Lecture Notes in Computer Science 8403

Commenced Publication in 1973
Founding and Former Series Editors:
Gerhard Goos, Juris Hartmanis, and Jan van Leeuwen

Alexander Gelbukh (Ed.)

Computational Linguistics and Intelligent Text Processing

15th International Conference, CICLing 2014
Kathmandu, Nepal, April 6-12, 2014
Proceedings, Part I

 Springer

Volume Editor

Alexander Gelbukh
National Polytechnic Institute
Center for Computing Research
Av. Juan Dios Bátiz, Col. Nueva Industrial Vallejo
07738 Mexico D.F., Mexico
E-mail: gelbukh@gelbukh.com

ISSN 0302-9743   e-ISSN 1611-3349
ISBN 978-3-642-54905-2   e-ISBN 978-3-642-54906-9
DOI 10.1007/978-3-642-54906-9
Springer Heidelberg New York Dordrecht London

Library of Congress Control Number: 2014934305

LNCS Sublibrary: SL 1 – Theoretical Computer Science and General Issues

Typesetting: Camera-ready by author, data conversion by Scientific Publishing Services, Chennai, India

Printed on acid-free paper

Springer is part of Springer Science+Business Media (www.springer.com)

Preface

CICLing 2014 was the 15th annual Conference on Intelligent Text Processing and Computational Linguistics. The CICLing conferences provide a wide-scope forum for discussion of the art and craft of natural language processing research as well as the best practices in its applications.

This set of two books contains four invited papers and a selection of regular papers accepted for presentation at the conference. Since 2001, the proceedings of the CICLing conferences have been published in Springer's *Lecture Notes in Computer Science* series as volume numbers 2004, 2276, 2588, 2945, 3406, 3878, 4394, 4919, 5449, 6008, 6608, 6609, 7181, 7182, 7816, and 7817.

The set has been structured into 17 sections, representative of the current trends in research and applications of natural language processing:

- Lexical Resources
- Document Representation
- Morphology, POS-tagging, and Named Entity Recognition
- Syntax and Parsing
- Anaphora Resolution
- Recognizing Textual Entailment
- Semantics and Discourse
- Natural Language Generation
- Sentiment Analysis and Emotion Recognition
- Opinion Mining and Social Networks
- Machine Translation and Multilingualism
- Information Retrieval
- Text Classification and Clustering
- Plagiarism Detection
- Style and Spelling Checking
- Speech Processing
- Applications

The 2014 event received submissions from 57 countries, a record high number in the 15-year history of the CICLing series. Exactly 300 papers (third highest number in the history of CICLing) by 639 authors were submitted for evaluation by the international Program Committee (see Figure 1 and Tables 1 and 2). This two-volume set contains revised versions of 85 regular papers selected for presentation; thus the acceptance rate for this set was 28.3%.

In addition to regular papers, the books feature invited papers by:

- Jerry Hobbs, ISI, USA
- Bing Liu, University of Illinois, USA
- Suresh Manandhar, University of York, UK
- Johanna D. Moore, University of Edinburgh, UK

Table 1. Number of submissions and accepted papers by topic[1]

Accepted	Submitted	% Accepted	Topic
19	45	42	Semantics, pragmatics, discourse
14	43	33	Lexical resources
12	31	39	Machine translation and multilingualism
12	33	36	Practical applications
12	35	34	Emotions, sentiment analysis, opinion mining
12	40	30	Clustering and categorization
12	56	21	Text mining
11	48	23	Information retrieval
10	29	34	Underresourced languages
8	26	31	Syntax and chunking
7	44	16	Information extraction
6	18	33	Social networks and microblogging
5	16	31	Natural language generation
4	11	36	Noisy text processing and cleaning
4	16	25	Summarization
3	4	75	Spelling and grammar checking
3	9	33	Plagiarism detection
3	12	25	Word sense disambiguation
3	16	19	POS tagging
2	5	40	Coreference resolution
2	7	29	Computational terminology
2	7	29	Other
2	9	22	Textual entailment
2	13	15	Formalisms and knowledge representation
2	17	12	Named entity recognition
2	20	10	Morphology
1	6	17	Speech processing
1	10	10	Natural language interfaces
1	11	9	Question answering
0	3	0	Computational humor

[1] As indicated by the authors. A paper may belong to several topics.

These speakers presented excellent keynote lectures at the conference. Publication of full-text invited papers in the proceedings is a distinctive feature of the CICLing conferences. Furthermore, in addition to presentation of their invited papers, the keynote speakers organized separate vivid informal events; this is also a distinctive feature of this conference series. In addition, Professor Jens Allwood of the University of Gothenburg was a special guest of the conference.

With this event we continued with our policy of giving preference to papers with verifiable and reproducible results. In addition to the verbal description of their findings given in the paper, we encouraged the authors to provide a proof of their claims in electronic form. If the paper claimed experimental results, we asked the authors to make available to the community all the input data necessary to verify and reproduce these results: if it claimed to introduce

Table 2. Number of submitted and accepted papers by country or region

Country or region	Authors Subm.	Papers[2] Subm.	Accp.	Country or region	Authors Subm.	Papers[2] Subm.	Accp.
Afghanistan	1	1	–	Japan	22	8.33	3
Algeria	2	0.67	–	Jordan	12	3.33	–
Australia	8	3	1	Kazakhstan	6	1.67	1.67
Bangladesh	9	3	–	Korea (South)	12	3.5	0.50
Belgium	3	2	–	Latvia	6	2	1
Brazil	18	6.17	2.17	Malaysia	4	1.67	–
Bulgaria	1	1	–	Mexico	19	12.42	2.67
Canada	13	7	4	Mongolia	1	0.5	0.5
China	57	21.1	7.35	Morocco	5	3	–
Christmas Isl.	1	0.2	0.2	Nepal	12	6	2
Colombia	3	1	1	Norway	1	0.2	–
Croatia	1	0.33	0.33	Pakistan	4	1.83	–
Czech Rep.	20	11.4	3	Poland	2	2	–
Denmark	3	0.38	–	Portugal	5	2.5	1
Egypt	12	7	1	Romania	10	5.67	–
Ethiopia	5	4	2	Russia	9	5.17	–
Finland	5	2	2	Singapore	9	2.78	1.78
France	29	12.42	9.67	Slovenia	2	0.67	0.67
Germany	19	7.33	4.33	Spain	13	3.7	0.67
Greece	1	0.33	0.33	Sweden	5	4	1
Hong Kong	4	2	1	Switzerland	6	5	2
Hungary	3	1	–	Taiwan	5	1	–
India	136	75.1	10.33	Thailand	2	1	–
Indonesia	3	1	–	Tunisia	20	8.83	1.83
Iran	4	2	–	Turkey	3	2.83	1.5
Iraq	0	1	–	UK	10	3.83	3.33
Ireland	0	0.5	–	USA	48	21.48	7.17
Israel	14	7	2	Vietnam	5	1.67	–
Italy	6	2.5	1	*Total:* 639	300	85	

[2] By the number of authors: e.g., a paper by two authors from the USA and one from UK is counted as 0.67 for the USA and 0.33 for UK.

an algorithm, we encouraged the authors to make the algorithm itself, in a programming language, available to the public. This additional electronic material will be permanently stored on the CICLing's server, www.CICLing.org, and will be available to the readers of the corresponding paper for download under a license that permits its free use for research purposes.

In the long run, we expect that computational linguistics will have verifiability and clarity standards similar to those of mathematics: In mathematics, each claim is accompanied by a complete and verifiable proof (usually much longer than the claim itself); each theorem's complete and precise proof—and not just a vague description of its general idea—is made available to the reader. Electronic

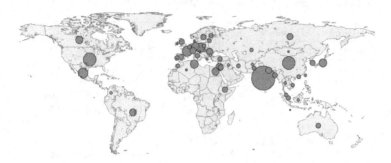

Fig. 1. Submissions by country or region. The area of a circle represents the number of submitted papers.

media allow computational linguists to provide material analogous to the proofs and formulas in mathematic in full length—which can amount to megabytes or gigabytes of data—separately from a 12-page description published in the book. More information can be found on www.CICLing.org/why_verify.htm.

To encourage providing algorithms and data along with the published papers, we selected a winner of our Verifiability, Reproducibility, and Working Description Award. The main factors in choosing the awarded submission were technical correctness and completeness, readability of the code and documentation, simplicity of installation and use, and exact correspondence to the claims of the paper. Unnecessary sophistication of the user interface was discouraged; novelty and usefulness of the results were not evaluated—instead, they were evaluated for the paper itself and not for the data.

The following papers received the Best Paper Awards, the Best Student Paper Award,[1] as well as the Verifiability, Reproducibility, and Working Description Award, respectively:

1st Place:	"A graph-based automatic plagiarism detection technique to handle artificial word reordering and paraphrasing", by Niraj Kumar, India
2nd Place:	"Dealing with function words in unsupervised dependency parsing," by David Mareček, Zdeněk Žabokrtský, Czech Republic
3rd Place:	"Extended CFG formalism for grammar checker and parser development," by Daiga Deksne, Inguna Skadiņa, Raivis Skadiņš, Latvia
and	"How preprocessing affects unsupervised keyphrase extraction," by Rui Wang, Wei Liu, Chris McDonald, Australia
Student:	"Iterative bilingual lexicon extraction from comparable corpora with topical and contextual knowledge," by Chenhui Chu, Toshiaki Nakazawa, Sadao Kurohashi, Japan

---

[1] The best student paper was selected among papers of which the first author was a full-time student, excluding the papers that received a Best Paper Award.

Verifiability: "How document properties affect document relatedness measures,"
by Jessica Perrie, Aminul Islam, Evangelos Milios, Canada

The authors of the awarded papers (except for the Verifiability award) were given extended time for their presentations. In addition, the Best Presentation Award and the Best Poster Award winners were selected by a ballot among the attendees of the conference.

Besides its high scientific level, one of the success factors of CICLing conferences is their excellent cultural program. The attendees of the conference had a chance to visit the wonderful historical and cultural attractions of the lesser-known country Nepal—the birthplace of the Buddha and the place where pagodas were invented before their spread to China and Japan to become an iconic image of East Asia. Of the world's ten highest mountains, eight are in Nepal, including the highest one Everest; the participants had a chance to see Everest during a tour of the Himalayas on a small airplane. They also attended the Seto MachindraNath Chariot festival and visited three historical Durbar squares of the Kathmandu valley, a UNESCO world cultural heritage site. But probably the best of Nepal, after the Himalayas, are its buddhist and hindu temples and monasteries, of which the participants visited quite a few. Even the Organizing Committee secretary and author of one of the best evaluated papers published in this set was the hereditary Supreme Priest of an ancient Buddhist temple!

I would like to thank all those involved in the organization of this conference. Firstly, the authors of the papers that constitute this book: it is the excellence of their research work that gives value to the book and sense to the work of all the rest. I thank all those who served on the Program Committee, Software Reviewing Committee, Award Selection Committee, as well as the additional reviewers, for their hard and very professional work. Special thanks go to Pushpak Bhattacharyya, Samhaa El-Beltagy, Aminul Islam, Cerstin Mahlow, Dunja Mladenic, Constantin Orasan, and Grigori Sidorov for their invaluable support in the reviewing process.

I would like to thank the conference staff, volunteers, and the members of the local Organizing Committee headed by Professor Madhav Prasad Pokharel and advised by Professor Jai Raj Awasthi. In particular, I am very grateful to Mr. Sagun Dhakhwa, the secretary of the Organizing Committee, for his great effort in planning all the aspects of the conference. I want to thank Ms. Sahara Mishra for administrative support and Mr. Sushan Shrestha for the website design and technical support. I am deeply grateful to the administration of the Centre for Communication and Development Studies (CECODES) for their helpful support, warm hospitality, and in general for providing this wonderful opportunity to hold CICLing in Nepal. I acknowledge support from the project CONACYT Mexico—DST India 122030 "Answer Validation through Textual Entailment" and SIP-IPN grant 20144534.

The entire submission and reviewing process was supported for free by the EasyChair system (www.EasyChair.org). Last but not least, I deeply appreciate the patience and help of Springer staff in editing these volumes and getting them printed in very short time—it is always a great pleasure to work with Springer.

February 2014   Alexander Gelbukh

Organization

CICLing 2014 was hosted by the Centre for Communication and Development Studies (CECODES), Nepal, and was organized by the CICLing 2014 Organizing Committee in conjunction with the CECODES, the Natural Language and Text Processing Laboratory of the CIC (Centro de Investigación en Computación) of the IPN (Instituto Politécnico Nacional), Mexico, and the Mexican Society of Artificial Intelligence (SMIA).

Organizing Chair

Madhav Prasad Pokharel

Organizing Committee

Madhav Prasad Pokharel (Chair)	Tribhuban University, Kathmandu
Jai Raj Awasthi (Advisor)	CECODES, Lalitpur
Sagun Dhakhwa (Secretary)	CECODES, Lalitpur
Bhim Narayan Regmi	CECODES, Lalitpur
Krishna Prasad Parajuli	CECODES, Lalitpur
Sandeep Khatri	CECODES, Lalitpur
Kamal Poudel	CECODES, Lalitpur
Bhim Lal Gautam	CECODES, Lalitpur
Krishna Prasad Chalise	CECODES, Lalitpur
Dipesh Joshi	CECODES, Lalitpur
Prajol Shrestha	NLP Engineer, Vision Objects

Program Chair

Alexander Gelbukh

Program Committee

Ajith Abraham
Rania Al-Sabbagh
Sophia Ananiadou
Marianna Apidianaki
Alexandra Balahur
Kalika Bali

Leslie Barrett
Roberto Basili
Pushpak Bhattacharyya
Nicoletta Calzolari
Nick Campbell
Sandra Carberry

Michael Carl
Hsin-Hsi Chen
Dan Cristea
Bruce Croft
Mike Dillinger
Samhaa El-Beltagy
Tomaž Erjavec
Anna Feldman
Alexander Gelbukh
Dafydd Gibbon
Gregory Grefenstette
Eva Hajicova
Sanda Harabagiu
Yasunari Harada
Karin Harbusch
Ales Horak
Veronique Hoste
Nancy Ide
Diana Inkpen
Hitoshi Isahara
Aminul Islam
Guillaume Jacquet
Sylvain Kahane
Alma Kharrat
Adam Kilgarriff
Valia Kordoni
Leila Kosseim
Mathieu Lafourcade
Krister Lindén
Bing Liu
Elena Lloret
Bernardo Magnini
Cerstin Mahlow
Suresh Manandhar
Diana Mccarthy
Alexander Mehler
Rada Mihalcea
Evangelos Milios
Dunja Mladenic
Marie-Francine Moens
Masaki Murata
Preslav Nakov
Costanza Navarretta
Roberto Navigli
Vincent Ng

Joakim Nivre
Attila Novák
Kjetil Nørvåg
Kemal Oflazer
Constantin Orasan
Ekaterina Ovchinnikova
Ivandre Paraboni
Saint-Dizier Patrick
Maria Teresa Pazienza
Ted Pedersen
Viktor Pekar
Anselmo Peñas
Octavian Popescu
Marta R. Costa-Jussà
German Rigau
Horacio Rodriguez
Paolo Rosso
Vasile Rus
Kepa Sarasola
Roser Sauri
Hassan Sawaf
Satoshi Sekine
Serge Sharoff
Grigori Sidorov
Kiril Simov
Vivek Kumar Singh
Vaclav Snasel
Thamar Solorio
Efstathios Stamatatos
Carlo Strapparava
Tomek Strzalkowski
Maosong Sun
Stan Szpakowicz
Mike Thelwall
Jörg Tiedemann
Christoph Tillmann
George Tsatsaronis
Dan Tufis
Olga Uryupina
Karin Verspoor
Manuel Vilares Ferro
Aline Villavicencio
Piotr W. Fuglewicz
Savas Yildirim

Software Reviewing Committee

Ted Pedersen
Florian Holz
Miloš Jakubíček

Sergio Jiménez Vargas
Miikka Silfverberg
Ronald Winnemöller

Award Committee

Alexander Gelbukh
Eduard Hovy
Rada Mihalcea

Ted Pedersen
Yorick Wiks

Additional Reviewers

Mahmoud Abunasser
Naveed Afzal
Iñaki Alegria
Hanna Bechara
Houda Bouamor
Janez Brank
Chen Chen
Víctor Darriba
Owen Davison
Ismaïl El Maarouf
Mahmoud El-Haj
Milagros Fernández-Gavilanes
Daniel Fernández-González
Corina Forascu
Kata Gabor
Mercedes García Martínez
Diman Ghazi
Rohit Gupta
Francisco Javier Guzman
Kazi Saidul Hasan
Radu Ion
Zahurul Islam
Milos Jakubicek
Antonio Jimeno
Olga Kolesnikova
Mohammed Korayem
Tobias Kuhn

Majid Laali
Yulia Ledeneva
Andy Lücking
Tokunbo Makanju
Raheleh Makki
Akshay Minocha
Abidalrahman Moh'D
Zuzana Neverilova
An Ngoc Vo
Mohamed Outahajala
Michael Piotrowski
Soujanya Poria
Francisco Rangel
Amir Hossein Razavi
Francisco Ribadas-Pena
Alvaro Rodrigo
Armin Sajadi
Paulo Schreiner
Djamé Seddah
Karan Singla
Vit Suchomel
Aniruddha Tammewar
Yasushi Tsubota
Francisco José Valverde Albacete
Kassius Vargas Prestes
Tim Vor der Brück
Tadej Štajner

Website and Contact

The webpage of the CICLing conference series is www.CICLing.org. It contains information about past CICLing conferences and their satellite events, including links to published papers (many of them in open access) or their abstracts, photos, and video recordings of keynote talks. In addition, it contains data, algorithms, and open-source software accompanying accepted papers, according to the CICLing verifiability, reproducibility, and working description policy. It also contains information about the forthcoming CICLing events, as well as contact options.

Table of Contents – Part I

Morphology, POS-tagging, and Named Entity Recognition

Syntax and Parsing

Anaphora Resolution

Recognizing Textual Entailment

Semantics and Discourse

Natural Language Generation

Table of Contents – Part II

Sentiment Analysis and Emotion Recognition

Opinion Mining and Social Networks

Machine Translation and Multilingualism

Information Retrieval

Text Classification and Clustering

Text Summarization

Plagiarism Detection

Style and Spelling Checking

Speech Processing

Applications

Using Word Association Norms
to Measure Corpus Representativeness

Reinhard Rapp

Aix-Marseille Université, Laboratoire d'Informatique Fondamentale
163 Avenue de Luminy – Case 901, 13288 Marseille Cedex 9, France
`reinhardrapp@gmx.de`

Abstract. An obvious way to measure how representative a corpus is for the language environment of a person would be to observe this person over a longer period of time, record all written or spoken input, and compare this data to the corpus in question. As this is not very practical, we suggest here a more indirect way to do this. Previous work suggests that people's word associations can be derived from corpus statistics. These word associations are known to some degree as psychologists have collected them from test persons in large scale experiments. The output of these experiments are tables of word associations, the so-called word association norms. In this paper we assume that the more representative a corpus is for the language environment of the test persons, the better the associations generated from it should match people's associations. That is, we compare the corpus-generated associations to the association norms collected from humans, and take the similarity between the two as a measure of corpus representativeness. To our knowledge, this is the first attempt to do so.

Keywords: Corpus linguistics, corpus representativeness, word association, association norms.

1   Introduction

Since the pioneering work on the Brown Corpus (Francis & Kuçera, 1989) it has often been tried to compile corpora comprising a balanced mix of subject areas and genres. Among the best known such efforts is the British National Corpus (Burnard & Aston 1998). This is a collection of samples of spoken and written language from a multitude of sources designed to represent present day British English. Biber (1993) describes a number of general design issues aiming at achieving *representativeness* in corpus design. These include a "definition of the target population, stratified versus proportional sampling of a language, sampling within texts, and issues related to the required sample size." McEnery and Wilson (1996: 87) define the term *corpus* in a way that implies representativeness: "a body of text which is carefully sampled to be maximally representative of a language or language variety". On the other hand, more recently Saldanha (2009) states: "The problem with making representativeness the defining characteristic of a corpus is that it is very difficult to evaluate". That is, despite an abundance of work in corpus linguistics, perhaps with the exception of

A. Gelbukh (Ed.): CICLing 2014, Part I, LNCS 8403, pp. 1–13, 2014.
© Springer-Verlag Berlin Heidelberg 2014

Brisbaert & New (2009) who validate corpus frequencies using lexical decision times, we still lack practical methods on how to measure corpus representativeness. Our goal here is to make an attempt to provide such a method.

We start by defining the term *representative* as reflecting the average (spoken and written) language use a native speaker typically encounters in everyday life over a longer period of time. But we cannot easily observe a person's language environment over years, so what information can we use? Our suggestion is to look at word associations as obtained from test persons. Such data has been collected from native speakers in large scale experiments, as exemplified in the *Edinburgh Associative Thesaurus* (*EAT*; Kiss et al., 1973) where English word associations for thousands of stimulus words are listed.

It is relatively easy to conduct such experiments: Typically, the subjects are given questionnaires with lists of stimulus words, and are asked to write down for each stimulus word the spontaneous association which first comes to mind. This leads to collections of associations, the so-called association norms, as exemplified in Table 1.

Table 1. Top ten associations to three stimulus words as taken from the Edinburgh Associative Thesaurus. The numbers of subjects responding with the respective word are given in brackets.

ABOVE	CONSTELLATION	FEMININE
below (59)	stars (39)	masculine (26)
high (4)	star (33)	girl (14)
over (4)	sky (5)	woman (8)
sky (4)	andromeda (2)	female (6)
all (3)	aquarius (2)	sex (3)
up (3)	plough (2)	beauty (2)
me (2)	aircraft (1)	bird (2)
under (2)	cancer (1)	girls (2)
us (2)	clear (1)	nice (2)
average (1)	dance (1)	pretty (2)

Association theory, which can be traced back to Aristotle in ancient Greece, has often stated that our associations are governed by our experiences. For example, more than a century ago William James (1890) formulated this in his book "The Principles of Psychology" as follows:

> "Objects once experienced together tend to become associated in the imagination, so that when any one of them is thought of, the others are likely to be thought of also, in the same order of sequence or coexistence as before. This statement we may name the law of mental association by contiguity."

This citation is talking of *objects*, but the question arose whether for words the same principles might apply, and with the advent of corpus linguistics it was possible to verify this experimentally by looking at the distribution of words in texts. Among the first to do so were Schvaneveldt et al. (1989), Wettler & Rapp (1989), and Church & Hanks (1990).

Their underlying assumption was that strongly associated words should often occur in close proximity in text corpora. This is actually confirmed by corpus evidence: Figure 1 assigns to each stimulus word position 0, and displays the occurrence frequencies of its *primary associative response* (most frequent response as produced by the test persons) at relative distances between -50 and +50 words. However, to give a general picture and to abstract away from idiosyncrasies, the figure is not based on a single stimulus/response pair, but instead represents the average of 100 English stimulus/response pairs as published by Jenkins (1970). The effect is in line with expectations: The closer we get to the stimulus word, the higher the chances that the primary associative response occurs. Only the distances plus and minus one and plus and minus three are exceptions, but this is an artefact because content words are typically separated by function words which carry not much content and are therefore of little interest here.

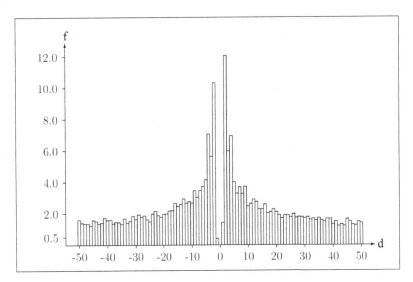

Fig. 1. Occurrence frequency f of a primary response at distance d from a stimulus word, averaged over the 100 stimulus/response pairs according to Jenkins (1970). At large distances from the stimulus word the average occurrence frequency of a primary response is 0.49 (adapted from Rapp, 1996).

Word associations derived from corpora are the basis underlying this work. Wettler & Rapp (1993) were probably the first to directly compare these to human associations and Wettler et al. (2005) describe a link to classical psychological learning theory, thereby providing strong evidence that human association learning is in essence corpus based. A later paper by Turney & Pantel (2010) nicely works out the relationship between frequency and meaning. In short, there is some evidence that the framework assumed here is sound.

The remainder of this paper is structured as follows: We first provide some relevant details on the EAT and show how to compute associations from a corpus. We then, by comparing human and corpus-based associations, provide a measure of corpus representativeness which we apply to a number of different corpora, thereby providing detailed computational results. We finally discuss these results and present some conclusions.

2   Language Resources

2.1   Association Norms

As discussed in the previous section, we assume that there is a relationship between word associations as collected from human subjects and word co-occurrences as observed in a corpus, and our hypothesis is that the strength of this relationship can be used as a measure of corpus representativeness. A corpus leading to simulated associations very similar to the ones collected from humans is likely to be a good surrogate for everyday language, although word associations constitute only one of many properties of a corpus, see the reflections on the shortcomings in Section 5. Nevertheless, for a corpus to be representative, it is a necessary (though not sufficient) condition that the word associations derived from it are similar to those collected from humans.

As our source of human data we use the EAT (Kiss et al., 1973) which is the largest classical collection of its kind.[1] The EAT comprises the associative responses as requested from around 100 British students for each of altogether 8400 stimulus words. As exemplified in Table 1, some of the responses to a particular stimulus word are given by many students, whereas others are given by only one or two. What is the reason for this? According to the theory, the associative response of a test person should reflect the language environment of this person, i.e. the text and speech this person previously encountered and whose statistical properties were stored in long term memory. As apparently there is some variation in each person's language history, some variation in the associative responses can be expected. This is a natural and desired effect.

However, there is also an undesired component in this variation: That test persons perceive only a single input word at a time is an idealization. In reality, they still have the previous input words in short term memory, and also the (e.g. classroom) environment provides lots of additional (e.g. visual) stimuli. So what the subjects actually come up with is based on a mix of all these stimuli plus of what they have in short (and possibly medium) term memory. For example, the responses on the stimulus word *artist* could be influenced by some artwork present in classroom, or of a

---

[1] An even larger, though possibly more noisy, association database has been collected via online gaming at www.wordassociation.org. Other collections of association data include, among others, the University of South Florida Free Association Norms, the Birkbeck Association Norms, the Minnesota Norms of Word Association, the Dutch Word Association Database of the KU Leuven, and Mathieu Lafourcade's Jeux de mots.

museum visit the day before. The responses to *transport* could be influenced by the means of transport the students used for arriving in class, such as car, bicycle, train, or bus. Or the response to *cinema* could depend on a recently watched movie.

As corpus representativeness should focus on long term averages and not on short term effects, such influences are undesired and should be avoided. We do so by assuming: Responses confirmed by many students are more likely to reflect long term averages (especially if the group of test persons is heterogeneous). And responses provided by only one or two test persons are more likely to be noise.

As the EAT is very large, we can afford to stay on the high quality side of this scale. For this reason, we decided to use only the primary associative response for each stimulus word, and to discard all other responses. Alternatively, we could have decided to take the top two or top five responses into account. However, preliminary experiments showed that this would "water down" our results. That is, the basic effects would be the same, but in a somewhat less salient fashion. See Rapp (2013) where such effects are quantified for the related task of multiword association.

The EAT uses uppercase characters only as at the time of its construction a distinction between uppercase and lowercase characters was not customary in computing. As this is only a minor shortcoming, we decided not to try to make up for it. We only converted the EAT to lowercase for easier readability. For reasons of consistency, we also converted all corpora (see section 2.2) to lowercase only.

Another problem with the EAT is that it contains some multiword units. In principle, our approach for computing word associations has no problem with multiword units if they are identified at the word segmentation stage. However, despite some attempts to come up with definitions, typologies, and classifications (e.g. Sag et al., 2002), there are no generally acknowledged principles on what should count as a multiword unit. Also, when computing associations for single word stimuli, it is not clear whether or not matching components of a multiword unit should be taken into account. For example, given *New Year*, should the occurrences of *Year* within this multiword unit nevertheless be used for computing the associations to *Year*? As for computing corpus representativeness it seemed not important to elaborate on this minor problem, we simply decided to remove all items from the EAT where either the stimulus word or the primary associative response happened to be a multiword unit.

Finally, as high frequency function words were considered of little relevance to our analysis and in order to keep our algorithm efficient, we decided not to take into account some of the most frequent words. Based on the word frequencies in the British National Corpus we compiled a list of words with frequencies 250,000 or higher.[2] If either the stimulus word or its primary response occurred in this list, we removed the respective item from the EAT.

In summary, of the 8400 items in the EAT we removed those involving multiword units and high frequency function words, thereby obtaining 7731 remaining items.

---

[2]   This list comprises the following words: the, of, and, to, a, in, that, it, is, i, was, for, s, on, you, he, with, as, be, at, by, are, this, but, have, not, had, from, his, they, she, or, which, we, t, an, there, her, were, one, all, been, if, their, has, will, so, what.

Whereas some previous studies involving the computation of word associations used lemmatization, we decided not to do so here. Firstly, in view of future work, we wanted to keep the basic algorithm for computing word associations as language independent as possible. Secondly, lemmatizing the EAT is problematic as it does not provide context for its words. Thirdly, lemmatization mainly has benefits at the evaluation stage, but this is not important in our setting. For example, given the stimulus word *table* and the primary associative response *chair*, if the simulation produced the plural form *chairs* this would count as incorrect without lemmatization. But as this is (at least for some of the most studied languages) a relatively infrequent phenomenon, and as it applies in a similar way across corpora, it is not very relevant for corpus comparisons where the emphasis is not on polishing individual results.

2.2   Corpora

Our corpus representativeness measure is to be applied to a number of well known corpora. These are:

1. Brown Corpus (balanced corpus of 1 million words; Francis & Kučera, 1989)
2. British National Corpus (BNC; balanced corpus of 100 million words; Burnard & Aston, 1998)
3. English Wikipedia (300 million words of encyclopaedic texts)[3]
4. ukWaC (British English web corpus of 2 billion words)[4]
5. English Gigaword Corpus 4th edition (4 billion words of newswire text)[5]

As the EAT is case-insensitive (see above), these corpora were converted to lowercase only. For the experiments described in the next sections we needed to cut off our corpora in order to provide results for subcorpora of particular sizes. Because corpus size was measured as the number of running words, and as these numbers depend somewhat on how word segmentation is conducted, let us briefly describe the procedure we used: In order to keep our algorithm as language independent as possible, we simply consider any uninterrupted sequences of alpha characters as words, but also any sequences of non-alpha characters except white space (blanks, tabulator, new line). That is, white space and transitions between the two types of characters (alpha and non-alpha) are considered as word separators.

3   Computing and Evaluating Word Associations

The five corpora were pre-processed as described above and either in full or in part used for computing word associations. Hereby the procedure was as follows: For all

---

[3]   We use the English part of the Wikipedia XML Corpus (Denoyer & Gallinary, 2006). Although this is smaller than current versions, it has the advantage that it is an offline copy so that our results can be replicated.

[4]   http://wacky.sslmit.unibo.it/doku.php?id=corpora

[5]   http://catalog.ldc.upenn.edu/LDC2009T13

7731 words occurring as stimuli in our EAT derived gold standard we computed co-occurrence vectors. That is, each vector contains the number of common occurrences of the stimulus word with other words belonging to a specific vocabulary. Hereby it counts as a co-occurrence if two words appear together within a distance of at most ten words, i.e. a text window of plus and minus ten words around the stimulus word is considered. The exact distance within the window is not taken into account. To reduce memory requirements and processing time, we restricted the vocabulary of co-occurring words to the primary associative responses relating to the 7731 stimulus words in our gold standard. As quite a few stimulus words lead to the same primary associative responses (e.g. *black* and *snow* both have *white* as their most frequent response) the number of distinct words in the vocabulary of primary associative responses is considerably lower than 7731, namely 2792.

Having completed the co-occurrence counting, in the next step an association measure was applied to the co-occurrence vectors. This is meant to account for the differences in absolute word frequencies. As our association measure of choice we used the log-likelihood ratio (Dunning, 1993) which is very well established for such purposes. It compares the observed co-occurrence counts with the co-occurrence counts expected from chance, thus strengthening significant word pairs and weakening incidental word pairs. The resulting vectors we call association vectors. Given these vectors, the strongest association to a given stimulus word can be determined by simply looking for the highest value within the respective association vector. The corresponding word is considered to be the associative response generated by the system. For the same stimulus words used in Table 1, Table 2 shows some sample associations as computed using the British National Corpus.

Table 2. Top ten corpus-derived associations for three stimulus words. The numbers of subjects from the EAT responding with the respective word (if larger than zero) are given in brackets.

ABOVE	CONSTELLATİON	FEMİNİNE
below (59)	stars (39)	masculine (26)
level	star (33)	women (2)
average (1)	southern	gender
high (4)	triangle	woman (8)
feet	bright	female (6)
water	planet (1)	men
head	rather	male (1)
see	south	more
ground	find	hair
left	map	soft

Let us briefly discuss the above choice of window size (±10 words). Some previous studies used smaller window sizes such as e.g. ±2 words in Wettler et al. (2005). However, other than in the present work this study had eliminated function words from the corpus, so that the effective window size might be around ±4 words. Also, it should be noted that the best window size depends on the size of the corpus: For small corpora the problem of data sparseness can be somewhat reduced by considering a larger window, whereas for larger corpora this is not necessary and a smaller window might possibly lead to a higher accuracy of the predictions (compare Figure 1).

Concerning evaluation, in principle the idea is to find matches between the human and the corpus-based associations. One possibility is to simply count the number of cases where the primary associative response matches the strongest corpus-based association. However, when it comes to very small corpus sizes of e.g. just 1000 words (see Section 4), the problem of data sparseness becomes so severe that a more tolerant evaluation method leads to more robust results less susceptible to statistical variation. This is why for measuring accuracy we count the number of cases where a human primary associative response is listed within the top ten corpus-based associations, rather than insisting on a match with the strongest association. This simple modification leads to improvements in reliability when measuring very low accuracies.

As readers familiar with the field will typically expect evaluations based on recall, precision, and/or f-measure, let us explain why we believe that these are not very appropriate here. In principle, it would be possible to e.g. look at the top ten human associations for a given stimulus word, and then find out how many of these occur in the top ten corpus-based associations. From the results recall, precision and f-measure could be computed. However, the problem is the following: These measures were developed in Information Retrieval under the assumptions that within the documents in a database two categories can be distinguished: Those relevant to a query and all others which are assumed to be irrelevant. However, what we have in the case of word associations is that the degree of relevance (here: association strength) is very important. For example, given the stimulus word *black*, 57 of 99 subjects answered with *white*, but each of the following 30 responses is given by at most three subjects. The problem is that an evaluation based on recall and precision would give such spurious responses the same weight that it gives the top response, which is clearly inappropriate. In short: We have chosen our straightforward evaluation methodology not because it is even simpler, but because it appears better suited for this particular purpose.

4   Results and Discussion

Concerning the representativeness of our five corpora, we tried to come up with some hypotheses before we started to compute the results. These were our predictions:

1. Representativeness should increase with corpus size.
2. The Brown corpus and the BNC should be more representative than unbalanced corpora of the same size.
3. The Brown corpus (1 million words) should be more representative than the first million words of the British National Corpus as the latter is balanced only over its full size (100 million words), but not over its first million words.
4. For same sizes, we would expect ukWaC to be more representative than Wikipedia as we think that corpus heterogeneity is a plus for representativeness. ukWaC is obviously more heterogeneous as, for example, it is multi genre multi topic whereas Wikipedia is single genre multi topic.

5. The Gigaword Corpus should be the least representative. Although, like Wikipedia, it is also single genre multi topic, the distribution of topics is not as wide because in newsticker texts there are strong foci e.g. on politics, culture, and sports.

The actual results are given in Table 3. There we find for each of the five corpora its size and the percentage of primary associative responses which ranked among the top ten in the corpus-based associations. These percentages we take as the *representativeness* of the respective corpus. The range of values can be between 0 and 100, whereby 0 denotes a complete lack of representativeness, and 100 denotes perfect representativeness. The values are also provided for partial corpora, whereby all parts have in common that they always start at the beginning of the respective corpus. Let us discuss these results by comparing them to the above expectations.

Hypothesis 1 said that representativeness should increase with corpus size. We can see in Table 3 that this is clearly the case: Representativeness is almost zero if only the first 100 words of a corpus are taken into account, and gradually increases to at least 14% for the full corpora. These increases can be better seen in Figure 2 which is a graphical representation of Table 3. Note that the horizontal axis has a logarithmic scale and that the curve for the Brown corpus is not well visible as it ends at 1 million words and thus has a range where all curves are closely together and overlapping.

Hypothesis 2, saying that the balanced corpora, namely the Brown corpus and the BNC, should be more representative for their sizes than non-balanced corpora, is also confirmed. At 1 million words, these two are the top performers. At 100 million words, the BNC performs best. Note, however, that for very small sizes, especially for 100 words, the results are less predictable as the sampling error increases for small corpora. For this reason it probably does not make much sense to compare the representativeness scores among the smaller partial corpora (e.g. below 100,000 words). We only presented them here to be able to verify hypothesis 1.

Hypothesis 3 (Brown better than BNC for 1 million words) is not confirmed, although with 13.22% (Brown) and 13.85% (BNC) both results are fairly close. In hindsight, our explanation for this is that the BNC better matches the language environment of the EAT students (experiments were conducted in Edinburgh) as it represents British English whereas the Brown corpus represents American English. The time periods when the text samples were produced could also play a role. The EAT experiments were conducted between June 1968 and May 1971. The BNC's text samples mainly date from 1975 to 1994 (only some imaginative texts date back earlier: 1960 to 1974). The materials in the Brown corpus were all published in the United States in 1961. This means: The EAT student's language environment was pre 1968 to pre 1971. The BNC authors' mainly pre 1975 to pre 1994. And the Brown authors' pre 1961.

Hypothesis 4, namely that ukWaC is better than Wikipedia, is confirmed for all corpus sizes above 100,000 words. As noted above, for smaller corpus sizes sampling errors are likely to be significant.

Table 3. Corpus representativeness scores for full and partial corpora

	Brown	BNC	Wiki-pedia	ukWaC	Giga-word
100 words	0.31%	0.22%	0.30%	0.47%	0.13%
1000 words	0.49%	0.67%	0.72%	0.61%	0.28%
10000 words	0.60%	0.91%	0.71%	0.69%	0.83%
100000 words	2.95%	3.09%	3.12%	2.90%	1.91%
1 million words	13.22%	13.85%	11.24%	11.87%	5.91%
10 million words	–	35.26%	26.83%	30.02%	14.20%
100 million words	–	52.67%	42.89%	48.40%	25.13%
1 billion words	–	–	–	55.92%	35.57%
full corpus	14.62%	53,63%	49.89%	57.07%	44.43%
size full corpus (million words)	1.18	117	313	2345	4371

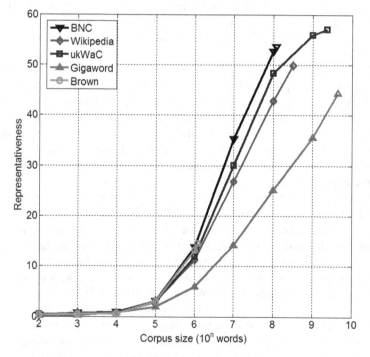

Fig. 2. Corpus representativeness scores depending on corpus size

Hypothesis 5, saying that the Gigaword corpus should be the least representative, is clearly confirmed for almost all corpus sizes except for 10,000 words where statistical variation is still very strong.

Overall we think that the hypotheses were reasonably well confirmed thereby providing evidence that the computed scores are actually related to what might sensibly be considered as the representativeness of a corpus. In particular, it is worth noting that our measure confirms the intuitions that it makes sense to balance a corpus and that corpus heterogeneity is a plus: With a score of 52.67% the balanced BNC is the top performer at 100 million words, and with 57.07% the very heterogeneous ukWaC is the top performer overall.

5   Summary and Outlook

In this work we defined the term *corpus representativeness* as the ability of a corpus to represent the average (spoken and written) language use a native speaker typically encounters in everyday life over a longer period of time. As we cannot easily observe test persons over years, our suggestion was to utilize word associations as obtained in the free association experiment. Word association data is available at least for some languages as it has been collected by psychologists from native speakers in large scale experiments.

Previous work has provided evidence that human word associations are based on the co-occurrences of words in perceived language, i.e. that our brain analyzes word co-occurrences and determines their saliency. Although this may still be controversial, in the current work we took this finding for granted but turned round the perspective. We said that a corpus is representative for the language environment of a group of persons if the word associations derived from it resemble these persons' word associations.

For five English corpora, namely the Brown Corpus, the British National Corpus, the English Gigaword Corpus, Wikipedia, and ukWaC, we computed the word associations for a test set of 7731 stimulus words. We then, for each corpus, compared the resulting associations to the associations as listed in the Edinburgh Associative Thesaurus, and computed a similarity score which we called corpus representativeness.

Our measure only aims for representativeness concerning an average person's language environment (rather than e.g. the sum of the language environments of all speakers of a language). Therefore it is by design that it does not take into account a corpus' comprehensiveness beyond the vocabulary and associations an average native speaker knows. Therefore it can be justified that the very large Gigaword corpus, despite its very comprehensive vocabulary, did not perform very well.

Much more severe appears the following shortcoming: Our measure is limited in so far as it only considers a particular aspect of corpus representativeness, namely word association. It does not explicitly consider higher level features e.g. concerning syntax, semantics, pragmatics, or style.

Let us therefore propose some extensions to be dealt with in future work: Whereas word associations are based on word co-occurrences and therefore on a type of first order statistics, we could extend the method to statistics of order zero (Rapp, in print) and of order two. Order zero would mean word frequency. Here human data is also readily available in the form of the so-called familiarity norms where test persons are asked to rate word familiarities. For example, the MRC psycholinguistic database (Coltheart, 1981) contains such data in large quantities. Second order statistics comprise word relatedness, thereby – in the spirit of Harris' distributional hypothesis – identifying words with common context. Here human data is also available: A prototypical example is the data from the well known TOEFL synonym test (Landauer & Dumais, 1997). But monolingual dictionaries listing synonyms could also be considered as human data, and in particular WordNet whose variants are available for many languages. That is, the quality of corpus-derived WordNet synsets could be taken as a measure of corpus representativeness.

After compiling a number of representativeness scores representing statistics of order zero, one and two, these scores might finally be combined into an overall score e.g. by computing their geometric mean. This would be in analogy to the BLEU score (Papineni et al., 2002) used in machine translation evaluation where matches (between machine translation and human reference translation) of various n-gram lengths are separately scored and then combined.

Acknowledgment. This research was supported by a Marie Curie Intra European Fellowship within the 7th European Community Framework Programme.

References

1. Biber, D.: Representativeness in Corpus Design. Literary and Linguistic Computing 8, 243–257 (1993)
2. Brisbaert, M., New, B.: Moving beyond Kučera and Francis: A critical evaluation of current word frequency norms and the introduction of a new and improved word frequency measure for American English. Behavior Research Methods 41(4), 977–990 (2009)
3. Burnard, L., Aston, G.: The BNC Handbook: Exploring the British National Corpus with Sara. University Press, Edinburgh (1998)
4. Church, K.W., Hanks, P.: Word association norms, mutual information, and lexicography. Computational Linguistics 16(1), 22–29 (1990)
5. Coltheart, M.: The MRC psycholinguistic database. Quarterly Journal of Experimental Psychology 33A, 497–505 (1981)
6. Denoyer, L., Gallinari, P.: The Wikipedia XML Corpus. SIGIR Forum 40(1), 64–69 (2006)
7. Dunning, T.: Accurate methods for the statistics of surprise and coincidence. Computational Linguistics 19(1), 61–74 (1993)
8. Francis, W.N., Kuçera, H.: Manual of Information to Accompany a Standard Corpus of Present-Day Edited American English, for Use with Digital Computers. Brown University, Department of Linguistics, Providence, R.I (1989)
9. James, W.: The Principles of Psychology. Holt, New York (1890), Reprinted Dover Publications, New York (1950)
10. Jenkins, J.J.: The 1952 Minnesota word association norms. In: Postman, L.J., Keppel, G. (eds.) Norms of Word Association, pp. 1–38. Academic Press, New York (1970)
11. Kiss, G.R., Armstrong, C., Milroy, R., Piper, J.: An associative thesaurus of English and its computer analysis. In: Aitken, A.J., Bailey, R.W., Hamilton-Smith, N. (eds.) The Computer and Literary Studies, pp. 153–165. University Press, Edinburgh (1973)

12. Landauer, T.K., Dumais, S.T.: A solution to Plato's problem: The latent semantic analysis theory of acquisition, induction, and representation of knowledge. Psychological Review 104(2), 211–240 (1997)
13. McEnery, T., Wilson, A.: Corpus Linguistics. Edinburgh University Press (1996)
14. Papineni, K., Roukos, S., Ward, T., Zhu, W.-J.: BLEU: A method for automatic evaluation of machine translation. In: Proceedings of the 40th Annual Meeting of the ACL, pp. 311–318. Philadelphia (2002)
15. Rapp, R.: Die Berechnung von Assoziationen. Olms, Hildesheim (1996)
16. Rapp, R.: From stimulus to associations and back. In: Proceedings of the 10th Workshop on Natural Language Processing and Cognitive Science, Marseille, France (2013)
17. Rapp, R.: Using word familiarities to measure corpus representativeness. In: Proceedings of the 48. Linguistics Colloquium, Alcala de Henares, Spain (September 2013) (in print)
18. Sag, I.A., Baldwin, T., Bond, F., Copestake, A., Flickinger, D.: Multiword expressions: A pain in the neck for NLP. In: Gelbukh, A. (ed.) CICLing 2002. LNCS, vol. 2276, pp. 1–15. Springer, Heidelberg (2002)
19. Saldanha, G.: Principles of corpus linguistics and their application to translation studies research. Tradumàtica 7, 1–7 (2009)
20. Schvaneveldt, R.W., Durso, F.T., Dearholt, D.W.: Network structures in proximity data. In: Bower, G. (ed.) The Psychology of Learning and Motivation: Advances in Research and Theory, vol. 24, pp. 249–284. Academic Press, New York (1989)
21. Turney, P.T., Pantel, P.: From frequency to meaning: Vector space models of semantics. Journal of Artificial Intelligence Research 37, 141–188 (2010)
22. Wettler, M., Rapp, R.: A connectionist system to simulate lexical decisions in information retrieval. In: Pfeifer, R., Schreter, Z., Fogelman, F., Steels, L. (eds.) Connectionism in Perspective, pp. 463–469. Elsevier, Amsterdam (1989)
23. Wettler, M., Rapp, R., Sedlmeier, P.: Free word associations correspond to contiguities between words in texts. Journal of Quantitative Linguistics 12(2), 111–122 (2005)

Optimality Theory as a Framework
for Lexical Acquisition

Thierry Poibeau

Laboratoire LATTICE
PSL*: Paris Sciences et Lettres*
1 rue Maurice Arnoux
92120 Montrouge France
thierry.poibeau@ens.fr

Abstract. This paper re-investigates a lexical acquisition system initially developed for French. We show that, interestingly, the architecture of the system reproduces and implements the main components of Optimality Theory. However, we formulate the hypothesis that some of its limitations are mainly due to a poor representation of the constraints used. Finally, we show how a better representation of the constraints used would yield better results.

1 Introduction

Natural Language Processing (NLP) aims at developing techniques for processing natural language texts using computers. In order to yield accurate results, NLP requires resources containing various information (sub-categorization frames, semantic roles, selection restrictions, etc.). Unfortunately, such resources are not available for most languages and are very costly to develop manually. A recent trend of research has tried to overcome these limitations through the development of automatic acquisition methods from corpora.

Automatic lexical acquisition is an engineering task aiming at providing comprehensive—even if not fully accurate—resources for NLP. As natural languages are complex, lexical acquisition needs to take into account a wide range of parameters and constraints. However, surprisingly, in the acquisition community, relatively few investigations have been done on the structure of the linguistic constraints themselves, beyond the engineering point of view (but note that this work has been extensively done for parsing, see [1]).

In this paper, we want to take another look at some experiments recently done on the automatic acquisition of lexical resources from textual corpora, more specifically on French. In a way, acquisition is converse to parsing: the task consists, from a surface form, in trying to find an abstract lexical-conceptual structure that justify the surface construction (taking into account the relevant set of constraints for the given language). Here, in order to get a tractable model, we limit ourselves to the acquisition of sub-categorization frames from corpora. The task is challenging since surface forms

* This work has received support of TransferS (laboratoire d?excellence, program "Investissements d'avenir" ANR-10-IDEX-0001-02 PSL* and ANR-10-LABX-0099).

incorporate adverbs, modifiers, interpolated clauses and some flexibility in the ordering of the arguments.

Most approaches, including ours, are based on simple filtering techniques. If a complement appears very rarely associated with a given predicate, the acquisition process will assume that this is an incidental co-occurrence that should be left out. However, as we will see, even if this technique is efficient for high frequency items, it leaves a lot of phenomena aside.

Following these observations, we get interested in Optimality Theory (OT). OT is based on a number of assumptions which are absolutely relevant for the lexical acquisition context [2,3,4]:

- Linguistic well-formedness is relative, not absolute. Perfect satisfaction of all linguistic constraints is attained rarely, and perhaps never.
- Linguistic well-formedness is a matter of comparison or competition among candidate output forms (none of which is perfect).
- Linguistic constraints are ranked and violable. Higher ranking constraints can compel violation of lower ranking constraints. Violation is minimal, however. And even low ranking constraints can make crucial decisions about the winning output candidate.
- The grammar of a language is a ranking of constraints. Ranking may differ from language to language, even if the constraints do not.

However, despite these observations, OT has been mainly applied to phonology, more rarely to morphology or syntax [5,1]. In this paper, we would like to show, on a precise example, that OT provides a very competitive framework for sub-categorization acquisition.

In order to apply OT to lexical acquisition, we first need to model all the language properties as constraints. The task consists then in identifying the relevant set of constraints that allow one to map a lexical structure to actual (surface) constructions. Note that the task is highly challenging since constraints interact with each other, must be ranked and can be violated.

2   From Corpus to Resources

2.1   OT and Syntax

OT has been mainly applied to syntax in the framework of the Principles and Parameters (P&P) theory developed by Chomsky [6] as part of his Minimalist Program. The central idea of P&P is that a person's syntactic knowledge can be modeled with two formal mechanisms:

- A finite set of fundamental principles that are common to all languages; e.g., a sentence must always have a subject, even if it is not overtly pronounced.
- A finite set of parameters that determine syntactic variability amongst languages; e.g., a binary parameter that determines whether or not the subject of a sentence must be overtly pronounced.

Within this framework, the goal of linguistics is to identify all the principles and parameters that are universal to human languages (i.e. what defines the Universal Grammar).

OT provides a nice framework to implement P&P since the formalism is constraint-based. The input is a set of (universal) abstract candidate forms[1]. Thus, principles and parameters just have to be translated into constraints (CON); then an evaluation function (EVAL) computes the best output given the input and the set of constraints (the principles and parameters) for a given language.

To summarize, here are the three main components of OT: GEN (+input), CON and EVAL.

- GEN takes a series of surface forms and generates an infinite number of candidates, or possible realizations of that input. A language's grammar (its ranking of constraints) determines which of the infinite candidates will be assessed as optimal by EVAL.
- CON includes the set of constraints to be used to determine which of the input candidates is the most likely to be accepted.
- EVAL determines the best analysis among input candidates, taking into account the set of constraints CON. Given two candidates, A and B, A is better than B on a constraint hierarchy if A incurs fewer violations than B. Candidate A is better than B on an entire constraint hierarchy if A incurs fewer violations of the highest-ranked constraint distinguishing A and B. A is optimal in its candidate set if it is better on the constraint hierarchy than all other candidates.

However, the task here is slightly different (converse) since we try to find the best underlying representation from the output (a given utterance), more precisely, we try to learn syntactic frames from data.

2.2   Learning Syntactic Frames from Raw Data

As already said, comprehensive and accurate lexical resources are key components of Natural Language Processing (NLP) systems. Hand-crafting lexical resources is difficult and extremely labour-intensive— particularly as NLP systems require statistical information about the behavior of lexical items in context, and this statistical information changes from one domain to the other. For this reason automatic acquisition of lexical resources from corpora has become increasingly popular.

One of the most useful lexical information for NLP is that related to the predicate-argument structure. The sub-categorization frames (SCFs) of a predicate capture the different combinations of arguments that a given predicate can take. For example, in French, the verb "*acheter*" (*to buy*) sub-categorizes for a subject, a direct object and an indirect object (a prepositional phrase governed by the preposition "*à*"). This can be formalized as follows: *N0 acheter N1 à N2*.

---

[1] This point, which is much controversial, is based on the assumption that linguistic principles— in P&P Theory—are supposed to be universal. There is a huge literature on this hypothesis that we will not address in this paper. We do not claim any universal feature in this work; we just use OT as an interesting framework for modeling the constraints used.

Sub-categorization lexicons can benefit many NLP applications. For example, they can be used to enhance tasks such as parsing [7,8] and semantic classification [9] as well as applications such as information extraction [10] and machine translation.

Several sub-categorization lexicons are available for many languages, but most of them have been built manually. For French these include the large French dictionary *"Le Lexique Grammaire"* [11] and the more recent *Lefff* [12] and *Dicovalence* (http://bach.arts.kuleuven.be/dicovalence/) lexicons.

Some work has been conducted on automatic sub-categorization acquisition, mostly on English [13,14,15,16] but also on other languages, from which German is just one example [17]. This work has shown that although automatically built lexicons are not as accurate and detailed as manually built ones, they can be useful for real-world tasks. This is mostly because they provide what manually built resources do not generally provide: statistical information about the likelihood of SCFs for individual verbs.

In what follows, we show that statistical information, in order to yield accurate results, must take into consideration a huge number of constraints. First experiments have given interesting results but the nature and the structure of constraints must be further explored in order to strengthen the existing results. We show that OT provides an interesting framework to identify and structure the set of relevant constraints.

2.3   Introducing Gradience in Lexical Acquisition

As for most linguistic questions, there is no well-established definition of what to include in a SCF, but everybody agrees that a SCF should minimally include the number and the type of the complements depending on the verb (or more generally on the predicative item considered, since adjectives and nouns can also have a SCF). Most authors agree on the fact that complements should be divided between arguments and adjuncts but the distinction between these two categories is far from obvious. Some linguistic tests exist (can the complement be deleted without changing the meaning of the sentence? Can it be moved easily? Can it be pronominalized? etc.) but none of these tests is sufficient or discriminatory enough.

As outlined by Manning [18] "rather than maintaining a categorical argument / adjunct distinction and having to make in/out decisions about such cases, we might instead try to represent SCF information as a probability distribution over argument frames, with different verbal dependents expected to occur with a verb with a certain probability". For example, from the analysis of a large news corpus, one can observe that the French verb *venir (to come)* accepts the frame *PP[de (from)]* with a relative frequency of 59.1% whereas it accepts the frame *PP[à (to)]* with a relative frequency of 5%. This phenomenon can be seen as a kind of selectional "preference" of certain verbs for certain SCFs; the link with more semantic information remains to be done.

It is well known that the evaluation of probability distributions is difficult, since it is by definition dependent on a given corpus. Hand-crafted dictionaries generally do not include any frequency information. Moreover, very few lexical acquisition frameworks currently integrate an efficient way to deal with various phenomena such as multiword expressions (especially light verb constructions and semi-idiomatic expressions), complement optionality, etc. Therefore, current approaches have a tendency to produce two many SCFs for a given items (semi-idiomatic expressions should be recognized as such and should not be added as new SCFs associated with head verbs, optionality should be handled to reduce the number of partial SCFs).

In the next section, we briefly present a state-of-the art system for French and its limitations; we show that the acquisition model corresponds to OT but does not take into consideration a precise enough set of constraints. We then make some proposals in order to get better results using a finer grain model of constraints.

3   ASSCI, A State-of-the Art Subcategorization Acquisition System for French

A system for the automatic acquisition of sub-categorization frames has recently been implemented for French. This system called ASSCI is capable of acquiring large scale lexicons from un-annotated corpora [19].

This system is close to other systems developed for example for English [15,20] in that it extracts SCFs from data parsed using a shallow dependency parser [21] and is capable of identifying a large number of SCFs. However, unlike most other systems that accept raw corpus data as input, it does not assume a list of predefined SCFs. The system is based on the assumption that the most relevant SCF corresponding to a given surface form will directly emerge from the application of the constraints on the various candidates, as postulated by OT.

ASSCI takes raw corpus data as input. Input text is first tagged and syntactically analyzed. Then, the system generates a list of candidate SCFs for each verb that occurs frequently enough in data (in the default setting, 200 occurrences of a given verb are necessary). *ASSCI* consists of three modules: a pattern extractor which extracts patterns for each target verb; a SCF builder which builds a list of candidate SCFs per verb (GEN), and a SCF filter (EVAL) which filters out SCFs deemed incorrect according to predefined parameters (CON). They are described briefly in the following sections. For a more detailed description of *ASSCI*, see [19].

3.1   Preprocessing: Morphosyntactic Tagging and Syntactic Analysis

The system first tags and lemmatizes corpus data using *TreeTagger* and then parses it thanks to *Syntex* [21]. *Syntex* is a shallow parser for French. It uses a combination of heuristics and statistics to find dependency relations between tokens in a sentence. It is a relatively accurate parser, e.g. it obtained the best precision and F-measure for written French text in the first EASY evaluation campaign (2006).

The below example illustrates the dependency relations detected by *Syntex* (2) for the input sentence in (1):

```
(1) La sécheresse s' abattit sur le Sahel en 1972-1973 .
(The drought came down on Sahel in 1972-1973.)
```

```
(2) DetFS|le|La|1|DET;2|
NomFS|sécheresse|sécheresse|2|SUJ;4|DET;1|
Pro|se|s'|3|REF;4|
```

```
VCONJS|abattre|abattit|4|SUJ;2,REF;3,PREP;5,PREP;8
Prep|sur|sur|5|PREP;4|NOMPREP;7
DetMS|le|le|6|DET;7|
NomMS|sahel|Sahel|7|NOMPREP;5|DET;6
Prep|en|en|8|PREP;4|NOMPREP;9
NomXXDate|1972-1973|1972-1973|9|NOMPREP;8|
Typo|.|.|10||
```

Syntex does not make a distinction between arguments and adjuncts - rather, each dependency of a verb is attached to the verb.

3.2   Producing the Input (The Pattern Extractor)

The pattern extractor collects the dependencies found by the parser for each occurrence of a target verb. Some cases receive special treatment in this module. For example, if the pronoun *"se"* is one of the dependencies of a verb, the system considers this verb like a new one. In (1), the pattern will correspond to *"s'abattre"* and not to *"abattre"*. If a preposition is the head of one of the dependencies, the module explores the syntactic analysis to find if it is followed by a noun phrase (+SN]) or an infinitive verb (+SINF]). (3) shows the output of the pattern extractor for the input in (1).

```
(3) VCONJS|s'abattre :
Prep+SN|sur|PREP__Prep+SN|en|PREP
```

3.3   GEN (The SCF Builder)

The SCF builder extracts SCF candidates for each verb from the output of the pattern extractor and calculates the number of corpus occurrences for each SCF and verb combination. The syntactic constituents used for building the SCFs are the following:

1. SN for nominal phrases;
2. SINF for infinitive clauses;
3. SP[*prep*+SN] for prepositional phrases where the preposition is followed by a noun phrase. *prep* is the head preposition;
4. SP[*prep*+SINF] for prepositional phrases where the preposition is followed by an infinitive verb. *prep* is the head preposition;
5. SA for adjectival phrases;
6. COMPL for subordinate clauses.

When a verb has no dependency, its SCF is considered as INTRANS.
(4) shows the output of the SCF builder for (1).

```
(4) S'ABATTRE+s'abattre ;;; SP[sur+SN]_SP[en+SN]
```

3.4 CON and EVAL (SCF Filter)

Each step of the process is fully automatic, so the output of the SCF builder is noisy due to tagging, parsing or other processing errors. It is also noisy because of the difficulty of the argument-adjunct distinction. The latter is difficult even for humans.

Many criteria that have been defined are not usable in our case because they either depend on lexical information which the parser cannot make use of (since the task is to acquire this information) or on semantic information which even the best parsers cannot yet learn reliably. The approach here is based on the assumption that true arguments tend to occur in argument positions more frequently than adjuncts. Thus many frequent SCFs in the system output are correct.

The strategy is then to filter low frequency entries from the SCF builder output. This is done using the maximum likelihood estimates [22]. This simple method involves calculating the relative frequency of each SCF (for a verb) and comparing it to an empirically determined threshold. The relative frequency of the SCF i with the verb j is calculated as follows:

$$rel\_freq(scf_i, verb_j) = \frac{|scf_i, verb_j|}{|verb_j|}$$

$|scf_i, verb_j|$ is the number of occurrences of the SCF i with the verb j and $|verb_j|$ is the total number of occurrences of the verb j in the corpus.

If, for example, the frequency of the SCF SP[sur+SN]_SP[en+SN] is below the empirically defined threshold, the SCF is rejected by the filter. The MLE filter is not perfect because it is based on rejecting low frequency SCFs. Although relatively more low than high frequency SCFs are incorrect, sometimes rejected frames are correct. The filter incorporates special heuristics for cases where this assumption tends to generate too many errors. With prepositional SCFs involving one PP or more, the filter determines which one is the less frequent PP. It then re-assigns the associated frequency to the same SCF without this PP.

For example, SP[sur+SN]_SP[en+SN] could be split to 2 SCFs : SP[sur+SN] and SP[en+SN]. In this example, SP[en+SN] is the less frequent prepositional phrase and the final SCF for the sentence (1) is (5).

(5) SP[sur+SN]

Note that SP[en+SN] is here an adjunct.

4 Some Limitations of This Approach

This approach is very efficient to deal with large corpora. However, some issues remain. As the approach is based on automatic tools (especially parsers) that are far from perfect, the obtained resources always contain errors and have to be manually validated. Moreover, the system needs to get enough examples to be able to infer relevant information. Therefore, there is generally a lack of information for a lot of low productivity items (the famous "sparsity problem").

More fundamentally, some constructions are difficult to acquire and characterize automatically. On the one hand, idioms are not recognized as such by most acquisition systems. On the other hand, some adjuncts appear frequently with certain verbs (eg. some verbs like *dormir – to sleep* – frequently appear with location complements). The system then assumes that these are arguments, whereas linguistic theory would say without any doubt that these are adjuncts. Lastly, surface cues are sometimes insufficient to recognize ambiguous constructions (cf. *...manger une glace à la vanille...* vs *...manger une glace à la terrasse d'un café... — to eat a vanilla ice-cream vs to eat an ice-cream at an outdoor cafe*).

In a traditional architecture, the filtering process incorporates in one modules the set of constraints (CON) and the evaluation function (EVAL). This makes the system less readable than if the constraints were modeled apart from the EVAL function. There is thus a need to refine the set of constraints

5   A Solution: Provide an Explicit Modeling of the Set of Constraints (CON)

We have shown in the previous section that a part of the errors produced were due to an over-simplification of the initial model. It is thus necessary to take other parameters into considerations in order to yield better results. This can be done by refining the set of constraints (CON).

5.1   Refining CON

The issues we have reported in the previous section do not mean that automatic methods are flawed, but they have a number of drawbacks that should be addressed. The acquisition process, based on an analysis of co-occurrences of the verb with its immediate complements (along with filtering techniques) makes the approach highly functional. It is a good approximation of the problem. However, this model does not take into account external constraints.

The analysis of the co-occurrences of the verb with its complement is meaningful but is not sufficient to fully grasp the problem. The fact that some phrasal complements (with a specific head noun) frequently co-occur with a given verb is most of the time useful, especially to identify idioms [23], colligations [24] and light verb constructions [25]. On the other hand, the fact that a given prepositional phrase appear with a large number of verbs may indicate that the preposition introduces an adjunct rather than an argument.

So, instead of simply capturing the co-occurrences of a verb with its complements, a number of important features should be taken into account:

- indicator of the dispersion of the prepositional phrases (PP) depending on the nature of the preposition (if a PP with a given preposition appears with a wide range of different verbs, it is more likely to be a modifier);
- indicator of the co-occurrence of the PP depending on the nature of the head noun (if a verb appears frequently with the same PP frame, it is more likely to form a semi-idiomatic expression);
- indicator of the complexity of the sentence to be processed (if a sentence is complex, its analysis is less reliable).

In order to do this, the pattern extractor has to be modified in order to keep most of the information that were previously rejected as not relevant. These indicators then need to be calculated so as to be taken into account by EVAL.

5.2   Modifying EVAL

All the constraints can be evaluated separately, so as to obtain for each of them an ideal evaluation of the parameter. There are two ways of doing this: *i*) by automatically inferring the different weights from a set of annotated data or *ii*) by estimating the results of various manually defined weights. We are currently using this last method since data annotation is very costly. However, the first approach would certainly lead to more accurate results.

The weight and the ranking of the different constraints must then be examined. A linear model can provide a first approximation but there are surely better ways to integrate the different constraints. Some studies provide some cues but they need to be proper evaluated in order to be integrated in this framework [5].

5.3   Manual Validation

Lastly, the approach requires a manual validation. Rather than leaving the validation process apart for further examination by a linguist, we propose to integrate it in the acquisition process itself. Taking into consideration the number of examples and the complexity of the sentences used for training, it is possible to associate confidence scores with the different constructions of a given verb: the linguist is then able to quickly focus on the most problematic cases. It is also possible to propose tentative constructions to the linguist, when not enough occurrences are available for training. In the end, when too few examples are available, the linguist can provide relevant information to the machine. However, with a well-designed and dynamic validation process, it is possible to obtain accurate and comprehensive lexicons, using only a small fraction of the time that would be necessary to manually develop a lexicon from scratch.

6   Conclusion

Tn this paper, we have proposed a new approach for the automatic acquisition of lexical knowledge from corpora using Optimality Theory. Using this model, it is possible to represent a large part of the language activity through constraints. We have shown that the individual evaluation of each constraint yields very accurate and precise results.

An implementation of this model is currently being done for Japanese [26]. The model provides a better integration of the linguistic contraints within the automatic processing system. First results were competitive with other approaches while providing a more accurate linguistic description.

References

1. Aarts, B.: Syntactic Gradience: The Nature of Grammatical Indeterminacy. Oxford University Press, Oxford (2008)

2. Kager, R.: Optimality Theory. Cambridge University Press, Cambridge (1999)
3. McCarthy, J.: Doing Optimality Theory. Blackwell, Oxford (2008)
4. Prince, A., Smolensky, P.: Optimality Theory: Constraint Interaction in Generative Grammar. Blackwell, Oxford (2004)
5. Blache, P., Prost, J.P.: A Quantification Model of Grammaticality. Constraints Solving and Language Processing (2008)
6. Chomsky, N.: The Minimalist Program. The MIT Press, Cambridge (1995)
7. John Carroll, G.M., Briscoe, T.: Can subcategorisation probabilities help a statistical parser? In: Proceedings of the 6th ACL/SIGDAT Workshop on Very Large Corpora, Montreal, Canada (1998)
8. Arun, A., Keller, F.: Lexicalization in crosslinguistic probabilistic parsing: The case of French. In: Proceedings of the 43rd Annual Meeting of the Association for Computational Linguistics (ACL 2005), pp. 306–313. Association for Computational Linguistics, Ann Arbor (2005)
9. Schulte im Walde, S., Brew, C.: Inducing German Semantic Verb Classes from Purely Syntactic Subcategorisation Information. In: Proceedings of the 40th Annual Meeting of the Association for Computational Linguistics, Philadelphia, PA, pp. 223–230 (2002)
10. Surdeanu, M., Harabagiu, S.M., Williams, J., Aarseth, P.: Using Predicate-Argument Structures for Information Extraction. In: Proceedings of the Association of Computational Linguistics (ACL), pp. 8–15 (2003)
11. Gross, M.: Méthodes en syntaxe. Hermann, Paris (1975)
12. Sagot, B., Clément, L., de La Clergerie, E., Boullier, P.: The Lefff 2 Syntactic Lexicon for French: Architecture, Acquisition, Use. In: Language Resource and Evaluation Conference (LREC), Genoa (2006)
13. Brent, M.R.: From Grammar to Lexicon: Unsupervised Learning of Lexical Syntax. Computational Linguistics 19, 203–222 (1993)
14. Manning, C.D.: Automatic Acquisition of a Large Subcategorization Dictionary from Corpora. In: Proceedings of the Meeting of the Association for Computational Linguistics, pp. 235–242 (1993)
15. Briscoe, T., Carroll, J.: Automatic Extraction of Subcategorization from Corpora. In: Proceedings of the 5th ACL Conference on Applied Natural Language Processing, Washington, DC, pp. 356–363 (1997)
16. Korhonen, A., Krymolowski, Y., Briscoe, T.: A Large Subcategorization Lexicon for Natural Language Processing Applications. In: Proceedings of the 5th International Conference on Language Resources and Evaluation, Genova, Italy (2006)
17. Schulte im Walde, S.: A Subcategorisation Lexicon for German Verbs induced from a Lexicalised PCFG. In: Proceedings of the 3rd Conference on Language Resources and Evaluation, Las Palmas de Gran Canaria, Spain, vol. IV, pp. 1351–1357 (2002)
18. Manning, C.D.: Probabilistic syntax. In: Bod, R., Hay, J., Jannedy, S. (eds.) Probabilistic Linguistics, pp. 289–341. MIT Press (2003)
19. Messiant, C.: ASSCI : A Subcategorization Frames Acquisition System For French. In: Proceedings of the Association for Computational Linguistics (ACL) Student Research Workshop. Association for Computational Linguistics, Colombus (2008)
20. Preiss, J., Briscoe, T., Korhonen, A.: A System for Large-Scale Acquisition of Verbal, Nominal and Adjectival Subcategorization Frames from Corpora. In: Proceedings of the Meeting of the Association for Computational Linguistics, Prague, pp. 912–918 (2007)
21. Bourigault, D., Jacques, M.P., Fabre, C., Frérot, C., Ozdowska, S.: Syntex, analyseur syntaxique de corpus. In: Actes des 12èmes Journées sur le Traitement Automatique des Langues Naturelles, Dourdan (2005)

22. Korhonen, A., Gorrell, G., McCarthy, D.: Statistical Filtering and Subcategorization Frame Acquisition. In: Conference on Empirical Methods in Natural Language Processing and Very Large Corpora, Hong Kong (2000)
23. Fabre, C., Bourigault, D.: Exploiter des corpus annots syntaxiquement pour observer le continuum entre arguments et circonstants. Journal of French Language Studies 18(1), 87–102 (2008)
24. Firth, J.R.: Descriptive Linguistics and the Study of English. In: Selected Papers of John R. Firth (1968)
25. Butt, M.: The Light Verb Jungle. Harvard Working Papers in Linguistics 9, 1–49 (2003)
26. Marchal, P., Poibeau, T., Lepage, Y.: Representing the Continuum between Arguments and Adjuncts within Predicate-Frames. In: NINJAL International Symposium on "Valency Classes and Alternations in Japanese", Tokyo (2012)

Verb Clustering for Brazilian Portuguese

Carolina Scarton[1,4], Lin Sun[2], Karin Kipper-Schuler[3], Magali Sanches Duran[4],
Martha Palmer[3], and Anna Korhonen[2]

[1] Department of Computer Science, University of Sheffield
Regent Court, 211 Portobello, Sheffield, S1 4DP, UK
c.scarton@sheffield.ac.uk
[2] Computer Laboratory, University of Cambridge
William Gates Building, 15 JJ Thomson Avenue, Cambridge CB3 0FD, UK
linsun84@gmail.com, alk23@cam.ac.uk
[3] Department of Linguistics, University of Colorado at Boulder
295 UCB Boulder, Colorado 80309-0295
karin_schuler@yahoo.com, martha.palmer@colorado.edu
[4] Interistitutional Center for Computational Linguistics, ICMC, University of São Paulo
Avenida Trabalhador são-carlense, 400 - Centro, 13566-590, São Carlos - SP
magali.duran@uol.com.br

Abstract. Levin-style classes which capture the shared syntax and semantics of
verbs have proven useful for many Natural Language Processing (NLP) tasks and
applications. However, lexical resources which provide information about such
classes are only available for a handful of worlds languages. Because manual development of such resources is extremely time consuming and cannot reliably
capture domain variation in classification, methods for automatic induction of
verb classes from texts have gained popularity. However, to date such methods
have been applied to English and a handful of other, mainly resource-rich languages. In this paper, we apply the methods to Brazilian Portuguese - a language
for which no VerbNet or automatic class induction work exists yet. Since Levin-
style classification is said to have a strong cross-linguistic component, we use
unsupervised clustering techniques similar to those developed for English without language-specific feature engineering. This yields interesting results which
line up well with those obtained for other languages, demonstrating the cross-
linguistic nature of this type of classification. However, we also discover and
discuss issues which require specific consideration when aiming to optimise the
performance of verb clustering for Brazilian Portuguese and other less-resourced
languages.

1 Introduction

Verbs are central to many Natural Language Processing (NLP) tasks. Typically the
main predicates of sentences, they provide the key syntactic and semantic information required for language understanding. Information regarding verbs is traditionally
obtained from lexical resources such as WordNet [1], FrameNet [2], PropBank [3] and
VerbNet [4].

In this paper we are particularly concerned with Levin-style verb classification. Beth
Levin [5] has classified verbs according to their participation in diathesis alternations.

A. Gelbukh (Ed.): CICLing 2014, Part I, LNCS 8403, pp. 25–39, 2014.

These are alternations in the syntactic realization of a verb that may sometimes also result in a slight change of meaning. For example, sentences (1) and (2) (from [5], p. 2) illustrate the locative alternation for verbs "spray" and "load":

1. (a) Sharon sprayed water on the plants.
   (b) Sharon sprayed the plants with water.
2. (a) The farmer loaded apples into the cart.
   (b) The farmer loaded the cart with apples.

Levin classes capture the regularity in verb meaning and behaviour at syntax-semantics interface. VerbNet [4], an extensive computational verb lexicon for English, extends Levin's original classification with additional classes and member verbs, and provides detailed syntactic-semantic information for the classes that span across the entire verb lexicon.

Because Levin classes capture useful generalisations about verb behaviour and meaning, they have proven useful for many NLP tasks and applications. Examples include information retrieval [6], semantic role labelling [7, 8], semantic parsing [9], word sense disambiguation [10–13], among others. Since many Levin-style classes are applicable across languages [14], they can also be useful for cross-linguistic tasks. However, their effective exploitation in the multi-lingual context has been limited to date because resources like VerbNet are currently only available for English and a handful of other languages, such as Spanish [15], Chinese [16] and Arabic [17].

Although manual development of VerbNets is under way for many languages, it is extremely time-consuming and cannot reliably capture domain variation. Therefore techniques that can automatically induce or update verb classifications from texts have gained popularity. Particularly attractive are unsupervised techniques such as clustering because they do not require manual annotations for training and are therefore easier port across NLP tasks.

For English, verb clustering approaches have been developed by Kingsbury and Kipper-Schuler [18], Sun and Korhonen [19] and Reichart and Korhonen [20], among others. The best techniques have been applied successfully to domains such as biomedicine [21] and they have produced promising results with demonstrated improvement on application tasks such as argumentative zoning [22] and metaphor identification [23].

For languages other than English, only a small number of clustering works exist that focus on Levin style syntactic-semantic classification, e.g. [24–26]. Interestingly, the recent experiment performed by Sun et al. demonstrates that it is possible to take an unsupervised clustering method developed for English [27] and apply it successfully to French [25], using French NLP tools for feature extraction, but without language specific feature engineering. If this approach was applicable to a wider range of languages, it could greatly support the development of VerbNets across languages and language domains.

In this paper, we explore this approach for a another language: Brazilian (Br.) Portuguese. Although Portuguese is a major world language (around 215 million speakers worldwide), no manually developed VerbNet or automatic verb clustering approach exists for it yet. While the language belongs to the family of Romance languages like

French, it is lexically more distant to English and is also less-resourced in terms of basic corpora and NLP tools than French, and therefore likely to be more challenging for clustering.

We develop and release the first gold standard Levin classification for Br. Portuguese and apply the state-of-the-art verb clustering approach developed for English [27] to this language. Using the NLP tools developed for Br. Portuguese for feature extraction, we experiment with the same basic features and the same clustering method as for English. The results are encouraging and support the hypothesis that Levin-style classes can indeed be cross-linguistically applicable: it was possible to obtain a gold standard largely via translation from English to Br. Portuguese, and the best performing features and clustering techniques matched with those for English and French. The level of clustering performance for Br. Portuguese lags behind that of resource-richer languages. We investigate reasons for this and discuss future work in this area.

2   Related Work

Several approaches have been developed for classifying English verbs into Levin-style classes in both supervised [28–30] and unsupervised [18–20, 27] manner. These approaches have employed a variety of different features for classification, ranging from shallow features (e.g. co-occurrences of verbs with other words) extracted from raw or part-of-speech (POS) tagged text to deeper features (e.g. grammatical dependencies) extracted from manually or automatically parsed data. Also lexical(-semantic) features which correspond more closely with the features Levin used for her manual classification have been used, such as verb subcategorization frames (SCFs), selectional preferences (SPs) and recently also diathesis alternation approximations [31]. These more sophisticated features have been learned from parsed data. Clustering approaches have ranged from the simple k-means [18] to more sophisticated techniques such as spectral clustering [27], hierarchical clustering using graph factorization [19] and determinantal point processes [20], among others.

For example, the state-of-the-art approach of Sun and Korhonen [27] which we plan to use as a starting point in our work, uses a variety of features based on co-occurrences, verb SCFs and lexical as well as selectional preferences of verbs on their argument heads, and yields promising performance when used with spectral clustering. When this approach was evaluated against gold standards based on Verbnet [30, 32], both containing hundreds of verbs in 15-20 classes, it achieved the highest performance (at around 80 F-measure) with deep linguistic features: SCFs refined with selectional preferences.

For languages other than English, few works exist. The most substantial related work focuses on German [33] where verb clustering has yielded promising results, but this work has emphasis on semantic rather than VerbNet style syntactic-semantic classification. Although the two classification types share many properties, the mapping between the two is only partial and many to many due to fine-grained nature of semantic classes based purely on synonymy [11].

The prior works most related to ours are those by Ferrer [24] for Spanish and Sun et al. [25] and Falk et al. [26] for French. Ferrer applied a simple hierarchical clustering approach developed for English to Spanish, and evaluated it against a manual classification of Vazquez [34] which is similar in nature (but not identical) to that of Levin's. The experiment included 514 verbs in 31 classes and produced results only slightly better than the random baseline.

Sun et al. [25] used the same features as Sun and Korhonen [27] for English to cluster 171 French verbs to 16 classes. The gold standard was obtained by translating the Levin-based gold standard of Sun et al. [32] from English to French, and a good correspondence was reported between the two gold standards. The authors reported the best results (64.5 F-measure) on high frequency verbs with the same combination of features (SCFs and selectional preferences) and the same clustering method (spectral clustering) as for English. Falk et al. [26] employed a neural clustering method for French verbs. They achieved 70 F-measure when evaluating on a slightly modified version of the Sun et al. 2010 gold standard for French. However, the method is not fully comparable to other works mentioned here because it uses features from lexical resources rather than those obtained solely by NLP.

The work reported on manual development of VerbNets for different languages [15–17] seems to support the hypothesis that Levin-style classes can be, to a considerable degree, cross-linguistically applicable. The experiment reported by Sun et al. [25] provides further evidence for this because it shows that it is possible to take an unsupervised technique developed for one language and apply it to another language without language specific tuning (other than use of language-specific corpora and basic NLP tools for feature extraction) and get promising results.

However, this experiment focused on French only. French shares some of its vocabulary with English, and like English, has large corpora, POS-taggers, parsers and lexical acquisition tools available. If this approach proves more widely applicable so that it can be successfully employed for other, including also less-resourced languages, verb clustering could offer a useful tool for hypothesizing Levin classes for other languages. We will take this line of research further and investigate whether the approach could be applied to Br. Portuguese which, like the majority of worlds languages, is less-resourced in terms of NLP than French and is thus likely to be a more challenging test case.

Portuguese is a Romance language which has its origins in Latin and is currently the seventh most spoken language in the world. From the 215 million people speaking Portuguese, 85% speak Br. Portuguese. Br. Portuguese differs from the European Portuguese largely in terms of lexicon. As we are dealing with the verb lexicon, we differentiate between the two variations of the language and focus on Br. Portuguese only. However, our work could be easily extended to accommodate European Portuguese as well.

Some major lexical resources are currently under development for both variants of Portuguese. The ones related to verbs include PropBank-Br [35] (based on PropBank), FrameNet Brasil [36] and FrameCorp [37] (based on FrameNet), WordNet.Br [38], WordNet.Pt [39, 40], one of the Wordnets in the MultiWordNet Project [41] (based on WordNet) as well as VerbNet.Br [42] based on English VerbNet. The latter project, which is most closely related to our work, provides alignments between English VerbNet, WordNet and WordNet.Br, and enables semi-automatically inferring Levin classes for Br. Portuguese from the alignment data. The classification created using this method is noisy and has not been manually validated.

3 Gold Standard for Brazilian Portuguese

As no VerbNet exists which could provide gold standard classes for our experiments we created the first gold standard including Levin classes for Br. Portuguese[1]. We used an approach similar to that earlier employed by Sun et al. [25] for building a gold standard for French. They took a gold standard frequently used for evaluating verb clustering for English [32] and translated its 204 verbs and 17 classes to French. The majority of verbs and classes could be translated successfully. To cover for the ones that could not, Sun et al. considered synonyms of known member verbs and added these in, where possible. French subcategorization frames (SCFs) and alternations, defined manually for each class, were used as evaluation criteria. The final gold standard included 171 verbs in 16 classes.

We employed a similar approach because it had proved successful for French and we were interested in exploring the cross-linguistic potential of Levin classification. We used a native speaker of Portuguese with expertise in VerbNet to develop the first version of the gold standard. She performed the translation and defined syntactic-semantic criteria for each class. We ended up with 203 verbs in 16 classes (12.69 verbs per class). The majority of verbs (including their synonyms) got translated successfully. Only one class in the English gold standard was deemed incompatible with Portuguese (peer-30.3), showing a strong cross-lingual element between English and Portuguese classifications, similar to that earlier observed with English and French.

Because many of the verbs in the resulting gold standard were quite low in frequency in our corpus, we supplemented the resource with additional member verbs from Verb-Net.Br – the resource recently developed by Scarton and Aluísio [42]. As the classifications in this resource are noisy, we used a native language expert to validate the class memberships according to the criteria we had developed during the translation of the first version of the gold standard. The resulting extended gold standard includes 540 verbs in 16 classes (c. 34 verbs per class).

Table 1 shows the resulting classes in the gold standard (indicated by original Levin class numbers) together with some example verbs.

4 Verb Clustering

4.1 Features

We employed a selection of syntactic and semantic feature sets that had proved promising for both English [27] and French [25]. To facilitate easy comparison of our results against earlier works we indicate each feature set using the same numbers as in [27] and [25]:

---

[1] We will release this gold standard together with a published version of this paper.

Table 1. Brazilian Portuguese gold standard classes with some example verbs

Number	Class	Portuguese Members
22.2	amalgamate	*alternar, contrastar, combinar, juntar, comparar,...*
31.1	amuse	*frustar, chatear, alegrar, decepcionar, encantar,...*
29.2	characterize	*diagnosticar, restabelecer, retratar, classificar,...*
36.1	correspond	*pechinchar, flertar, simpatizar, colidir, cooperar,...*
13.5.1	get	*arranjar, colher, reservar, adquirir, obter,...*
18.1	hit	*martelar, esmagar, espancar, bater,...*
43	light emission	*resplandecer, raiar, cintilar, piscar, brilhar,...*
37.3	manner of speaking	*cochichar, rosnar, sussurrar, berrar, ...*
47.3	modes of being with motion	*boiar, flutuar, vibrar, oscilar,...*
40.2	nonverbal expression	*bocejar, roncar, solućar, suspirar, sorrir,...*
45	other cos	*encurtar, afrouxar, alargar, estreitar, derreter,...*
9.1	put	*cravar, posicionar, mergulhar, situar, inserir,...*
10.1	remove	*erradicar, subtrair, descarregar, remover,...*
51.3.2	run	*marchar, nadar, passear, voar, correr,...*
37.7	say	*segredar, reportar, dizer, proclamar, exprimir,...*
11.1	send	*despachar, transportar, remeter, enviar,...*

- F1: SCFs and their relative frequencies with individual verbs (without parameterising for prepositions).
- F2: F1 with SCFs parameterized for the tense (i.e. POS tag) of the verb.
- F3: F1 with SCFs parameterized for specific prepositions.
- F7: Collocations (COs) extracted from the window of 6 words, with the relative word position recorded. We followed the work of Li and Brew [29] where COs were extracted from the window of words immediately preceding and following a POS-tagged and lemmatized verb (stop words were removed before the extraction).
- F13: All Lexical Preferences (LPs) in argument head positions: the type and frequency of words acting as prepositions (PREP), subjects (SUBJ), indirect objects (IOBJ) and direct objects (OBJ) in dependency-parsed data were considered.
- F16: F3 parameterized for LPs.
- F17: F3 refined with Selectional Preferences (SPs).

The extraction of these features requires POS-tagging and, with the exception of F7, also parsing data, and using additional technology to extract SCFs and SPs from the parsed data. We used the three publicly available corpora for Brazilian Portuguese to ensure that as much data as possible was available for clustering. These were (i) Lácio-Ref [43] which includes legal, news, scientific and literary texts - approximately 9 million words in total, (ii) PLN-BR-FULL [44] which provides 29M words of news texts and (iii) Revista Pesquisa FAPESP corpus [45] which contains 6M words of scientific text.

These corpora were POS-tagged and parsed using the rule-based PALAVRAS [46] parser which outputs grammatical relations. According to the evaluation performed by the authors, this rule-based parser achieves 99% of correctness for POS and 97% for syntax. We used the system of Zanette et al. [47] to extract SCFs from the resulting parsed data. Similar to the system of [48] for French, the system generates SCFs from the dependency relations associated with individual verbs in parsed data. According to the evaluation of Zanette [49], this system performs at around 50.6% F-measure.

We considered all the dependencies of interest and all the SCFs with frequency higher than 5 in our experiments. This yielded 3,779 verb lemmas, 408 basic SCF types and 3,578 preposition-parameterized SCF types. For SP acquisition, we used the method proposed by Sun and Korhonen (2009) (without the automatic definition of best number of clusters in Sun et al. (2010)). The method involves (i) taking the SUBJ, OBJ and IOBJ relations associated with verbs in parsed data, (ii) extracting all the argument heads in these relations, and (iii) clustering the resulting N most frequent argument heads into M classes. We considered frequency higher than 5 and N {200, 500} most frequent argument heads and M {10, 20, 30, 80} classes. Finally, all feature vectors were normalized by the sum of the feature values before clustering.

4.2   Clustering Algorithms

We used two clustering algorithms in our work: the MNCut spectral clustering algorithm (SPEC) which produced the best results in both English and French [25, 27] and a recent Data-Cluster-Data (DCD) algorithm [50], not previously employed for verb clustering. We wanted to experiment with DCD because it had been shown to work together with SPEC to reduce problems of data or feature sparsity which a less-resourced language, in particular, will suffer from.

In DCD, SPEC is first used to perform dimensionality reduction using measures of distributional similarity. The resulting feature space tends to be dense and the infrequent (and potentially unreliable) features become less important when distributional similarity measures are used. DCD takes the output of SPEC as the initial guess and performs further optimization. In the experiments performed by [50] the method further improved the performance of SPEC on varied datasets (consisting of text, images and other material).

We introduce the two clustering approaches in the below sections, respectively.

Spectral Clustering. Spectral clustering (SPEC) has proved promising in several previous verb clustering experiments, e.g. [27, 51]. Following [27] we used the MNCut spectral clustering [52].

The similarity matrix A is normalized into a stochastic matrix P.

$$P = D^{-1}A \tag{1}$$

The degree matrix D is a diagonal matrix where $D_{ii} = \sum_{j=1}^{N} A_{ij}$.

It was shown by [52] that if P has the K leading eigenvectors that are piecewise constant[2] with respect to a partition I^* and their eigenvalues are not zero, then I^* minimizes the multiway normalized cut which is the sum of transition probabilities across different clusters.

---

[2] The eigenvector v is piecewise constant with respect to I if $v(i) = v(j) \forall i, j \in I_k$ and $k \in 1, 2...K$.

In practice, the leading eigenvectors of P are not piecewise constant. However, we can extract the partition by finding the approximately equal elements in the eigenvectors using a clustering algorithm like KMeans. KMeans is a simple clustering method that iteratively partitions data in order to minimize the within-cluster sums of point-to-cluster-centroid distance.

Data-Cluster-Data. In DCD[3] given a similarity matrix A of the n verbs, the clustering task is to divide the verbs into r disjoint subsets. The pairwise similarity is measured using Jensen Shannon Divergence as in [27]. The aim of the clustering is to find the probability of assigning the ith verb to the kth cluster $p(k|i)$.

The similarity matrix can be seen as an undirected similarity graph where each node corresponds to a verb. If we augment the similarity graph by r cluster nodes, the connection weight between the verb and the cluster is (assuming uniform prior $p(i) = 1/n$):

$$p(i|k) = \frac{p(k|i)p(i)}{\sum_v p(k|v)p(v)} = \frac{p(k|i)}{\sum_v p(k|v)}$$

The similarity between two verbs can be defined as two-step random walks from ith verb to jth verb via all clusters:

$$p(i|j) = \sum_k p(i|k)p(k|j) = \sum_k \frac{p(k|i)p(k|j)}{\sum_v p(k|v)}$$

The objective of the clustering is to find a good approximation between the input similarity matrix A and the random walk probabilities $\hat{A}$. The difference between the two matrices is measured using the Kullback-Leibler divergence. The learning target can be formulated as the following optimization problem:

$$\min_{w \geq 0} D_{KL}(A||\hat{A}) = \sum_{ij} \left(A_{ij} \log \frac{A_{ij}}{\hat{A}_{ij}} - A_{ij} + \hat{A}_{ij} \right)$$

$$\text{s.t.} \sum_k W_{ik} = 1, i - 1, ..., n, \tag{2}$$

where we define $W_{ik} = p(k|i)$, and thus

$$\hat{A}_{ij} = \sum_k \frac{W_{ik} W_{ij}}{\sum_v W_{vk}} \tag{3}$$

By dropping the constant terms, the optimization problem with dirichlet prior on W is equivalent to minimising:

$$J(W) = -\sum_{ij} A_{ij} \log \hat{A}_{ij} - (\alpha - 1) \sum_{ik} \log W_{ik} \tag{4}$$

---

[3] For more detailed information about DCD that that we are able to provide in this section, please see [50].

where α is the parameter of the dirichlet prior.

Taking the constraint in equation 2 into account, the optimization object becomes:

$$L(W, \lambda) = J(W) + \sum_i \lambda_i (\sum_k W_{ik} - 1) \tag{5}$$

[50] proved that the L is non-increasing under the update rule of W and λ detailed in algorithm 1.

Algorithm 1. Optimization Algorithm for DCD [50]

Require: similarity matrix A, number of clusters r, nonnegative initial guess of W

  repeat

    $Z_{ij} = (\sum_k \frac{W_{ik} W_{jk}}{\sum_v W_{vk}})^{-1} A_{ij}$

    $s_k = \sum_v W_{vk}$

    $\nabla_{ik}^- = 2(ZW)_{ik} s_k^{-1} + \alpha W_{ik}^{-1}$

    $\nabla_{ik}^+ = (W^T ZW)_{kk} s_k^{-2} + W_{ik}^{-1}$

    $a_i = \sum_l \frac{W_{il}}{\nabla_{il}^+}, b_i = \sum_l W_{il} \frac{\nabla_{il}^-}{\nabla_{il}^+}$

    $W_{ik} \leftarrow W_{ik} \frac{\nabla_{ik}^- a_i + 1}{\nabla_{ik}^+ + b_i}$

  until W is unchanged

  return cluster assigning probabilities W

The initial guess of W can be produced from the result of another clustering algorithm. As suggested by [50], the result of the spectral clustering can be used for the initialisation of W. Thus, by using the result of spectral clustering as a starting point, we can expect DCD to further improve the clustering result. We convert the result of SPEC to an $n \times r$ binary indicator matrix, and add a small positive random number to all entries. This matrix is used as the input W matrix for the algorithm 1.

4.3   Evaluation Metrics

We evaluate the results of the clustering against the gold standard using F-Measure as in [27] and [25] to facilitate meaningful comparison against previous works. F-measure provides the harmonic mean of precision (P) and recall (R). P is calculated using modified purity – a global measure which evaluates the mean precision of clusters. Each cluster ($k_i \in K$) is associated with the gold-standard class to which the majority of its members belong. The number of verbs in a cluster (k_i) that take this class is denoted by $n_{prevalent}(k_i)$.

$$P = \frac{\sum_{k_i \in K : n_{prevalent(k_i)} > 2} n_{prevalent}(k_i)}{|verbs|}$$

R is calculated using weighted class accuracy: the proportion of members of the dominant cluster DOM-CLUST$_i$ within each of the gold-standard classes $c_i \in C$.

$$R = \frac{\sum_{i=1}^{|C|} |\text{verbs in DOM-CLUST}_i|}{|\text{verbs}|}$$

We calculate the random baseline as follows: 1/number of classes. We also calculate the statistical significance of the results by using the one-tailed McNemar's test [53], with the extension of [54]. We considered p-value lower than 0.05.

5 Results

Table 2 shows the F-measure results for both clustering algorithms with different featuresets (see Section 4.1 is for legends of the different featureset codes). We can see that SPEC performs better than DCD with nearly all the featuresets. This difference is statistically significant for the majority of features and becomes more pronounced as we move towards more sophisticated featuresets. The only featureset on which DCD seems outperform SPEC is F1 which consists plain SCFs, but this difference in performance is not statistically significant. While the poor performance of DCD may seem surprising in the light of the good results of [50], our dataset, focusing solely on natural language, is different in nature and also multiple times smaller than the data employed by Yang and Oja.

Table 2. Results for Br. Portuguese verb clustering, considering 16 clusters (number of classes in the gold standard) – * means no statistically significant difference between the algorithms

Feature	Spectral Cluster	DCD
F1*	33.62	**35.08**
F2*	**39.16**	36.79
F3	**42.27**	40.94
F7	**35.79**	32.23
F13	**39.77**	37.13
F16	**41.23**	37.55
F17 (N=200, M=10)	**38.66**	35.99
F17 (N=200, M=20)	**41.15**	35.62
F17 (N=200, M=30)	**39.70**	34.37
F17 (N=200, M=80)	**39.34**	39.26
F17 (N=500, M=10)	**38.54**	35.66
F17 (N=500, M=20)	**42.06**	34.33
F17 (N=500, M=30)	**42.77**	35.92
F17 (N=500, M=80)	**38.51**	35.33

Regarding features, the best individual featureset is F17 with N=500 and M=30 which yields F-measure of 42.8 with SPEC. This is the most sophisticated featureset which incorporates preposition-prameterized SCFs with SPs. Also F3 (SCFs parameterized for prepositions) and F16 (F3 parameterized for LPs) yield F-measure which is

Table 3. The results for Br. Portuguese (BP), English and French (* inducates the best results for each language) using SPEC

Feature	BP	French	English
F1	33.62	42.4	57.8
F2	39.16	45.9	46.7
F3	42.27	50.6	63.3
F7	35.79	55.1	-
F13	39.77	52.7	74.6
F16	41.23	53.4	73
F17	42.77*	54.6*	80.4*

clearly above 40, but there is a statistically significant difference between the performance of these featuresets and that of F17.

To provide an idea of how this performance compares with results obtained for resource-richer languages, Table 3 shows the previously reported results for English [27] and French [25]. Although the same feature sets and the same SPEC algorithm were used here for all the languages, it is important to note that the results are not directly comparable due to differences in data, NLP tools and gold standards (which are not identical, even though they were derived from the same gold standard). However, this table serves to give an idea of the general performance level and the best performing features for the three languages.

We can observe that clustering performs clearly the best for English (with top performance at 80.5F) which has the largest corpus data and the highest quality NLP tools among the three languages. French does not perform equally well (with top performance at 54.6F), with errors reported due to the poor quality of parsing and data sparsity [25] [4]. Br. Portuguese falls behind English and French in performance. Yet, the best results for this language, obtained without any language specific feature engineering, are (in contrast to the earlier verb clustering experiment for Spanish [24]) well beyond the random baseline. This together with the fact that similar feature sets tend to obtain the highest and lowest results among the three languages is encouraging and also demonstrates the cross-linguistic potential of Levin's classification.

The lower quality NLP tools are likely to be one explanation for the lower performance for Br. Portuguese. For example, the SCF system of Zanette, used for many of the featuresets, was reported to perform only at around 50.6% F-measure. The featuresets which were created using multiple NLP tools are likely to suffer from this problem the most due to error propagation.

Another explanation is the small corpus size (our corpus is e.g. four times smaller than that used in the French experiment) since verb clustering is known to be sensitive to data sparsity. We therefore conducted another experiment on the full set of verbs where we investigated the effect of instance filtering on the performance of the best features sets: F3, F13, F16 and F17.

The results shown in Table 4 demonstrate the strong effect of data size of the performance. The results for high frequency verbs are considerably better than those for

---

[4] Note that higher performance at 65.4 F was reported for high frequency French verbs.

lower frequency verbs. SPEC, in particular, is able to take advantage of big data size. While DCD and SPEC perform quite similarly for for the full set of verbs, the difference between the two methods gets more pronounced when the scope is restricted to high frequency verbs (in particular those that have 2000 or more occurrences). For the highest frequency group (verbs with 4000 or more occurrences), SPEC performs considerably better than DCD for all featuresets.

The most sophisticated featureset F17 performs clearly the best with SPEC, obtaining its top performance for verbs which have 4000 or more occurrences at 75.2 F. The second best featureset is F16 (at 68.3 F) – another linguistically sophisticated featureset which refines preposition-parameterized SCFs with lexical preferences. All the four featuresets obtain results of over 63 F at the highest frequency group.

The results for high frequency verbs approach those obtained for all verbs in English (with the majority of verbs having frequency more than 1000 in English), showing that big data size can compensate for the lower quality NLP tools.

Table 4. The effect of verb frequency on clustering performance

Freq.	Verbs	F3		F13		F16		F17	
		DCD	SPEC	DCD	SPEC	DCD	SPEC	DCD	SPEC
50	454	40.82	41.79	36.50	35.44	37.23	39.44	32.06	35.20
100	371	37.74	43.72	37.59	41.43	37.84	41.11	32.41	37.66
150	321	39.41	42.27	35.62	41.62	38.80	42.97	32.38	37.83
200	290	39.53	41.42	36.64	39.17	36.29	39.37	30.84	37.03
400	222	43.67	43.98	36.86	44.77	41.42	39.89	31.80	40.51
1000	131	41.11	44.52	42.13	44.37	45.43	46.85	40.99	47.62
2000	82	42.63	55.54	46.20	50.26	49.02	59.12	40.20	52.08
3000	63	44.32	53.21	55.60	57.48	50.32	62.03	50.50	57.50
4000	46	45.32	64.66	53.77	63.29	48.24	68.31	43.90	75.21

6   Conclusion

In this paper we have presented the first work of clustering verbs into Levin-style classes in Br. Portuguese. We have explored, in particular, the cross-linguistic potential of Levin style classification and how this could be exploited in the creation of a gold standard as well as in the development of a verb clustering approach for this language.

We first created a gold standard for evaluation of clustering by translating it from English to Br. Portuguese, showing that the two gold standards share nearly all of their classes and the majority of member verbs. The gold standard was extended further mainly because many member verbs were low in frequency. We then used existing Br. Portuguese NLP tools to extract similar feature sets as previously used for English and French, and clustered them using similar clustering algorithms. The results for different feature sets were in line with those obtained earlier for English and French. In particular, the most sophisticated features – the ones which are in the closest agreement with Levin's original features – produced the best results across the three languages.

The level of clustering performance for Br. Portuguese was considerably lower than that for English and French when the full set of verbs was considered. However, when the scope was restricted to high frequency verbs the results were substantially better, demonstrating that big data size is important for this task. The top performance was, again, obtained using linguistically sophisticated features.

In the future, to improve the results for Br. Portuguese, we plan to use larger corpus data (e.g. supplement existing corpora with text from the web), to improve the accuracy of feature extraction (e.g. SCF acquisition), and to investigate whether it is possible to refine feature sets with language specific constraints.

We have shown that it is possible to adopt a verb clustering method developed for resource-rich language and apply it to a less-resourced language without language specific feature engineering, and obtain a useful result. Future work should investigate the applicability of this approach to other, more distant languages and language families. Such investigations can be highly valuable for the majority of world's languages that suffer from the lack of NLP resources, and could greatly benefit from techniques for (semi-)automatic lexicon development.

Acknowledgements. This work was supported by FAPESP/Brazil (No. 2010/03785-0 and No. 2011/22882-0), EXPERT (EU Marie Curie ITN No. 317471) project and the Royal Society University Research Fellowship (UK).

References

1. Fellbaum, C.: WordNet: An electronic lexical database. MIT Press, Cambridge (1998)
2. Baker, C.F., Fillmore, C.J., Lowe, J.F.: The Berkeley Framenet Project. In: 36th Annual Meeting of the Association for Computational Linguistics and 17th International Conference on Computational Linguistics, University of Montréal, Canadá, pp. 86–90 (1998)
3. Palmer, M., Gildea, D., Kingsbury, P.: The Proposition Bank: A Corpus Annotated with Semantic Roles. Computational Linguistics 31(1), 71–106 (2005)
4. Kipper-Schuler, K.: Verbnet: A broad coverage, comprehensive verb lexicon. Doctor of philosophy, University of Pennsylvania (2005)
5. Levin, B.: English Verb Classes and Alternation, A Preliminary Investigation. The University of Chicago Press, Chicago (1993)
6. Crouch, D., King, T.H.: Unifying Lexical Resources. In: Interdisciplinary Workshop on the Identication and Representation of Verb Features and Verb, Saarbruecken, Germany, pp. 32–37 (2005)
7. Swier, R., Stevenson, S.: Unsupervised Semantic Role Labelling. In: EMNLP 2004, Barcelona, Spain, pp. 95–102 (2004)
8. Yi, S., Lopper, E., Palmer, M.: Can Semantic Roles Generalize Across Genres? In: NAACL HLT 2007, Rochester, NY, USA, pp. 548–555 (2007)
9. Shi, L., Mihalcea, R.: Putting Pieces Together: Combining Framenet, Verbnet and Wordnet for Robust Semantic Parsing. In: 6th International Conference on Computational Linguistics and Intelligent Text Processing, Mexico City, Mexico, pp. 99–110 (2005)
10. Girju, R., Roth, D., Sammons, M.: Token-level Disambiguation of Verbnet Classes. In: Interdisciplinary Workshop on the Identification and Representation of Verb Features and Verb Classes, Saarbruecken, Germany (2005)
11. Abend, O., Reichart, R., Rappoport, A.: A Supervised Algorithm for Verb Disambiguation into Verbnet Classes. In: LREC 2008, Manchester, UK, pp. 9–16 (2008)

12. Chen, L., Eugenio, B.D.: A Maximum Entropy Approach to Disambiguating Verbnet Classes. In: Proceedings of the 2nd Interdisciplinary Workshop on Verbs, The Identification and Representation of Verb Features, Pisa, Italy (2010)
13. Brown, S.W., Dligach, D., Palmer, M.: Verbnet Class Assignment as a WSD Task. In: IWCS 2011, Oxford, UK, pp. 85–94 (2011)
14. Jackendoff, R.: Semantic Structures. MIT Press, Cambridge (1990)
15. Taulé, M., Martí, M.A., Borrega, O.: Ancora-net: Mapping the spanish ancora-verb lexicon to verb-net. In: The Workshop on Verbs. The Identification and Representation of Verb Features, Pisa, Italy (2010)
16. Liu, M.C., Chiang, T.Y.: The construction of mandarim verbnet: A frame-based study of statement verbs. Language and Linguistics 9(2), 239–270 (2010)
17. Mousser, J.: Classifying arabic verbs using sibling classes. In: International Workshop on Computational Semantics, Oxford, UK (2011)
18. Kingsbury, P., Kipper-Schuler, K.: Deriving Verb-Meaning Clusters from Syntactic Strucutres. In: The Workshop on Text Meaning, in Conjunction with NAACL HLT 2003, Edmonton, Canad (2003)
19. Sun, L., Korhonen, A.: Hierarchical Verb Clustering Using Graph Factorization. In: EMNLP 2011, Edinburgh, UK, pp. 1023–1033 (2011)
20. Reichart, R., Korhonen, A.: Improved lexical acquisition through dpp-based verb clustering. In: ACL 2013, Sofia, Bulgaria (2013)
21. Korhonen, A., Krymolowski, Y., Collier, N.: The choice of features for classification of verbs in biomedical texts. In: COLING 2008, Manchester, UK (2008)
22. Guo, Y., Korhonen, A., Poibeau, T.: A weakly-supervised approach to argumentative zoning of scientific documents. In: EMNLP 2011, Edinburgh, UK (2011)
23. Shutova, E., Sun, L.: Unsupervised metaphor identification using hierarchical graph factorization clustering. In: NAACL 2013, Atlanta, USA (2013)
24. Ferrer, E.E.: Towards a semantic classification of spanish verbs based on subcategorisation information. In: The Workshop on Student Research, in Conjunction with ACL 2004, Barcelona, Spain, pp. 163–170 (2004)
25. Sun, L., Korhonen, A., Poibeau, T., Messiant, C.: Investigating the cross-linguistic potential of Verbnet-style classification. In: The 23rd International Conference on Computational Linguistics, Beijing, China, pp. 1056–1064 (2010)
26. Falk, I., Gardent, C., Lamirel, J.C.: Classifying french verbs using french and english lexical resources. In: ACL 2012, Jeju, Republic of Korea, pp. 854–863 (2012)
27. Sun, L., Korhonen, A., Krymolowski, Y.: Improving verb clustering with automatically acquired selectional preferences. In: EMNLP 2009, Singapore, pp. 638–647 (2009)
28. Merlo, P., Stevenson, S.: Automatic verb classification based on statistical distributions of argument structure. Computational Linguistics 27(3), 373–408 (2001)
29. Li, J., Brew, C.: Which Are the Best Features for Automatic Verb Classication? In: ACL 2008 (2008)
30. Joanis, E., Stevenson, S., James, D.: A General Feature Space for Automatic Verb Classication. Natural Language Engineering (2008)
31. Sun, L., McCarthy, D., Korhonen, A.: Diathesis alternation approximation for verb clustering. In: ACL 2013, Sofia, Bulgaria, pp. 736–741 (2013)
32. Sun, L., Korhonen, A., Krymolowski, Y.: Verb class discovery from rich syntactic data. In: The 9th International Conference on Computational Linguistics and Intelligent Text Processing, Haifa, Israel, pp. 16–27 (2008)
33. Schulte im Walde, S.: Experiments on the Automatic Induction of German Semantic Verb Classes. Computational Linguistics 32(2), 159–194 (2006)
34. Vázquez, G., Fernández, A., Castellón, I., Martí, M.A.: Clasificasión verbal: Alternancias de diátesis. Quaderns de Sintagma, Universitat de Lleida (2000)

35. Duran, M.S., Aluisio, S.M.: Propbank-br: A brazilian treebank annotated with semantic role labels. In: LREC 2012, Istanbul, Turkey (2012)
36. Salomao, M.M.: Framenet Brasil: Um trabalho em progresso. Revista Calidoscópio 7(3), 171–182 (2009)
37. Bertoldi, A., Chishman, R.: Frame semantics and legal corpora annotation: Theoretical and applied challenges. Linguistic Issues in Language Technology 7(9) (2012)
38. da Dias Silva, B.C., Felippo, A.D., Nunes, M.G.V.: The Automatic Mapping of Princeton Wordnet lexical-conceptual relations onto the Brazilian Portuguese Wordnet database. In: Proc. LREC 2008, Marrakech, Morocco, pp. 1535–1541 (2008)
39. Marrafa, P.: Portuguese wordnet: General architecture and internal semantic relations. DELTA 18, 131–146 (2002)
40. Marrafa, P., Amaro, R., Chaves, R.P., Lourosa, S., Martins, C., Mendes, S.: Wordnet.pt new directions. In: The Third Global WordNet Association Conference, Jeju, Republic of Korea, pp. 319–320 (2008)
41. Bentivogli, L., Pianta, E., Girardi, C.: Multiwordnet: Developing an aligned multilingual database. In: The First International Conference on Global WordNet Conference, Mysore, India, pp. 293–302 (2002)
42. Scarton, C., Aluísio, S.M.: Towards a cross-linguistic Verbnet-style lexicon to Brazilian Portuguese. In: The Workshop on Creating Cross-language Resources for Disconnected Languages and Styles, in Conjunction with LREC 2012, Istanbul, Turkey (2012)
43. Aluísio, S.M., Pinheiro, G.M., Manfrim, A.M.P., Genovês Jr., L.H.M., Tagnin, S.E.O.: The Lácio-web: Corpora and Tools to Advance Brazilian Portuguese Language Investigations and Computational Linguistic Tools. In: LREC 2004, Lisbon, Portugal, pp. 1779–1782 (2004)
44. Muniz, M., Paulovich, F.V., Minghim, R., Infante, K., Muniz, F., Vieira, R., Aluísio, S.: Taming the tiger topic: An xces compliant corpus portal to generate subcorpus based on automatic text topic identification. In: CL 2007, Birmingham, UK (2007)
45. Aziz, W., Specia, L.: Fully automatic compilation of a Portuguese-English parallel corpus for statistical machine translation. In: STIL 2011, Cuiabá, MT (October 2011)
46. Bick, E.: The Parsing System Palavras: Automatic Grammatical Analysis of Portuguese in a Constraint Grammar Framework. Doctor of philosophy, University of Aarhus (2005)
47. Zanette, A., Scarton, C., Zilio, L.: Automatic extraction of subcategorization frames from corpora: An approach to Portuguese. In: PROPOR 2012 - Demo Session, Coimbra, Portugal (2012)
48. Messiant, C.: A subcategorization acquisition system for French verbs. In: NAACL HLT 2008, Columbus, OH, pp. 55–60 (2008)
49. Zanette, A.: Aquisiçao de Subcategorization Frames para Verbos da Língua Portuguesa. Projeto de diplomação, Federal University of Rio Grande do Sul (2010)
50. Yang, Z., Oja, E.: Clustering by low-rank doubly stochastic matrix decomposition. In: ICML (2012)
51. Brew, C., Schulte im Walde, S.: Spectral clustering for german verbs. In: EMNLP 2002, pp. 117–124 (2002)
52. Meila, M., Shi, J.: A random walks view of spectral segmentation. In: AISTATS (2001)
53. McNemar, Q.: Note on the sampling error of the difference between correlated proportions or percentages. Psychometrika 12(2), 153–157 (1947)
54. Dietterich, T.: Approximate statistical tests for comparing supervised classification learning algorithms. Neural Computation 10(7), 1895–1923 (1998)

Spreading Relation Annotations
in a Lexical Semantic Network Applied to Radiology

Lionel Ramadier[1,2], Manel Zarrouk[1], Mathieu Lafourcade[1], and Antoine Micheau[2]

[1] LIRMM, University Montpellier 2, France
[2] IMAIOS, Montpellier, France

Abstract. Domain specific ontologies are invaluable but their development faces many challenges. In most cases, domain knowledge bases are built with very limited scope without considering the benefits of including domain knowledge to a general ontology. Furthermore, most existing resources lack meta-information about association strength (weights) and annotations (frequency information like *frequent*, *rare* ... or relevance information like *pertinent* or *irrelevant*). In this paper, we are presenting a semantic resource for radiology built over an existing general semantic lexical network (JeuxDeMots). This network combines weight and annotations on typed relations between terms and concepts. Some inference mechanisms are applied to the network to improve its quality and coverage. We extend this mechanism to relation annotation. We describe how annotations are handled and how they improve the network by imposing new constraints especially those founded on medical knowledge.

Keywords: relation inference, lexical semantic network, relation annotation, radiology.

1 Introduction

For more than two decades, medical practice and bio-medical research have benefited from the availability of biomedical ontologies (Bodenreinder, 2008). These resources are used for semantic analysis such as entity recognition (i.e., the identification of biomedical entities in texts as name of genes, disease, etc.), and relation extraction (i.e., the identification of semantic relationships among biomedical entities like for instance interaction between proteins). In the framework of the UMLS project, which interrelates some 60 controlled vocabularies, an upper-level ontology, the UMLS semantic network (Lomax, 2004) has been built. In the field of radiology, such a semantic network is used to facilitate or automate the analysis of radiologist reports in order to extract recommended courses of action or to trigger warning systems to improve patient management (Yetisgen-Yildiz and al., 2013). There exist reference ontologies in biomedical domain (UMLS), but they might not be suited to a particular domain like radiology because result sets are too large and too complex (Mejino 2008). To solve this problem, the Radiology Society of North America (RSNA) has created reference ontology for radiology RadLex (Rubin, 2008). RadLex and its derivatives rely on English and are not considered medically complete (Hong, 2012).

A. Gelbukh (Ed.): CICLing 2014, Part I, LNCS 8403, pp. 40–51, 2014.

There is a German version of RadLex (Gertsmair, 2012) but none exist in French, at our knowledge. More importantly, in the domain of radiology, the relationship between terms is crucial and the ontology model might not capture this information as well as a semantic network. The ontology indicates generally only the hierarchy between terms and lacks specific relations relevant either to medicine or how doctors express their knowledge in reports. When making clinical diagnosis based on a radiologist report it is crucial for the medical practitioner to be presented with information from many different non-hierarchical sources but not so important to know the exact hierarchy of a term (as this information is already known beforehand). For example, it is important to give an exhaustive list of symptoms that the medical practitioner should look for regardless of the ontological hierarchy associated with each term. These terms can be better linked when modeled by a semantic network and even better a lexical semantic network taking into account facts of medical language. While general purpose semantic networks will certainly help, they need to be extended to specific domains such as radiology.

The combined method of modeling is important for radiology reports because they contain several distinct sections. In the History section for example, there are typically descriptive texts written in everyday language while in other sections, such as Findings, the language changes to specialized terms. The goal of the construction of this lexical network is to analyze radiological report in order to extract terms and semantic relations between them. Another aim of such research is to carry out a semantic annotation of medical images in order to improve their retrieval.

Lexical-semantic networks can be manually constructed or generated by algorithmic analysis of texts. For instance, the ConceptNet, a freely available general knowledge base, is generated automatically from the 700 000 sentences of the Open Mind Common Sense Project (H Liu and al, 2004). But fully automated generation are generally limited to term co-occurrences as extracting precise semantic relations between terms from corpora remains difficult.

In our combined general purpose-specialized network, we decided to use JeuxDeMots (Lafourcade 2007) as a basis for the general purpose network. What we wish to have is a general knowledge base of a very broad scope, in the spirit of Wikipedia but under the machine tractable form of a lexical-semantic network. JeuxDeMots relies on crowdsourcing to manually construct a knowledge base. For this purpose, JeuxDeMots provides a contributive tool called *Diko*. This tool is important because we can use it to improve the network completeness in specific areas where the game approach is not suitable (relation too complicated, not lexicalized enough). Diko also exploits an inference mechanism (Zarrouk, 2013) to automatically propose relations (between terms) on the basis of what already exists in the network. This approach of inference is strictly endogenous as it does not rely on any external resources. JeuxDeMots uses crowdsourcing to incrementally attribute weight to relations between terms. If a large number of users/players associate two given nodes, the weight will be higher than a link that was only mentioned by fewer users. While this user provided weight is adequate for general purposes, it fails in the diagnostic purpose of radiology reports because the overall frequency of a symptom is not a good indication of its relevance. In a clinical situation, many patients complain of a headache and almost none report *arm drift* before suffering from a stroke, but *arm drift* is the most important term. Generally, there is not always a correlation between the associative strength and its importance between two terms. The arc weight indeed implements the associative strength but it correlates neither to the truth nor to the frequency. The medical significance of the relationship should be indicated to generate faithfully this specialist radiology semantic network. The goal of our current work is to develop the cost function that best captures this medical significance and then to train the semantic network through inference mechanisms. We introduce annotation between some relations in the field of radiology in the semantic lexical network. The goal of the relation annotation is to guide the process of inference and semantic analysis.

The rest of the paper is organized as follows. In section 2 we describe the principles behind of lexical network construction and illustrated it with JeuxDeMots. We discuss also about the building of a network specialized in radiology. We present also one type of inference: the deduction scheme. In section 3 we turn to describing the annotation of the relation between medical terms. Section 4 is devoted to describing our experiments and commenting on their results. Section 5 concludes, pointing at avenues for future research.

2   Lexical Networks

The type of lexical network where are working with is a *graph with lexical items or concepts as nodes connected through arcs interpreted as relations between items.* Those relations are semantically typed and represent (typical) lexical or ontological relationship possible between terms (hypernym, synonym, antonym, part of, cause, consequence, typical location, telic role, semantic role and so on). Besides being typed, relations are weighted and directed (no automated symetrisation is undertaken). The contributive approaches for building such a network are more and more popular because they are both cheap to set up and efficient in quality. In recent years, there is an increasing trend of using on-line GWAPs (game with a purpose) (Thaler and al, 2011) for feeding such resource. The JDM lexical network is constructed through a set of on-line associate games and contributive tools. We briefly describe it in the following section.

2.1   The JDM Game Model

JeuxDeMots is a two player blind game based on agreement on term associations. At the beginning of a game session the player is given an instruction related to a target term (for example: *give any term that is related to disease*). The user has a limited time to enter as many propositions as possible. At the end of the allowed time, player proposals are compared to those of another player for the same game, and points are earned on the basis on the common proposals. Terms in agreement are added to the lexical network with the relation corresponding of the game instruction. If the relation already exists, its weight is increased, otherwise the relation is added. This game is adequate for general common sense knowledge but may be not very efficient for specialized domain. For our project - building a lexical network for radiology, we use a contributive tool, compatible with the JDM lexical network, named *Diko* that we explain the principle below.

2.2   The Contributive Model of Diko and Relation Annotations

Diko is a web based tool for displaying information contained in the JDM lexical network but that can also be used for contribution. The necessity to not be only dependent on the JDM game for the construction of the lexical network comes from the fact that many relation types of JDM are either difficult to grasp for a casual player or not very productive (not possible many answers). In order to build a specialized knowledge we use Diko to propose new relations between terms relevant to the domain at hand. The principle of the contribution process is that a proposition made by a user will be voted pro or con by an expert validator in radiology. In the field of medicine, we added some relations like *symptom* or *diagnostic*. This contributive work is needed to build a knowledge substrate for radiology and eventually, the purpose of the project is to extract in a semi-automatic way words and relations from the radiology reports to enhance the specialized network.

To improve the quality of the network, we add more medical significance of relationships between terms thanks to *annotations*. For instance, for the following relation *measles (target) children* we can add the annotation "frequent" regardless of the weight of the relation (Fig.1, and another example is given in Fig.2). In section 3, we will detail the concept of *annotations* and their utility.

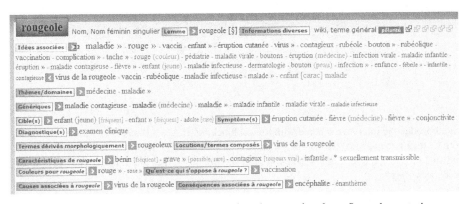

Fig. 1. Example of term "measles" with annotations between brackets. Several annotations are possible for a given relation like *frequent*.

In order to increase the number of relations in the JDM network an inference engine has been proposed. This latter proposes relations as if it was a contributor, to be validated by other human contributors or expert in the case of specialized knowledge. In this paper we describe one type of inference: the deduction scheme.

This deductive scheme is based on the transitivity of the ontological relation is-a (hypernym). If a term A is a kind of B and B has some relation R with C (the premises),

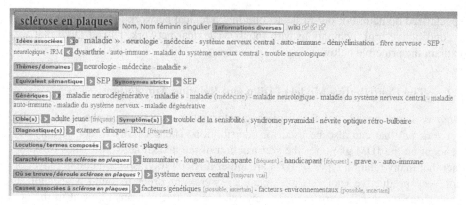

Fig. 2. Example of term "multiple sclerosis" which has for example as causes *genetic factors* and *environmental factors* annotated *possible* but uncertain

then we expect that A holds the same relation with C (the conclusion). The inference engine can be applied on terms having at least one hypernym. If a term has a set of weighted hypernym, the inference engine deduces a set of inferences. These hypernyms are classified according a hierarchical order. The weight of an inference proposed is the incremental geometric mean of each occurrence. In fact, this scheme is too simple, in effect the term B may be polysemous and ways to avoid probably wrong inference can be done by a logical blocking (fig.3). This mechanism has been described in a previous work (Zarrouk 2013).

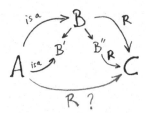

Fig. 3. Deductive inference scheme with logical blocking. If A is a B and B has some relation R with C, then it is expected that A has the same relation R with C. However, if B is polysemous, and two different refinement (B' and B") hold the premises, then the relation A R C is most probably wrong.

In case of invalidation of an inferred relation, a reconciliator is invoked to try to assess why the inferred relation is wrong. The reconciliation allows us to identify the cause of the wrong inference: an exception, an error in the premises or transitivity confusion due to polysemy with the identification of the proper word senses at stake.
In what follows, we are going to consider this type of inference for being annotated. Nevertheless, there are two other types of inference: the induction (from specific to general) and abduction (imitation from examples).

3   Relation Annotations

Generally, above all in specialized knowledge, the correlation between the weight of the relation and its importance is not strict. In the case of *hepatocellar carcinoma* the relation with *wash-out* is specific of radiology so the weight of the relation will be low but for the radiologist this relation will be important. This is why it appears interesting to introduce annotations for some relations as they can be of a great help in the medical area. In the lexical network, a relation is represented by a 3-uple:

$<Node_{start}$, Relation type/annotation, $Node_{end}>$ formally written
$Node_{start}$ (Relation type/ annotation) $Node_{end}$.

For the field of radiology, the most useful relations are shown in table 1. In radiological ontology like RadLex, there are not many relation types or occurence which can be really useful for the analysis of radiological reports. In an information retrieval, this annotation can be helpful to the users. As often, they want to know if a characteristic of one pathology is rare or frequent. This kind of information is generally absent from a network or ontology. For example, the relation between *hepatocellular carcinoma* and *hypervascular* are frequent and this information will be directly available in the network.

Table 1. Relevant relations in the radiology field with explanation, examples and their annotations

Relation type	Explanation, examples and annotation
is-a	Hypernym, *MRI* is-a *medical imaging (possible)*
has-parts	Element of the term, *liver* has part *segment I (always true)*
characteristic	*Hepatocellular carcinoma* carac *hypervascular (frequent)*
typical location	Typical place where can be the term/object in question, *multiple sclerosis* typ location *central nervous system (always true)*
target	Population affected by the term, *measles* target *children (frequent)*
diagnosis	Examen, *multiple sclerosis* diag *MRI (frequent, crucial)*
symptom	Symptom, *measles* symptom *fever (frequent)*
against	What the start term opposes/fight/prevents, *malignant tumor* against *chemotherapy (frequent)*
cause	B(that you have to give) is a cause of A, *cirrhosis* cause *alcoholism (frequent)*
consequence	The end term is a possible consequence of the start term, *stroke* consequence *hémiplegia (possible)*

These annotations will have a filter function in the inference scheme. The types of annotations are of several natures (frequency and relevance information). Below, we presented the different main annotation labels.

- frequency annotations : very rare, rare, possible, frequent, always true
- usage annotations :
  - o   often believed true
  - o   language misuse
- quantifier : any number, like 1, 2, 4 etc. or many, few
- qualitative: pertinent, irrelevant, inferable, potential.

Concerning **language misuse**, a doctor can use the term *flu* (illness) instead of *virus of influenza*: it's a misuse of language as the doctor just makes use of a language shortcut.   The annotation **often believed true** applies for a wrong relation (with a negative weight) which is very often considered as true, for example *spider (*is-a/often believed true) insect*. This kind of annotations could be used to block the inference scheme.

Qualitative annotation relates to the inferable status of a relation, especially concerning inference.   The **pertinent** annotation refers to a proper ontological level for a given relation. For example: *living being (carac/pertinent) alive* or *living being (can/pertinent) die*. The **inferable** annotation is supposedly to be put when a relation is inferable (or has been inferred) from already existing relation, for example :   *dog (carac/inferable) alive* because *dog (isa) living being*. A **potential** annotation may be put for terms above pertinent ones in the ontological hierarchy, for examples : *bird (haspart/always true) wings* and *animal (haspart/potential) wings*. Finally, the **irrelevant** annotaion is put for true relation which are considered as too far below the pertinent level, for example *animal (haspart/irrelevant) atoms*.

The **quantifier** represents the number of part of a object. Each human has two lungs so quantifier will be 2. This kind of annotation is not necessarily a numeral, but can be of more or less subjective value, like *few*, *many*, etc.

The **frequency** annotations are of five types (*always true, frequent, possible, rare and exceptional*) and qualitative are two types (*pertinent and irrelevant*). We have attributed empirical values to each annotation's label like 4 to always true, 3 to frequent, 2 to possible, 1 to rare and 0 to the rest of the annotations. These allow us to select some annotation to facilitate or block the inference scheme.

The first annotations have been made by hand but with the help of inference scheme they will spread through the network. To improve the quality of the network and to prevent some incoherent inferences some kind of annotation will block the potentially absolute relations. For instance, the annotation *language misuse* or *irrelevant* will block the inference scheme.

Moreover, to have the most accurate annotation, we need to order the central terms from the most specific to the less specific. That is to say, to reconstitute the taxonomic order related to the hypernym relation (is-a). For the term *hepatocellular carcinoma* the (several) order of hyponyms will be:

hepatocellular carcinoma
< malignant tumors of liver < tumor of liver < liver pathology < pathology

hepatocellular carcinoma
< malignant tumors of liver < tumor of liver < tumor < pathology

According to the term the annotation will be different. To choose the right annotation of the new inferred relation, this order plays an important role. The annotation of the most specific term is more crucial (important) than the less specific. We must take into account this fact for the inference mechanism with annotations.

In the inference mechanism, the term B (central term cf. fig. 2) plays a crucial role. We look at the hierarchy of the terms B according to which a specific relation was inferred many times and we keep the most specific. If we end up with two or more terms, we apply the max rules to the values corresponding to each annotation. The result will be the value of the annotation we will give to the inferred relation (Fig 4).

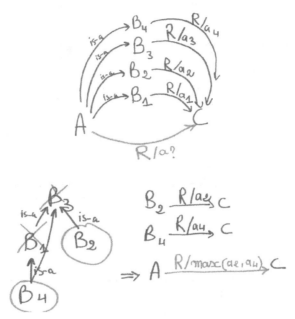

Fig. 4. Approach based on hierarchy used to choose the most accurate annotation to give to an inferred relation via several central terms

4   Experimentation

In the previous experiments conducted in (Zarrouk 2013), the deduction engine was applied to the whole lexical network to prove the efficiency of the approach. However, in this paper we unleashed the experiment on a subset of the lexical network JDM which contains all the hypernmy relations (is-a) in which is based the deduction scheme and all the manually annotated relations and that is in order to reduce the search space.

4.1   Unleashing Relation Inference

To increase the result's accuracy and to avoid to infer noisy relations , we blocked inferences on relations which are annotated as irrelevant or exception. Moreover, more detailed results and experiments about the deduction engine are provided in (Zarrouk 2013 (1)). The deduction inference engine applied on **146 934** relations produces a total of **1 825 933** relations with **573 613** distinct ones which make the average of 3 occurrences per relation (Table.2)

Table 2. Number of inferred relations from those already existing ones

Existing relations	146 934
Inferred relations	1 825 933
Distinct inferred relations	573 613

4.2   Spreading Relation Annotation

The annotations inference engine is the second part of the system. It will be unleashed over the relations (the lexical network) previously enriched with the use of the deduction engine. The relation annotation system runs only on the inferred relations. It takes into consideration the annotations of the premises used to infer a certain relation as mentioned. If there is just one available premise, the annotation of this premise, if any, is affected to the relation inferred. If there are many premises, the system will rebuild the hierarchy between these ones and will keep the annotation of the nearest premise for being the most accurate. In case of having some premises with the same level in the hierarchy, a maximum rule is applied between them and the annotation having the strongest number (always true: 4, frequent: 3, possible:2,.. etc.) will be affected to the inference. This system guarantees a good accuracy of the annotation spreading.

As noticed, contrary to the original deduction engine, we allowed redundancy in because it increases the accuracy of the relation annotation spreading system's results. To clarify, we propose the following example:

Premises: *stroke (is-a) cerebral infraction* **&** *cerebral infraction (diagnosis/frequent) MRI*
→ **inferred relation:**   *stroke (diagnosis/frequent) MRI*   *(1)*

Premises: *stroke (is-a) cerebrovascular disease* **&** *cerebrovascular disease (diagnosis/possible) MRI*
→ **inferred relation:** *stroke (diagnosis/possible) MRI*   *(2)*

The annotation system having these two occurrences **(1)** and **(2)** of the same relation *stroke (diagnosis) MRI*, annotated differently (possible, frequent) will decide to keep the strongest one (frequent). It is informed about the annotation's strength by empirical values we have attributed to each annotation's label according to their frequency like 4 to "always true", 3 to "frequent", 2 to "possible", 1 to "rare" and 0 to the rest of the annotations.

The annotation's inference system applied on the relations base stemmed of the deduction engine run, annotated **10 085** relations starting from only **72** ones (Table.3).

Table 3. Number of annotations inferred after the application of the relation annotation system on the existing ones

Annotation's Label	Existing annotation	Inferred annotation
Frequency: frequent &always true	38	8 093
Frequency: possible	16	150
Frequency: rare & very rare	7	35
Qualifier: often believed true	1	7
Qualifier: irrelevant	5	1 604
Quantifier	5	178
Total	**72**	**10 085**

In this experiment, we have not considered *potential* and *inferable* annotations (more than 43 000 distinct annotations for one unique run, 97% are correct and 3% false) because they are more utility annotations than semantically relevant in the context of radiology. Instead, we focused here on the annotations illustrating frequency since it is a very important information in the radiological area.

The number of annotated relations per annotation's label does not depend on the number initially existing as noticed from Table.2, but simply on the number of the ongoing hypernym relations of the central term of the scheme as in the simplified example:

1) The basic inference scheme is the following:

$$A \text{ (is-a) } B \ \& \ B \text{ (R/annot) } C \rightarrow A \text{ (R/annot) } C$$

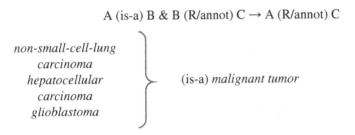

non-small-cell-lung carcinoma
hepatocellular carcinoma
glioblastoma

⎫ (is-a) *malignant tumor*

& *malignant tumor* (carac/frequent) *poor pronastic*

→3 relations annotated as *frequent (non-small-cell-lung carcinoma /hepatocellular carcinoma/ glioblastoma (carac/frequent) poor pronastic)*

The larger the number of hypernym relations toward the term B (*malignant tumor*) which has an outgoing relation annotated *(malignant tumor (carac/frequent) poor pronastic)*, the larger is the number of annotated relations.

2) However, for the existing annotated relations which do not contribute a lot in the inferring process, as the annotation frequent (Table.2), they are attributed to relations which are ineligible to the annotations scheme which is a deductive like for example: *Hepatocellular carcinoma (carac/frequent) hypervascular*

The term *Hepatocellular carcinoma* does not have any ongoing hypernym relation (*x (is-a) Hepatocellular carcinoma*), so in this case the annotation frequent will not generate other annotations.

We statistically evaluated the produced annotation, and it appears than 87% of them have been evaluated as "correct", 5 % as "incorrect" and the rest (8 %) as "debatable" (that is to say that experts might discuss not its validity but rather if the frequency value should be modify). The evaluation has been done manually by three *experts* on random sample of at least 100 annotations up to 10% for each annotation values. Each evaluator had to choose between the three possible values above: correct, incorrect and debatable. The Cohen's kappa coefficient was equal to 0.83.

A debatable result is one felt by the evaluator as not incorrect but where points of view can be in conflict. Most of the cases are between *frequent* and *possible*, or to a lesser extend between *rare* and very *rare*.

In this experiment we applied the relation/annotation system through a single run. But naturally, the system which is actually running iteratively along the contributions and the games uses the new terms and annotations added and the previously inferred ones to continue annotating more relations.

5   Conclusion

Annotations viewed as information added to typed relations between terms add a new dimension in the knowledge contained in lexical networks. Even when weighted, relation strength not always relates to its confidence. Something can be rare but very important, and conversely a relation can be so obvious that its intensity is low.

We presented some issues related to building a lexical semantic network with games and user contributions and about inferring new annotated relations from existing ones. To be able to enhance the network quality and coverage, we proposed a consolidation approach based on a relations and annotations inference engine. The annotation system we presented in this paper is a complement for the lexical network consolidation system presented in (Zarrouk, 2013 (2)). This enhanced consolidation approach can provide, thanks to the annotation system, a crucial information which can be used not only in radiology as shown previously but also in other specialized domains, and certainly for common sense reasoning.

It seems to us interesting to develop knowledge in a specialized domain inside a general lexical network. Further research must improve the spreading relation annotation and also the specialized lexical in radiology with the help of expert but also non expert contributors.

References

1. Bodenreider, O.: Biomedical ontologies in action: Role in knowledge management, data integration and decision support. Yearb Med. Inform. 47, 67–79 (2008)
2. Chamberlain, J., Fort, K., Kruschwitz, U., Lafourcade, M., Poesio, M.: Using games to create language resources: Successes and limitations of the approach. In: The People's Web Meets NLP, pp. 3–44. Springer, Heidelberg (2013)
3. Gala, N., Lafourcade, M.: NLP lexicons: Innovative constructions and usages for machines and humans. In: Proc. of Electronic Lexicography in the 21st Century: New Applications for New Users (eLEX 2011), Bled, Slovenia, November 10-12, p. 12 (2011)
4. Gerstmair, A., Daumke, P., Simon, K., Langer, M., Kotter, E.: Intelligent image retrieval based on radiology reports. European Radiology 22(12), 2750–2758 (2012)
5. Hong, Y., Zhang, J., Heilbrun, M.E., Kahn Jr., C.E.: Analysis of RadLex coverage and term co-occurrence in radiology reporting templates. Journal of Digital Imaging 25(1), 56–62 (2012)
6. Lafourcade, M.: Making people play for Lexical Acquisition. In: Proc. SNLP 2007, 7th Symposium on Natural Language Processing, Pattaya, Thailande, December 13-15, p. 8 (2007)
7. Liu, H., Singh, P.: ConceptNet—a practical commonsense reasoning tool-kit. BT Technology Journal 22(4), 211–226 (2004)
8. Lomax, J., McCray, A.T.: Mapping the gene ontology into the unified medical language system. Comparative and Functional Genomics 5(4), 354–361 (2004)
9. Mejino Jr., J.L., Rubin, D.L., Brinkley, J.F.: FMA-RadLex: An application ontology of radiological anatomy derived from the foundational model of anatomy reference ontology. In: AMIA Annual Symposium Proceedings, vol. 2008, p. 465. American Medical Informatics Association (2008)
10. Rubin, D.L.: Creating and curating a terminology for radiology: ontology modeling and analysis. Journal of Digital Imaging 21(4), 355–362 (2008)
11. Thaler, S., Siorpaes, K., Simperl, E., Hofer, C.: A survey on games for knowledge acquisition. Rapport Technique, STI, 26 (2011)
12. Yetisgen-Yildiz, M., Gunn, M.L., Xia, F., Payne, T.H.: A text processing pipeline to extract recommendations from radiology reports. Journal of Biomedical Informatics (46), 354–362 (2013)
13. Zarrouk, M., Lafourcade, M., Joubert, A.: Inference and Reconciliation in a Crowdsourced Lexical-Semantic Network. In: CICLING 2013: International Conference on Intelligent Text Processing and Computational Linguistics, March 24-30, p. 13. University of the Aegean, Samos (2013)
14. Zarrouk, M., Lafourcade, M., Joubert, A.: Proc of 9th International Conference on Recent Advances in Natural Language Processing (RANLP 2013), Hissar, Bulgaria, September 7-13, p. 6 (2013)

Issues in Encoding the Writing of Nepal's Languages

Pat Hall[1], Bal Krishna Bal[1], Sagun Dhakhwa[2], and Bhim Narayan Regmi[2]

[1] Language Technology Kendra, Patan Dhoka, Lalitpur, Nepal
{pavhall,bkbal}@ltk.org.np
www.ltk.org.np
[2] Centre for Communication and Development Studies, Nagbahal-16, Lalitpur, Nepal
{rite2sagun,bhimregmi}@gmail.com
www.desceco.org

Abstract. The major language of Nepal, known today as Nepali, is spoken as mother tongue by nearly half the population, and as a second language by nearly all of the rest. A considerable volume of computational linguistics work has been done on Nepali, both in research establishments and commercial organizations. However there are another 94 languages indigenous to the country, and the situation for these is not good. In order to apply computational linguistics methods to a language it must first be represented in the computer, but most of the languages of Nepal have no written tradition, let alone any support by computers. It is the written form that is needed for full computational processes, and it is here that we encounter barriers or at best inappropriate compromises. We will look at the situation in Nepal, ignoring the 17 cross-border languages where the major speaker population lies outside Nepal. We are left with only three languages with written traditions: Nepali which is well served, Newari with over 1000 years of written tradition but which so far has been frustrated in attempts to encode its writing, and Limbu which does have its writing encoded though with defects. Many of the remaining languages may be written in Devanagari, but aspire to something different that relates to their languages and has a more visually distinctive writing to mark their identity. We look at what can be done for these remaining languages and speculate whether a common writing system and encoding could cover all the languages of Nepal. Inevitably we must focus on the current standard for the computer encoding of writing, Unicode, but we find that while language activists in Nepal do not adequately understand what is possible with the technology and pursue objectives within Unicode that are not necessary or helpful, external experts only have limited understanding of all the issues involved and the requirements of living languages and their users and instead pursuing scholarly interests which offer limited support for living users.

Keywords: Nepal, Nepali, Tibeto-Burman languages, Indo-Aryan languages, Newari, phonetics, writing systems, encoding, Unicode.

A. Gelbukh (Ed.): CICLing 2014, Part I, LNCS 8403, pp. 52–67, 2014.
© Springer-Verlag Berlin Heidelberg 2014

1   Introduction

We are concerned about the use of languages in Nepal by the community of living speakers and writers, so that they can access and share knowledge in their own languages. For this to happen, those languages must be written with the writing encoded in the computer and supported by a range of computational linguistics resources. While our concerns focus on Nepal, we realize that the issues are global.

Ethnologue [18] record 7,105 living languages worldwide, of which 2,387 (33.6%) are viewed as endangered. This edition of Ethnologue includes indications of the vitality of languages, using a system of levels called EGIDS [17]. EGIDS contain 11 levels with two divided levels, languages at levels 0 to 4 are actively written and deemed not in danger, while levels 6b to 10 are considered endangered with the language not being passed on to the next generation. In between are levels 5 and 6a:

> 5: "The language is in vigorous use, with literature in a standardized form being used by some though this is not yet widespread or sustainable."
> 6a: "The language is used for face-to-face communication by all generations and the situation is sustainable."

An important aspect of language vitality is the language's ability to be written, captured in levels 5 and above in the EGIDS framework. This is only 31.2% of the world's languages, 2,216 languages. We need to be circumspect about these figures, because what constitutes being written depends on what you value in writing.

The EGIDS analysis for Nepal shows that only 20 of the 95 indigenous languages are in levels 5 and above (21%), with most languages totally unsupported by writing let alone computer encodings. In section 2 we discuss the general context of South Asian writing, and then in section 3 we look at the three languages indigenous to Nepal that do have written traditions: Nepali, Limbu and Newari (also known as Nēpāl bhāṣā). In Section 4 we look at the remaining languages of Nepal.

In Section 5 we discuss three factors relevant to the encoding of writing: the multidisciplinary requirements of encoding, the multiple stakeholder interests involved, and perverse incentives that encourage unnecessary encodings. These issues are not specific to Nepal and this leads us to consider encoding problems worldwide, and what could be done about this on the international stage.

But we start with a short subsection setting the background for Nepal.

1.1   Nepal Background

Nepal's topography varies from the Himalayan mountains in the north to the Gangetic plains in the south. Linguistic communities have migrated in from both directions, those from the north brought languages of the Tibeto-Burman family, while communities migrating from the south mostly brought languages of the Indo-Aryan family. Once in Nepal communities remained completely isolated by steep valleys and high mountains and by thick forest, leading to the evolution of many distinct languages. The 2011 census of Nepal [8] listed 123 languages, while Ethnologue [18] lists 124 languages for Nepal. Ethnologue includes 4 deaf sign languages and 8 languages that have no speakers and of the remaining 112 languages, 17 languages have much larger speaker populations in neighbouring countries; removing these languages leaves 95 languages that belong predominantly to Nepal. The two sources differ in detail significantly with Ethnologue dividing the large Tibeto-Burman languages Tamang and Gurung and others into several sub-languages while the census does not make these distinctions but does list a number of sublanguages or dialects of Nepali as distinct. Because the compilers of Ethnologue are linguists we will use their analysis in this paper.

Among the languages with EGIDS status of 5 or better, some languages with large populations of speakers such as Magar (770,000) and Gurung (352,000) are marked as endangered at 6b, while some languages of very few speakers such as Helambu Sherpa (3,990) and Koi (2,640) are marked as developing 5. While most of the 17 cross-border languages have strong written traditions, only three of the 95 indigenous languages have any tradition of being written: Nepali, Newari, and Limbu.

2   The Context of South Asia

Nearly all writing in South Asia and South-East Asia has evolved from the original Brahmi system over the past 2,000 years, apart from the Arabic-derived writing which was introduced into the region with Islam (eg [9], [10], [31]). While originally hand-written, Indic writing has been printed in movable type since around 1800, with the type evolving and being simplified over the centuries (eg [32]). When computers became used for writing and publishing, the encoding of Devanagari and other Indic scripts was undertaken in India, leading to ISCII, the Indian Script Code for Information Interchange [7]. Devanagari was planned for inclusion in ISO 8859 as part 12, expecting to adopt ISCII's codes. However ISO 8859 was superseded in 1990 by Unicode [37] which included code blocks for Devanagari and other major Indic scripts from the start, adapted from a 1988 version of ISCII. One significant difference between ISCII and Unicode was that in ISCII all the scripts of India had been unified within a single table, with the different scripts selected by appropriate font, whereas in Unicode these were dis-unified into separate code blocks for reasons that are not clear, though presumably because they looked very different. The importance of language considerations in encodings has recently been emphasized for us by the draft of an Indian Standard "Devanagari Script Behaviour for Hindi" [36].

Unification is an important issue in encoding – should a range of scripts and languages share a common encoding, or should each be encoded separately? The extreme in unification is for the Latin or Roman script, where all the languages of western Europe share a common code block, with extra blocks for extensions of the script for other languages. But the move has been away from unification, as seen in the move from ISCII to Unicode. The problem with dis-unification is that if a language is written in a number of scripts, with digitized documents in these separate scripts, and these scripts have been separately encoded, then bringing these resources together requires special transformation processes to reconcile them. Why make this necessary when a simple font change could switch between external presentations of the texts?

The transition from hand-writing to typing at a keyboard requires an addition to the writing system, because what had been an implicit choice about the writing of conjunct ligatures, now has to be made explicit. In the abugida system of Brahmi-derived writing all consonants carry with them an implicit /a/ vowel unless this is overridden by an alternative vowel or the special virama which suppresses the vowel, or if the consonant forms part of a conjunct consonantal cluster or ligature. For typewriters the choice was limited by keyboard size, and only a few selected consonants were given half-forms which when typed before a consonant (or another half-consonant) looked acceptably like a conjunct, though these horizontal conjuncts may not have been a common practice before. In handwriting vertically compressed half forms were frequently used for vertically stacked conjuncts, but typewriter technology could not handle these subjoined forms. Figure 1 illustrates these half forms, both horizontal and vertical subjoined.

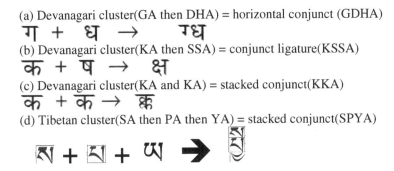

(a) Devanagari cluster(GA then DHA) = horizontal conjunct (GDHA)

ग + ध → ग्ध

(b) Devanagari cluster(KA then SSA) = conjunct ligature(KSSA)

क + ष → क्ष

(c) Devanagari cluster(KA and KA) = stacked conjunct(KKA)

क + क → क्क

(d) Tibetan cluster(SA then PA then YA) = stacked conjunct(SPYA)

ས + ས + ཡ → སྤྱ

Fig. 1. Consonant clusters as Conjuncts in handwriting copied from Unicode [37]

With computers able make key-mappings, simple transformations at input, system designers were able to make more natural extension of Devanagari so that a virama after a consonant followed by another consonant signaled a ligature conjunct:

(a) input (GA, virama, DHA) later output as ligature GDHA

(b) input (KA, virama, SSA) later output as ligature KSSA

(c) input (KA, virama, KA) later output as ligature KKA

By contrast in Tibetan, there are two forms of each consonant, and head form and a subjoined form, and thus:

(d) input (head SA, subjoined PA, subjoined YA) later output as stacked SPYA

When creating computer codes, the practice was to replicate one-for-one the keyboard sequences (or vice versa, the keyboard follows the encoding), leading to two encoding conventions – the virama model, and the subjoined model. Most South Asian encodings followed the virama model while Tibetan followed the subjoined model. It is not clear what dictates the choice, though clearly the virama model requires fewer code points, and fits more comfortably on a conventional keyboard. We are left puzzled as to why Tibetan adopted the subjoined model.

From a computational perspective there is a much more natural encoding, to move the encoding away from the abugida practice of all consonants carrying an implicit /a/ vowel to making the vowels always explicit. This makes the internal coding much closer to the spoken form, emphasizing the link to language; we see this as a possible approach to the writing of currently unwritten languages of Nepal.

In support of this link between the written form and the spoken form, we note the decision in ISCII, and later followed by Unicode, to make the sequence of codes for the consonant combined with a matra-vowel (Unicode call this a dependent vowel sign) always follow the spoken sequence with vowel following the consonant, even though in some cases such as the matra I (f◯) in Devanagari is written before the consonant, as shown in Figure 2.

Fig. 2. The written syllable /ksshi/ on left is written /i/ /kssh/ but its storage form on the right is the consonant cluster followed by the matra vowel. (from Unicode Ch 2 p 4.[37])

Unicode introduced two further codes, the zero-width joiner (ZWJ) and zero width non-joiner (ZWNJ); while these may be useful for typography, they have no use in everyday creation of written content and will not be discussed further in this paper.

3   Nepal's Written Languages

Apart from the 17 cross-border languages shared with India and China, only three languages of Nepal have written traditions:

- Nepali, until recently the only permitted language for formal use in Nepal;
- Newari, known as Nēpāl bhāṣā within the linguistic community, has 846,557 speakers, and has been written for over a thousand years in a number of scripts;
- Limbu with 343,603 speakers, has a number of written sacred texts dating back several hundred years.

While Ethnologue records Nepali at EGIDS level 1, it only reports Newari at level 3 and Limbu at level 5. This is not surprising since these and all languages other than Nepali were suppressed by successive governments of Nepal from the late 18th century until 1990. While the writing of Limbu was probably only ever used for special cultural and religious texts, Newari writing was used for a wide range of purposes until the overthrow of their regime by the Gorkhas in the mid 18th century. The cross-border language Maithili has its own mature literature with its own distinctive script - Mithilaksha or Tirhuta.

3.1   Nepali Written in Devanagari

Nepali, historically known as Khas, Parbatiya and Gorkhali, with 11,826,953 mother-tongue speakers in 2011, has been written for at least 800 years ([1], [6], [30]). It became the national language of Nepal following the conquest of the country by the

Gorkha king Pritivi Narayan Shah in the mid 1700s. It only acquired the name "Nepali" in the 1930s, reportedly as a rallying point for nationalism against the then ruling oligarchy in Nepal.

Nepali is the national language, it is written in Devanagari with this requirement enshrined in law. Around 1998 a proposal was submitted to Unicode by a committee in Nepal for a distinct encoding for the writing of Nepali, to include three common consonant clusters (conjuncts) - <tra>, <ksha>, and <gya> - that often collate separately and are taught as part of the basic alphabet for Nepali. This proposal was rejected on the basis that these conjuncts did not define a different writing system and they should continue to be treated as conjuncts with the collation differences handled through collation algorithms, a view we agree with.

In 1998 a Nepali Unicode CD working with the Unicode Devanagari code block was produced containing fonts and keyboard drivers, followed a few years later by development projects aimed at producing versions of Linux and Windows localized to Nepali. A national committee agreed around 3,000 computer terms, and then teams translated millions of words of interface text, documentation and help text. Localized versions of Linux and Windows were released within a couple of months of each other at the end of 2005. These included spell checkers initially based purely on dictionary lookup, with the Linux spell-checker then moving to the more sophisticated HunSpell system.

There has been considerable activity in computational linguistics focused on Nepali [4,5], including Nepali stemmers and morphological analyzers aiming at full computational support for Nepali, necessary for grammar checkers and rule-based translation. The Dobhase English to Nepali rule-based translator was released in 2006. A TTS system was produced as part of the Linux system, using Festival/Festvox, but the size of Festival meant that a screen reader was not possible.

As part of supporting work a number of corpora for Nepali were produced, recently released through ELDA: text corpora of 1 million annotated words from representative genre following FLOB plus 13 million opportunistically collected words, a limited collection of English-Nepali parallel texts, speech corpora of 260,000 words collected in everyday settings, and a spoken corpus for TTS of 1,200 sentences containing all 1764 diphones of Nepali. The gathering of the speech corpus included a considerable amount of video recording, but because of concern about identification of the speakers, this material is not generally available. An 8,000 word corpus-based dictionary has been produced.

Much of this work has been funded by external agencies, IDRC, EuropeAid, UNESCO, Microsoft. On-line translation for Nepali has recently been introduced by Google. Nepali is now a relatively well resourced language, though much work has only been partially completed and must continue.

3.2   Limbu Language and Sirijanga Script

Limbu was added to the Unicode Standard in April 2003. Limbu's traditional script is claimed to have been invented in the 9th century and then revived in the 17th century by Te-ongsi Sirijonga, and was then revived again in 1925 and named "Sirijanga".

Limbu standardisation began in 1999, with a proposal in 2002 [21] jointly between a field linguist and a Unicode expert.

In 2011 two additional conjunct characters, <gya> and <tra> were proposed [24]. In Devanagari these would be encoded simply as a three code sequence, the two consonants with a virama in between. This new proposal exposed a failure in the initial encoding: even though the Limbu writing system Sirijanga is a Brahmi-derived writing system, the coding had not adopted either a virama model or a subjoined model. However there is a diacritic /sa-i/ which suppresses the implicit vowel, just as the original virama in Devanagari did, and clearly /sa-i/ could be interpreted as a virama.

3.3   Newari Language and Multiple Scripts

The Newars had been the rulers of the Kathmandu valley for many centuries before they were conquered by the Gorkhas from a neighbouring Himalayan kingdom. They call the Kathmandu valley "Nepal", their language "Nēpāl bhāṣā" (which we refer to as "Newari" here) and their writing "Nepal Lipi" or "Nepaalalipi", the language and writing of the Kathmandu valley. You can see Newar writing carved into stone or wood, or embossed in brass or other metals, in temples around the valley. There are two distinct styles of writing, an ornate style with many long downward diagonal strokes called "Ranjana", and a more rounded style "Prachalit". Both these styles are still in use today, as seen in the extract from a monthly newspaper shown in Figure 3.

Fig. 3. Newar writing from Lipi Pau monthly – Ranjana headline, Prachalit body

Other styles of writing have also appeared. Shakya [34] identifies a third style Bhujimmola, while Shakyavansha [35] identified 9 styles. Different styles appear to have been used for different purposes – Ranjana for sacred and religious texts, Prachalit for everyday secular writings and Bhujimmola for administrative purposes. Much current writing of Nēpāl bhāṣā is done in Devanagari, though this is deprecated by the Newar community. Hack fonts are available for both Ranjana and Prachalit (eg [34]), but what is needed is a proper encoding in Unicode of Nepal Lipi.

In 2001 a Unicode block named "Newari" illustrated with Ranjana characters was proposed [12] along with another code block named "Nepali" illustrated with Prachalit characters. These drafts included the same three conjuncts, <tra>, <ksha>, and <gya>, of that earlier failed proposal for Nepali.

Shakya [34] gives tables of the basic characters of the writing for Ranjana, Prachalit, and Bhujimmola. Most characters have Devanagari equivalents, but there are also a number of distinct aspirated or breathy consonants, supported by the phonology of

Hale and Shrestha [15] as part of the spoken language. These glyphs must be viewed as distinct letters of the Newar alphabet.

In 2009 a proposal to encode Ranjana [13] advocated:

> "Since Rañjana is visually and structurally similar to the Lañtsa & Wartu scripts used for Buddhist Sanskrit documents in Tibet (China), Bhutan, Mongolia, Nepal, Sikkim & Ladakh (India) it has been considered would be practical to merge these two scripts (Lañtsa and Wartu) with Rañjana for encoding purposes." (p1)

A strong lobby of Newars wanted Prachalit to be encoded first, as the style of writing used in the daily life of the Newars, and after much a second document [14] proposed unifying a number of scripts used for Newari, concluding:

> "**Encoding considerations**. It should first be said that some members of the user community have criticized the idea of unifying these "scripts". It may be that this is a misunderstanding of the UCS; the analogy of the Latin script with its 𝔊𝔞𝔢𝔩𝔦𝔠and 𝕱𝖗𝖆𝖐𝖙𝖚𝖗 variants, however, is probably applicable, which is why the recommendations here have been made." (p2)

We agree with the second of these because the unification is of very similar "scripts" for a single language. However we are unsure about the first because the unification is across significantly different languages, though we must bear in mind the examples given in the second quotation. Significantly different languages such as Icelandic, French and Finnish all use the common Roman script, albeit with some notable differences.

A meeting of the Nepal Lipi Guthi in July 2008 explored the idea that each style of writing should be separately encoded while a meeting in March 2010 considered that a standard should cover all the variants for writing Newari within a single code block.

In 2011 a proposal for Prachalit [23] was made, soon to be replaced by a very similar proposal for "Newar" [25]. This second proposal also included "additional consonantal forms", shown in Figure 4, exactly those extra aspirated consonants included by Shakya [34], but arguing that these should not be separately encoded, but should be viewed as conjuncts which have been written in the wrong order.

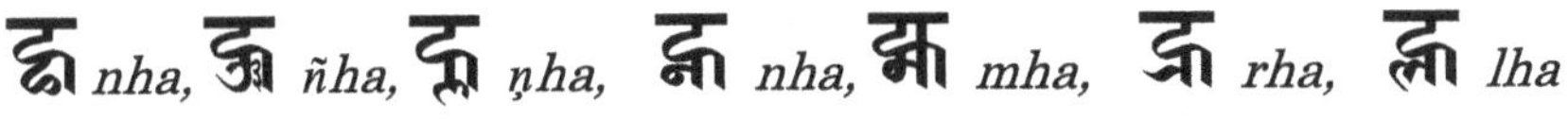

रॆ *nha,* ञ्ह *ñha,* ण्ह *ṇha,* न्ह *nha,* म्ह *mha,* र्ह *rha,* ल्ह *lha*

Fig. 4. Newar characters – are these consonants or conjuncts?

In 2012 a group in Nepal submitted an alternative proposal [19] to the Unicode Technical Committee (UTC) describing Newar writing from a Newar perspective. They included the breathy consonants of Figure 4, but apart from that and what the Unicode block should be called, there is strong agreement. It is clear that the breathy consonants are present in the spoken language, and should be encoded, they are currently written as a digraph not a conjunct, they are distinct from the consonant followed by a /ha/.

Some rapprochement between these two proposals is needed, though Unicode experts in north and west clearly favour the "Newar" proposal [25] because of its grounding visually in samples of written material, discounting the perspective from Nepal [19] because it is supported by linguistic arguments. The semi-official Unicode proposer and author of [19] has been planning to travel to Nepal to meet with the

Newar community since early 2012, but so far he has never yet made that trip. A Nepal national standard might be an alternative way forward.

The relationship between Ranjana and these proposals needs to be resolved, noting that slots for both Newar and Ranjana appear in the forward planning Roadmap of Unicode and in the forward planning of the Script Encoding Initiative [3]. The group in Nepal has also recently put forward a proposal for Ranjana [20] distinct from their earlier Nepaalalipi proposal. A reader of Devanagari, a reviewer of this paper, commented that he could read Prachalit but not Ranjana posing this as evidence that they should not be unified, but we wonder whether as a reader of Roman writing he could immediately read the 𝔖𝔞𝔢𝔩𝔦𝔠 or 𝕱𝖗𝖆𝖐𝖙𝖚𝖗 fonts in the quotation above, or the many exotic Roman fonts available such as 𝕕𝕖𝕧𝕒𝕟𝕒𝕘𝕒𝕣𝕚𝕤𝕙 , 𝒢𝒾𝒹𝒹𝓎𝓊𝓅, MESQUITE, 𝔭𝔯𝔞𝔨𝔯𝔱𝔞, or 𝔷𝔞𝔪𝔞𝔯𝔨𝔞𝔫 ?

3.4 Cross-Border Languages

Maithili is the second largest linguistic community in Nepal with 2.8 million mother tongue speakers, while in India there are over 30 million. When the Indian constitution first scheduled its official languages Maithili was viewed as a dialect of Hindi but this was vigorously contested and eventually led to the inclusion of Maithili as a distinct scheduled language in 2004. It is written in Devanagari, while their traditional style of writing, known as Mithilakshar or Tirhuta, has been treated as an exotic for use in wedding invitations and similar. In 2008 a Unicode compliant Tirhuta font Janaki was produced in Nepal, mapped to the Devanagari code block. Then in 2011 a separate encoding of Tirhuta [26] was proposed, arguing that it could not be unified with Bengali, while not discussing Devanagari at all. This would seem to be a retrograde step, splitting the corpus of writing into two encodings, though of a course a simple program could move the corpus into the new coding, but why require that?

Other Indo-Aryan languages shared between India and Nepal do not have traditional scripts, and are well served by Devanagari, though some scholars say that the Kaithi script which entered Unicode in 2009 was used.

There is also a small community of Tibetan speakers in Nepal, 5,280 in the 2011 census, with just over a million in China. Tibetan has long been encoded in Unicode, but uses the subjoined model rather than the virama model for its coding. The only other cross-border language that is written is Lepcha, which has its own distinctive Rong script encoded in Unicode, though we understand that older hack fonts are used in preference to Unicode compliant open type fonts which may indicate a problem in Unicode.

4 The Unwritten Languages of Nepal

The UNESCO process for creating a writing system for a language [30] begins with a phonological analysis of the spoken language. Field linguists aiming to document the languages that they study have for many years follow this approach and improvise means of writing the languages, in Nepal usually based on Devanagari. For

Indo-Aryan languages this works without problems, but for Tibeto-Burman languages new characters had to be created or combinations used; Noonan [22] describes these and reports a similar problem to that found in Newari of aspirated consonants in Chantyal handled by digraphs in a similar manner. Basing the writing on Devanagari is very attractive because most people in Nepal will be familiar with the script through schooling in the medium of Nepali written in Devanagari. Field studies of languages has meant that many of them have basic documentation, a useful basis for further development.

In 2008 Regmi [28] proposed a modified and simplified Devanagari for writing all the languages of Nepal, removing conjuncts and adding the vowel /a/ and further vowels, consonants and tone marks for the Tibeto-Burman languages. A seminar in 2011 brought together a number of scholars who concluded that 85 symbols would be sufficient, with representatives of 30 language communities being interested in adopting such a system.

Fig. 5. The Sikkim Herald in 11 languages (with permission from Mark Turin.)

Subsequently Regmi and colleagues [29] consolidated the phonemic inventory of 38 languages of Nepal based on secondary sources. These languages include Bankariya, Bantawa, Bhojpuri, Bhujel, Bote, Byansi, Chepang, Dhimal, Dumi, Dura, Gurung, Hayu, Kaike, Kirati-Koinch, Kisan, Koyu/Koyi, Kulung, Kumal, Kusunda, Lhomi, Limbu, Lohorung, Magar, Maithili, Meche, Mewahang, Nepali, Newari, Pahari, Raji, Raute, Santhali, Sherpa, Tamang, Thakali, Thulung, Uranw, and Wambule. Describing these languages in Nepali or English causes difficulty and confusion because there is no proper representation of some of the phonemes of the languages of Nepal while describing and comparing the languages. This study recommends a unified writing system for the languages where glottal stop, velar, and implosive consonant sounds, high and mid central vowels, front and back low vowels, front rounded and back unrounded vowels, and supra segmentals like length, tone and stress as well as syllabic consonants can be written.

These ideas have been applied to the writing of the Tibeto-Burman Lohorung language with 4,970 speakers [8] in a crowd-sourced multimedia dictionary project [2], though some members of the community have not liked the adoption of Devanagari-based writing and some even preferred Roman writing.

The rejection of Devanagari by language activists is to be expected given the history of linguistic persecution by the state and its Devanagari script. In response activists of the Dhimal, Bantawa, Gurung, Magar and Sunawar communities have created their own distinctive writing, with proposals that have reached discussion towards standardization [3]. Many of these languages are also spoken in Sikkim where the Sikkim Herald is published in 11 languages with distinctive scripts and typography, as seen in Figure 5.

This pursuit for visually distinctive writing is one way of marking the writers identity [33]. However this can be achieved by distinctive fonts, and does not require separate encodings as suggested by the many ISO proposals. This observation led to a proposal to unify all these proposals with Nepaalalipi [16] basing the unification on phonological arguments, but this proposal has not been discussed at ISO or UTC meetings, and the few comments made have been very negative – unifying based on spoken languages is not favoured by Unicode "experts". An even more broadly based unification along the lines of Regmi et al [29] discussed above encouraging separate linguistic communities to have distinct fonts may be the way forward, though we must anticipate problems with any proposal that reaches the ISO and Unicode committees.

5   The Barriers and How to Circumvent Them

We need to understand why there have been difficulties in the incorporation of Nepal's languages into Unicode, so that standardization can move forwards. There appear to be three critical factors, discussed below.

5.1   The Multiple Disciplines of Encoding

Deciding on the encodings for the writing of a language requires knowledge of such a wide range of disciplines that no single individual can be expected to encompass them, encoding essentially must be a team activity respecting all opinions. The disciplines involved are:

- computer technology:
  - input-output systems: the way input mechanisms from keyboard and stylus tracing and OCR are transformed into internal codes, and the way sequences of internal codes are rendered in print and on screen;
  - text processing: collation and matching, the separation of concerns of styles (including fonts) from content and the encodings involved, typical computational processes that are undertaken on texts;
  - localization and internationalization methods;
  - standards: the importance of standards and the role they play in ensuring interoperation of computer systems.

- writing systems and orthography:
  - variety of writing systems and how they work, ranging from ideographic systems to syllabaries to alphabets in various forms, and their inter-relationships;
  - spelling and dictionaries: compromises between word relationships and how they are spoken.
- linguistics:
  - language structure: word morphology, syntactic structures;
  - phonology: the elements of spoken languages;
  - computational linguistics: important specific processes carried out on written texts;
  - social linguistics: how languages and their writing are used in everyday life.

While it is perfectly acceptable that individuals are ignorant of some or most of these areas of knowledge, what is not excusable is the denial that such knowledge is important and the refusal to take into account opinions about encoding derived from these other knowledge perspectives. Regrettably we have come across too many instances of claims based upon partial knowledge.

5.2   Multiple Stakeholders

There is a variety of different groups of people who wish to benefit from the encoding of writing in different ways. In coming to an agreement about encoding we need to understand these different interests and accommodate them.

Language activists are concerned about their own language community maintaining their own identity marked by their languages, social customs, and indigenous knowledge. In debates about encoding they can mistakenly focus on elements that seem to mark identity, such as the name of the code block in Unicode. However these code block names are not visible outside the standard, and their interest may best be served by distinctive fonts with their language or ethnic group marked in the name of the font.

Missionary groups wish to see their sacred texts translated into a particular language with no real interest in the wider use of the writing of the language. While they might adopt the locally dominant writing system, they may compromise on the orthography and produce a limited dictionary. It has been suggested that the missionaries' interest in mother tongue education has been so that the children concerned can then read the missionaries' sacred texts.

Librarians and archivists have been active in encoding for many years, and until the 1980s sought to establish their own computer codes. Their interest was originally to enable cataloguing of document collections, but this interest has since grown to include the digitization of complete documents. People interested in **antiquarian texts** and in the **history** of such texts also hold common cause with librarians and archivists, focusing on archaic forms of the writing and seeing commonality in visual appearance rather than underlying use. Visually different scripts are seen as different writing systems warranting different encodings, rather than seeing common ground in common usages and marking the differences simply by different fonts. This group of stakeholders is unlikely to write new documents in their chosen script, and may simply use the font produced as a focus for internet-based discussions, where a photographic image might have sufficed.

Software producers want to see their technology used as widely as possible – the WWW consortium and Unicode consortium must be counted in here, as well as the many people involved in software localization and the national standards committees of nations who are members of ISO. These are the people who nominally determine the standards, taking advice from the experts they have chosen.

The claim for expertise within Unicode appears to have been captured by the librarians and archivists, at least in the encodings for Nepal. Everson argued for the unification of Ranjana and similar looking scripts in Tibet and Bhutan (2009a), and that Prachalit and Ranjana were different because their headlines were different (2009b), though he has also pointed to the visually distinctive Fraktur and Gaelic scripts. Pandey, a historian, focuses on visual commonality of scripts across languages, rather than on linguistic commonality across scripts.

5.3   Perverse Incentives

Coding proposals are written by a surprisingly small set of people. These proposals are listed in the document registers on the ISO WG2 website. Table 1 shows the number of proposals and other standards documents authored or co-authored by seven Unicode authors.

Table 1. Authorship of ISO WG2 proposals

Person	number of documents authored in period			total
	Sept 08 – Oct 10	Oct 10 – June 11	June 11 – Feb 12	
star 1	42	41	15	98
star 2	19	41	21	81
star 3	9	20	15	44
star 4	19	17	4	40
other 1	9	5	0	14
other 2	6	7	1	14
other 3	2	5	0	7

It clearly helps in proposal writing to have experienced people who know what needs to be done, so the top four people would be valuable assets to Unicode and ISO WG2. They receive special accolades from Unicode, and in one instance an author was featured in the New York Times [11]. Some of the authors get paid for their contribution through the Script Encoding Initiative in Berkeley who in turn receives funding from the US government. Unfortunately all this has the perverse consequence that the more scripts they can successfully encode, the higher the rewards, and instead of seeking to unify scripts, the writers of proposals are incentivized to see differences and encode scripts separately. We have heard other stories where the financial incentives have led the authors of alternative proposals for encoding a particular script to behave as competitors rather than seeking to collaborate to produce the best encoding for the community of users. This perverse incentive can also affect language activists who can receive development funding to champion their particular favoured writing and it's encoding.

6   Conclusions

It seems clear that some reorientation of the encoding standardisation process towards languages and the living users of those languages is necessary if we are prevent the needless proliferation of encodings for scripts which are essentially the same. If this proliferation continues we will end up with a situation not unlike that encountered in Asia with hack fonts and their accidental encoding requiring that if people are to share documents across the internet they must all possess the same font with its own unique encoding. Unicode had seemed the means of saving South Asian languages from hack fonts, it now seems likely to perpetuate the same situation with hack encodings.

Meanwhile those of us concerned about the application of computers to Nepal's languages must aim to develop encoding standards that are agreed within Nepal. Hopefully such proposals would obtain agreement within the wider international community in ISO WG2. However we should not sacrifice the interests of Nepal's language communities to the interests of remote scholars in the west and north and an intermediate position would be to enact Nepal national standards.

Beyond this we are concerned that hack encodings may already have proliferated within Unicode, having in mind the N'Ko writing system of West Africa that looks so like a variant of Arabic writing from which it has clearly been derived. How many more are there like these, should a major review be undertaken?

There is a deep social injustice in the current situation that favours a few hundred scholars of an extinct writing over the interest of hundreds of thousands in a community of living users of a language. Is this a case for formal UNESCO involvement, to extend their concern for languages as cultural expressions to concern for the support of those languages by computer technology?

References

1. Adhikari, S.: History of Nepali Language. Bhundi Puran Prakashan, Kathmandu (2056 BS) (in Nepali)
2. Allwood, J., Dhakhwa, S., Regmi, B.N., Shrestha, P.: Multimodal corpus using multimodal dictionary in Lohorung. In: Proceedings of the International Conference on Speech Database and Assessments - Poster Sessions, Hsinchu, Taiwan, pp. 109–114. National Chiao Tung University (October 2011), The dictionary is viewable at http://desceco.org/lohorung
3. Anderson, D.: Liaison Report, Script Encoding Initiative, UC Berkeley. Document ISO/IEC JTC1/SC2/WG2 N4220 (February 12, 2012)
4. Bal, B.K.: Towards Building Advance Natural Language Applications – An Overview of the Existing Primary Resources and Applications in Nepali. In: 7th Workshop on Asian Language Resources, ACL-IJCNLP 2009, pp. 165–170 (2009)

5. Bal, B.K., Gurung, S., Hall, P.: Towards Universal Access to ICTs in Nepal. In: Computer Society of India Conference, Kolkata, India (2006)
6. Bandhu, C.M.: Origin of Nepali, 5th edn. Sajha Prakashan, Lalitpur (2052 BS) (in Nepali)
7. BIS: Indian Standard Indian Script Code for Information Interchange – ISCII. IS 12194: 1991. Bureau of Indian Standards, New Delhi (1991)
8. Central Bureau of Statistics: National Population and Housing Census 2011 (National Report). Central Bureau of Statistics, Kathmandu (2012)
9. Coulmas, F.: The Writing Systems of the World. Blackwell (1989)
10. Diringer, D.: Writing. Thames and Hudson (1962)
11. Erard, M.: For the World's A B C's, He Makes 1's and 0's. New York Times (September 26, 2003)
12. Everson, M.: Newari code table and names list (2000), http://www.evertype.com/standards/tai/newari.pdf
13. Everson, M.: Preliminary proposal for encoding the Rañjana script in the SMP of the UCS. ISO/IEC JTC1/SC2/WG2 N3649 (2009), http://std.dkuug.dk/jtc1/sc2/wg2/docs/n3649.pdf
14. Everson, M.: Roadmapping the scripts of Nepal. ISO/IEC JTC1/SC2/WG2 N3692 (2009), http://std.dkuug.dk/jtc1/sc2/wg2/docs/n3692.pdf
15. Hale, A., Shrestha, K.P.: Newar. Languages of the World/Meterials, vol. 256. Lincom Europa, Munich and Newcastle (2005)
16. Hall, P.: Proposal to Encode Nepal Himalayish Scripts in ISO/IEC 10646, ISO/IEC JTC1/SC2/WG2 N4347 (2012)
17. Lewis, M.P., Simons, G.F.: Assessing Endangerment: Expanding Fishman's GIDS. Revue Roumaine de Linguistique (RRL) LV(2), 103–120 (2010), http://www.lingv.ro/RRL-2010.html
18. Lewis, M.P., Simons, G.F., Fennig, C.D. (eds.): Ethnologue: Languages of the World, 17th edn., SIL International, Dallas, Texas (2013), http://www.ethnologue.com
19. Manandhar, D.D., Karmacharya, S., Chitrakar, B.: Proposal for the Nepaalalipi script in the UCS. ISO/IEC JTC1/SC2/WG2 N4322 (2012), http://std.dkuug.dk/jtc1/sc2/wg2/docs/n4322.pdf
20. Manandhar, D.D., Karmacharya, S., Chitrakar, B.: Proposal to Encode Ranjana Script in ISO/IEC 10646 (draft submitted to ISO on December 31, 2013)
21. Michailovsky, B., Everson, M.: Revised proposal to encode the Limbu script in the UCS. ISO/IEC JTC1/SC2/WG2 N2410 (2002), http://std.dkuug.dk/jtc1/sc2/wg2/docs/n2410.pdf
22. Noonan, M.: Recent Adaptations of the Devanagari Script for the Tibeto-Burman Languages of Nepal (2003), http://pantherfile.uwm.edu/noonan/www/Papers.html
23. Pandey, A.: Preliminary Proposal to Encode the Prachalit Nepal Script in ISO.IEC 10646. ISO/IEC JTC1/SC2/WG2 N4038 (May 3, 2011), http://std.dkuug.dk/jtc1/sc2/wg2/docs/n4038.pdf
24. Pandey, A.: Proposal to Encode the Letters GYAN and TRA for Limbu in the UCS. ISO/IEC JTC1/SC2/WG2 N3975 (January 14, 2011)
25. Pandey, A.: Proposal to Encode the Newar Script in ISO.IEC 10646. ISO/IEC JTC1/SC2/WG2 N4184 (January 5, 2012), http://std.dkuug.dk/jtc1/sc2/wg2/docs/n4184.pdf
26. Pandey, A.: Proposal to Encode the Tirhuta Script in ISO/IEC 10646. ISO/IEC JTC1/SC2/WG2 N4035 (May 5, 2011)

27. Pokharel, B.K.: Five Hundred Years, 4th edn. Sajha Prakashan, Lalitpur (2050 BS) (in Nepali)

28. Regmi, B.N.: Developing a Devanagari-based multi-language orthography for Nepalese languages. In: Second International Conference on Language Development, Language Revitalization, and Multilingual Education in Ethnolinguistic Communities, Bangkok, Thailand, July 1-3 (2008)

29. Regmi, B.N., Regmi, D.R., Acharya, M., Mahato, H.N., Lamichhane, B.: Typological Study of the Languages of Nepal. Report Submitted to Second Higher Education Project University Grant Commission, Nepal (2012) (in Nepali)

30. Robinson, C., Gadelii, K.: Writing Unwritten Languages. UNESCO website, search for title (2003)

31. Rogers, H.: Writing Systems. A Linguistic Approach. Blackwell Publishing (2005)

32. Ross, F.: The printed Bengali character and its evolution. Curzon Press (1999)

33. Sebba, M.: Sociolinguistic approaches to writing systems research. Writing Systems Research, vol. 1(1). Oxford University Press (2009)

34. Shakya, R.: Alphabet of the Nepalese Script. Motiraj Shakya and Sanunani Shakya, Patan, Nepal (2002)

35. Shakyavansha, H.: Nepalese Alphabets = Nepāl lip i saṁgraha, 7th edn (1985)

36. TDIL: Devanagari Script Behaviour for Hindi Draft issued for comment by the Government of India's Technology Development for Indian Languages Programme (2013)

37. Unicode Consortium: http://www.unicode.org includes Unicode Version 6.3 (2014)

Compound Terms
and Their Multi-word Variants:
Case of German and Russian Languages

Elizaveta Clouet and Béatrice Daille

University of Nantes, LINA
{elizaveta.loginova,beatrice.daille}@univ-nantes.fr

Abstract. The terminology of any language and any domain continuously evolves and leads to a constant term renewal. Terms undergo a wide range of morphological and syntactic variations which have to be handled by any NLP applications. If the syntactic variations of multiword terms have been described and tools designed to process them, only a few works studied the syntagmatic variants of compound terms. This paper is dedicated to the identification of such variants, and more precisely to the detection of synonymic pairs that consist of "compound term - multi-word term". We describe a pipeline for their detection, from compound recognition and splitting to alignment of the variants with original terms, through multi-word term extraction. The experiments are carried out for two compound-producing languages, German and Russian, and two specialised domains: wind energy and breast cancer. We identify variation patterns for these two languages and demonstrate that the transformation of a morphological compound into a syntagmatic compound mainly occurs when the term denomination needs to be enlarged.

1 Introduction

Terms are not always employed in their basic form, a large number of term occurrences in texts are variants [1]. Grouping the variants of a term is useful in various NLP applications such as machine translation, computer-assisted translation [2], Information Retrieval [3], filling of terminological databases and question-answering systems. The set of variants reported for a term could reach 30% of its occurrences [4,5] and even more, depending on genre and domain.

In this paper we focus on identifying the variants of compound terms. By compound we mean "a lexeme that consists of more than one stem" [6]. We admit that a lexical unit is a compound when its components are concatenated or hyphen-separated. Compounding is productive in German, Greek, Finnish and many other languages. Specialised domains are particularly conductive for compounds.

To detect variants, we follow a definition given by Daille [5]: "a variant of a term is an utterance which is semantically and conceptually related with an original term". This means that a variant could have a certain semantic distance from

A. Gelbukh (Ed.): CICLing 2014, Part I, LNCS 8403, pp. 68–78, 2014.

the original term. We examine pairs made of a compound and a syntagmatic variant. We distinguish between synonymic pairs, e.g.

DE[1] Brustdrüsenentfernung, 'mammary gland removing' -
Entfernung der Brustdrüse, 'removing of mammary gland',

and quasi-synomymic pairs that can replace each other only in certain contexts when the term denomination needs to be detailed, e.g.

RU энергоисточник, 'energy source' -
источник тепловой энергии, 'source of heat energy'.

The syntagmatic variant often (but not mandatory) reveals a more specific concept than the compound as shown in the previous example.

The present work aims at detecting compound terms and their multi-word variants in the corpora. Section 2 presents our approach toward compound splitting, term extraction and variant alignment. Section 3 describes experimental data and tool parameters. Section 4 analyses the detected variants. Section 5 gives a short outlook of related works. Section 6 concludes our experiments and suggests some future work.

2   Method

Our approach to identify "compound term - MWT" variation is general and can be applied to any language, to the extent that this language forms compounds. To identify variants, we first extract the compound terms from a specialised corpus. The detected compounds are split into their component parts. Second, we extract a list of MWTs from the same corpus. Finally, we match components of compounds (or more precisely their lemmas) with the MWTs from the list, and if a MWT matches all components of a compound term, we align them. The alignment step is language-independent, whereas the two previous steps include some language-specific rules (see Fig. 1).

2.1   Compound Splitting

German is known to be a highly compounding language. German compounding is well-described in the literature (for example see [7]). Compounding in Russian is less regular than in German, but also productive, particularly in specialised domains. When two words form a compound, a "linking morpheme" is often inserted - s (es) or n (en) in German, o or e in Russian, - and word inflection may be omitted:

DE Blindleistungs<u>bedarf</u>, 'reactive power requirement'
= <u>Blindleistung</u>, 'reactive power' + <u>Bedarf</u>, 'requirement';
RU водоснабжение, 'water supply'
<u>vodo</u><u>snabzhenie</u>[2] = <u>voda</u>, 'water' + <u>snabzhenie</u>, 'supply'.

---

[1] DE denotes to German language, RU - to Russian language.

[2] Here and further transliteration is given for Russian examples.

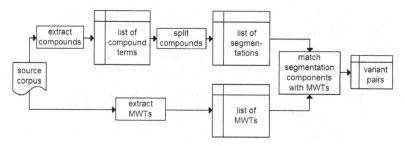

Fig. 1. From term extraction to variant alignment

To recognize and to split compound words, we use a multilingual splitting system combining language independent and language dependent features. The candidate compound is lemmatised. Splitting starts with the generation of all possible two-part segmentations. The component length varies, beginning with a minimum length of 3 characters (it is a common convention in compound splitting, for instance used by Koehn and Knight[8]).

DE Reihenuntersuchung[3], 'mass screening':
reihenuntersuchung → rei + henuntersuchung
reihenuntersuchung → reih + enuntersuchung
. . .
reihenuntersuchung → reihen + untersuchung
. . .
reihenuntersuchung → reihenuntersuch + ung

For each candidate segmentation, the rules which transform components into their independent lemmas are first applied (e.g. DE ”n” → ””: *Reihen* → *Reihe*). If the rules are not available or not sufficient to cover all transformation cases, the prospective lemmas are proposed using a string similarity measure (Levenshtein distance).

Second, the lemmas for both components are matched with a lexicon (for example, a monolingual dictionary) and with a monolingual corpus. The usage of a lexicon allows us to deal with neoclassical compounds [9], i.e. compounds including elements of Latin or Greek origin (e.g. DE *Histopathologie*, 'histopathology'). Such elements are not independent words and thus are not separately employed in a corpus. The usage of a corpus instead helps to recognise the components formed by highly specialised words, neologisms or loan words that are not listed in general language dictionaries. The segmentation score (from 0 to 1) depends on the following features for each component: similarity between the component and its lemma, presence of the component lemma in a lexicon and its frequency in a corpus. The features are combined by a linear interpolation, and the feature weights are learnt during separate experiments on the training data.

---

[3] Splitter input and output are in lower case.

Table 1. Examples of syntactic patterns for MWT extraction

	Pattern	MWT	English translation
German	ADJ N	*Genetische Veränderung*	*Genetic modification*
	N S:p N	*Netzintegration von Windenergie*	*Grid integration of wind energy*
Russian	ADJ N	*Биологическое топливо*	*Bio fuel*
	N:gen N	*Потребление тепла*	*Consumption of heat*

The right side component is split further in a recursive manner in order to process compounds consisting in more than two parts:

RU килоэлектронвольт ("kiloelectronvolt"):
kiloelektronvolt → kilo + elektronvolt
elektronvolt → elektron + volt

Finally, the system returns a top N of the best segmentations ordered by their score. For DE *Reihenuntersuchung* the output is:

reihe untersuchung 0.9
reihe untersuchend 0.85
reiben untersuchung 0.85
reifen untersuchung 0.85
reißen untersuchung 0.85

In this example, the best-ranked segmentation Reihe 'series, number' + Untersuchung 'screening' is correct.

In order to recognize whether a candidate is compound or not using this system, a threshold should be defined. This score differs according to the language and to the target application. If a very precise result and only reliable compounds are needed, the score should be close to 1, but some compounds would be missing. If a high recall and a large number of compounds are the goal, this minimal score should be decreased.

2.2 Multi-word Term Extraction

MWT candidates are extracted from the corpora using TermSuite[4]. This tool allows identification of single-word terms as well as multi-word terms, but for the task we address in the present paper, only MWTs are needed. MWTs are defined in this tool as noun phrases consisting in 2 or 3 content words, and are detected by means of hand-crafted syntactic patterns. For some examples of frequent patterns[5] in DE and RU see Table 1.

2.3 Alignment between Compounds and MWTs

Alignment between compound terms and their multi-word variants is performed by simple matching of compound element lemmas with the words forming MWTs.

---

[4] version 1.4, http://code.google.com/p/ttc-project
[5] POS notation: ADJ - adjective, N - noun, N:gen - noun in genitive case, S:p - preposition.

We aim at detecting the variants with different structures, thus we deliberately do not limit variant identification by any syntactic rules. The fact that a list of extracted units and not a raw corpus is used for matching ensures that the variants are well-formed terms. We allow up to a 15-symbol window between compound elements in order to cover quasi-synonymic variants. The order of elements inside a multi-word variant may differ from the order inside a compound:

> RU ветропоток - поток ветра
> *translit.* vetropotok - potok vetra
> 'airflow' - 'flow of air';

> DE Häufigkeit-Ergebnis-Beziehung -
> Beziehung von Häufigkeit und Ergebnis
> 'frequency-result-correlation' -
> 'correlation between frequency and result (of treatment).'

For three-component terms, all possible permutations are considered[6].

3   Data Pre-processing and Settings

We applied this method of variant extraction to three specialised corpora of comparable size: Russian and German corpora related to the wind energy domain, and German corpus from the medical field (breast cancer domain). The corpora were collected from the web. German and Russian wind energy corpora are comparable, i.e. the texts deal with the same topic, but are not translations from a language to the other. For corpora statistics, see Table 2.

Table 2. Experimental Setup

	RU WIND	DE WIND	DE CANCER
Initial corpus size	323 946	358 602	378 474
Lexicon size after filtering	5970	3815	3739
Extracted compounds	1114	2281	2092
Extracted terms (MWT & SWT)	19 292	15 709	16 840

The corpora were previously lemmatised and annotated with part-of-speech (POS) tags by TreeTagger[7]. To reduce the data to be processed by the splitting module, a lexicon was produced from each corpus. The lexicon includes only lemmas, and it is filtered by lemmas POS, frequency and length. Only nouns

---

[6] The scripts for variant alignment, as well as for compound splitting, are available on http://www.CICLing.org/2014/data/57

[7] http://www.cis.uni-muenchen.de/~schmid/tools/TreeTagger/

and adjectives were kept, because other POS seldom form compounds. The lemmas which appear less than 5 times in the corpus were excluded to ensure keeping only correct forms. We also excluded lemmas shorter than 6 characters.

TreeTagger is a probabilistic tool trained on a large amount of textual data from general language. However domain-specific compounds seldom appear in the general language texts, so the tagger often fails performing their lemmatisation. Both languages we treat are morphologically rich languages with large case systems, and proper lemmatisation is important for accurate compound splitting. Thus we added some basic rules to improve lemmatisation for German (dropping plural and case inflections *es, en, er, em, ns, s, n, e*). Lemma correction for Russian would be necessary too, but the rules are not straightforward, and it was not performed in the present work.

As regards compound splitting module, only the best-ranked candidate segmentation was used for each compound. To choose an optimal score threshold for our task, we previously trained the module on a subset of corpora lexicons manually annotated with a category (compound or not) and with the correct segmentation(s) for compounds. As expected, the splitting precision grows with the threshold, whereas recall gets lower (cf. Fig. 2).

Splitting precision is calculated as the ratio between the number of words properly split by the module, and the number of words that have been split:

$$SplittingPrecision = \frac{nbCorrectSplits}{nbSplits} \tag{1}$$

Recall is calculated as the ratio between the number of words correctly split, and the total number compounds to be analysed:

$$SplittingRecall = \frac{nbCorrectSplits}{nbCompounds} \tag{2}$$

The challenge is to choose a threshold allowing a good balance between precision and recall. For the present work, we set a minimal threshold of 0.8 for German and of 0.75 for Russian so as to privilege recall but not to affect too much precision.

When extracting MWTs, a frequency threshold of 2 was defined to avoid hapax legomena and to reduce extraction errors. We did not put this threshold to a higher value because a variant is a rare occurrence and only occasionally appears in the texts. The line *Extracted terms* of the Table 2 shows the number of terms, both single and multi-word, detected by TermSuite.

4   Results

The process described above resulted in the extraction of a number of candidate multi-word variants comprised between 87 and 153 depending on language and domain. For details see Table 3.

For our task, the evaluation of recall is difficult to carry out because it requires to previously enumerate all variants of all possible structures appearing in the corpora. So we limited the evaluation of the results by assessment of precision.

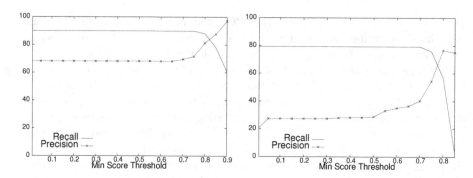

Fig. 2. Recall and precision of compound recognition and splitting for the training data from DE breast cancer corpus (left) and RU wind energy corpus (right)

Table 3. Variant Detection Results

	RU WIND	DE WIND	DE CANCER
Extracted variant pairs	109	87	153
Correct variant pairs	79	67	133
Precision (%)	72	77	87

Precision is estimated as the number of correct variant pairs divided by the total number of extracted variant pairs:

$$Precision = \frac{nbCorrectVariants}{nbVariants} \tag{3}$$

A variant pair "compound - MWT" is considered to be correct if two terms are synonyms or quasi-synonyms. According to the language and domain, precision is between 72% and 87% (cf. Table 3).

4.1 Error Analysis

Since we did not set pre-determined variation patterns, we did not expect a high precision of results. Some extracted multi-word candidates are not related to the original terms:

RU парогенератор - параметр генератора
translit. parogenerator - parameter generatora
'steam generator' - 'generator parameter'.

Other candidates are related to the original term, but do not refer to the same concept:

DE Genveränderung, 'genetic mutation' -
Gentest auf Veränderungen, 'genetic screening for the mutations'.

The head of the compound term, *Veränderung*, plays the role of modifier in the MWT. It could be convenient to extract such type of variants, but in this study we considered them as incorrect.

A considerable number of errors are due to incorrect lemmatisation, especially for the Russian corpus since a lemma correction was not applied to it. Among 31 variant pairs considered as incorrect, 17 are actually correct but include a term that has already been extracted in another inflected form.

Some errors were introduced during MWT extraction. Incomplete terms were sometimes extracted, for instance DE *Risiko an Brustkrebs*, a part of the phrase *Risiko an Brustkrebs zu erkranken*, 'risk to fall ill with a breast cancer'. This incomplete term candidate was then aligned with compound *Brustkrebsrisiko*, 'breast cancer risk'.

Interestingly, we noticed that no error was generated by inaccurate compound splitting: all incorrect segmentation candidates were filtered during the alignment step.

4.2 Variant Structures

Among the extracted variant pairs, we observed two most productive variation patterns, and both can be expanded by additional element(s). Let us formulate these patterns using the following annotation: A and B - lexical units forming a compound, A' and B' - the same stems as A and B but appearing in the right part of the rule, prime means that the parts of MWT do not always match exactly the elements of compound, X - additional expanding element. A, B and X are the content words of any POS. We use + as the concatenation operator, − means hyphen, ? is the optional operator, [] defines a set of symbols.

1. A+B → B' (X[+-])?A'
   Morpho-syntactic realisation differs for German MWT: $N\ S{:}p\ N$, and for Russian MWT: $N\ N{:}gen$.

   DE Mammakarzinom-Patientin, 'breast cancer patient' -
   Patientin mit Mammakarzinom, 'patient with breast cancer';

   RU энергобаланс - баланс энергии
   translit. energobalans - balans energii
   'energy balance' - 'balance of energy'.

For both languages, additional element X may be space-separated (independent word, usually modifier) or concatenated to one of the MWT components (by juxtaposition or by hyphen):

   RU энергоресурс - ресурс <u>ветровой энергии</u>
   translit. energoresurs - resurs vetrovoj energii
   'energy source' - 'source of wind energy';

   DE Krebs-Früherkennung, 'cancer early diagnosis' -
   Früherkennung von <u>Brust</u>krebs, 'early diagnosis of breast cancer'.

2. A+B → A′(X[+-])? B′

   Morpho-syntactic structure for MWT is the same for both languages: *ADJ N*, and the expansion is also possible.

   DE Altanlage - alte Anlage, 'old plant'.

   This pattern has already been identified for multi-word expressions in English [10].

Expansion can be only graphical as in the example just below, or also semantic if the additional element has its proper meaning (cf. the second example):

   DE Magnetfeld - magnet<u>ische</u> Feld, 'magnetic field';

   RU фотоэлемент - <u>фотоэлектрический</u> элемент
   translit. fotoelement - fotoelektricheskiy element
   'photocell' - 'photoelectric cell'.

Variants with semantic expansion are more likely to be quasi-synonyms, but full synonymy is also possible:

   DE Küstenregion, 'coastal area' -
   küstennahe Region, lit. 'coast-near area'.

In RU wind energy corpus the second pattern dominates, in DE wind energy corpus the distribution is more comparable with a slight domination of the first pattern, whereas in DE breast cancer domain the first pattern overcomes the second.

When analysing a pair gathering a compound and a multi-word term, we consider the compound as the base term and the noun phrase as the variant. Indeed, for highly specialised units, compound forms have been inserted into dictionaries and seem to dominate in the texts. However many compounds are etymologically formed from multi-word expressions. Our work advocates for the handling of compounds and MWTs as units of the same level.

5   Related Work

Variant Extraction. Variant detection task in a multilingual context was addressed for English [11,10], French [11,2], German [2] and Japanese [3] languages. Jacquemin [11] designed a tool for variant recognition in the corpora, Fastr, which is based on morpho-syntactic rules. These rules have been enlarged for English language by defining the left and right context of the noun phrase which forms a variant [10]. This led to the overall precision of 83% on the variants pairs "MWT-MWT".

Yoshikane et al.[3] adapted Fastr for Japanese variant detection. The morpho-syntactic rules required by Fastr were established after investigation of the aligned pairs of, on the one hand, the MWTs issued from a terminology database and, on the other hand, the corpus sentences. A large spectrum of variation types involving MWTs was considered, including a type "compound - MWT". For this type the approach achieved a precision of about 94% for 806 extracted variants.

Weller et al. [2] discussed variants of compound terms which have various structures, including the variants with expansion. For variant extraction, the authors focused on a single multi-word variant pattern, $N\ S{:}p\ N$. For 100 analysed German variants, precision of 74% was attested using a predefined set of variation patterns. They also evoked a non-symbolic approach to variant extraction, i.e. without the usage of variation patterns.

Compound Splitting. As regards compound splitting, which is necessary to treat variation involving compound terms, several methods were proposed, from language-specific (e.g. morphological analyser SMOR [12] for German) to probabilistic and thus fully language-independent [13,14,6]. The first corpus-driven approach proposed by Koehn and Knight [8] became a state-of-the art one. The authors estimated probability for a segmentation from the geometric mean of the components frequencies in the corpus. Some splitters match the components with monolingual dictionary (BananaSplit [15]). Even statistical methods currently include some linguistic knowledge, i.e. a list of linking morphemes, in order to improve performance. The approach we used for this work combines corpus frequency, dictionary matching and transformation rules, with addition of string similarity when the rules are not sufficient.

6   Conclusions and Future Work

In the present work we addressed multi-word variants of compound terms on the example of German and Russian languages. To identify the variants, we applied a simple language-independent method based on string matching of component lemmas with the list of previously extracted MWT.

The matching with the term list instead of a raw corpus filtered many inaccurate or incomplete variants, even if some errors have occurred during MWT extraction. Our method gave although lower precision than the one that could be obtained using a large set of pre-defined variation patterns. Instead it allowed us to investigate the variants of diverse syntactic structures.

Error analysis showed that some pre-processing mistakes (compound lemmatisation, term extraction) were propagated on the variant identification, which means that precision could be increased by improving the quality of pre-processing. Compound splitting has automatically been accomplished, but all splitting errors were filtered on the variant alignment step.

After observation of identified variant pairs, we formulated the two most common variation patterns, and we paid a particular attention to variation with expansion. It would be interesting to analyse distribution of expansion subtypes (left or right expansion) according to language and domain.

Similar experiments could be carried out for other languages in order to compare variation patterns. The variation involving a compound term and a MWT occurs even in languages which are generally considered as non-compounding, for instance in French: *insulino-résistance - résistance à l'insuline,* 'insulin resistance'. Another variation type was not addressed in this work, i.e. a coordination variation: DE *Östrogenrezeptor, Progesteronrezeptor - Östrogen- und Progesteronrezeptoren,* 'oestrogen and progesterone receptors'. This variation type is also common for several languages.

References

1. Jacquemin, C.: Syntagmatic and paradigmatic representations of term variation. In: Proceedings of 37th Annual Meeting of the Association for Computational Linguistics (ACL 1999), pp. 341–348 (1999)
2. Weller, M., Blancafort, H., Gojun, A., Heid, U.: Terminology extraction and term variation patterns: A study of french and german data. In: Proceedings of German Society for Computational Linguistics and Language Technology (GSCL 2011), Hamburg, Germany (2011)
3. Yoshikane, F., Tsuji, K., Kageura, K., Jacquemin, C.: Detecting japanese term variation in textual corpus. In: Proceedings of 4th International Workshop on Information Retrieval with Asian Languages (IRAL 1999), Taipei, Taiwan, pp. 97–108 (1999)
4. Jacquemin, C.: Spotting and Discovering Terms through Natural Language Processing. MIT Press, Cambridge (2001)
5. Daille, B.: Variations and application-oriented terminology engineering. Terminology 11, 181–196 (2005)
6. Macherey, K., Dai, A., Talbot, D., Popat, A., Och, F.: Language-independent compound splitting with morphological operations. In: Proceedings of ACL 2011, Portland, Oregon, pp. 1395–1404 (2011)
7. Langer, S.: Zur Morphologie und Semantik von Nominalkomposita. In: Proceedings of KONVENS 1998, Bonn, pp. 83–97 (1998)
8. Koehn, P., Knight, K.: Empirical methods for compound splitting. In: Proceedings of EACL 2003, Budapest, Hungary (2003)
9. Namer, F.: Morphologie, Lexique et Traitement Automatique des Langues. Lavoisier, Paris (2009)
10. Ville-Ometz, F., Royauté, J., Zasadzinski, A.: Enhancing in automatic recognition and extraction of term variants with linguistic features. Terminology 13, 61–84 (2007)
11. Jacquemin, C.: Fastr: A unification-based front-end to automatic indexing. In: Proceedings of Intelligent Multimedia Information Retrieval Systems and Management (RIAO 1994), pp. 34–47 (1994)
12. Schmid, H., Fitschen, A., Heid, U.: SMOR: A german computational morphology covering derivation, composition, and inflection. In: Proceedings of LREC 2004, Lisbon, Portugal, pp. 1263–1266 (2004)
13. Dyer, C.: Using a maximum entropy model to build segmentation lattices for mt. In: Proceedings of HLT-NAACL 2009 (2009)
14. Hewlett, D., Cohen, P.: Fully unsupervised word segmentation with bve and mdl. In: Proceedings of ACL 2011, Portland, Oregon, pp. 540–545 (2011)
15. Ott, N.: Measuring semantic relatedness of german compounds using germanet (2005),
   http://niels.drni.de/n3files/bananasplit/Compound-GermaNet-Slides.pdf

A Fully Automated Approach for Arabic Slang Lexicon Extraction from Microblogs

Hady ElSahar and Samhaa R. El-Beltagy

Center of Informatics Sciences, Nile University, Cairo, Egypt
hadyelsahar@gmail.com, samhaa@computer.org

Abstract. With the rapid increase in the volume of Arabic opinionated posts on different social media forums, comes an increased demand for Arabic sentiment analysis tools and resources. Social media posts, especially those made by the younger generation, are usually written using colloquial Arabic and include a lot of slang, many of which evolves over time. While some work has been carried out to build modern standard Arabic sentiment lexicons, these need to be supplemented with dialectical terms and continuously updated with slang. This paper proposes a fully automated approach for building a dialectical/slang subjectivity lexicon for use in Arabic Sentiment analysis using lexico-syntactic patterns. Since existing Arabic part of speech taggers and other morphological resources have been found to handle colloquial Arabic very poorly, the presented approach does not employ any such tools, allowing the presented approach to generalize across dialects with some minor modifications. Results of experiments, that targeted Egyptian Arabic, show the approach's ability to detect subjective internet slang represented by single words or by multi-word expressions, as well as classifying the polarity of these with a high degree of precision.

1 Introduction

Recent years, have seen an enormous increase in the use of microblogging services and social media sites across the world and especially in the Arab World. A study prepared and published by Semiocast in 2012 has revealed that Arabic was the fastest growing language on Twitter in 2011, and was the 6th most used language on Twitter in 2012 [1]. A recent breakdown of Facebook(FB) users by country, places Egypt with 16 million users as the Arabic speaking country with the largest number of FB users, and ranks it at 17 among all countries of the world [2]. This represents a growth of 41% in terms of users, from the previous year [2]. Both Twitter and Facebook are characterized by having a high percentage of highly opinionated posts. The presence of such a large volume of opinionated data highlights the need for sentiment analysis tools which can make use of these whether in marketing, politics or other areas. Since slang and dialectical expressions are very commonly used for expressing sentiments and opinions on social media, augmenting sentiment lexicons with slang terms and expressions can directly impact the sentiment analysis process. The

A. Gelbukh (Ed.): CICLing 2014, Part I, LNCS 8403, pp. 79–91, 2014.
© Springer-Verlag Berlin Heidelberg 2014

aim of this work is to present an approach for capturing dialectical or slang terms or compound expressions that are highly indicative of subjectivity as well as assigning polarity to these. The rest of this paper is organized as follows: section 2, describes the different aspects of the problem that this works aims to address, section 3 briefly reviews related work, section 4 provides an overview of the proposed system, section 5 presents the experiments carried out to evaluate the work and their results, and finally section 6 concludes this paper.

2   Problem Statement

Sentiment analysis of Arabic social media is a challenging task not only because social media language is rich with colloquial Arabic, compound terms and idioms as well as a lack of resources as detailed in [3], but also because the language used in social media, and twitter in particular, has been shown to be of a highly dynamic and evolving nature [4]. Creative expressions that imply subjectivity are often created on the fly by popular tweeps (twitter users) and then quickly propagated and widely employed by other social media users; they then become strong subjective clauses. Subjective terms often emerge from peculiar exchanges observed on TV shows, advertisements, or trending YouTube videos. For example, the word "حبيبح", which the presented work was able to learn and which has no meaning in colloquial or MSA Arabic, is now commonly used to indicate a positive sentiment. The word is the written form of a sigh made by a famous Egyptian TV presenter in a positive context. Political situations and public figures are yet another source of inspiration for the creation of new expressions the usage of which indicates high subjectivity. For example, the adjective "نكسجي" ("someone who creates setbacks") which was created after the 30th of June events in Egypt, is used as a negative reference to a set of people with a specific political opinion. Also the term "بتاع الاستبن" ("affiliated to the spare tire"), which doesn't make much sense unless understood in context, is a negative term referring to those who support former Egyptian president Muhammad Morsi. The term "الاستبن" which means "the spare tire" was widely used for former president Muhammad Morsi during the 2012th presidential elections, by those who opposed him.

Another problem that this work tries to address is the wide use of transliterated English to reflect sentiment.   For example, the presented work is capable of picking up words like "كيوت" and "قيوط" both of which are Arabic transliterations of the English word "cute" and which reflect a positive sentiment. "اوفر" is another example of a commonly used transliterated English term, which is "over", and which is used in the social media context to indicate "exaggeration" or to something that is over the top. The presented work also tries to capture subjective slang and dialectical expressions that are well established within a certain culture or country.

The goal of this work is not only to try to detect slang and informality in a fully automatic way but to assign polarity to extracted terms. It's important to re-iterate that the terms of interest can be single words or compound phrases.

3   Related Work

There hasn't been much work that directly targets slang detection, let alone assigning polarity to slang expressions. However, if we consider slang as a special case of lexicon learning, then a lot of work becomes relevant. Research has been carried out to address the question of how to build a polarity lexicon for many languages including but not limited to German, Dutch, Spanish, Chinese, and Japanese, but most notably, English. This section will focus primarily on relevant work carried out for Arabic, which is the focus of this work, and on English where most efforts have been condensed,   as well as on approaches that claim language independence.

Turney [5] proposed   an unsupervised algorithm for inferring the polarity of phrases that have adjectives and adverbs within a POS-tagged corpus. Polarity is calculated based on Pointwise Mutual Information between unknown phrases and the words "excellent and poor". The approach suffers from several drawbacks if considered for application within an Arabic social media context: 1. it relies on the existence of a huge POS-tagged corpus, which in the case of Arabic social media is particularly challenging as the language used within this media, and especially in microblogs, is highly unstructured and a POS tagger that can actually work on these with any acceptable degree of accuracy, is yet to be developed. 2. The approach only targets adjectives and adverbs.

Banea el al [6] presented an approach which they claim can work for building subjective lexicons for languages with scarce resources. The approach requires a small seed of subjective words, and with the aid of a dictionary, uses those to generate a set of other candidate subjective terms. The candidate terms are ranked using Latent Semantic Analysis as a similarity measure between a candidate and the original seed. The approach was applied on Romanian. Based on the description provided for the approach it can be deduced that the approach is incapable of handling slang, which is very common in a social media settings as slang can rarely be found in language dictionaries, and completely ignores multiword expressions, which are very commonly used in Arabic to convey sentiment.

Abdul-Mageed & Diab [7] proposed an approach for building a large scale Arabic sentiment lexicon. To do so, they expand on a Modern standard Arabic (MSA) polarity lexicon of 3225 adjectives which was built manually, by using a number of existing English lexicons including SentiWordNet [8]. The authors report having problems with both the coverage and the quality of some of the entries. They also state that they have not tested the system for the task of sentiment analysis. But even without the reported limitations, this approach will still be incapable of encompassing slang, dialectical Arabic,   and multiword expressions.

Velikovich & Blair-Goldensohn [9]   proposed yet another approach for building a polarity lexicon using graph propagation over a phrase similarity graph built using 4 billion unlabeled web documents and a set of seed terms. Results of experiments carried out by the authors show that the derived lexicon improves the accuracy for the sentence polarity classification task. The advantage of using this method is that it is capable of learning slang and multi-word expressions. It is not clear however, how well it will perform if the graph is built using a smaller corpus.

Volkova et al. [4], proposed an approach for bootstrapping subjectivity clues from Twitter without relying on language-dependent tools. To do so, strong subjective terms were selected from the MPQA lexicon (the number of which was not specified) and used to annotate a set of 1M Tweets in English. To show that the presented approach can generalize across languages, the selected seed terms were also translated to Russian and Spanish using bilingual dictionaries. Translations were used to annotate a set of Tweets for corresponding languages. The polarity of new terms was determined based on the probability of the term appearing in positive or negative tweets. It was not clear how objective terms were handled. For this approach to work for other languages, either the target language must have a large lexicon that can be used as a seed, or a bilingual dictionary that can be used to translate from the original English seed lexicon to the target language must exist. The work does not address idioms and multi-word expressions.

4   The Proposed Approach

As stated before, the goal of this work is to present an approach capable of detecting commonly used dialectical or slang terms that reflect subjectivity. A term in the context of this work can be made up of a single word or multiple words. Furthermore, the work aims to classify the polarity of detected terms. So, the presented approach is a two phase one. In the first phase, candidate subjectivity terms are detected, and in the second phase, they are classified.

For detecting highly subjective words/expression, we have identified a set of lexico-syntactic patterns indicative of subjectivity. The use of handcrafted patterns is not a novel idea in the context of information extraction, for example, Hearst [10] has used such patterns for acquiring hyponyms from large text corpora, while Klaussner and Zhekova [11] have used them for ontology learning.

The tags used within the proposed patterns do not require the use of a part of speech tagger; in fact each tag has a list of finite possible values that is dialect dependant. So while the presented patterns themselves, are mostly dialect independent, the range of possible values from which tags in a pattern can be derived, depends on the dialect that is being targeted. Our work and experiments have focused on Egyptian Arabic; and more specifically, the Cairene dialect.

In the first phase, the extraction patterns are applied on a large corpus of text obtained from twitter to yield a set of subjective terms. In the second phase, the extracted terms, are assigned polarity based on the normalized point wise mutual information score between them and positive and negative terms derived from an existing polarity lexicon. Details of the process are presented in the following subsections. Figure 1 summarizes the entire process.

4.1   Extraction Patterns

When identifying extraction patterns, we had to make sure that they satisfy the following set of constraints: 1) patterns shouldn't have to rely on any part of speech taggers or parsers 2) defined patterns should be as general as possible so as to increase recall, 3) patterns should exhibit a very high rate of accuracy so as to be able to learn subjectivity in a totally automated way. The last point requires individual evaluation of identified patterns.

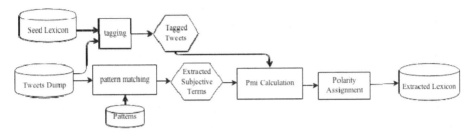

Fig. 1. Overview of the Lexicon Extraction System

Table 1 shows the tags that we have employed in our extraction patterns. Each tag is accompanied by a brief description and a set of Arabic and English examples as well as the count of the tokens that were used as a value pool for that tag. While most tags are fairly standard, the "Person Reference" [PR], "Strong subjective" [SS], and "Subjective Expression" {SE} are not. The [PR] tag refers to any word that can be made to reference a human being. Egyptian dialectical terms that do so include "راجل" (man), "بت" and "بنت" both of which correspond to "young lady" the former being a rather vulgar form for the term, "بني ادم" (human being), "كائن" (creature) and "انسان" (human).The [SS] tag refers to a set of strong subjective positive and negative terms generalized to their different masculine, feminine and plural forms. As can be seen from the table, only a very small seed of 25 terms was used. Terms that are learned at the end of our extraction process, can be made to augment this list, to relearn other terms iteratively. The {SE} tag, is simply a place holder for the expression that will be extracted from any pattern

Since our work is mainly concerned with slang, tag lists were built to match with slang rather than with Modern Standard Arabic (MSA). For example in the list of "Demonstrative Pronouns", terms such as "ده" which corresponds to masculine "this" and "دي" which is the feminine form of "this" as well as "دول" which means "those" were included. The same was applied to all other lists. The slang intensifiers list for example included "خالص", "موت", "السنين" and "بغباوه" which all corresponds to "a lot", and affirmations like "صحيح" and "بجد" which mean "indeed". The personal pronouns list included only personal prounouns that can be used for addressing someone rather than all personal pronouns.

The initial set of candidate patterns is shown in **Table 2**. As can be seen from the table, specifically from examples and their translations, the structure of Arabic sentences is quite different from English ones. For example, in English, intensifiers often precede the terms they intensify, while in Arabic they often follow them. So, for pattern 1 to work in English for example, it should become [DP] [PR] [Ints] {SE} and {SE} should be limited to a single word to avoid noise. A literal translation of the matched examples in the table would have not made much sense, so instead we include as an accurate a translation as possible (which does not always correspond to the Arabic structure) and highlight English intensifiers with italics.

Table 1. Tags used in extraction patterns

Tag	Description	Arabic Examples	English Examples	Count
[Neg]	Negator	لايمكن, مش, لا	Not	46
[DP]	Demonstrative Pronoun	دي,ده	This, that	15
[Ints]	Intensifier	جدا, طحن, اوي	Very much	40
[PR]	Person Reference	إنسان, راجل , ست	Person, man, woman	40
[PP]	Personal Pronoun	إنت, يا	You	5
[Conj]	Conjunction	و, او	And, or	3
[SS]	Strong subjective	جميل,حقير	Pretty, vile	25
{SE}	Subjective Expression to be extracted			N/A

The set of patterns listed in **Table 2**, are based on observing how people commonly express sentiments about people or things in Arabic. Before carrying out any rigorous evaluation, each of these patterns was used to extract a limited set of subjective terms from a subset of the used twitter corpus (described in details in the next section). A quick examination revealed that patterns P9, P10 and P11, often result in noisy matches. For example, one of the expressions captured by pattern P9 was " سكره ايه اللي حصل" which translates to "sugar what happened". While "sugar" is indeed a subjective term, the rest of the expression "what happened" is simply noise. What all three patterns have in common is being open ended, meaning that the captured slang expression can match with any words that follow the initial pattern with no constraints. This has the potential of introducing too much noise and should be avoided which is why these patterns were omitted from the final pattern set. A thorough examination of how well the remaining patterns perform is presented in section 5.

All remaining patterns, except for patterns P2 and P5 are dialect independent. To adapt those to other dialects, the text hardwired in these patterns needs to be translated to the target dialect. The intensifier at the end of pattern 1, and the personal pronoun at the end of pattern 3, simply ensure that the extracted term is of high subjectivity and that multi-word sentiments are extracted as {SE} can match with one or more words. Patterns P4, P6, P7 and P8 capitalize on that fact positive or negative terms usually occur together by using of conjunctions or using multiple "Personal Pointers" [PP].

4.2   Polarity Classification

After extracting candidate subjective terms, there is a need for assigning polarity to these. In order to do so, this work proposes the use of co-occurrence statistics between each of candidate term and known positive and negative terms in a large collection of microblogs or tweets. The reason a microblog or tweet is favored as a

Table 2. Initial set of candidate patterns with examples

ID	Pattern	Example of a Match
P1	[DP] [PR] {SE} [Ints]	ده عيل {مستفز} بغباوه This boy is *incredibly* {irritating}
P2	" ايه ال" {SE} [DP]	ايه ال{شياكه} دي what {elegance } this is
P3	"ال"[PR] [DP] {SE} [Ints]	الراجل ده {حظه وحش} جدا This man has *incredibly* {bad luck}
P4	"ال"[PR] [DP] {SE} [Conj] [SS]	الست دي { مبدعه دائما } و فنانه This lady is {always creative} and artistic
P5	"اما انك" {SE} [Ints]	اما انك {فرفوور} صحيح You are such a {wimp} *indeed*

ID	Pattern	Example of a Match
P6	[PP] [SS] [PP] {SE} [PP]	يا رائع ياللي {بتفتح النفس} ياللي ... You wonderful you who {motivates} who...
P7	[PR] {SE} [Conj] [SS]	راجل {صاحب مبادئ} و محترم A man {with principles}and respectable
P8	[SS] [Conj] {SE} [Ints]	محترم و {مؤدب} جدا Respectable and *very*{polite}
Excluded Patterns		
P9	[PP] "لامؤاخذه" {SE}	انت لامؤاخذه {غبي} you, excuse me, are {dumb}
P10	[DP] "حتت" {SE}	دي حتت {سكره } She's a piece of {Sugar}
P11	"راجل" "ده" {SE}	ده راجل { مش متربي} This man {has no manners}

unit of information instead of a document or a longer posting, is that the short nature of tweets increases the likelihood that only sentiments with similar polarity will co-occur in a single tweet. Tweets are also rich with slang. By tagging individual tweets in a large corpus using an existing sentiment lexicon, co-occurrence can be calculated using normalized point wise mutual information $nPmi$ [12] as represented by equation (1), where x represents the candidate subjective term and y is the polarity class which can be positive or negative .

$$nPmi(x, y) = \left(\ln \frac{P(x,y)}{P(x)P(y)} \middle/ -\ln P(x,y) \right) \tag{1}$$

5   Experiments and Results

To determine whether or not the proposed approach is capable of achieving its goals, the proposed patterns had to be applied on a large corpus in order to extract terms. The extracted terms then had to be annotated with polarity.   Both the system's ability to extract subjectivity and slang and its ability to assign polarity were individually evaluated. Section 5.1 describes the dataset used in the experiments, while sections 5.2 and 5.3 describe the experiments carried out to evaluate the system's ability to detect subjectivity or slang and its ability to assign polarity respectively.

5.1   The Used Data Set

To build the corpus on which to apply the patterns described in the previous section, a set of approximately 11 million Arabic tweets was collected using the twitter API [13] and a set of trending hashtags as search terms. A preprocessing step was then made on individual tweets to remove unwanted features and noise. The process included:

1) Removal of hyperlinks
2) Removal of the hash letter '#' to capture subjective text in hash tags if available and also the replacement of underscores by spaces in hash tags to convert them into regular words
3) Text normalization according to the rules described in [14] as Arabic speakers usually mix between characters like "ي"and "ى" or "أ" and "ا" or "ه" and "ة"
4) Removal of redundant tweets, usually resulting from re-tweets. Instead of checking for exact matches, the cosine similarity function [15] with a threshold value of 0.7 was used. This guarantees uniqueness of instances in the dataset and excludes quoted re-tweets that may give a false indication for the high occurrence of a specific term.

The pre-processing step reduced the size of the corpus to 7.5M unique normalized tweets.

5.2   Evaluation of Slang Extraction

To evaluate the process of detecting and extracting slang terms, the proposed set of eight predefined patterns was applied to the dataset consisting of 7.5M preprocessed tweets. The pattern matching process resulted in the extraction of a set of 633 unique terms.

Three graduate students, who are also Arabic native speakers, were asked to manually annotate the set of 633 terms extracted from the pattern matching process using positive, negative, or not-a-sentiment tags. Each term was assigned the tag which had the most consensuses from the three judges. In addition, out of dictionary terms were labeled as *Slang*, swear terms and profanity were labeled as *Profanity*, and multi-word terms were labeled as *Compound*. The tagged dataset was then treated as a ground truth against which obtained results were compared. For each of the eight predefined patterns subjectivity precision was measured by calculating the percentage of terms that were assigned a subjective label (positive or negative) from the total number of extracted terms. **Table 3** summarizes the results.

Table 3. Summary of patterns and their corresponding statistis

	# of extracted terms	Profanity	Slang	Compound	Precision
P1	144	0.23	0.56	0.23	0.92
P2	14	0.43	0.93	0.14	1
P3	135	0.24	0.7	0.27	0.81
P4	4	0	0.5	0	1
P5	21	0.86	0.91	0.48	1
P6	222	0.53	0.59	0.24	0.9
P7	123	0.33	0.49	0.19	0.93
P8	98	0.1	0.44	0.27	0.84
Total	**633**	**0.32**	**0.6**	**0.29**	**0.89**

By analyzing the output of the different patterns, as reflected by Table 3, we find that P6 which was based on the detection of multiple personal pointers [PP] showed relatively higher performance than the rest of the patterns with 222 extracted terms and 90% precision. On the other hand P4, which is considered to be a mixed pattern, was able to extract only 4 terms due to having relatively more restrictions. Having said that, the count of the extracted terms would have been significantly higher if we could have overcome the limitation of the Twitter API by querying it directly for tweets that match our patterns instead of querying over a downloaded corpus. The overall pattern matching technique resulted in a total precision of 88.6%.

Some examples of extracted terms are shown in **Table 4**. The examples illustrate the system's ability to detect slang terms (both established and newly evolved) as well as subjective expressions. For example the term "امنجي" (social security secret agent), extracted by pattern 3, is considered a relatively new term with a negative connotation, as it is often used to refer to squealers. The expression "دمه خفيف"[1] (has a good sense of humor), picked up by pattern 7, is a well established expression that refers to funny people and is usually used to reflect a positive sentiment. In isolation, the individual words that make up this expression carry no sentiment. The same applies for the expression "تفتح النفس" (whets the appetite), also picked up by the same pattern.

---

[1] The literal translation of this tern is "has light blood".

The fact that the system was capable of detecting these compound expressions, shows that it can add value to existing sentiment lexicons.   Some of the detected words, are terms transliterated from other languages like "كيوت" ("cute") and "برنس" ("prince"). It's also interesting to see that the system has picked up the English word "cool" (pattern 1) as it is often used in its original English form within Arabic text, to mean cool.

Table 4. Examples of terms that were detected for each pattern

	Examples	**Respective translations**
P1	ذو قيمه , cool	Of value, Cool
P2	مشكوك فيه, لاسع	Insane, Suspicious
P3	هيييييح ,امنجي ,اتشحور	Was viciously beaten, Security state agent, sigh
P4	هتموت ,برنس	Prince, You will die
P5	برجوازي ,اعمي القلب والنظر متسلق	Sightless, Social climber
P6	بتاع ,ارهابي ,اشطه ,ابن الخروفه شاغل بالي ,الاستبن الطرطور	Son of a sheep, Fine, Terrorist Affiliated to the spare tire the clown on my mind
P7	قميل ,دمه خفيف , تغريداته مفيده	Posts useful tweets, Funny, Beautiful
P8	كيوت ,تفتح النفس ,احساسه عالي	Is very sensitive, Motivating, Cute

5.3   Evaluation of Polarity Assignment

A seed lexicon L_{seed} of 2K strongly subjective terms was selected from the Egyptian dialect lexicon L_{total} created by El-Beltagy and Ali [16][3] and was used for automatically labeling the preprocessed tweets dataset. Each tweet in the tweets dataset was considered to have positive polarity if it contained one or more positive terms from our seed lexicon and was considered to have negative polarity if it contained one or more negative terms from the seed lexicon. Tweets containing both positive and negative terms were excluded from our calculations and so were tweets containing negations and disjunctive conjunctions. Handling negations could have been done by converting subjective words preceded by a negation word to the opposite polarity but this method always leaves some unhandled noise especially when the negation word does not directly precede the subjective word. Since disjunctive conjunctions such as "بس" (but) and "لكن" (however) usually reverse the polarity of subsequent terms, tweets containing those, were also omitted. The annotation process using the selected seed lexicon resulted in 1.3M positively annotated tweets and 1.5M negatively labeled ones. To examine how the size of the lexicon affects the results, the same tweet dataset was annotated with the entire lexicon developed by El-Beltagy and Ali [3]. Table 5 reports the label specific statistics resulting from annotating the tweets dump using both L_{seed} and L_{total}.

Table 5. Labels resulting from the tagging process

	POS=1	NEG=1	POS $\geq$ 2	NEG$\geq$2	Mixed	Neutral	Total
L$_{seed}$	1M	1.17M	0.3M	0.35M	0.45M	4M	7.5M
L$_{total}$	1.18M	1.43M	0.4M	0.57M	0.76M	3M	7.5M

Following the tagging process, *nPmi* [12] was used to measure the co-occurrence of a term with positive and negative labeled tweets in the tagged dataset using equation (1). For each term x in the selected candidate terms, not in L$_{seed}$ and that occurs at least Ω times (we used Ω =10) in the labeled dataset we calculate it's *nPmi* with the positive labeled tweets $nPmi(x, POS)$ and also with negative labeled tweets $nPmi(x, NEG)$. Polarity is then assigned according to the nPmi value which has the highest value and that has an absolute difference α of a value greater than a specific threshold θ, where α is calculated using equation 2.

$$\alpha = |nPmi(x, POS) - nPmi(x, NEG)| \qquad (2)$$

α is thus used as an indication of the confidence of the process of polarity assignment. Terms that have a value of $\alpha < \theta$, are simply ignored. So selecting a relatively high θ value will affect the overall recall by excluding terms with relatively lower confidence.

Fig. 2 shows how precision of the results is affected by the changes in the confidence threshold α where precision is calculated by comparing the terms with assigned polarity against the polarity of their counterparts in the ground truth dataset described in the previous section. The value of θ was set to equal 0.001, which resulted in a total precision of 84.5% with 344 terms out the total 377 labeled correctly. The new terms were then added to the original 2K lexicon Lseed. This addition constitutes an increase of 18.85% of the size of the lexicon.

To keep a lexicon updated with new terms, the process of building a twitter corpus and applying patterns to it, should be done periodically.

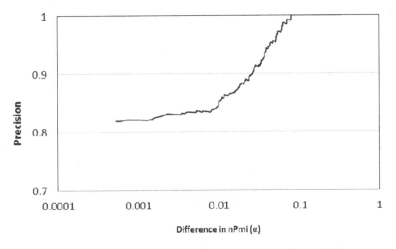

Fig. 2. Graph showing precision against difference in nPmi (α)

6   Conclusion and Future Work

In this paper we presented an approach for detecting subjective slang terms with the aim of building slang Arabic lexicons without depending on any language-dependent part of speech taggers or parsers. The proposed technique showed through experimentation, an ability to deal with informality which was illustrated by the precentages of extracted slang, profanity and compound-terms. The approach also showed that it was able to deal with the evolving nature of the language used in social media by picking up slang subjective terms that only trended very recently.

In the future, we plan on measuring the effect of augmenting a slang lexicon to one or more existing lexicons, on the task of sentiment analysis. Future work will also target extending the number of extracted terms by both increasing the number of defined patterns, and continuously feeding our system with recent tweets from which to extract terms. We are also looking into ways for increasing the precision of the obtained results without adversely affecting the recall. Automatically detecting extraction patterns, rather than having handcrafted ones, is yet another area of future work. We would also like to assign positive and negative weights to extracted terms rather than absolute polarity as many subjective terms can be used in positive, negative or neutral contexts.

References

1. Semiocast, Geolocation analysis of Twitter accounts and tweets by Semiocast (2012), http://bit.ly/1kwY9OZ
2. Farid, D.: Egypt has the largest number of Facebook users in the Arab world. Daily News Egypt (September 2013)
3. El-Beltagy, S.R., Ali, A.: Open Issues in the Sentiment Analysis of Arabic Social Media: A Case Study. In: Proceedings of 9th International Conference on Innovations in Information Technology (IIT), pp. 215–220 (2013)
4. Volkova, S., Wilson, T., Yarowsky, D.: Exploring Sentiment in Social Media: Bootstrapping Subjectivity Clues from Multilingual Twitter Streams. In: Proceedings of the 51st Annual Meeting of the Association for Computational Linguistics, pp. 505–510 (2013)
5. Turney, P.: Thumbs up or thumbs down?: Semantic orientation applied to unsupervised classification of reviews. In: Proceedings of the 40th Annual Meeting on Computational Linguistics (ACL), pp. 417–424 (July 2002)
6. Banea, C., Mihalcea, R., Wiebe, J.: A Bootstrapping Method for Building Subjectivity Lexicons for Languages with Scarce Resources. In: Proceedings of the Sixth International Conference on Language Resources and Evaluation (LREC 2008), pp. 215–220 (2008)
7. Abdul-Mageed, M., Diab, M.: Toward Building a Large-Scale Arabic Sentiment Lexicon. In: Proceedings of the 6th International Global WordNet Conference, pp. 18–22 (2012)
8. Esuli, A., Sebastiani, F.: Determining the semantic orientation of terms through gloss classification. In: Proceedings of the 14th ACM International Conference on Information and Knowledge Management, pp. 617–624 (2005)
9. Velikovich, L., Blair-Goldensohn, S.: The viability of web-derived polarity lexicons. In: Proceedings of the 11th Annual Conference of the North American Chapter of the Association for Computational Linguistics (ACL), pp. 777–785 (2010)

10. Hearst, M.: Automatic Acquisition of Hyponyms from Large Text Corpora. In: Proceedings of the 14th Conference on Computational Linguistics, COLING 1992, vol. 2, pp. 539–545 (1992)
11. Klaussner, C., Zhekova, D.: Lexico-Syntactic Patterns for Automatic Ontology Building. In: Proceedings of the Student Research Workshop Associated with RANLP, pp. 109–114 (2011)
12. Xu, J., Croft, W.B.: Corpus-Based Stemming using Co-occurrence of Word Variants 1 Introduction. ACM Trans. Inf. Syst. 16(1), 61–81 (1998)
13. Twitter REST API version 1.1, https://dev.twitter.com/docs/api/1.1
14. Larkey, L.S., Ballesteros, L., Connell, M.E.: Light Stemming for Arabic Information Retrieval. In: Arabic Computational Morphology, pp. 221–243 (2007)
15. Singhal, A.: Modern Information Retrieval: A Brief Overview. In: Bulletin of the IEEE Computer Society Technical Committee on Data Engineering, pp. 35–43 (2001)
16. El-Beltagy, S.R., Ali, A.: unWeighted Opinion Mining Lexicon, Egyptian Arabic (2013), http://bit.ly/MGtMqU

Simple TF·IDF Is Not the Best You Can Get for Regionalism Classification[*]

Hiram Calvo

Centro de Investigación en Computación
Instituto Politécnico Nacional,
Av. J. D. Bátiz s/n esq. M.O. de Mendizábal, México, D.F., 07738
hcalvo@cic.ipn.mx

Abstract. In broadly spoken languages such as English or Spanish, there are words akin to a particular region. For example, there are words typically used in the UK such as *cooker*, while *stove* is preferred for that concept in the US. Identifying the particular words a region cultivates involves discriminating them from the set of common words to all regions. This yields the problem where a term's frequency should be salient enough to be considered of importance, while being a common term tames this salience. This is the known problem of Term Frequency versus the Inverse Document Frequency; nevertheless, typical TF·IDF applications do not include weighting factors. In this work we propose several alternative formulae empirically, and then we conclude that we need to dig in a broader search space; thereby, we propose using Genetic Programming to find a suitable expression composed of TF and IDF terms that maximizes the discrimination of such terms given a reduced bootstrapping set of examples labeled for each region (400). We present performance examples for the Spanish variations across the Americas and Spain.

Keywords: Regionalisms, Genetic Programming, TF·IDF, Bootstrapping.

1 Introduction

Regionalism classification consists in identifying words in a certain language, *v.gr.* Spanish, and determining the region, or country where they are used the most. Differently to other words, regionalisms are used only in certain parts, and, despite belonging to a global language, they remain mostly not understood in other parts. For example, the term "pibe" is clearly a regionalism used in Argentina, while "chavo" is mostly used in Mexico. Regionalisms might include different names for local food, or ways of addressing people (cf. "Pana", in Ecuador).

There are several dictionaries covering regionalisms, particularly for Mexico, for example, we have the dictionary of Mexicanisms, by the Mexican Language Academy, 2010, which seeks to update the commonly used Dictionary of

---

[*] Work done with support from CONACyT-SNI, Mexico, and SIP project IPN 20121202.

A. Gelbukh (Ed.): CICLing 2014, Part I, LNCS 8403, pp. 92–101, 2014.

Mexicanisms edited in 1959 by Francisco Javier Santamaría. Another dictionary worth mentioning is the Dictionary of Spanish from Mexico, edited by the College of Mexico in 2010.

The process of compiling regionalisms by lexicographers and specialists is a process that consumes a great amount of time; that is why it is convenient to adopt as possible, several tools to allow them to refine their work swiftly. For example, there are tools known as *concordancers*, which are designed for finding occurrences of a particular word in several contexts, like the Corpus of Contemporary Mexican Spanish (1921-1974)[1][1], or the multilingual Sketch Engine[2][2], with billions of words. These tools usually provide users with statistical information about the usage of words and their contexts.

The use of certain words is a dynamic process that changes quickly. In contrast, regionalism dictionaries tend to be static, until a new edition is issued. To meet the demand of this dynamic change, we think of a tool that is able to suggest that, given the local usage of a word, it is a good candidate to be studied. As far as we know, there is not such computer tool, and that is why we address this problem. In order to do so, we propose the following hypothesis:

Regionalisms related to different countries can be identified automatically using the Web to identify which set of words are exclusive or particular to a certain nation.

We will use the bootstrapping technique, which, starting from a few words previously classified for some countries, seeks to obtain more words for each one of these. In this stage, it is necessary to establish criteria to separate common words from local words. Finally, this will allow us to extend the original dictionaries with proposed words that were identified as locally used ones. In the next section we will show details about how this is accomplished.

2   Discovering Americanisms

In order to discover new Americanisms, our method consists of the following steps:
1. Collecting resources
2. Web access
3. Analysis of discovered related words
4. Selection of the model for Americanism discrimination
5. Evaluation

We will use a dictionary for bootstrapping that contains 400 terms, manually classified as belonging to several countries such as Cuba, Bolivia, Mexico, Peru, etc.

For accessing the Web, Google's API is used. This tool allows requesting up to 1,000 queries daily. For each Americanism previously existent in our bootstrapping dictionary, we will obtain 50 text fragments (snippets). These fragments usually contain from 10 to 20 words. We store them in a database, next to the country identifier of the domain name where they were found. See Figure 1. In order to avoid

---

[1] Available at http://www.corpus.unam.mx:8080/cemc

[2] http://www.sketchengine.co.uk

domain names that do not have a country identifier, we filter them out in our query. For the example shown in Figure 1 the query was the following:

```
abogángster -site:com -site:org -site:net -site:info -site:tv
-site:edu -site:gov -site:biz
```

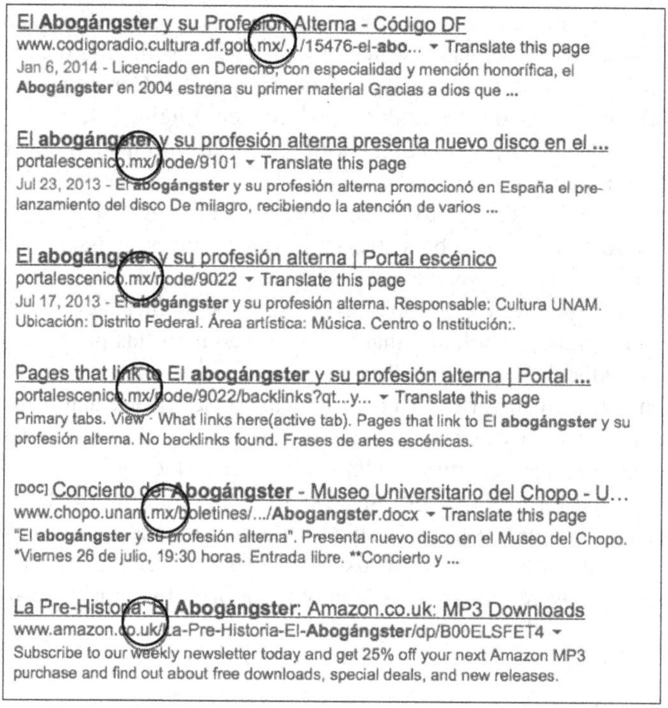

Fig. 1. Google results for the word "abogángster" (a Mexicanism meaning a mixture between lawyer and gangster), listed in the bootstrapping dictionary. The country identifiers of the domain name are marked in circles.

For the example shown in Figure 1, all words that appear in the first fragment (*Licenciado, en, Derecho, con, especialidad, y, mención honorífica, el, en,* 2004, *estrena,* su, *primer, material, Gracias, a, dios, que*) will be labelled as found in the .mx domain, that is, Mexico. The same will happen with those of the second snipped, and so on, until completing 50 snippets. Note that the last fragment contains words in English, and that those will be labelled as belonging to the .uk domain, that is, United Kingdom.

As a result of this process, new words, as those shown in Table 1, are found. Both *canchanchán* and *achichincle* are Mexicanisms, and mean something like "general purpose helper person". The counts of both words and their related domains reflect the fact that they appear the most in the .mx domain.

Nevertheless, in this case we performed a directed search of these particular words, but we already knew they were akin to Mexican Spanish. But, what happens if we delve for new words? Let us examine a fragment of words extracted from the Mexico and Argentina domains, respectively. See Table 2.

Table 1. Obtained words for specific domains (br, cc, de, es) and the number of times they appear for each domain

canchanchan,br:6	achichincle,ca:3
canchanchan,cc:1	achichincle,de:1
canchanchan,de:4	achichincle,es:4
canchanchan,do:5	achichincle,gt:1
canchanchan,es:3	achichincle,hn:2
canchanchan,fr:1	achichincle,mx:33
canchanchan,gs:1	achichincle,ro:1
canchanchan,it:2	achichincle,tk:2
canchanchan,ms:1	achichincle,vg:2
canchanchan,mx:20	achichincle,ws:1
canchanchan,uk:5	
canchanchan,us:1	

Table 2. Fragment of new words obtained for Mexico (.mx) and Argentina (.ar), beginning with "bol"

mx,bolsa:50	ar,bolso:2
mx,bolsa.:3	ar,bolso,:1
mx,bolsas:124	ar,bolson:1
mx,bolsas,:3	ar,bolsos:4
mx,bolsas.:1	ar,bolsón,:5
mx,bolsillo.:1	ar,bolsón.:1
mx,bolsita:4	ar,boludas:2
mx,bolsitas:1	ar,boludeces:52
mx,bolso:1	ar,boludeo:2
mx,bolsos:2	ar,boludez:1
mx,bolsos,:1	ar,boludez.:1
mx,bolton:1	ar,boludo:7
mx,bólón,:2	
mx,bolívar:5	

In Table 2, we see that the word *bolsas* (general word for bags) appears 124 times for Mexico, and *boludeces* (foolish things – a word used most notably in Argentina). Appears 52 times. With this, we could infer that, if *boludeces* is a regionalism from Argentina, then *bolsas* (bags) is indeed a Mexicanism, but this is not the case! Then, if this is not a regionalism, *boludeces* is neither the case, but it is! Then, we are in need of a more elaborate criterion for selecting relevant regionalism. For this we will use a commonly used measure, IDF: Inverse Document Frequency, proposed originally by Spärck-Jones in 1972 [3]:

Let TF=Term Frequency, the frequency of words for a particular country, and DF=Document Frequency; the number of countries where this word appears.

For a particular word w, we seek to consider its frequency for a certain country, while reducing its span in several countries, that is, we seek for a low DF. That is, TF/DF, written more often as TF·IDF, where IDF=1/DF (Inverse Document Frequency). Then we have

$$TF = \frac{\text{number of occurrences of } w \text{ in a class}}{\text{total number of ocurrences in a class}}, \text{ and}$$

$$IDF = \log\left(\frac{\text{total number of classes}}{\text{number of different classes that contain } w}\right)$$

However, when we directly apply this formula, we find that IDF is not enough for reducing the relevance of very common words. Note that we are not performing any stop words removal process. See the first column of Table 3.

As IDF should be used with a higher weight, we are tempted to empirically propose the formula TF·IDF·IDF. The results of this modification can be seen in the second column of Table 3. In that table, words that are Mexicanisms are highlighted (manually) in bold. In the last row, the total number of true Mexicanisms found amongst the first 27 words is shown. As can be seen, by adding weight to IDF, the number of detectable Mexicanisms increases for this example. The third column shows results for the formula TF·IDF·IDF·IDF, for which, 19 of the first 27 results are true Mexicanisms; that is, 70% of them.

Can we continue modifying this formula to obtain only pure Mexicanisms? Is this phenomenon extendable for other Americanisms?

Until now, we have shown only results for Mexicanisms, does this scheme work for detecting regionalisms of other countries as well? On the other hand, our modification consisted in only adding more IDF terms in the denominator but, are there other suitable modifications that yield better results?

In order to answer these questions, we decided to set up an experiment to explore a wide space of different formulas for detecting Americanisms. This is described in the following section.

3   Experiment

To carry out an adequate evaluation of the reach of different formulae to recognize regional words, we extracted a set of 1,137 Americanisms from the Anaya Dictionary (2010 edition). The experiment will consist in applying different formulae based on the TF and IDF terms, and finding how many words within the first 200 words found by the corresponding formula are actually Americanisms, *i.e.*, they are found in the extracted set of the 1,137 Americanisms from the Anaya Dictionary. These Americanisms were not included in the original 400 seed Americanisms previously used for bootstrapping. See Figure 2 for a fragment of the Americanisms extracted from the Anaya Dictionary.

Table 3. First position of the most relevant words for the domain .mx (Mexico), using several empirically found formulas that combine TF and IDF with regard to other countries. True relevant words are highlighted with bold.

TF·IDF		TF·IDF·IDF		TF·IDF·IDF·IDF	
palabra	pos.	palabra	pos.	palabra	pos.
de	0.0200	cosmos	0.0218	cosmos	0.1012
y	0.0199	bolsas	0.0190	bolsas	0.0884
en	0.0155	vender	0.0161	universal	0.0610
el	0.0142	universal	0.0154	andaba	0.0513
que	0.0128	comprar	0.0150	**chachalaca**	**0.0478**
a	0.0122	méxico	0.0119	quintana	0.0435
la	0.0114	andaba	0.0111	**xochimilco**	**0.0385**
los	0.0096	portal	0.0108	**chamarra**	**0.0364**
del	0.0095	**chachalaca**	**0.0103**	vender	0.0362
para	0.0095	alta	0.0097	méxico	0.0362
con	0.0094	y	0.0097	**chiluca**	**0.0356**
se	0.0087	quintana	0.0094	**tameme**	**0.0356**
un	0.0083	calzado	0.0091	**xochitepec**	**0.0349**
por	0.0079	**xochimilco**	**0.0083**	**chamuco**	**0.0349**
las	0.0074	**chamarra**	**0.0078**	**calmecac**	**0.0349**
vender	0.0072	**chiluca**	**0.0077**	**xotepingo**	**0.0342**
comprar	0.0069	**tameme**	**0.0077**	**pochteca**	**0.0335**
no	0.0069	información	0.0077	**esquites**	**0.0335**
es	0.0067	paloma	0.0076	**cachirul**	**0.0328**
al	0.0059	en	0.0075	comprar	0.0323
una	0.0058	**xochitepec**	**0.0075**	**ziranda**	**0.0321**
o	0.0054	**chamuco**	**0.0075**	**tecorral**	**0.0321**
como	0.0054	**calmecac**	**0.0075**	**piloncillo**	**0.0321**
lo	0.0054	**xotepingo**	**0.0074**	**xoloescuintle**	**0.0321**
su	0.0050	**pochteca**	**0.0072**	**xola**	**0.0321**
cosmos	0.0047	**esquites**	**0.0072**	**chapopote**	**0.0314**
bolsas	0.0041	para	0.0071	**achichincle**	**0.0314**
0/27	**0%**	**11/27**	**40.7%**	**19/27**	**70.4%**

mochilear. intr. Chile. To go hiking carrying things in a backpack (*mochila.*)

molonquear. tr. El Salv. shake (|| move violently). 2. El Salv. shake (|| beat.)

monga. f. P. Rico. Strong flu.

monito. m. Ur. A game where two players throw each other a ball, avoiding that a third player could grab it. || 2. Ur. The player located in the middle of this game (the third player.)

moñita. f. Ur. A kind of necktie (|| cravate which is tied by the front side using a knot shape.)

Fig. 2. Fragment of Americanism extracted from the Anaya dictionary

For this evaluation we will only test if the newly found Americanisms are also found in the Anaya Dictionary, regardless of their country identification. For example, *mochilear* is an Americanism widely used in Mexico, but the Anaya Dictionary lists it only for Chile. If our system finds this Americanism and tags it as a salient word for Mexico, then it will be accounted as good.

Results of applying the empirically motivated formulae appear in Table 4. From this table we can see that, if we keep adding IDF terms indefinitely, the ability of finding new Americanisms reaches a limit. On the other hand, it is important to note that despite the third column of Table 4 shows the number of words that the system extracted, but were not found in the Anaya Dictionary, this does not mean that they are not regionalisms. Amongst found words, we can find words such as "delegación" (county), that has a particular sense and usage in Mexico for its capital; or places and proper nouns such as Xochicalco, Morelos, or Veracruz, that are not listed as Americanisms in the Anaya Dictionary, but have a salient usage for a particular country. However, amongst the found words we find also *cosmos, vender* (to sell), and *paloma* (pidgeon), that are common Spanish words, and thus, they are not regionalisms.

Until now, we have manually proposed some empirical formulae following some intuition; however, we are interested in finding a TF and IDF combination that allows us to discern as many Americanisms as possible from a reduced set of seeds. In the next section we will present a method for automatically exploring several formulae in order to automatically find the best solution.

Table 4. Results of several formulae for detecting Americanisms amongst the first 200 words extracted by our system from the Web

Formula	Found in the Anaya Dictionary	Not found	%
TF-IDF	6	194	3.00%
TF	2	198	1.00%
TF·IDF	23	177	11.50%
TF·IDF·IDF	46	154	23.00%
TF·IDF·IDF·IDF	62	138	**31.00%**
TF·IDF·IDF·IDF·IDF	58	142	29.00%
IDF	4	196	2.00%

4   Genetic Programming

The method we will use for generating different formulae and improve them gradually, trying to find the best one, is Genetic Programming (GP). It was invented by John Koza [4,5]. In GP, a population consisting in different functions (each individual is a function) is evolved. In our case, functions are formulae consisting on a combination of TF and IDF terms with operators such as addition, subtraction, multiplication, etc. GP stochastically seeks to transform population in new and better functions. This stochastic process does not guarantee to achieve the best formula; essentially, GP is a controlled random method, but it is able to avoid falling into traps where other deterministic method could stall. As in nature, genetic programming is able to evolve function sets and to find new ways to solve problems. In figure 3 a scheme of genetic programming workflow is shown.

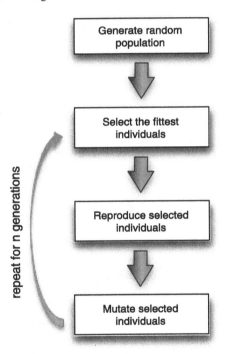

Fig. 3. Genetic programming workflow

For implementing our tests, the genetic programming algorithm was implemented with the following operators: addition, subtraction, multiplication, protected division (returns 0 if the denominator is zero) and logarithm. We previously tested with the exponential function as well, but we did not obtain good results –we obtained very big numbers that resulted in over and underflows. Exponents appear in formulae as a result of simplifying the multiplication of an operand by itself. TD and IDF were the only operands allowed.

After 3,521 iterations, the genetic programming algorithm obtained the following formula:

$$\frac{1}{tf^5} \log \left(idf^2tf^3 \left(idftf^2 + 5tf + 2\log\left(idf\right) + \log\left(tf\right)\right)\right)$$

This formula was capable of finding 68 Americanisms in the Anaya Dictionary amongst the first 200 words returned by our system. This represents a 34% of precision, and an improvement of 4% versus the best empirically found function. We performed several experiments, each one of them resulting in different formulae combining TF and IDF such as in the formula shown above, but none of them obtained more than 68 elements found amongst the first 200 as previously described. The formulae were similar between them, for example compare:

$$\frac{1}{tf^5} \log \left(idf^2tf^3 \left(5tf + 2\log\left(idf\right) + \log\left(tf\right)\right)\right)$$

$$\frac{1}{tf^5} \log \left(idf^2tf^3 \left(tf\left(tf + \log\left(tf\right)\right)\left(idftf + tf\right) + 3tf + 2\log\left(idf\right) + \log\left(tf\right)\right)\right)$$

$$\frac{1}{tf^5} \log \left(idf^2tf^3 \left(idftf^2 + 5tf + 2\log\left(idf\right) + \log\left(tf\right)\right)\right)$$

More importantly, all formulae found exactly the same 68 Americanisms, with only a slight variation in their fitness, but keeping the same ranking.

5   Conclusions and Future Work

In this work we have shown that it is possible to automatically distinguish Americanisms from common words, by filtering those words that particularly appear in a certain specific Internet domain, that do not appear in other countries' domains. In turn, it is necessary to discard words that are common to several countries. For doing this, we began implementing the widely used formula TF·IDF, and we found that the results this formula provides were not good enough. In order to perform an objective evaluation, we set up an experiment in which we verified that the newly found Americanisms were found as well in a list previously gathered from a Spanish Dictionary from which we extracted 1,137 Americanisms.

Because our algorithm produces an undetermined number of Americanism proposals, we limited ourselves to study only the first 200. These appear ranked by a formula based on the TF and IDF measures, so that studying the first 200 ones, means to study the best 200 ones. Using several formulae, we empirically found that when appending more IDF terms, the system's performance for automatically finding Americanisms increased, but this condition ceased after adding the third IDF —The best formula was TF·IDF·IDF·IDF, with 62 Americanisms found.

We have proposed using Genetic Programming for exploring a wide space of all possible formulae involving TF and IDF, and by this means try to find a better solution for identifying Americanisms. We found this sought formula, and it allowed us to automatically find 68 Americanisms with several proposed formulae. We ran our experiments several times, but a better solution than 68 Americanisms was not found, and the set of the found Americanisms was kept the same. As a future work we plan to use other classification techniques to attest whether it is possible to improve results.

References

1. Sánchez, F., Porta, J., Sancho, J.L., Nieto, A., Ballester, A., Fernández, A., Gómez, J., Gómez, L., Raigal, E., Ruiz, R.: La anotación de los corpus CREA y CORDE. In: Proceedings of SEPLN, vol. 99 (1999)
2. Kilgarriff, A., Rychly, P., Smrz, P., Tugwell, D.: ITRI-04-08 The Sketch Engine. Information Technology 105, 116 (2004)
3. Jones, K.S.: A statistical interpretation of term specificity and its application in retrieval. Journal of Documentation 28, 11-21, 60, 493–502 (1972, 2004)
4. Koza, J.R.: Non-Linear Genetic Algorithms for Solving Problems. United States Patent and Trademark Office (1988)
5. Koza, J.R.: Genetic evolution and co-evolution of computer programs. In: Artificial Life II, pp. 603–629 (1990)

Improved Text Extraction from PDF Documents for Large-Scale Natural Language Processing*

Jörg Tiedemann

Department of Linguistics and Philology
Uppsala University, Uppsala, Sweden
`jorg.tiedemann@lingfil.uu.se`

Abstract. The inability of reliable text extraction from arbitrary documents is often an obstacle for large scale NLP based on resources crawled from the Web. One of the largest problems in the conversion of PDF documents is the detection of the boundaries of common textual units such as paragraphs, sentences and words. PDF is a file format optimized for printing and encapsulates a complete description of the layout of a document including text, fonts, graphics and so on. This paper describes a tool for extracting texts from arbitrary PDF files for the support of large-scale data-driven natural language processing. Our approach combines the benefits of several existing solutions for the conversion of PDF documents to plain text and adds a language-independent post-processing procedure that cleans the output for further linguistic processing. In particular, we use the PDF-rendering libraries pdfXtk, Apache Tika and Poppler in various configurations. From the output of these tools we recover proper boundaries using on-the-fly language models and language-independent extraction heuristics. In our research, we looked especially at publications from the European Union, which constitute a valuable multilingual resource, for example, for training statistical machine translation models. We use our tool for the conversion of a large multilingual database crawled from the EU bookshop with the aim of building parallel corpora. Our experiments show that our conversion software is capable of fixing various common issues leading to cleaner data sets in the end.

Keywords: noisy text processing, text normalization, parallel corpora.

1 Introduction

Data-driven technique dominate modern natural language processing. Much progress has been reported in various fields of NLP due to advances in machine learning and growing data sets. However, the availability of clean data sets is still a serious bottleneck for most languages of the world. It is common practice to crawl the World Wide Web to extend data sets (see, e.g., [7,8]) and to find resources not only for medium and low density languages but also for the

---

* This research is supported by the Swedish Research Council (Vetenskapsrådet) through the project on Discourse-Oriented Machine Translation (2012-916).

A. Gelbukh (Ed.): CICLing 2014, Part I, LNCS 8403, pp. 102–112, 2014.
© Springer-Verlag Berlin Heidelberg 2014

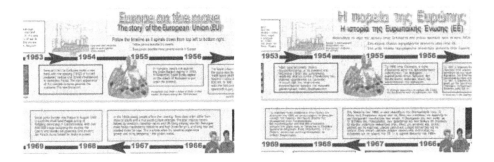

Fig. 1. A screenshot of translated PDF documents (English and Greek) from the EU bookshop with advanced layout

most common languages. The demand for training data is ever growing and domain adaptation problems require additional resources even for the dominating languages on the Internet. Crawling data, however, leads to a serious problem of noise which is unavoidable with the diversity of material available on-line. A large portion of the public documents is not available in clean textual formats that can easily be pushed through common NLP pipelines. In this paper we address the problem of converting PDF documents which is a very challenging task due to the flexibility of that format.

PDF is a file format that is optimized for printing. It encapsulates a complete description of the layout of a document including text, fonts, graphics and other elements. PDF files can be created by a large variety of tools and software products and all of them have their own way of encoding document information in the layout commands defined by the various PDF standards. It is easy to see that the generic conversion of PDF to plain text is a tough problem and probably never solvable with complete satisfaction. However, many documents are available exclusively in that format and, therefore, an extraction of text in the best possible way is an important task for data-driven techniques that rely on this kind of data.

In our work, we are interested in the conversion of documents published by the European Union through the EU bookshop website. Most of the data is completely free and many of the documents are translated into a variety (mostly European) languages, which makes it a valuable resource for many purposes, for example, training machine translation models [5].

One may think that the publications by the EU strictly follow certain standards and style sheets but we have realized that the database is very diverse similar to general web-crawling data. The bookshop includes not only proceedings of the European Parliament but all kinds of genres ranging from treaties, guidelines, and surveys to comics and children's books. Figure 1 shows a typical example of a translated document with advanced layout and structure. Many documents are full of tables, figures and various boxes that make the conversion a tedious task. Design and layout differ a lot and the extraction of text became the largest challenge when working with the crawled data set. Note that we aim at a batch conversion of massive amounts of data. Manual work is not an option; not even semi-automatic approaches would work on that scale.

After testing a variety of existing solutions we discovered that the final result was still not satisfactory, which lead to our own development of a tool that fixes some of the major issues we found when inspecting the results of preliminary runs. In the following we will first discuss the issues that we address and then present our solutions. In the end we will also describe our data sets collected and converted in our project with statistics and evaluations to emphasize the use of our tool. Note that data sets and tools are freely available through our websites (http://opus.lingfil.uu.se and http://bitbucket.org/tiedemann/pdf2xml).

2   Converting PDF to XML

Our main goal is to build linguistic resources out of PDF documents. We, therefore, opted for a conversion from PDF directly to a useful XML format that includes proper linguistic boundaries in a standardized and consistent format. However, our conversion tool relies on other software packages as described below which partially produces other types of output (plain text, for example) which we need to handle internally. Several command-line options are available in our practical implementation to control the behavior of our software. In the following, we first discuss the basic tools and libraries we integrate in our package and, then, we introduce the filtering procedures and post-processing features that we have implemented.

2.1   Basic Tools and Parameters

We rely on three public implementations of PDF rendering and conversion software. Firstly, we use the Apache Tika library[1] [6] that comes with various conversion tools not only for PDF documents. Secondly, we integrate the PDF tools provided by the Poppler Developers,[2] a PDF rendering library based on the xpdf 3.0 code base.[3] Finally, we also apply the PDF Extraction Toolkit pdfXtk [2], which is a Java framework built upon PDFBox[4] for document analysis and content extraction.

Apache Tika is a well-known Java-based content analysis toolkit that has a large community within the open-source projects hosted by the Apache Software Foundation and was formerly a sub-project of the information retrieval package Lucene. Apache Tika comes with well-documented API's and command-line tools which makes it easy to integrate the library in other software packages. The Poppler PDF rendering library is a fork of the xpdf toolkit, which is a de-facto

---

[1] http://tika.apache.org
[2] http://poppler.freedesktop.org
[3] http://www.tamirhassan.com/pdfxtk.html
[4] http://pdfbox.apache.org

standard tool in the GNU/Linux world for viewing and converting PDF documents. It includes the text extraction tool *pdftotext*, which we use extensively in our experiments below. Its main advantage is its speed that makes it well suited for large batch processes. The final package, pdfXtk is less known but provides more advanced document analysis techniques based on graph-based wrapping [1], which enables a reliable detection of text boxes in the layout-oriented PDF format. This enables better recognitions of text boundaries based on geometric structure and content attributes such as fonts, styles and font sizes.

All of these tools are able to extract plain text content from arbitrary PDF documents and they all produce reasonable results for most documents. However, we found out that there are still quite a lot of remaining issues that distort the original text and produce garbled results in some cases, which we will discuss in more detail in the following sections.

2.2 Identify Word Boundaries

One problem we identified in the output of common conversion tools is the recognition of proper word boundaries.[5] In some cases it is difficult to interpret the spacing between characters and conversion results may look like the sample text in Figure 2. This kind of noise is quite common and distorts the text substantially. For our purposes this kind of output is not acceptable even if the majority of the remaining parts are converted correctly.

As illustrated in Figure 2, the problem of additional spacing is not consistent throughout a document and its appearance is hard to predict. Furthermore, different tools and various modes may result in quite different results as we can see in the lower part of Figure 2, in which the "raw-mode" of *pdftotext* produces a more readable but still not perfect output. Note that the opposite phenomenon also frequently appears; word boundaries are missing and the conversion tools produce concatenated strings of multiple words or entire lines from the document.

Our strategy for handling these problems is based on a data-driven merging and splitting strategy. We run the document through several tools and modes and read the vocabulary from the output of their conversion. Alternatively, word lists can be given to define accepted tokens. Both sources can also be combined. From the data, we then create simple unsmoothed unigram models and use them to process the output of one of the integrated tools to better match the on-the-fly language model. Command-line tools can be used to adjust the conversion modes considered and to select the base tool used for starting the post-processing step. The default settings use *pdftotext* in "raw" and "standard" mode for language modeling and pdfXtk for producing the base conversion. For the latter, Apache Tika can be used as an alternative. Both of them produce XML markup that ensure that we have proper paragraph boundaries which are, otherwise, difficult to detect from raw text output.

---

[5] We focus here on languages that use spaces to mark word boundaries. For languages without explicit orthographic boundaries, these issues are not apparent.

Original PDF:

PRESENTATION ET RAPPEL DES PRINCIPAUX RESULTATS 9

CHAPITRE 1 - LE CHOIX DES SECTEURS ETUDIES 15

1. Le principal élément du choix : la concentration des besoins en vapeur 15
2. Les critères de choix : la consommation de combus-tibles et leur modalité d'utilisation d'une part, la concentration d'autre part 16

Converted to text using *pdftotext*:

```
P R E S E N T A T I O N ET R A P P E L DES P R I N C I P A U X R E S U L T A T S

9

C H A P I T R E 1 - L E CHOIX DES S E C T E U R S ETUDIES
1. Le p r i n c i p a l é l é m e n t du choix : la c o n c e n t r a t i o n d e s
b e s o i n s en v a p e u r
2. L e s c r i t è r e s de choix : la c o n s o m m a t i o n de
c o m b u s t i b l e s et l e u r m o d a l i t é d ' u t i l i s a t i o n d'une
p a r t , la concentration d'autre part
```

Converted to text using *pdftotext* in "raw mode":

```
PRESENTATION ET R A P P E L DES PRINCIPAUX RESULTATS 9
CHAPITRE 1 - LE CHOIX DES SECTEURS ETUDIES 15
1. Le principal élément du choix : la concentration des
besoins en vapeur 15
2. Les c r i t è r e s de choix : la consommation de combus-
tibles et leur modalité d'utilisation d'une part, la
concentration d'autre part 16
```

Fig. 2. Problems with word boundary detection in *pdftotext*

Note that we always use Unicode UTF8 to be as flexible as possible and the standard for language modeling is based on lowercased text. Our re-segmentation procedure uses the unigram language model from above together with an efficient inference algorithm based on dynamic programming. We record the longest word in our vocabulary to restrict the history that needs to be considered and run through the string of space separated segments to find possible units that need to be merged. Within this loop we also try de-hyphenation to further improve the results. The highest scoring word sequence according to our language model is then returned as the best output of the post-processing procedure.

Several parameters and heuristics can be used to influence the procedure. First of all, the input sequence can be split into sequences of single character to force the segmentation to rely entirely on the language model (ignoring the given segmentation provided by the basic conversion tool). This is in general not a good idea and may lead to a decreased conversion quality. However, such a splitting strategy is necessary to handle cases in which words are erroneously concatenated with each other, a problem that frequently appears as well. Therefore, we apply the following heuristics to enable both, splitting and merging: (i) We split strings into single characters if no space character is included in the entire string on one line. (ii) We split tokens that are suspiciously long into single character sequences (i.e. words that are longer than the longest word in the language model). (iii) We split tokens into single character sequences if they contain lower-case letters followed by upper-case letters.[6]

For conversions based on pdfXtk we apply our language-model-based re-segmentation only to those strings that have been split into single characters using the heuristics above. We add another loop that concatenates adjacent words if they exist in the vocabulary. For conversions based on Apache Tika we apply the re-segmentation based on language models on space separated text units. Global splitting into character sequences can be switched on on demand.

2.3 Ligatures

Another problem in the automatic conversion from PDF is the handling of ligatures. The tools we use differ in their capabilities of managing these contracted character sequences. Some of them manage to recognize them correctly and produce single character ligatures as an output. In some cases, ligatures are split and in other cases one (usually the second) character is missing. In our implementation, we normalize existing ligatures using a fixed substitution list. It includes the ligatures for the letter combinations 'IJ', 'ij', 'ff', 'fi', 'fl', 'ffi', 'ffl' and 'st'. Especially pdfXtk has an issue with swallowing some letters if they happen to be part of some specific ligatures. This creates problems especially when applying the merging heuristics described in the previous section. We, therefore, added a test that checks if a character of a known ligature needs to be inserted in order to create a string that is known to the language model. This additional heuristics is quite efficient despite its simplicity and takes care of most of the problems that occur.

2.4 De-hyphenation

Another important issue is hyphenation. In NLP, we usually do not want to have hyphenated words preserved as they appear in formatted texts. We, therefore, add further heuristics to take care of such cases. For this, our line-based post-processing procedure considers two adjacent lines and checks whether the last word of the first line ends with a hyphen. If this is the case then we consider two version in our re-segmentation procedure; one that removes the hyphen and

---

[6] This heuristic applies to languages that make a distinction between upper-case and lower-case letters as defined by Unicode characters sets.

one that leaves it in place. Our merging heuristics explained earlier then decide whether to concatenate the two strings (last word of the first line and first word of the second line) or not. The software generally prefers the version that allows a concatenation based on having a hyphen or not.

We also apply de-hyphenation heuristics when reading through pre-converted texts when building our language models. In this way, we also obtain words in their de-hyphenated form in our vocabulary and in the language model. It is also possible to test de-hyphenation for all words in the text and evaluate this test based on our on-the-fly vocabulary.

2.5 Paragraph Boundaries

Detecting paragraph boundaries is another important issue that influences subsequent linguistic processing. As we can see in Figure 2, plain text produced by common tools such as *pdftotext* is difficult to work with. Empty lines are simple indicators of new paragraphs. However, many subsequent lines contain paragraph boundaries and without marking them as such, subsequent sentence boundary detection can easily fail. Looking at the example in the figure again, it is not straightforward to identify the start of a new sentence if explicit punctuation and other common linguistic clues are missing. Alternatively, a layout-oriented text format could be chosen. *pdftotext* provides this mode as another possibility and in many cases, this format is much more suited for the recognition of textual units such as lists, headers and paragraphs. However, this mode makes it much harder to handle columns, tables and other formated text that interrupt the normal text flow.

Fortunately, tools such as Apache Tika and pdfXtk address this problem by adding explicit markup for text boxes. Therefore, we use those tools to produce the basic segmentation of texts into coherent segments. In particular, pdfXtk is able to detect very fine-grained boundaries due to its graph-based wrapping algorithm. It is, therefore, our choice for the base conversion. However, pdfXtk tends to over-generate paragraphs and, in this way, it splits many sentences and other coherent elements into pieces. For this reason, we add another simple heuristic to repair common issues and to restore paragraphs based on a simple linguistic rule. Basically, we observe the end of each paragraph and the beginning of the following paragraph in order to make a decision whether to merge them into one or not. In our current implementation we simply merge paragraphs if the first one does not end with a sentence-final punctuation characters and the next one starts with a lower-case letter as defined by the appropriate Unicode character class. This simple rule is very effective and seems to work well in most cases. It can also be switched off on demand.

2.6 Language Detection

Another property that we often require for NLP is that a corpus is homogenous with respect to the language used. PDF documents coming from the European Union, however, are often a mix and may include text written in other languages.

Table 1. The number of documents for the ten largest languages in the EU bookshop collection

language	nr. of doc's
English (en)	37,664
French (fr)	17,260
German (de)	15,585
Italian (it)	9,151
Spanish (es)	7,715
Dutch (nl)	7,687
Danish (da)	7,081
Greek (el)	6,486
Portuguese (pt)	6,380
Finnish (fi)	4,055

This is certainly also the case in other web-crawled data and automatic language identification is a common task that needs to be performed to clean up the data. In our approach, we added an existing language identifier to our software, the Google Compact Language Detector library[7] and its integration into the blacklist classifier for language identification [9]. This feature is optional and can be enabled while converting PDF documents. When switched on, the software runs each paragraph through the language classifier and rejects it if the detected language does not match the given language. This feature is a very useful tool that largely removes unwanted content from our corpora. It actually also helps to remove a lot of garbage that comes from non-text included in many PDF documents or garbled output produced by the PDF rendering libraries. It can be enabled to either remove non-matching text or to just mark each paragraph with the language detected. The latter is useful if subsequent processes need access to language detection information but still require the complete content of the document. This can be the case, for example, for automatic sentence alignment where text removal may cause serious problems.

3   Building a Multilingual EU Bookshop Corpus

In this section, we report our on-going efforts on creating a multilingual parallel corpus from the public documents provided by the EU bookshop. Our collection contains 135,849 PDF documents taken from the official website and many of them are available in various translations. Table 1 lists the ten largest languages represented in the current collection.

We used our tool to convert these documents and we also performed a conversion based on *pdftotext* (in standard mode) as a baseline reference. In our discussion, we focus on four languages: English (en), French (fr), German (de)

---

[7] https://code.google.com/p/chromium-compact-language-detector/

Table 2. Statistics of four disjoint data sets selected from the converted PDF documents. *pdftotext* refers to data sets created using a standard PDF conversion tool and standard linguistic pre-processing (paragraph and sentence boundary detection). *pdf2xml* refers to data created with our PDF conversion tools. The table lists the number of sentences (*sents*), the number of words and the average lengths of sentences in each data set (*w/s*).

	data set A		data set B		data set C		data set D	
lang	pdftotext	pdf2xml	pdftotext	pdf2xml	pdftotext	pdf2xml	pdftotext	pdf2xml
de sents	2.79M	2.93M	0.40M	0.49M	6.90M	7.54M	0.37M	0.39M
words	70.34M	70.46M	10.59M	10.14M	141.76M	134.62M	9.25M	8.92M
w/s	25.224	24.074	26.519	20.762	20.552	17.843	25.334	22.745
en sents	3.65M	3.95M	0.53M	0.65M	33.00M	36.25M	0.60M	0.59M
words	95.96M	95.45M	12.56M	12.01M	621.52M	584.35M	13.71M	13.20M
w/s	26.260	24.189	23.489	18.355	18.833	16.122	23.033	22.224
es sents	1.76M	1.84M	0.25M	0.30M	2.54M	2.74M	0.15M	0.17M
words	54.12M	53.87M	8.52M	8.09M	74.21M	71.07M	6.20M	6.03M
w/s	30.698	29.342	34.073	26.720	29.259	25.954	41.830	36.515
fr sents	1.90M	1.98M	0.46M	0.47M	8.45M	8.83M	0.32M	0.36M
words	57.02M	56.59M	13.36M	12.57M	213.16M	200.80M	12.13M	11.38M
w/s	29.997	28.580	29.270	26.898	25.217	22.737	38.168	31.181

and Spanish (es). We selected four data sets of different sizes based on file name patterns (without implying anything about their contents) to study the differences between the baseline conversion and our improved conversion. In both cases, we run sentence boundary detection after the basic conversion and tokenize the text with the same standard tools. For the *pdftotext* baseline, we also used a simple heuristic rule to improve paragraph detection. Short lines that start with an upper-case letter or a digit are treated as headers which has a great positive effect on subsequent sentence boundary detection. Our length threshold is set to 40 characters. Without this heuristic we would end up with many large text units that would be hard to split later. We also applied the same language detection filter (on the sentence level) in both versions to make the comparison fair.

Let us first look at some statistics from the data selected. Table 2 lists the sentence and token counts for each sub-corpus and each language. Here, we can see some interesting differences between the baseline conversion and our improved conversion. In general, the number of sentences is larger with our tool but the token counts are comparable. This leads to smaller average sentence length which indicates that our character merging and string splitting strategies have a clear effect together with the paragraph boundary detection heuristics.

Certainly, this does not prove that the changes actually improve the data even if manual inspection seems to verify this. Evaluating the general conversion quality is tricky as we do not have any gold standards, and large-scale manual evaluations are too expensive. One possibility is to test the data in a down-stream

Table 3. The test set perplexity of language models trained on 4 disjoint data sets created by a standard PDF conversion tool (*pdftotext*) and our implementation (*pdf2xml*)

lang	tool	data set A	data set B	data set C	data set D
de	pdftotext	467.033	639.193	588.271	553.339
	pdf2xml	464.785	620.899	574.903	530.933
en	pdftotext	314.590	599.654	390.718	556.243
	pdf2xml	312.888	580.897	384.214	541.568
es	pdftotext	256.589	439.685	341.331	415.313
	pdf2xml	256.840	424.248	332.031	401.819
fr	pdftotext	198.377	381.100	233.022	348.049
	pdf2xml	197.642	366.404	226.964	333.529

application. A typical application is the use of such data for language modeling, which is an essential part of many applications. A common metric for showing the appropriateness of a language model given some data is perplexity. Our data is especially interesting for machine translation due to its multilingual contents. Therefore, we selected the news test sets of the SMT evaluation campaign from the annual workshop on machine translation (WMT) from the year 2013. Table 3 lists the test set perplexities measured on these data sets using standard trigram language models trained on our converted PDF documents. We estimated the LM probabilities with KenLM [4] with standard settings (using modified Kneser-Ney smoothing without pruning) and used the KenLM tools [3] for querying the language models.

Concluding from the table, we can see a consistent perplexity reduction on unrelated test data when using our improved PDF conversion as the basis for training language models. The only exception is Spanish on data set A but here, the perplexity is almost identical in both cases. Test set A seems to be the easiest collection as all perplexity scores are very similar anyway. Otherwise, the reduction is quite substantial given that most of the documents are successfully converted with the standard tools as well and only a smaller proportion of the data is actually influenced by the additional post-processing steps.

4   Conclusions

In this paper, we present a new tool for improved text extraction from arbitrary PDF documents. Our approach combines the benefits of several PDF rendering libraries and fixes common problems using several post-processing steps and heuristics. Its main purpose is the creation of large-scale data sets for empirical NLP from noisy and diverse document collections. Our tool manages to improve the detection of word boundaries using on-the-fly language models and efficient re-segmentation procedures. It also normalizes ligatures and removes hyphenations if necessary. The approach does not require external linguistic resources and is completely open-source and freely available.[8] We used the tool for creating a multilingual corpus of documents published by the European Union. The data sets are also freely available from OPUS[9] [10] and will be useful for cross-lingual applications such as statistical machine translation. In our experiments, we could show that the improved conversion techniques lead to cleaner data sets that reduce the perplexity of unseen test data when measured with a standard n-gram language model trained on the automatically converted documents. In future work, we would like to use the data in machine translation. For this, we need to align all documents pairwise in order to create parallel training data that is applicable for the SMT training pipelines.

References

1. Hassan, T.: Graphwrap: A system for interactive wrapping of pdf documents using graph matching techniques. In: ACM Symposium on Document Engineering, pp. 247–248 (2009)
2. Hassan, T.: Object-level document analysis of PDF files. In: ACM Symposium on Document Engineering, pp. 47–55 (2009)
3. Heafield, K.: Kenlm: Faster and smaller language model queries. In: Proceedings of the Sixth Workshop on Statistical Machine Translation, pp. 187–197. Association for Computational Linguistics, Edinburgh (July 2011), http://www.aclweb.org/anthology/W11-2123
4. Heafield, K., Pouzyrevsky, I., Clark, J.H., Koehn, P.: Scalable modified kneser-ney language model estimation. In: Proceedings of the 51st Annual Meeting of the Association for Computational Linguistics, vol. 2, pp. 690–696. Association for Computational Linguistics, Sofia (2013), http://www.aclweb.org/anthology/P13-2121
5. Koehn, P.: Statistical Machine Translation. Cambridge University Press (2010)
6. Mattmann, C.A., Zitting, J.L.: Tika in Action. Manning Publications Co. (2011), http://manning.com/mattmann/
7. Resnik, P.: Mining the web for bilingual text. In: Proceedings of the 37th Annual Meeting of the Association for Computational Linguistics, pp. 527–534. Association for Computational Linguistics, College Park (June 1999), http://www.aclweb.org/anthology/P99-1068
8. Resnik, P., Smith, N.A.: The Web as a parallel corpus. Computational Linguistics 29(3), 349–380 (2003); special Issue on the Web as Corpus
9. Tiedemann, J., Ljubešić, N.: Efficient discrimination between closely related languages. In: Proceedings of COLING 2012, pp. 2619–2634. The COLING 2012 Organizing Committee, Mumbai, India (2012), http://www.aclweb.org/anthology/C12-1160
10. Tiedemann, J.: Parallel data, tools and interfaces in OPUS. In: Proceedings of the Eight International Conference on Language Resources and Evaluation (LREC 2012), European Language Resources Association (ELRA), Istanbul, Turkey (May 2012)

---

[8] http://bitbucket.org/tiedemann/pdf2xml
[9] http://opus.lingfil.uu.se

Dependency-Based Semantic Parsing
for Concept-Level Text Analysis

Soujanya Poria[1,4,6], Basant Agarwal[2], Alexander Gelbukh[3],
Amir Hussain[4], and Newton Howard[5]

[1] School of Electrical & Electronic Engineering, Nanyang Technological University, Singapore
[2] Department of Computer Engineering, Malaviya National Institute of Technology, India
[3] Centro de Investigación en Computación, Instituto Politécnico Nacional, Mexico
[4] Department of Computing Science and Mathematics, University of Stirling, UK
[5] MIT Media Laberotory, MIT, USA
[6] The Brain Sciences Foundation, Cambridge, MA 02139, USA
sporia@ntu.edu.sg, {sp47,ahu}@cs.stir.ac.uk, basant@mnit.ac.in,
www.gelbukh.com, nhmit@mit.edu

Abstract. Concept-level text analysis is superior to word-level analysis as it preserves the semantics associated with multi-word expressions. It offers a better understanding of text and helps to significantly increase the accuracy of many text mining tasks. Concept extraction from text is a key step in concept-level text analysis. In this paper, we propose a ConceptNet-based semantic parser that deconstructs natural language text into concepts based on the dependency relation between clauses. Our approach is domain-independent and is able to extract concepts from heterogeneous text. Through this parsing technique, 92.21% accuracy was obtained on a dataset of 3,204 concepts. We also show experimental results on three different text analysis tasks, on which the proposed framework outperformed state-of-the-art parsing techniques.

1 Introduction

Concept-level text analysis [24,26,25] focuses on a semantic analysis of text [12] through the use of web ontologies or semantic networks, which allow the aggregation of conceptual and affective information associated with natural language opinions. By relying on large semantic knowledge bases, such approaches step away from blind use of keywords and word co-occurrence count, but rather rely on the implicit features associated with natural language concepts. Unlike purely syntactical techniques, concept-based approaches are able to detect also sentiments that are expressed in a subtle manner, e.g., through the analysis of concepts that do not explicitly convey any emotion, but which are implicitly linked to other concepts that do so. The bag-of-concepts model can represent semantics associated with natural language much better than bags-of-words [4]. In the bag-of-words model, in fact, a concept such as `cloud computing` would be split into two separate words, disrupting the semantics of the input sentence (in which, for example, the word `cloud` could wrongly activate concepts related to `weather`).

The analysis at concept-level allows for the inference of semantic and affective information associated with natural language opinions and, hence, enables a comparative

A. Gelbukh (Ed.): CICLing 2014, Part I, LNCS 8403, pp. 113–127, 2014.

fine-grained feature-based sentiment analysis. Rather than gathering isolated opinions about a whole item (e.g., iPhone5), users are generally more interested in comparing different products according to their specific features (e.g., iPhone5's vs Galaxy S3's touchscreen), or even sub-features (e.g., fragility of iPhone5's vs Galaxy S3's touchscreen). In this context, the construction of comprehensive common and common-sense knowledge bases is key for feature-spotting and polarity detection, respectively. Common-sense, in particular, is necessary to properly deconstruct natural language text into sentiments— for example, to appraise the concept small room as negative for a hotel review and small queue as positive for a post office, or the concept go read the book as positive for a book review but negative for a movie review [2]. Common-sense knowledge describes basic understandings that people acquire through experience. In cognitive science, building conceptual representations is a fundamental ability to understand and handle objects and actors of an operating environment [15].

To this end, the proposed concept parser aims to break text into clauses and, hence, deconstruct such clauses into concepts, to be later fed to a vector space of common-sense knowledge. For applications in fields such as real-time human-computer interaction and big social data analysis, in fact, deep natural language understanding is not strictly required: a sense of the semantics associated with text and some extra information (e.g., affect) associated with such semantics are often enough to quickly perform tasks such as emotion recognition and polarity detection. Common-sense reasoning is often performed through common-sense ontologies and the employment of reasoning algorithms, such as predicate logic and machine learning, to reach a conclusion.

In this paper, we propose a novel concept parser based on the semantic relationship between words in natural language text and on the semantics of the ConceptNet ontology. The paper is organized as follows: Section 2 describes related works in semantic parsing; Section 3 discusses the proposed algorithm; Section 4 offers a summary of the novelty of our work; Section 5 presents experimental results and a comparative evaluation against the state of the art; Section 6 proposes three possible applications of the proposed concept parser; finally, Section 7 concludes the paper.

2 Related Work

Automatic knowledge mining from text is a popular research field and concept extraction is one of its key steps. [5] used domain specific ontologies to acquire knowledge from text. Using such ontologies the authors extracted 1.1 million common-sense knowledge assertions. Concept mining is useful for tasks such as information retrieval [29], opinion mining [3], text classification [35].

State-of-the-art approaches mainly exploit term extraction methods to obtain concepts from text. The approaches can be classified into two main categories: linguistic rules [7] and statistical approaches [36] [1]. [36] used term frequency and location of the words and, hence, employed a non-linear function to calculate term weighting. [1] mined concepts from the Web by using webpages to construct topic signatures of concepts and, hence, built hierarchical clusters of such concepts (word senses) that lexicalize a given word. [9] and [34] combined linguistic rules and statistical approaches to enhance the concept extraction process.

Other relevant works in concept mining focus on concept extraction from documents. Gelfand et al. have developed a method based on the Semantic Relation Graph to extract concepts from a whole document [10]. They used the relationship between words, extracted on a lexical database, to form concepts. Our approach also exploit the relationship between words but it obtain the semantic relationship between words based on dependency parsing. We gather more conceptual information of a concept using the ConceptNet ontology. Concepts extracted from text are sent as a query to ConceptNet to extract their semantics.

Nakata has described a method to index important concepts described in a collection of documents belonging to a group for sharing them [20].

Lexicon syntactic patterns is also one of the popular techniques for concept extraction. [14] extracted hyponomy relations from text from Grolier's Encyclopedia by matching 4 given lexicon-syntactic patterns. Her theory explored a new direction in the concept mining field. She claimed existing hyponomy relations can be used to extract new lexical syntactic patterns. [17] and [18] used the "isa" pattern to extract Chinese hyponymy relations from unstructured Web corpus and obtained promising results.

2.1 Part Of Speech Based Concept Parsing Model

Rajagopal et al. 2013 [28] proposed a novel Part Of Speech based approach to extract concepts. This is the only state of the art approach which tried to understand the meaning of the text. Later, we compare our approach with [28]. Below, we briefly present the POS algorithm proposed in [28].

First, the semantic parser breaks text into clauses. Each verb and its associated noun phrase are considered in turn, and one or more concepts is extracted from these. As an example, the clause "I went for a walk in the park", would contain the concepts go walk and go park. The Stanford Chunker [8] is used to chunk the input text. A sentence "I am going to the market to buy vegetables and some fruits" would be broken into "I am going to the market" and "to buy vegetables and some fruits". A general assumption during clause separation is that, if a piece of text contains a preposition or subordinating conjunction, the words preceding these function words are interpreted not as events but as objects.

The next step of the algorithm then separates clauses into verb and noun chunks, as suggested by the parse trees shown in Fig. 1. Next, clauses are normalized in two stages. First, each *verb* chunk is normalized using the Lancaster stemming algorithm [21]. Second, each potential *noun* chunk associated with individual verb chunks is paired with the stemmed verb in order to detect multi-word expressions of the form 'verb plus object'. Objects alone, however, can also represent a common-sense concept. To detect such expressions, a POS-based bigram algorithm checks noun phrases for stopwords and adjectives. In particular, noun phrases are first split into bigrams and then processed through POS patterns, as shown in Algorithm 1.

POS pairs are taken into account as follows:

1. ADJ + NOUN : The adj+noun combination and noun as a stand-alone concept are added to the objects list.
2. ADJ + STOPWORD : The entire bigram is discarded.

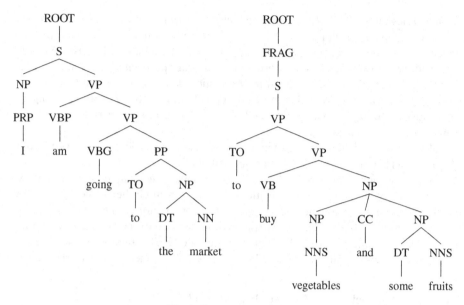

Fig. 1. Example parse trees

3. NOUN + ADJ : As trailing adjectives do not tend to carry sufficient information, the adjective is discarded and only the noun is added as a valid concept.
4. NOUN + NOUN : When two nouns occur in sequence, they are considered to be part of a single concept. Examples include *butter scotch, ice cream, cream biscuit*, and so on.
5. NOUN + STOPWORD : The stopword is discarded, and only the noun is considered valid.
6. STOPWORD + ADJ: The entire bigram is discarded.
7. STOPWORD + NOUN : In bigrams matching this pattern, the stopword is discarded and the noun alone qualifies as a valid concept.

The POS-based bigram algorithm extracts concepts such as *market, some fruits, fruits*, and *vegetables*. In order to capture event concepts, matches between the object concepts and the normalized verb chunks are searched. This is done by exploiting a parse graph that maps all the multi-word expressions contained in the knowledge bases. Such an unweighted directed graph helps to quickly detect multi-word concepts, without performing an exhaustive search throughout all the possible word combinations that can form a commonsense concept.

Single-word concepts, e.g., *house*, that already appear in the clause as a multi-word concept, e.g., *beautiful house*, in fact, are pleonastic (providing redundant information) and are discarded. In this way, the algorithm 2 is able to extract event concepts such as *go market, buy some fruits, buy fruits*, and *buy vegetables*, representing the concepts to be fed to a common-sense reasoning algorithm for further processing.

Data: NounPhrase
Result: Valid object concepts
Split the NounPhrase into bigrams ;
Initialize concepts to Null ;
for *each NounPhrase* **do**

>   **while** *For every* bigram *in the NounPhrase* **do**
>   >   POS Tag the Bigram ;
>   >   **if** adj noun **then**
>   >   > | add to Concepts: noun, adj+noun
>   >
>   >   **else if** noun noun **then**
>   >   > | add to Concepts: noun+noun
>   >
>   >   **else if** stopword noun **then**
>   >   > | add to Concepts: noun
>   >
>   >   **else if** adj stopword **then**
>   >   > | continue
>   >
>   >   **else if** stopword adj **then**
>   >   > | continue
>   >
>   >   **else**
>   >   > | Add to Concepts : entire bigram
>   >
>   >   **end**
>   >   repeat until no more bigrams left;
>
>   **end**

end

Algorithm 1. POS-based bigram algorithm

3 Algorithm

First, we extract dependency relations between the words of a sentence. Then, those relations are used to formulate complex concepts. Once, these concepts are extracted we obtain related common-sense knowledge of the concepts from ConceptNet. Below, we first describe the use of the dependency relations to form concepts and latter we discuss how related common-sense knowledge can be inferred from ConceptNet.

3.1 Formation of Concepts Using Dependency Relations

Subject Noun Rule

Trigger: when the active token is found to be the syntactic subject of a verb.
Behavior: if a word h is in a subject noun relationship with a word t then the concept t-h is extracted.
Example: In (1), *movie* is in a subject relation with *boring*.

(1)   The movie is boring.

Here the concept (boring-movie) is extracted.

Data: Natural language sentence
Result: List of concepts
Find the number of verbs in the sentence;
for *every clause* **do**
> extract VerbPhrases and NounPhrases;
> stem VERB ;
> **for** *every NounPhrase with the associated verb* **do**
> > find possible forms of *objects* ;
> > link all *objects* to stemmed verb to get *events*;
>
> **end**
> repeat until no more clauses are left;

end

Algorithm 2. Event concept extraction algorithm

Joint Subject Noun and Adjective Complement Rule

Trigger: when the active token is found to be the syntactic subject of a verb and the verb is on adjective complement relation with an adverb.

Behavior: if a word *h* is in a subject noun relationship with a word *t* and the word *t* is with adjective complement relationship with a word *w* then the concept *w-h* is extracted.

Example: In (2), *flower* is in a subject relation with *smells* and *smells* is in adjective complement relationship with *bad*.

(2)   The flower smells bad.

Here the concept (`bad-flower`) is extracted.

Direct Nominal Objects. This complex rule deals with direct nominal objects of a verb.

Trigger: when the active token is head verb of a direct object dependency relation.

Behavior: if a word *h* is in a direct nominal object relationship with a word *t* then the concept *h-t* is extracted.

Example: In (3) the system extracts the concept (`see,movie`).

(3)   Paul saw the movie in 3D.

(`see,in,3D`) is not treated at this stage since it will later be treated by the standard rule for prepositional attachment.

Adjective and Clausal Complements Rules. These rules deal with verbs having as complements either an adjective or a closed clause (i.e. a clause, usually finite, with its own subject).

Trigger: when the active token is head verb of one of the complement relations.

Behavior: if a word *h* is in a direct nominal object relationship with a word *t* then the concept *h-t* is extracted.

Example: in (4), *smells* is the head of a clausal complement dependency relation with *bad* as the dependent.

(4)   This meal smells bad.

In this example the concept (smell,bad) is extracted.

Negation. Negation is also a crucial components of natural language text which usually flips the meaning of the text. This rule is used to identify whether a word is negated in the text.

Trigger: when in a text a word is negated.

Behavior: if a word *h* is negation by a *negation marker t* then the concept *t-h* is extracted.

Example: in (5), *like* is the head of the negation dependency relation with *not* as the dependent. Here, *like* is negated by the negation marker *not*.

(5)   I do not like the movie.

Based on the rule described above the concept (not, like) is extracted.

Open Clausal Complements. Open clausal complements are clausal complements of a verb that do not have their own subject, meaning that they (usually) share their subjects with that of the matrix clause. The corresponding rule is complex in the same way as the one for direct objects.

Trigger: when the active token is the head of the relation

Behavior: as for the case of direct objects, the algorithm tries to determine the structure of the dependent of the head verb. Here the dependent is itself a verb, therefore, the system tries to establish whether the dependent verb has a direct object or a clausal complement of its own. In a nutshell, the system is dealing with three elements: the head verb(h), the dependent verb(d), and the (optional) complement of the dependent verb (t). Once these elements have all been identified, the concept (h,d,t) is extracted

Example: in (6), *like* is the head of the *open clausal complements* dependency relation with *praise* as the dependent and the complement of the dependent verb *praise* is *movie*.

(6)   Paul likes to praise good movies.

So, in this example the concept (like,praise,movie) is extracted.

Modifiers

Adjectival, Adverbial and Participial Modification. The rules for items modified by adjectives, adverbs or participles all share the same format.

Trigger: these rules are activated when the active token is modified by an adjective, an adverb or a participle.
Behavior: if a word w is modified by a word t then the concept (t,w) is extracted.
Example: in (7) the concept *bad, loser* is extracted.

(7)   a.   Paul is a bad loser.

Prepositional Phrases. Although prepositional phrases do not always act as modifiers we introduce them in this section as the distinction does not really matter for their treatment.

Trigger: the rule is activated when the active token is recognized as typing a prepositional dependency relation. In this case, the head of the relation is the element to which the PP attaches, and the dependent is the head of the phrase embedded in the PP.
Behavior: instead of looking for the complex concept formed by the head and dependent of the relation, the system uses the preposition to build a ternary concept.
Example: in (8), the parser yields a dependency relation typed `prep_with` between the verb *hit* and the noun *hammer* (=the head of the phrase embedded in the PP).

(8)   Bob hit Marie with a hammer.

Therefore the system extracts the complex concept (`hit, with, hammer`).

Adverbial Clause Modifier. This kind of dependency concerns full clauses that act as modifiers of a verb. Standard examples involve temporal clauses and conditional structures.

Trigger: the rule is activated when the active token is a verb modified by an adverbial clause. The dependent is the head of the modifying clause.
Behavior: if a word t is a adverbial clause modifier of a word w then the concept $(t-w)$ is extracted.
Example: in (9), the complex concept (`play,slow`) is extracted.

(9)   The machine slows down when the best games are playing.

Noun Compound Modifier

Trigger: the rule is activated when it finds a noun composed with several nouns. A noun compound modifier of an NP is any noun that serves to modify the head noun.
Behavior: if a noun-word w is modified by another noun-word t then the complex concept $(t-h)$ is extracted.
Example: in (10), the complex concept (`birthday,party`) is extracted.

(10)   This is a birthday gift for you.

Single Word Concepts. Words having part-of-speech VERB, NOUN, ADJECTIVE and ADVERB are also extracted from the text. Single word concepts which exist in the multi-word-concepts are discarded as they carry redundant information. For example, concept *party* that already appears in the concept *birthday party* so, we discard the concept *party*.

3.2   Obtaining Common-Sense Knowledge from ConceptNet

ConceptNet [13] represents the information from the Open Mind corpus as a directed graph, in which the nodes are concepts and the labeled edges are common-sense assertions that interconnect them. For example, given the two concepts person and cook, an assertion between them is CapableOf, i.e. a person is capable of cooking [13].

After obtaining concepts from the text we send them as queries to ConceptNet. From ConceptNet we find the common-sense-knowledge related to the query concepts. For example, when we send the concept *birthday party* as a query to ConceptNet we get related concepts such as *cake, buy present*. From ConceptNet we find the following relations

- cake – AtLocation ↝ birthday party.
- buy present – UsedFor ↝ birthday party.

These common-sense concepts are used to gather more knowledge about the concepts as they have direct connections with *birthday party*. From ConceptNet we get *cake* is used in *birthday party* and people *buy present* for the *birthday party*. So, this process help us to acquire more knowledge about the concepts we extract by the methodology described in Section 4.1. Hence, the joint exploitation of the extracted concepts and ConceptNet offer machine a better understanding of the natural language text. Our approach enables computer to understand the topic of the text as well as the meaning conveyed by the text.

4   Novelty of Our Work

Existing approaches mainly discuss on the automatic extraction of concepts based on the hyponomy and hypernomy relationship of words in a text. The concepts extracted by their methods can easily identify on which topic the text is all about but cant describe the meaning inferred by the text i.e. using those methods we are unable to know what the text tells about the topic. Such information are often found to be crucial for several cognitive tasks such as sentiment analysis, emotion analysis, opinion mining etc where both topic and meaning of the text are important. Our method is able to extract concepts which carry the meaning expressed by the text as well as our method also extracts the concepts which tells about the topic or theme of the text. The difference between our approach and state of the art can be explained using a simple example (11-a). For (11-a) existing approaches can only extract concepts related to Coffee and Starbucks based on the ontologies the methods use. However, our approach extracts the concepts: *like-coffee, coffee-of-Starbucks, coffee, Starbucks* as well as concepts related to *like-coffee, coffee, coffee-of-Starbucks* and *Starbucks*. Concepts related to *like-coffee, coffee, coffee-of-Starbucks* and *Starbucks* are extracted from the ConceptNet ontology. Clearly, the concepts extracted by our approach carry the meaning (here the sentiment of the speaker) expressed by (11-a), while the state of the art approaches fail to do it.

(11)   a.   I like the coffee of Starbucks.

Readers may be confused our approach with the syntactic ngrams proposed by [32]. Here, we first describe syntactic n-grams and then show the differences between our concept parser and syntactic n-gram. By dependency syntactic n-gram (sn-gram) we understand a subtree of the dependency tree of a sentence that contains n nodes [30]. Sn-grams can be used as features to represent sentences in the same scenarios as conventional n-grams [31]; more specifically, sn-grams represent dependency trees as vectors in the same way as conventional n-grams represent strings of words. However, unlike conventional n-grams [6], sn-grams represent linguistic entities and are thus much more informative and less noisy. While sn-grams go a long way towards linguistically meaningful representation, numerous phenomena from the presence of functional words to synonymous expressions to insignificant details still introduce noise in this representation and prevent semantically similar constructions to be mapped to identical feature vectors. In this work we present near-paraphrasic rules that simplify and normalize the dependency trees in order to reduce synonymous variation and remove insignificant details and thus improve similarity between feature vectors of semantically similar expressions and reduce data sparseness. Another difference is that syntactic n-grams convey all characteristics of basic n-gram whereas our concept parser extracts semantic from the text. Lets discuss the differences between syntactic n-gram and our proposed concept parser through an example [32].

(12)   a.   I can even now remember the hour from which I dedicated myself to this
             great enterprise.

Here, extracted syantactic n-grams are [*remember now, now even, remember hour, remember dedicated, dedicated enterprise, enterprise great, remember now even, remember hour dedicated, hour dedicated enterprise, dedicated enterprise great*].

Whereas, extracted concepts by our concept parser are [*even now, even now remember, remember hour, hour, remember from dedicate, dedicate which to enterprise, dedicate myself to enterprise, dedicate to enterprise, great enterprise*].

After sending these concepts as query to conceptnet in order to acquire more common-sense knowledge we obtain the concept list [*even now, even now remember, remember hour, hour, remember from dedicate, dedicate which to enterprise, dedicate myself to enterprise, dedicate to enterprise, great enterprise, still, sixty minute*]. Here, from conceptnet we find commonsense knowledge *still, sixty minute* related to the concepts *even now* and *hour* respectively.

Clearly from above examples we see the proposed concept parser is able to extract more semantic. *even now, even now remember* extracted by proposed concept parser express more semantic compare to *now even* and *remember now even* extracted by syntactic n-grams.

In (13). our concept parser extracts *food, food smell, bad food, smell bad*. But, syntactic n-gram method extracts *smell bad food, smell bad*. From this example, our concept parser is able to extract good semantic conveyed by *bad food*.

(13)   a.   The food smells bad.

5   Experiments and Results

To calculate the performance, we selected 300 sentences from the *Stanford Sentiment Dataset* [33] and extracted the concepts manually. This process yielded 3204 concepts. Below in Table 1 we show the accuracy of concept mining process using approach with the POS based approach described in Section 2. concepts in them manually.

Table 1. Results obtained using different algorithms on the dataset

Algorithm	Precision
Part-of-Speech Approach	86.10%
Proposed Approach	92.21%

6   Applications of the Proposed Concept Parser

We used the proposed concept parser in many applications and found it to perform superior to the existing concept parsers. As, to the best of our knowledge Part Of Speech based concept parser has the highest accuracy till now in extracting concepts from text so we compare the result obtained using our concept parser with the Part Of Speech based concept parser. This section also shows the proposed concept parser out-performs Syntactic N-grams [32] technique in these tasks. Syntactic N-grams method uses dependency tree of a text and by following the paths in the tree it extracts ngrams. It is called syntactic because it carries syntactic information of words i.e. information on word relations in a text. But, the method consists all characteristics of the ngrams.

   We treated each application as classification task. As discussions on feature extraction process and classification method are out of the scope of this paper, we do not present those details in this paper. Please find those details in [23][22][27].

6.1   Sentiment Analysis of Text

For experiments on detecting positive and negative sentiment in texts, we used Stanford Twitter dataset[11]. We cast this task as a classification task. For sentiment analysis experiment,this was binary classification.We report the results obtained with the Extreme Learning Machine (ELM)[16] as the classifier. Concept parser was used to extract concepts from a text and those concepts were used to form feature vector. Details of the feature formulation is skipped in this paper as this is not the focus of the paper. Table 2 shows the experimental results and comparison between the performance of proposed concept parser and POS based concept parser and Syntactic N-grams in the task.

Table 2. Sentiment analysis on Stanford Twitter dataset

Algorithm	Precision
Syntactic N-grams	83.23%
Part-of-Speech Approach	82.20%
Proposed Approach	85.05%

6.2 Emotion Recognition from Text

As a dataset for the emotion detection experiment, we used the ISEAR dataset. We cast the task as a six-way classification, where the six classes were anger, sadness, disgust, fear, surprise, and joy. This experiment was also based on the concept extraction process from text and the extracted concepts were used to form feature vector to learn the Emotion Recognition classifier. ELM was used as a classifier for this task. Table 3 shows the significant improvement in the accuracy of the Emotion Recognition task when proposed concept parser is used instead of POS based concept parser and syntatic N-grams are used for the task.

Table 3. Emotion detection on the ISEAR dataset

Algorithm	Precision
Syntatic N-grams	61.25%
Part-of-Speech Approach	62.10%
Proposed Approach	63.25%

6.3 Personality Recognition from Text

For experiments on detection personality from text,we used five-way classification according to the five personality traits described by Mathews et al. (2009), which are-openness, conscientiousness, extraversion,agreeableness, and neuroticism, sometimes abbreviated as OCEAN by their first letters. To experiment, we used the dataset provided by [19]. We treated this task as a classification. For this task also, we used concept parser to extract concepts from the text and later they were used to form the features to train the classifier. As a classifier, we used ELM. Table 4 shows the experimental results.

Table 4. Personality detection on the essays dataset for personality detection

	Extraversion	Neuroticism	Agreeableness	Conscientiousness	Openness
Syntatic N-grams	0.532	0.561	0.502	0.566	0.592
Part-of-Speech Approach	0.546	0.557	0.540	0.564	0.604
proposed method	0.634	0.637	0.615	0.633	0.661

7 Conclusion and Future Work

In this work, we use the dependency relation between words to extract concepts from text. The joint exploitation of these concepts and conceptnet help to acquire more knowledge thus it enable a better understanding of the text. Experiment shows how well it performs and it outperforms state of the art model. Future work involves to discover more useful dependency relationship to mine the concepts. Also, removing the concepts which do not carry good semantic rather carry noise is a challenging task. Along with using conceptnet, how other ontologies can help to enrich the concept mining process is also a big task to deal with. We also aim to use extracted concepts for cognitive tasks such as opinion mining, sentiment analysis, personality detection etc.

Acknowledgements. This work was partially supported by the Governments of Mexico and India (grant CONACYT 122030) and Government of Mexico (grant IPN SIP 20144534; COFAA-IPN).

References

1. Agirre, E., Ansa, O., Hovy, E., Martínez, D.: Enriching very large ontologies using the www. arXiv preprint cs/0010026 (2000)
2. Cambria, E., Gastaldo, P., Bisio, F., Zunino, R.: An ELM-based model for affective analogical reasoning. Neurocomputing (2014)
3. Cambria, E., Hussain, A., Havasi, C., Eckl, C., Munro, J.: Towards crowd validation of the uk national health service. In: WebSci, Raleigh (2010)
4. Cambria, E., White, B.: Jumping NLP curves: A review of natural language processing research. IEEE Computational Intelligence Magazine 9(2) (2014)
5. Cao, C., Feng, Q., Gao, Y., Gu, F., Si, J., Sui, Y., Tian, W., Wang, H., Wang, L., Zeng, Q., et al.: Progress in the development of national knowledge infrastructure. Journal of Computer Science and Technology 17(5), 523–534 (2002)
6. Çelebi, A., Özgür, A.: N-gram parsing for jointly training a discriminative constituency parser. Polibits 48, 5–12 (2013)
7. Chen, W.L., Zhu, J.B., Yao, T.S., Zhang, Y.X.: Automatic learning field words by bootstrapping. In: Proc. of the JSCL, vol. 72. Tsinghua University Press, Beijing (2003)
8. De Marneffe, M.-C., Manning, C.D.: The stanford typed dependencies representation. In: Coling 2008: Proceedings of the Workshop on Cross-Framework and Cross-Domain Parser Evaluation, pp. 1–8. Association for Computational Linguistics (2008)
9. Du, B., Tian, H.F., Wang, L., Lu, R.Z.: Design of domain-specific term extractor based on multi-strategy. Computer Engineering 31(14), 159–160 (2005)
10. Gelfand, B., Wulfekuler, M., Punch, W.F.: Automated concept extraction from plain text. In: AAAI 1998 Workshop on Text Categorization, pp. 13–17 (1998)
11. Go, A., Bhayani, R., Huang, L.: Twitter sentiment classification using distant supervision. CS224N Project Report, Stanford, pp. 1–12 (2009)
12. Harlambous, Y., Klyuev, V.: Thematically reinforced explicit semantic analysis. International Journal of Computational Linguistics and Applications 4(1), 79–94 (2013)
13. Havasi, C., Speer, R., Alonso, J.: Conceptnet 3: A flexible, multilingual semantic network for common sense knowledge. In: Recent Advances in Natural Language Processing, pp. 27–29 (2007)

14. Hearst, M.A.: Automatic acquisition of hyponyms from large text corpora. In: Proceedings of the 14th Conference on Computational Linguistics, vol. 2, pp. 539–545. Association for Computational Linguistics (1992)
15. Howard, N., Cambria, E.: Intention awareness: Improving upon situation awareness in human-centric environments. Human-centric Computing and Information Sciences 3(9) (2013)
16. Huang, G.-B., Zhu, Q.-Y., Siew, C.-K.: Extreme learning machine: Theory and applications. Neurocomputing 70(1), 489–501 (2006)
17. Liu, L., Cao, C., Wang, H.: Acquiring hyponymy relations from large chinese corpus. WSEAS Transactions on Business and Economics 2(4), 211 (2005)
18. Liu, L., Cao, C.-G., Wang, H.-T., Chen, W.: A method of hyponym acquisition based on "isa" pattern. Journal of Computer Science, 146–151 (2006)
19. Mairesse, F., Walker, M.A., Mehl, M.R., Moore, R.K.: Using linguistic cues for the automatic recognition of personality in conversation and text. J. Artif. Intell. Res. (JAIR) 30, 457–500 (2007)
20. Nakata, K., Voss, A., Juhnke, M., Kreifelts, T.: Collaborative concept extraction from documents. In: Proceedings of the 2nd Int. Conf. on Practical Aspects of Knowledge Management (PAKM 1998). Citeseer (1998)
21. Paice, C.D.: Another stemmer. ACM SIGIR Forum 24(3), 56–61 (1990)
22. Poria, S., Cambria, E., Hussain, A., Bin, G.-H.: Big multimodal data analysis. Elsevier Neural Networks (2014)
23. Poria, S., Gelbukh, A., Agarwal, B., Cambria, E., Howard, N.: Common sense knowledge based personality recognition from text. In: Castro, F., Gelbukh, A., González, M. (eds.) MICAI 2013, Part II. LNCS, vol. 8266, pp. 484–496. Springer, Heidelberg (2013)
24. Poria, S., Gelbukh, A., Cambria, E., Das, D., Bandyopadhyay, S.: Enriching senticnet polarity scores through semi-supervised fuzzy clustering. In: 2012 IEEE 12th International Conference on Data Mining Workshops (ICDMW), pp. 709–716. IEEE (2012)
25. Poria, S., Gelbukh, A., Cambria, E., Yang, P., Hussain, A., Durrani, T.: Merging senticnet and wordnet-affect emotion lists for sentiment analysis. In: 2012 IEEE 11th International Conference on Signal Processing (ICSP), vol. 2, pp. 1251–1255. IEEE (2012)
26. Poria, S., Gelbukh, A., Hussain, A., Howard, N., Das, D., Bandyopadhyay, S.: Enhanced senticnet with affective labels for concept-based opinion mining. IEEE Intelligent Systems 28(2), 31–38 (2013)
27. Poria, S., Winterstein, G., Cambria, E., Gelbukh, A., Hussain, A., Bin, G.-H.: Dependency-based rules for concept-level sentiment analysis. Elsevier Knowledge-Based Systems (2014)
28. Rajagopal, D., Cambria, E., Olsher, D., Kwok, K.: A graph-based approach to commonsense concept extraction and semantic similarity detection. In: Proceedings of the 22nd International Conference on World Wide Web Companion, pp. 565–570. International World Wide Web Conferences Steering Committee (2013)
29. Ramirez, P.M., Mattmann, C.A.: Ace: Improving search engines via automatic concept extraction. In: Proceedings of the 2004 IEEE International Conference on Information Reuse and Integration (IRI 2004), pp. 229–234. IEEE (2004)
30. Sidorov, G.: Non-continuous syntactic n-grams. Polibits 48, 67–75 (2013)
31. Sidorov, G.: Syntactic dependency based n-grams in rule based automatic English as second language grammar correction. International Journal of Computational Linguistics and Applications 4(2), 169–188 (2013)
32. Sidorov, G., Velasquez, F., Stamatatos, E., Gelbukh, A., Chanona-Hernández, L.: Syntactic n-grams as machine learning features for natural language processing. Expert Systems with Applications 41(3), 853–860 (2014)

33. Socher, R., Perelygin, A., Wu, J.Y., Chuang, J., Manning, C.D., Ng, A.Y., Potts, C.: Recursive deep models for semantic compositionality over a sentiment treebank. In: Conference on Empirical Methods in Natural Language Processing (EMNLP). EMNLP (2013)
34. Velardi, P., Fabriani, P., Missikoff, M.: Using text processing techniques to automatically enrich a domain ontology. In: Proceedings of the International Conference on Formal Ontology in Information Systems, vol. 2001, pp. 270–284. ACM (2001)
35. Yuntao, Z., Ling, G., Yongcheng, W., Zhonghang, Y.: An effective concept extraction method for improving text classification performance. Geo-Spatial Information Science 6(4), 66–72 (2003)
36. Zheng, J.H., Lu, J.L.: Study of an improved keywords distillation method. Computer Engineering 31(18), 194–196 (2005)

Obtaining Better Word Representations via Language Transfer

Changliang Li[*], Bo Xu, Gaowei Wu, Xiuying Wang, Wendong Ge, and Yan Li

Institute of Automation of the Chinese Academy of Sciences (CASIA), Beijing, China
{changliang.li,xubo,gaowei.wu,xiuying.wang,wendong.ge}@ia.ac.cn,
liyan.startup@gmail.com

Abstract. Vector space word representations have gained big success recently at improving performance across various NLP tasks. However, existing word embeddings learning methods only utilize homo-lingual corpus. Inspired by transfer learning, we propose a novel language transfer method to obtain word embeddings via language transfer. Under this method, in order to obtain word embeddings of one language (target language), we train models on corpus of another different language (source language) instead. And then we use the obtained source language word embeddings to represent target language word embeddings. We evaluate the word embeddings obtained by the proposed method on word similarity tasks across several benchmark datasets. And the results show that our method is surprisingly effective, outperforming competitive baselines by a large margin. Another benefit of our method is that the process of collecting new corpus might be skipped.

1 Introduction

Vector space models represent word meanings with vectors that capture syntactic and semantic information of words. The representations can be introduced to similarity measure and have recently demonstrated outstanding results across a variety of NLP tasks, such as Name Entity Recognition (NER), document classification, relation extraction and so on [1]. So work to improve word representations is very worthwhile.

An effective approach to word representation is to learn distributed representations. A distributed representation is dense, low dimensional, and real-valued. Distributed word representations are also called word embeddings. Each dimension of the embeddings represents a latent feature of the word, hopefully capturing useful syntactic and semantic properties [2]. Many approaches have been proposed to learn good performance word embeddings, such as RNN [3], HLBL [4], Morphological RNNs [5], SENNA [6].

However, despite much work on improving the learning of word embeddings, all the work shares a common problem that only utilizes homo-lingual corpus to learning word embeddings.

---

[*] Corresponding Author.

A. Gelbukh (Ed.): CICLing 2014, Part I, LNCS 8403, pp. 128–137, 2014.

Transfer learning is the improvement of learning in a new task through the transfer of knowledge from a related task that has already been learned. Common machine learning algorithms, in contrast, traditionally address isolated tasks. Transfer learning attempts to change this by developing methods to transfer knowledge learned in one or more source tasks and use it to improve learning in a related target task. Techniques that enable knowledge transfer represent progress towards making machine learning as efficient as human learning [7,8].

In this paper, inspired by transfer learning, we propose a novel language transfer method to obtain word embeddings via language transfer. We use source language word embeddings to represent target language word embeddings according to some certain rule. Because one target word can translated into more than one similar word in source languages, the target word embeddings could capture more semantic information via our method.

We evaluate the word embeddings obtained through our method on word similarity task across many benchmark datasets. The experiments results show that the word embeddings learned through our method outperform several competitive baselines by a large margin, at most 32.3 Spearman's rank correlation ($\rho \times 100$).

Similar to transfer learning, another benefit of our method is that it could make full use of existing corpus. If we want obtain a new language word embeddings, the process of collecting new corpus might be skipped.

2   Related Work

A word representation is a mathematical object associated with each word, often a vector. Each dimension's value corresponds to a feature and might even have a semantic or grammatical interpretation, so we call it a word feature [2]. One common approach to inducing word representation is to use clustering, perhaps hierarchical. This technique was used by a variety of researchers [9,10,11,12,13]. This leads to a one-hot representation over a smaller vocabulary size.

Neural language models [14,15], on the other hand, induce dense real-valued low-dimensional word embeddings using unsupervised approaches. Historically, training and testing of neural language models has been slow, scaling as the size of the vocabulary for each model computation. Many approaches have been proposed to eliminate that linear dependency on vocabulary size and allow scaling to very large training corpora.

Collobert and Weston presented a neural language model that could be trained over billions of words, because the gradient of the loss was computed stochastically over a small sample of possible outputs, in a spirit similar to Bengio [16]. This neural model of Collobert and Weston was refined and presented in greater depth in [17,18].

It was found that word representations could capture meaningful syntactic and semantic regularities in a very simple way, such as the singular/plural relation that V_{apple} - V_{apples} $\approx$ V_{car} - V_{cars}. The regularities are observed as constant vector offsets between pairs of words sharing a particular relationship. It is also true for a variety of semantic relations, as measured by the SemEval 2012 task of measuring relation similarity [19].

Vector space word representations have been successfully at improving performances across a variety of NLP tasks. Much work has been focused on improving word embeddings. Socher et al., recently proposed several kinds of recursive neural networks language models, such as RNN, MV-RNN, RNTN [20,21,22,23,24]. Mikolov et al., presented language model based on recurrent neural networks [25,26]. Most of the work focuses on improvement of language model. Meanwhile, corpus also plays an important role in training word embeddings. There are several corpora publicly available, most of them are English.

Many benchmark datasets, such as WordSim-353 [27], RG [28], MC [29], SCWS* [1] and RW [5], have been widely used to evaluate word representations. All these benchmark datasets are scored with human similarity ratings on pairs of words. In this paper, we evaluated the word embeddings obtained by our method on these datasets in word similarity task.

3　Language Transfer

In this section, we firstly described the details of language transfer method. And then we introduced a global context-aware neural language model [1] to train source language word embeddings.

3.1　Language Transfer Method

All common languages share concepts that are grounded in the real world (such that cat is an animal smaller than a dog) [30]. Meanwhile, one word often corresponds to more than one word in another different language. For example, the English word "computer" is corresponding to three Chinese words "计算机;电脑;计算装置". Similarly, the Chinese word "电脑" is corresponding to two English words "PC; computer". This is not limited between English and Chinese. It also works among other languages.

Based on the linguistic knowledge and phenomenon, we propose a novel language transfer method inspired by transfer learning. Suppose we have had source language word embeddings, our method could provide better target word embeddings than training language models on target language corpus from scratch. The details are described as follows.

For each target language word W_{targ}^i, which means the ith word in target language vocabulary, we translate it into source language word(s). Each corresponding source word is represented as $W_{sour}^{i,n}$, which means the nth source word corresponding to target word W_{targ}^i.

We attempt to represent the target language word embeddings $V(W_{targ}^i)$ through using the corresponding source language word embeddings $V(W_{sour}^{i,n})$. It is illustrated by formula (1). N is the number of source words for the target word.

$$V(W_{targ}^i) = Func\left\{\sum_{n=1}^{n=N} V(W_{sour}^{i,n})\right\} \tag{1}$$

$Func(.)$ is normalized function. Figure 1 illustrates the process of language transfer method. Some target word corresponds to only one source word. In this case, $V(W_{targ}^{i}) = V(W_{sour}^{i,1})$.

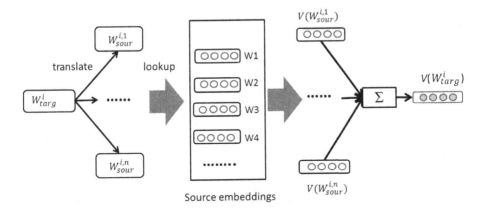

Fig. 1. Process of language transfer method

For example, if we select English as target language and Chinese as source language, the embeddings of target word "computer" is obtained through the following equation (3).

$$V(computer) = Func\{V(计算机) + V(电脑) + V(计算装置)\} \tag{3}$$

Figure 2 shows the example in vector space. The vectors of both English word and corresponding Chinese words have been projected down to two dimensions using PCA algorithm.

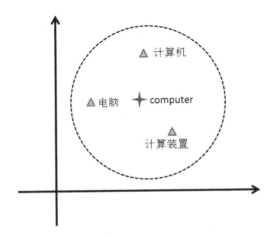

Fig. 2. Example of language transfer method

From this example, we can see that English word "computer" embeddings are obtained through combining three corresponding Chinese words embeddings. And it is supposed that the ability of word embeddings to capture semantic information is strengthened.

3.2   Neural Language Model

As our main work focuses on the language transfer method, in this paper, we do not research on a new language model. Instead, we choose to employ a global context-aware neural language model [1] to train source language word embeddings.

The model defines two scoring components that contribute to the final score of a (word sequence, document) pair. The scoring components are computed by two neural networks, one capturing local context and the other global context. Figure 3 shows the context-aware neural language model.

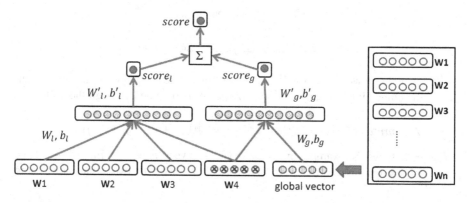

Fig. 3. An overview of global context-aware neural language model. The model can make full use of both local and global context.

The neural language model assigns a score for each N-gram consisting of words $x_1, x_2 \ldots x_n$ as formula (4).

$$s(ngram_i) = W'_l f(X_l W_l + b_l) + b'_l + W'_g f(X_g W_g + b_g) + b'_g \qquad (4)$$

Where X_l represents the concatenation of the input word embeddings; X_g is the concatenation of the weighted average document vector and the vector of the last word in sequence; b_l, b'_l, b_g, b'_g are biases; W'_l, W_l, W'_g, W_g are weights of the neural network.

We employ a ranking-type cost as objective function to minimize as described by formula (5).

$$J(\theta) = \sum_{i=1}^{M} \max \{0, 1 - (s(ngram_i) - s(ngram_i'))\} \qquad (5)$$

M is the number of all available N-grams in the training corpus, whereas $(ngram_i')$ is a N-gram obtained from $(ngram_i)$ by replacing its last word with a randomly chosen different word. This model alternate between two stages, forward pass and back-propagation pass [1]. The model assigns higher scores to valid N-gram than invalid ones and has been demonstrated to be both efficient and effective in learning word representations [5].

In learning process, we take the derivative of the ranking loss with respect to the parameters: weights of the neural network and the embedding matrix. These weights are updated via back propagation algorithm. The embedding matrix is the word representations (see [1] for more details about the global context-aware neural language model).

4   Experiments

In order to illustrate the effectiveness of our method, we evaluated the word embeddings obtained through our method on several benchmark datasets in word similarity task. In our experiment, we selected English as target language, Chinese as source language.

4.1   Obtaining Target Language Word Embeddings

We selected BaiduBaike [31] documents as the corpus, due to its wide range of topics and word usages, and its clean organization of document by topic. In preprocessing, the numbers were mapped to a NUMBER token. Meanwhile some other rare tokens were mapped to an UUUNKKK token.

For the neural language model training, we used 10-word windows of text as the local context, 100 hidden units, and no weight regularization for both neural networks. In order to compare with baselines results, we used 50-dimensional embeddings.

After training and simple process, we have obtained final Chinese word embeddings, including NUMBER and UUUNKKK tokens. And then we used formula (1) to build target language word embeddings for 100232 English words built by C&W [17]. If some English word or token could not be translated into Chinese, we replaced it with UUUNKKK in formula (1).

4.2   Benchmark Datasets

We evaluated word embeddings obtained through our method in word similarity task on many benchmark datasets. Theses benchmark datasets includes SCWS* [1], WordSim-353 (WS for short) dataset [27], RG dataset [28], and MC dataset [29]. Each dataset contains a list of word pairs rated by multiple human annotators (WS: 353 pairs, 13 annotators; MC: 28 pairs, 38 annotators; RG: 65 pairs, 36 annotators; SCWS*: 2003 pairs, 10 annotators).

And also we believed that good performance word representations should be able to learn useful representations for both frequent words and rare words. So we also evaluated our method on Rare Word dataset [5], RW for short. RW concludes 2034 pairs of rare words, and each pair is scored by human judgment.

4.3   Results

We evaluated word embeddings obtained by our method, referred as LTM, on word similarity task. The results are evaluated using standard metrics, Spearman's rank correlation ρ, which is used to gauge how well the relationship between two variables, the similarity scores given by a variety of language models   and the human annotators.

Detailed performance of word embeddings obtained through our method (LTM) is given in Table 1. We also reported baseline result provided by [1] (referred as HSMN). It employed the same language model as this paper on English corpus. Furthermore, we reported three more competitive baselines HSMN+stem, HSMN+cimRNN, and HSMN+csmMRNN [5]. They all are based on the same language model as this paper too. However, HSMN+stem is based on the morphological segmentation of unknown words: instead of using a universal vector presentation representing the stems of unknown, it used vectors representing the stems of unknown words; HSMN+cimRNN is context-insensitive combined with morphology; and HSMN+csmMRNN is context-sensitive combined with morphology.

Table 1. Spearman's rank correlation (ρ × 100) between similarity scores assigned by various models or methods and by human annotators

	WS	MC	RG	SCWS*	RW
HSMN	62.58	65.44	62.81	32.11	1.97
HSMN+stem	62.58	65.90	62.81	32.11	3.40
HSMN+cimRNN	62.81	65.90	62.81	32.97	14.85
HSMN+csmRNN	**64.58**	71.72	65.45	43.65	22.31
LTM	60.1	**75.90**	**72.44**	**50.49**	**34.2**

Compared to baseline HSMN, from the result, we can see that the obtained word embeddings by our method outperforms baseline HSMN by a good margin for all datasets (except for WS).   Especially for RW and SCWS* dataset, our method outperforms the baseline by 32.3 and 18.3 Spearman's rank correlation (ρ × 100) respectively.

Even compared to the competitive baseline with best performance, HSMN+csmRNN, our method also outperforms by a large margin on all datasets (except for WS), especially by 11.9 Spearman's rank correlation (ρ × 100) on RW dataset.

We notice that our method outperforms the baselines on RW datasets by a larger margin than on other datasets. The property of RW dataset is that it consists of rare and complex words. According to our analysis, the reasonable explanation to this performance could be described as follows.

Rare and complex English words are sparse in English corpus. While training on language model, the sparsity would reduce embeddings' ability to capture semantic properties. This might be the reason that the baseline performs badly on RW dataset. However, most of rare and complex English words can be translated into common Chinese words. The combination of these common Chinese words embeddings are supposed to capture more useful semantic properties. This is exactly the exemplification of our method's idea.

5   Discussions

The results highlight the fact that our method is reliable. This might be explained by two reasons: 1) all common languages share concepts which are grounded in the real world; 2) one word often corresponds to more than one word in another different language. From the result, we also could infer that it is possible to skip the process of collecting target language corpus, instead to utilize existing source language corpus to train word embeddings. This is exactly like the idea of transfer learning.

However, on WS dataset the method performs not as good as on other datasets. The possible reason is that the transliteration way employed while translating person and place name between English and Chinese. Transliteration only strives to represent the characters accurately without considering word meaning. There are several pairs of person name and place name words in WS, which would be translated into Chinese by the transliteration way. These translated words did not capture any corresponding semantic information. So the final effect of language transfer method on WS dataset is corrupted.

The quality of translation plays a significant role in our method. Except for transliteration, the translation of past tense might also impact the final result of our method. When comes to longer phase representations, the translation quality becomes more important. This also implies significance of another NLP task, machine translation. Good performance word representation could improve machine translation, and vice versa.

The language transfer method might play a significant role in many NLP tasks such as information retrieval, parsing and so on. In this paper, we took English as target language, Chinese as source language. However, our method is not limited to the two languages. And it can be generalized to other languages.

6   Conclusion

This paper has presented a novel language transfer method to obtain word embeddings. Unlike previous method to use homo-lingual, the main property of language transfer method is to use source language word embeddings to represent target language word embeddings. The results have shown that language transfer method outperforms competitive baselines by a large margin in similarity task on many benchmark datasets.

Therefore we could infer that the word embeddings obtained by our method could capture more semantic information. Furthermore, another benefit of our method is that the process of collecting new language corpus might be skipped. However, we have recognized that the transliteration of person name and place name might corrupt the final effect of our method. An important direction for our further work might be to reduce the impact brought by transliteration among languages. And another promising direction is to extent the language transfer method to more NLP tasks as well as to more languages.

Acknowledgements. This work is partly supported by National Program on Key Basic Research Project (973 Program) under Grant 2013CB329302, National Natural Science Foundation of China (NSFC) Grants No.61203281 and No.61175050.

References

1. Huang, E.H., Socher, R., Manning, C.D., Ng, A.Y.: Improving word representations via global context and multiple word prototypes. In: Annual Meeting of the Association for Computational Linguistics, ACL (2012)
2. Turian, J., Ratinov, L., Bengio, Y.: Word representations: A simple and general method for semisupervised learning. In: ACL (2010)
3. Mikolov, T., Karafiát, M., Burget, L., Cernocký, J., Khudanpur, S.: Recurrent neural network based language model. In: INTERSPEECH (2010)
4. Mnih, A., Hinton, G.E.: A scalable hierarchical distributed language model. In: NIPS, pp. 1081–1088 (2009)
5. Luong, M., Socher, R., Manning, C.: Better word representations with recursive neural networks for morphology. In: CONLL (2013)
6. Bengio, Y., Louradour, J., Collobert, R., Weston, J.: Curriculum learning. In: ICML (2009)
7. Torrey, L., Shavlik, J.: Transfer learning. In: Soria, E., Martin, J., Magdalena, R., Martinez, M., Serrano, A. (eds.) Handbook of Research on Machine Learning Applications. IGI Global (2009)
8. Asadi, M., Huber, M.: Effective control knowledge transfer through learning skill and representation hierarchies. In: International Joint Conference on Artificial Intelligence (2007)
9. Huang, F., Yates, A.: Distributional representations for handling sparsity in supervised sequence labeling. In: ACL (2009)
10. Ratinov, L., Roth, D.: Design challenges and misconceptions in named entity recognition. In: CoNLL (2009)
11. Koo, T., Carreras, X., Collins, M.: Simple semi-supervised dependency parsing. In: ACL, pp. 595–603 (2008)
12. Miller, S., Guinness, J., Zamanian, A.: Name tagging with word clusters and discriminative training. In: HLT-NAACL, pp. 337–342 (2004)

13. Liang, P.: Semi-supervised learning for natural language. Master's thesis, Massachusetts Institute of Technology (2005)
14. Bengio, Y.: Neural net language models. Scholarpedia 3, 3881 (2008)
15. Bengio, Y., Ducharme, R., Vincent, P.: A neural probabilistic language model. In: NIPS (2001)
16. Bengio, Y., Ducharme, R., Vincent, P., Jauvin, C.: A neural probabilistic language model. Journal of Machine Learning Research 3, 1137–1155 (2003)
17. Collobert, R., Weston, J.: A unified architecture for natural language processing: Deep neural networks with multitask learning. In: ICML (2008)
18. Morin, F., Bengio, Y.: Hierarchical probabilistic neural network language model. AISTATS (2005)
19. Mikolov, T., Yih, W.-T., Zweig, G.: Linguistic regularities in continuous space word representations. In: NAACL-HLT (2013)
20. Miller, S., Guinness, J., Zamanian, A.: Name tagging with word clusters and discriminative training. In: HLT-NAACL, pp. 337–342 (2004)
21. Socher, R., Pennington, J., Huang, E.H., Ng, A.Y., Manning, C.D.: Semi-supervised recursive autoencoders for predicting sentiment distributions. In: EMNLP (2011)
22. Socher, R., Manning, C., Ng, A.: Learning continuous phrase representations and syntactic parsing with recursive neural networks. In: NIPS*2010 Workshop on Deep Learning and Unsupervised Feature Learning (2010)
23. Socher, R., Lin, C.C., Ng, A., et al.: Parsing natural scenes and natural language with recursive neural networks. In: Proceedings of the 28th International Conference on Machine Learning (ICML 2011), pp. 129–136 (2011)
24. Socher, R., Bauer, J., Manning, C.D., et al.: Parsing with compositional vector grammars. In: Proceedings of the ACL Conference (2013)
25. Mikolov, T., Kombrink, S., Burget, L., Cernocký, J., Khudanpur, S.: Extensions of recurrent neural network language model. In: ICASSP (2011)
26. Mikolov, T., Zweig, G.: Context dependent recurrent neural network language model. In: SLT (2012)
27. Finkelstein, L., Gabrilovich, E., Matias, Y., Rivlin, H., Solan, Z., Wolfman, G., Uppin, E.: Placing search in context: The concept revisited. ACM Transactions on Information Systems 20(1), 116–131 (2002)
28. Rubenstein, H., Goodenough, J.B.: Contextual correlates of synonymy. Commun. ACM 8(10), 627–633 (1965)
29. Miller, G., Charles, W.: Contextual correlates of semantic similarity. Language and Cognitive Processes 6(1), 1–28 (1991)
30. Mikolov, T., Le, Q.V., Sutskever, I.: Exploiting Similarities among Languages for Machine Translation. arXiv preprint arXiv:1309.4168 (2013)
31. http://baike.baidu.com/

Exploring Applications of Representation Learning in Nepali

Anjan Nepal and Alexander Yates

Temple University
Broad St. and Montgomery Ave.
Philadelphia, PA 19122
{anjan.nepal,yates}@temple.edu

Abstract. We explore the applications of representation learning in Nepali, an under-resourced language. Using distributional similarity on a large amount of unlabeled Nepali text, we induce clusters of different sizes. The use of these clusters as features significantly improves the performance compared to the baseline on two standard NLP tasks. In a part-of-speech (PoS) tagging experiment where the train and test domain are the same, the accuracy on the unknown words increased by up to 5% compared to the baseline. In a named-entity recognition (NER) experiment in domain adaptation setting with a small training data size, the F1 score improved by up to 41% compared to the baseline. In a setting where train and test domain are the same, the F1 score improved by 13% compared to the baseline.

Keywords: Domain Adaptation, Representation Learning, Nepali Language, PoS Tagging, NER.

1 Introduction

Nepali is the lingua franca of Nepal and the mother tongue of almost half the population of the country [10], yet only a few NLP resources exist to make use of the language processing techniques for practical applications. One reason for this is because the manual annotation is usually an expensive task. However, unlabeled data is usually cheap to find and can be used to improve the performance by using semi-supervised learning approach. Semi-supervised learning by using representations learned from a large unlabeled data is well studied in English and a few other languages. This approach needs fewer labeled training examples to reach the same performance compared to the traditional system. If same amount of training examples are used it can boost the performance of the system [24]. In a domain adaptation setting, this approach has been shown to be useful to learn a system that has a target domain performance comparable to the source domain [17,36]. These important properties suggest that this approach will also be useful for under-resourced languages. In this paper, we explore semi-supervised representation learning in Nepali. We show that this approach out-performs the baseline system by 5% in accuracy on the unknown

A. Gelbukh (Ed.): CICLing 2014, Part I, LNCS 8403, pp. 138–150, 2014.

words on the PoS tagging task. In an NER experiment, we show an improvement of 13% in F1 score compared to the baseline reaching 76% even when the labeled training examples used are in the amount of just a few hundreds.

In the next section we briefly describe about the Nepali language. In section 3, we discuss the previous work on representation learning and the Nepali language. Section 4 describes the representation learning models used in this paper. In section 5, we provide the results and discussion for the experiments. Section 6 concludes.

2   Background on Nepali Language

Nepali is an Indo-Aryan language and linguistically closely related to Hindi. It is written in Devanagari script. There are 11 vowels and 33 consonants. There is no notion of capitalization in the language which makes the NER task challenging. It is a highly inflectional language. Nouns, adjectives, verbs and adverbs can have inflected forms and pronouns, coordinating conjunctions, subordinating conjunctions, postpositions, interjections, vocatives and nuance particles have uninflected forms. Some of the inflection causing cases are: number (singular/plural), gender, status, person and tense. Postpositions are parallel to prepositions in English and occur after the nouns. A detailed report on the Nepali language and grammar can be found in Bal *et al.* [3].

3   Previous Work

Representation learning is well studied in NLP and the techniques can be broadly divided into four categories: 1) vector space models of meaning based on document-level lexical co-occurrence statistics [34,37,33]; 2) dimensionality reduction techniques for vector space models [13,16,21,32,7,38]; 3) using clusters that are induced from distributional similarity [8,30,26] as non-sparse features [24,9,22,41]; 4) and recently, language models [5,27] as representations [12,39,11,4], some of which have already yielded state of the art performance on domain adaptation tasks [17,19,36,18] and information extraction tasks [1,15].

The research in Nepali natural language processing is in its early stage. Bista [6] builds an English to Nepali translation system by producing the parse tree for the source language, applying translation rules on the constituent phrases, applying rules to change the syntax into the target language and adding morphological rules to inflect the root of the words to finally produce the target sentence. Bal *et al.* [2] develop rule-based morphological analyzer and stemmer for Nepali and also provide a tokenizer software. Shahi *et al.* [35] build a PoS tagger model based on SVM. Our work is different than the previous work and the contribution is three-fold: 1) We provide the results for the experiments on language models trained on a large corpus which can be helpful for machine translation systems and to induce representations; 2) We use the representations from these language models and cluster based methods to improve the performance of the supervised sequence labeling tasks; 3) We present a new dataset and experimental results on the NER task for the Nepali language.

4 Models for Representation Learning

We use two representation learning models: hidden markov model (HMM) [31] and Brown Clustering algorithm [8] to induce representations of the data in our experiments.

4.1 Hidden Markov Model

An HMM is a generative probabilistic model that generates each word x_t in the corpus conditioned on a latent variable Y_t which in turn is conditioned on the previous latent variable Y_{t-1} in the sequence. Each Y_t takes a value between 1 and K, where K is the state size of the model. The joint probability of the words $\mathbf{x} = (x_1, x_2, ..., x_T)$ and the latent variables $\mathbf{y} = (y_1, y_2, ..., y_T)$ is given by

$$P(\mathbf{x}, \mathbf{y}) = \prod_{t=1}^{T} P(x_t|y_t)P(y_t|y_{t-1})$$

The parameters of the model $P(x_t|y_t)$ and $P(y_t|y_{t-1})$ are estimated using Expectation-Maximization (EM) algorithm [14]. We can find the optimal sequence of latent states $\hat{\mathbf{y}} = \arg\max_{\mathbf{y}} P(\mathbf{y}|\mathbf{x})$ using the Viterbi algorithm. We can also find the soft clusters, which is the posterior probability distribution over the hidden states. The training time for an HMM is $O(K^2 \cdot T \cdot I)$ where I is the number of iterations of the EM algorithm.

4.2 Brown Clustering Algorithm

Brown Clustering algorithm is a hierarchical clustering algorithm which maximizes the mutual information of the bigrams in the data. Given the words in the corpus, $\mathbf{x} = (x_1, x_2, ..., x_T)$, the objective function of the model that we want to maximize is given by:

$$P(\mathbf{x}) = \prod_{t} P(x_t|C(x_t))P(C(x_t)|C(x_{t-1}))$$

where C is a deterministic function that maps a word in the vocabulary of size V to a cluster of size K. The training time for the algorithm is $O(K^2 \cdot V)$.

Both of these model are based on the idea that the words which have similar distribution of words in their context are similar. However, these two models differ in a few aspects. HMM makes use of the whole sequence as context and can give different cluster to the same word depending on the context. Brown algorithm uses a local context (the bigram statistics) and assigns a single cluster for a word in all contexts. Additionally, we can choose between hard-clusters or soft-clusters in an HMM whereas the representations from Brown algorithm are always hard clusters. The choice of representation to use usually depends on the property of the supervised data and the supervised task.

The representation models are trained on a large amount of unlabeled text. Then the latent states from these models are induced for each token of the supervised training and test data. These latent states can be used as features in any structured classifier, and we use a linear-chain Conditional Random Fields (CRF) [23] which is widely used in information extraction.

4.3   Supervised Classifier: CRF

Conditional Random Fields (CRF) is a discriminative undirected probabilistic graphical model with the objective function:

$$P(y|\mathbf{x}, \boldsymbol{\lambda}) = \frac{1}{Z} \exp \sum_{k=1}^{K} \lambda_k f_k(y, \mathbf{x})$$

where, y is a label sequence; $\mathbf{x}$ is an input sequence (word and representation features); f_k are the feature combination for the labels y and inputs $\mathbf{x}$; $\boldsymbol{\lambda}$ are the weights for the features and Z is a normalization constant that ensures that P is a proper probability distribution.

During the training, both the input $\mathbf{x}$ and the label y from the supervised training data are available and we estimate the weights that maximize the probability $P(y|\mathbf{x}, \boldsymbol{\lambda})$ for all the training examples, $i = 1$ to N. Maximizing the probability is equivalent to maximizing the log of the probability and the objective function can be written as:

$$l(\boldsymbol{\lambda}) = \sum_{i=1}^{N} \sum_{t=1}^{T} \sum_{k=1}^{K} \lambda_k f_k(y_t^{(i)}, y_{t-1}^{(i)}, \mathbf{x}_t^{(i)}) - \sum_{i=1}^{N} \log Z(\mathbf{x}^{(i)})$$

The feature functions are functions of consecutive labels at each time-step due to the linear-chain structure of the CRF to ensure tractable inference, but they can include the whole input $\mathbf{x}$. All these features are collected for all the time-steps $t = 1$ to T of the sequence. The optimization however does not have a closed-form solution and numerical optimization methods like gradient descent are used. The gradient is given by:

$$\frac{\partial l}{\partial \lambda_k} = \sum_{i=1}^{N} \sum_{t=1}^{T} f_k(y_t^{(i)}, y_{t-1}^{(i)}, \mathbf{x}_t^{(i)}) - \sum_{i=1}^{N} \sum_{t=1}^{T} \sum_{y,y'} P(y, y'|\mathbf{x}_t^{(i)}) f_k(y, y', \mathbf{x}_t^{(i)})$$

The first term is the expected value of f_k under the empirical distribution and the second term is the expected value of f_k under the model distribution. By setting the derivative to zero, we are trying to find the weights which try to make these expected values equal to each other. During the test, we find the most probable sequence of the label y which has the maximum probability given the learned weights $\boldsymbol{\lambda}$ and the inputs $\mathbf{x}$.

5   Experiments

We expect that the use of the representations improves the performance of a supervised classifier in both in-domain and domain adaptation setting. We also expect that the representations with a larger number of latent states give better performance. To test these hypotheses, we train representation models on a large unlabeled corpus, induce representations using these models for the supervised classifier data and compare the results in the supervised experiments.

5.1   Experimental Setup

We used the unlabeled data from the Nepali National Corpus [40] which contains 14 million tokens. The corpus contained a lot of noise which might have been mainly introduced in the data collection and tokenization stages. We applied the following preprocessing steps to clean the corpus. We manually added more than 600 rules to fix the tokenization by looking at the most frequent errors first. After that, we removed the sentences which satisfy one or more of these conditions: has a word with character length greater than 30; number of words is less than 5 and greater than 50 and has total number of characters greater than 500. We also added rules to remove the sentences containing some pattern of jumbled text that were most frequent in the corpus and probably were a result of the data collection error. After the preprocesing, we randomly selected 10,000 sentences each for the test and dev set and the remaining 622,555 sentences for the training of the representation systems. The number of tokens in the training corpus was 10.5 million. We collapsed all words occurring less than or equal to 3 times into a special *unk* symbol and all numbers into a special *num* symbol which resulted in the training vocabulary size of 65,106. For representation learning, we built HMM language models of state size 25, 50, 100 and 200. We also ran Brown Clustering algorithm with cluster size 100, 200, 400 and 800. We chose larger size for Brown because it can be trained relatively faster than an HMM of the same latent state size.

To evaluate the performance of the representations, we used them in two standard sequence labeling tasks: PoS tagging and NER. PoS tagging is the task of assigning the part-of-speech tags like noun and verb to the words in a sentence. NER is the task of finding the names of the entities like Person (PER), Organization (ORG) and Location (LOC) in a sentence. We used the free implementation of the CRF package called CRFSuite[1]. For each token, we used the following features in the CRF:

- Words: $w_{i-2}, w_{i-1}, w_i, w_{i+1}, w_{i+2}$
- Prefixes: length 1, 2, 3, 4
- Suffixes: length 1, 2, 3, 4
- Representations (if applicable): $r_{i-2}, r_{i-1}, r_i, r_{i+1}, r_{i+2}$
- hasNumber?, hasASCII?, hasHyphen? (features specific to PoS)

5.2   PoS Tagging

For the supervised PoS tagging, we used the standard labeled dataset which has the translated sentences of sections 00-02 of the WSJ corpus and are manually

---

[1] http://www.chokkan.org/software/crfsuite/

annotated by Rupakheti *et al.* [29]. The annotation of the tagset is similar to the Penn-treebank [25] and contains 42 tags. We used sections 00 and 01 as the training set (99,870 tokens) and section 02 as the test set (11,524 tokens).

5.3 Named Entity Recognition

We developed a new dataset for the NER experiment in Nepali. We selected 896 sentences (22,436 tokens) from the WSJ translated sentences as the source domain for the domain adaptation settings. From now on, we will refer to this dataset as WSJ. We randomly selected 705 sentences (18,003 tokens) from the Nepali local business news text and separated 505 sentences (12,979 tokens) as the test set for all the experiments. From now on, we will refer to these 505 sentences as NEPALITEST and to the remaining 200 sentences as NEPALITRAIN. We manually annotated the sentences for person, organization and location following the guidelines from Chinchor *et al.* [28].

Our domain adaptation setting is a challenging task. In the NEPALITEST corpus, 29% of the words are out of vocabulary of the WSJ corpus. Out of 800 total entities in the NEPALITEST, there are no matching entities for the person and the organization and out of the 384 locations, only 33 exactly match in the WSJ. When the NEPALITRAIN is combined with the WSJ sentences for training, the entities in the test that match to the training are: 13 out of 284 persons, 83 out of 366 organizations and 211 out of 384 locations.

5.4 Results and Discussion

We represent the baseline system as BASELINE, which uses all the features except the representation features in the CRF. Also, we represent the HMM model with state size K as HMM-K and Brown Clusters with K number of clusters as BROWN-K and use the representation features in addition to the baseline system features. We report the performance for PoS tagging by accuracy for all words, out-of-vocabulary (OOV) words and the sentences. We report the precision (P), recall (R) and F1 score for the evaluation of the NER system.

The results for the PoS tagging experiment are shown in Table 1. For the NER task when trained only on in-domain 200 sentences from NEPALITRAIN we only report the important results in the Table 2. Both results show that the systems using representations as features generally out-perform the baseline system. In all NER experiments, we perform the significance test only on the recall because the performance gain is mainly due to the improvement in the recall. The best performing representation BROWN-400 on PoS tagging experiment improved the accuracy on the OOV words by 5% and best performing representation BROWN-200 on the NER task improved the F1 score by more than 14%. Including both representations as features lowered the score in the PoS tagging experiment while it pushed the score further up by 2% in the NER experiment. This suggests that HMM representations do not provide more information than the Brown representations for the PoS tagging task for our dataset.

Table 1. The accuracy of the supervised system on PoS tagging experiment using different representations reported on all, out-of-vocabulary (OOV) words and sentences. The differences in OOV accuracy between the BASELINE and HMM-100+BROWN-400 is statistically significant at $p < 0.01$ using McNemar's test with chi-squared distribution with 1 degree of freedom.

Model	All Acc %	OOV Acc %	Sentence Acc %
BASELINE	94.53	80.39	34.51
HMM-25	94.77	80.49	34.29
HMM-50	94.63	81.16	34.95
HMM-100	94.60	81.73	35.60
HMM-200	94.67	81.09	35.38
BROWN-100	95.22	84.13	38.46
BROWN-200	95.31	84.22	39.12
BROWN-400	95.21	**85.46**	37.58
BROWN-800	**95.53**	85.26	**39.78**
HMM-100+BROWN-400	95.24	84.65	38.46

In the domain adaptation NER experiment with WSJ as the training data, the results of various representation systems are reported in Table 3. For comparison, we include all our representations. All representations improved the score compared to the BASELINE. Among the HMMs, HMM-100 performed the best and among the Brown clusters, BROWN-400 performed the best. Using both of these representations improved the performance further, reaching F1 score of 59.88. This score is nearly equal to the the baseline score of 60.02 when the in-domain NEPALITRAIN is used as the training data. (see Table 2)

We breakdown the scores for each entity type for the best performing representation systems and show them in Table 4. We see that the HMM representation

Table 2. Precision(P), Recall(R) and F1 score for the supervised NER experiment using different representations when only NEPALITRAIN is used for training. The difference between the recall values of the baseline and the best representation system is statistically significant at $p < 0.01$ using paired t-test.

Model	P	R	F1
BASELINE	84.20	46.62	60.02
HMM-100	82.32	59.38	68.99
BROWN-200	81.28	68.38	74.27
BROWN-400	**84.35**	64.00	72.78
HMM-100+BROWN-200	83.51	**70.25**	**76.31**

Table 3. Precision(P), Recall(R) and F1 score for the supervised NER experiment using different representations when WSJ is used for training. The difference between the recall values of the baseline and the best representation system is statistically significant at $p < 0.01$ using paired t-test.

Model	P	R	F1
BASELINE	47.72	11.75	18.86
HMM-25	57.94	18.25	27.76
HMM-50	57.41	19.38	28.97
HMM-100	66.67	31.50	42.78
HMM-200	69.01	29.50	41.33
BROWN-100	64.06	32.75	43.34
BROWN-200	67.68	39.00	49.48
BROWN-400	69.38	35.12	46.64
BROWN-800	69.08	17.88	28.40
HMM-100+BROWN-200	**72.26**	**51.12**	**59.88**

gives best performance on ORG and PER entities while the Brown cluster gives best performance on LOC. This result suggested for using a combination of these two representations in our tasks. The combination improved the score than using them individually. We can see that the use of representations significantly improve the F1 score of PER and LOC entity types.

Table 4. Breakdown of the scores for each entity type using best performing representations when only WSJ is used for training in the NER experiment. We see that HMM is doing well on ORG and PER while Brown is doing well on LOC. A combination of these two representations improves the performance on all entity types.

Model	LOC			ORG			PER		
	P	R	F1	P	R	F1	P	R	F1
BASELINE	60.00	10.16	17.37	45.74	16.54	24.29	31.58	7.69	12.37
HMM-100	67.52	20.57	31.54	**50.69**	28.08	36.14	85.47	64.10	73.26
BROWN-200	82.63	45.83	58.96	41.36	25.77	31.75	80.23	44.23	57.02
HMM-100+BROWN-200	**82.85**	**51.56**	**63.56**	45.73	**35.00**	**39.65**	**93.75**	**76.92**	**84.51**

Table 5 shows the result for the domain adaptation setting where a few labeled data from the test domain is available. We ran the training on WSJ and NEPALITRAIN dataset. Again, HMM-100 performed the best among the HMM models. In contrast to previous NER experiments, BROWN-400 performed best compared to BROWN-200 when we compared all the Brown clusters. When both HMM and Brown representations are used, we get the highest F1 score among all the experiments.

Table 5. Precision(P), Recall(R) and F1 score for the supervised NER experiment using different representations when WSJ and NEPALITRAIN is used for training. The difference between the recall values of the baseline and the best representation system is statistically significant at $p < 0.01$ using paired t-test.

Model	P	R	F1
BASELINE	77.22	54.25	63.73
HMM-100	79.35	63.88	70.78
BROWN-200	78.61	68.00	72.92
BROWN-400	**82.50**	67.75	74.40
HMM-100+BROWN-400	81.77	**71.75**	**76.43**

Table 6. The perplexlity of a word in unlabeled train, dev and test corpus from HMM lanugage models trained for representations. As the state space size increases, the perplexity of the data decreases.

Model	Train	Dev	Test
HMM-25	783.12	799.75	798.47
HMM-50	663.15	686.53	689.42
HMM-100	613.05	651.50	652.89
HMM-200	**479.98**	**533.19**	**535.00**

The results from PoS and NER experiments support our first hypothesis that representation learning improves the performance of the system than the baseline in both the in-domain and the domain adaptation setting. Our second hypothesis that with higher number of states the performance improves is not fully supported by the experiment: the maximum performance is achieved in the mid range, usually 100 for HMM and 200 for Brown representations. Previous work has shown that generally higher number of latent states in HMM decreases the perplexity of the data and increases the performance in the supervised task [1]. In our experiments with HMM as language model, we also found that the perplexity goes down with the increase in number of states (see Table 6). However, the HMM representation model with 100 states performed better than with 200 states. Previous work on Brown clustering has also shown improvement in performance as the number of latent states increases [36]. However, our Brown representation model with 200 and 400 clusters usually did better than with 800 clusters. We think these are mainly due to sparsity introduced due to fewer supervised training samples. Generally, higher number of latent states give better representations of the words, but it also increases the feature size, which in turn requires a larger number of training examples to accurately estimate the feature weights. Our experiments with small supervised training examples support this hypothesis. For example, the fraction of the tokens in the NEPALITEST dataset whose Brown clusters did not appear in the training dataset halved when NEPALITRAIN and WSJ were used as the training set compared to when only NEPALITRAIN was used for the training set. This can be the reason why BROWN-200 performed better when NEPALITRAIN is used alone for training (see Table 2) compared to NEPALITRAIN and WSJ are used for training (see Table 5). Huang *et al.* [17] have shown that HMM performs better than other representations on polysemous words. We believe that our data does not have has has many polysemous words after seeing that the performance of the HMM on PoS tagging is lower than the Brown clusters. The results from both the PoS and NER experiments show that representation learning is useful to improve the performance of the supervised systems but the choice of the representation depends on the property of the supervised data and the task.

6   Conclusion and Future Work

In this paper, we showed that semi-supervised learning by representation learning on unlabeled data can be helpful for the under-resourced Nepali language. The experiments on PoS tagging showed that the use of representations improved the accuracy on the unknown words. The domain adaptation experiments on NER showed that this approach can also be used to easily improve the performance of the system on a dataset of different domains. The experiments with same training and test domain showed that even with few hundred examples, we can achieve significantly high performance with the use of representations. In the future, we would like to build a model that cleans and fixes the tokenization of the corpus automatically, a work which took a significant amount of our time in this project. Also, we would like to explore the use of representation learning techniques to other NLP tasks like parsing and translation.

Acknowledgments. We acknowledge Center for Research in Urdu Language Processing (www.crulp.org), National University of Computer and Emerging Sciences (www.nu.edu.pk), Pakistan for the PoS labeled data. We would like to thank Bal Krishna Bal for providing us with the Nepali corpus. This research was supported in part by NSF Award 1065397.

Appendix: Some Nepali word clusters from HMM

1. ठूलो (big) कम (little) राम्रो (beautiful) महत्वपूण (important) सजिलो (easy)
2. *num* एक (one) दुई (two) दुइ (two) तीन (three) नौ (nine) पहिलो (first) धेरै (many)
3. कृष्ण (Krishna) गिरिजाप्रसाद (Girija Prasad) मन्त्री (minister) डा. (Dr.) श्री (Mr.)
4. श्रेष्ठ (Shrestha) थापा (Thapa) नेपाल (Nepal) शर्मा (Sharma) कोइराला (Koirala)
5. म (I) सबै (all) तिमी (you) हामी (we) त्यो (that) तपाईं (you) आमा (mother)
6. माघ (Magh, a Nepali month) रु. (Rs.) करीब (approximately) सय (hundred)
7. काठमाडौँ (Kathmandu) भारत (India) अमेरिका (America) चीन (China) समाज (society)
8. मा (in) बाट (from) सँग (with) माथि (above) पछि (behind) भन्दा (than) बीच (between)

We can find some semantic and syntactic similarity among the words in a cluster (approximate translation of the words in English are in parenthesis). For example, the first and second cluster mainly contain adjective and numbers. Third and fourth clusters mainly contain the first name and the last names of people. These kinds of clusters generally improve the performance in tasks like NER even when the name of the person in the test data is never seen in the training data.

References

1. Ahuja, A., Downey, D.: Improved extraction assessment through better language models. In: Proceedings of the Annual Meeting of the North American Chapter of the Association of Computational Linguistics, NAACL-HLT (2010)
2. Shrestha, P., Bal, B.K.: A morphological analyzer and stemmer for nepali. In: Workshop on Morpho-Syntactic Analysis, School of Asian Applied Natural Language Processing for Linguistics Diversity and language Resource Development (2007)
3. Bal, B.K.: Structure of nepali grammar (2004–2007)
4. Bengio, Y., Louradour, J., Collobert, R., Weston, J.: Curriculum learning. In: International Conference on Machine Learning, ICML (2009)
5. Bengio, Y.: Neural net language models. Scholarpedia 3(1), 3881 (2008)
6. Bista, S.K.: Interim report on dobhase: Online english to nepali machine translation system. Technical Report, Kathmandu University (2005)
7. Blei, D.M., Ng, A.Y., Jordan, M.I.: Latent dirichlet allocation. Journal of Machine Learning Research 3, 993–1022 (2003)
8. Brown, P.F., de Souza, P.V., Mercer, R.L., Pietra, V.J.D., Lai, J.C.: Class-based n-gram models of natural language. Computational Linguistics, 467–479 (1992)
9. Candito, M., Crabbé, B.: Improving generative statistical parsing with semi-supervised word clustering. In: IWPT, pp. 138–141 (2009)
10. Nepal Central Bureau of Statistics. Major highlights. Census (2011)
11. Collobert, R., Weston, J.: A unified architecture for natural language processing: Deep neural networks with multitask learning. In: International Conference on Machine Learning, ICML (2008)
12. Collobert, R., Weston, J., Bottou, L., Karlen, M., Kavukcuoglu, K., Kuksa, P.: Natural language processing (almost) from scratch. Journal of Machine Learning Research 12, 2493–2537 (2011)
13. Deerwester, S.C., Dumais, S.T., Landauer, T.K., Furnas, G.W., Harshman, R.A.: Indexing by latent semantic analysis. Journal of the American Society of Information Science 41(6), 391–407 (1990)
14. Dempster, A.P., Laird, N.M., Rubin, D.B.: Maximum likelihood from incomplete data via the em algorithm. Journal of the Royal Statistical Society, Series B 39(1), 1–38 (1977)
15. Downey, D., Schoenmackers, S., Etzioni, O.: Sparse information extraction: Unsupervised language models to the rescue. In: ACL (2007)
16. Honkela, T.: Self-organizing maps of words for natural language processing applications. In: Proceedings of the International ICSC Symposium on Soft Computing (1997)
17. Huang, F., Ahuja, A., Downey, D., Yang, Y., Guo, Y., Yates, A.: Learning Representations for Weakly Supervised Natural Language Processing Tasks. Computational Linguistics 40(1) (2014)
18. Huang, F., Yates, A.: Distributional representations for handling sparsity in supervised sequence labeling. In: Proceedings of the Annual Meeting of the Association for Computational Linguistics, ACL (2009)

19. Huang, F., Yates, A.: Exploring representation-learning approaches to domain adaptation. In: Proceedings of the ACL 2010 Workshop on Domain Adaptation for Natural Language Processing, DANLP (2010)
20. Huang, F., Yates, A.: Open-domain semantic role labeling by modeling word spans. In: Proceedings of the Annual Meeting of the Association for Computational Linguistics, ACL (2010)
21. Kaski, S.: Dimensionality reduction by random mapping: Fast similarity computation for clustering. In: IJCNN, pp. 413–418 (1998)
22. Koo, T., Carreras, X., Collins, M.: Simple semi-supervised dependency parsing. In: Proceedings of the Annual Meeting of the Association of Computational Linguistics (ACL), pp. 595–603 (2008)
23. Lafferty, J., McCallum, A., Pereira, F.: Conditional random fields: Probabilistic models for segmenting and labeling sequence data. In: Proceedings of the International Conference on Machine Learning (2001)
24. Lin, D., Wu, X.: Phrase clustering for discriminative learning. In: ACL-IJCNLP, pp. 1030–1038 (2009)
25. Marcus, M.P., Marcinkiewicz, M.A., Santorini, B.: Building a large annotated corpus of English: The Penn Treebank. Computational Linguistics 19(2), 313–330 (1993)
26. Martin, S., Liermann, J., Ney, H.: Algorithms for bigram and trigram word clustering. Speech Communication 24, 19–37 (1998)
27. Mnih, A., Hinton, G.E.: A scalable hierarchical distributed language model. In: Neural Information Processing Systems (NIPS), pp. 1081–1088 (2009)
28. Ferro, L., Chinchor, N., Brown, E., Robinson, P.: 1999 named entity recognition task definition (1999)
29. Bal, B.K., Rupakheti, P., Khatiwada, L.P.: Report on nepali computational grammar. Technical Report, Madan Puraskar Pustakalaya, Lalitpur, Nepal
30. Pereira, F., Tishby, N., Lee, L.: Distributional clustering of English words. In: Proceedings of the Annual Meeting of the Association for Computational Linguistics (ACL), pp. 183–190 (1993)
31. Rabiner, L.R.: A tutorial on hidden Markov models and selected applications in speech recognition. Proceedings of the IEEE 77(2), 257–285 (1989)
32. Sahlgren, M.: An introduction to random indexing. In: Methods and Applications of Semantic Indexing Workshop at the 7th International Conference on Terminology and Knowledge Engineering, TKE (2005)
33. Sahlgren, M.: The word-space model: Using distributional analysis to represent syntagmatic and paradigmatic relations between words in high-dimensional vector spaces. PhD thesis, Stockholm University (2006)
34. Salton, G., McGill, M.J.: Introduction to Modern Information Retrieval. McGraw-Hill (1983)
35. Balami, B., Shahi, T.B., Dhamala, T.N.: Support vector machines based part of speech tagging for nepali text. International Journal of Computer Applications (2013)
36. Turian, J., Ratinov, L., Bengio, Y.: Word representations: A simple and general method for semi-supervised learning. In: Proceedings of the Annual Meeting of the Association for Computational Linguistics (ACL), pp. 384–394 (2010)
37. Turney, P.D., Pantel, P.: From frequency to meaning: Vector space models of semantics. Journal of Artificial Intelligence Research 37, 141–188 (2010)
38. Väyrynen, J.J., Honkela, T., Lindqvist, L.: Towards explicit semantic features using independent component analysis. In: Proceedings of the Workshop Semantic Content Acquisition and Representation, SCAR (2007)

39. Weston, J., Ratle, F., Collobert, R.: Deep learning via semi-supervised embedding. In: Proceedings of the 25th International Conference on Machine Learning (2008)
40. Lohani, R.R., Regmi, B.N., Gurung, S., Gurung, A., McEnery, T., Allwood, J., Yadava, Y.P., Hardie, A., Hall, P.: Construction and annotation of a corpus of contemporary nepali. Corpora 3, 213–225 (2008)
41. Zhao, H., Chen, W., Kit, C., Zhou, G.: Multilingual dependency learning: A huge feature engineering method to semantic dependency parsing. In: CoNLL 2009 Shared Task (2009)

Topic Models Incorporating Statistical Word Senses

Guoyu Tang[1,2,3], Yunqing Xia[1,2,3], Jun Sun[4],
Min Zhang[5], and Thomas Fang Zheng[1,2,3]

[1] Center for Speech and Language Technologies, Division of Technical Innovation and
Development, Tsinghua National Laboratory for Information Science and Technology
[2] Center for Speech and Language Technologies, Research Institute of Information
Technology
[3] Department of Computer Science and Technology, Tsinghua University, Beijing,
China
[4] Institute for Infocomm Research, A-STAR, Singapore
[5] Soochow University, China
sweetyuer@gmail.com, {yqxia,fzheng}@tsinghua.edu.cn,
sunj@i2r.a-star.edu.sg, mzhang@suda.edu.cn

Abstract. LDA considers a surface word to be identical across all documents and measures the contribution of a surface word to each topic. However, a surface word may present different signatures in different contexts, i.e. polysemous words can be used with different senses in different contexts. Intuitively, disambiguating word senses for topic models can enhance their discriminative capabilities. In this work, we propose a joint model to automatically induce document topics and word senses simultaneously. Instead of using some pre-defined word sense resources, we capture the word sense information via a latent variable and directly induce them in a fully unsupervised manner from the corpora. Experimental results show that the proposed joint model outperforms the classic LDA and a standalone sense-based LDA model significantly in document clustering.

Keywords: topic modeling, word sense induction, document representation, document clustering.

1 Introduction

Latent Dirichlet Allocation (LDA) was developed as a powerful unsupervised algorithm in analyzing topic distribution for a document collection [1].

The classic LDA model relies on the co-occurrences of surface words to capture their semantic relations. In reality, a surface word is likely to be highly associated to more than one topic and presents different word senses in different topics. LDA considers the surface word to be identical in both contexts and leverages on its co-occurrences with other words in the context to differentiate those two topics. Ideally, if a model is able to differentiate word senses in different contexts, the sense disambiguated words can contribute more probability masses to the

A. Gelbukh (Ed.): CICLing 2014, Part I, LNCS 8403, pp. 151–162, 2014.

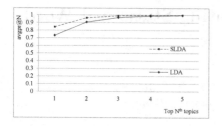
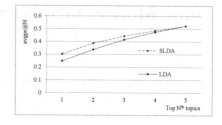

Fig. 1. Averaged per word (sense) topic distribution on the top-5 topics where the cumulative curve presents *avpgr* over the top-k topics

Fig. 2. Averaged per document topic distribution on the top-5 topics. where the cumulative curve presents the *avpgr* over the top-k topics.

corresponding topic than the surface words alone. In other words, word senses if applicable can serve as additional features for topic models and enhance the discriminative capability of topic models.

The intuition can be further verified with an empirical study on the average probability mass over the top N topics ($argpr@N$). We perform the surface word LDA model and the sense-based LDA model (SLDA) model (see details in Section 3.1) on Reuters20 data and compare the probability mass over the top N topics (N=5 in our analysis). We study two pairs of conditional probabilities: (1) The first pair is $p(z|w)$ and $p(z|s)$, which refer to the topic distribution given a surface word and a word sense respectively. The quantities are averaged over all word (sense) types in the data set. (2) The second pair is $p_w(z|d)$ in LDA and $p_s(z|d)$ in SLDA models, which refer to the topic distribution given a surface word based document and a word sense based document respectively. The quantities are averaged over all documents in the data set. The first pair is presented in Fig. 1 and the second is in Fig. 2, where the sense based model is referred as SLDA. All figures are drawn based on the experiments in Section 4.2. From Fig. 1, we can find that SLDA is above LDA in the cumulative curves. This suggests that word sense is a more indicative signature to describe the topic preferences for documents than surface word. From Fig. 2, we find that SLDA concentrates a document more on the top topics and provides sharper posterior topic estimation than LDA. This indicates that SLDA offers more confidence on the posterior estimation by means of the indicative word senses. Details of this analysis can be further found in Section 4.2.

In this paper, we will not only verify that word sense features provide topic models with more confidence in the posterior estimation, but also propose appropriate approaches to verify that the reinforced confidence is meaningful and helpful to improve the quality of the induced topics. The major contributions in this paper thus can be compiled in two perspectives. First, we incorporate the word sense information in the LDA generative story and construct a joint model to infer word senses for words and topics for documents simultaneously. Rather than applying the word sense information as an external feature or isolate the word sense induction as a pre-processing step [2] the proposed model is more generic by incorporating the word sense feature as a latent variable in the graphical model. Second, our model is completely unsupervised and is able to work with external resources minimized. Previous researches[3,4] attempted to introduce word senses from WordNet to topic models. However, their models rely on the external knowledge source, i.e. WordNet, to construct a pool of word senses for a given word. Alternatively, we induce word senses automatically from the corpora. This is especially advantageous for resource poor languages that are short of available pre-defined word senses as well as domain specific documents that may contain terms beyond the general resources.

Specifically, we employ Hierarchical Dirichlet Processes (HDP) [5] as a non-parametric prior for word sense induction, because HDP can prevent us from explicitly bounding the number of word senses for each word. Two models are proposed in this paper: Standalone SLDA (SA-SLDA) considers word sense induction and document representation as standalone modules; Collaborative SLDA (CO-SLDA) takes the topics of senses from SLDA as the pseudo feedback for Word Sense Induction (WSI) and iteratively infers both topics and word senses.

The reminder of this paper is organized as follows: we first present some background for this work in Section 2. After that, we describe the approaches to incorporate statistical word senses for the LDA topic models. Experimental results and discussions are presented in Section 4. We conclude this paper in Section 5.

2   Related Work

2.1   Semantic Interpretation of Documents

In Vector Space Model (VSM) [6] , it is assumed that terms are independent of each other and the semantic relations between terms are ignored.Recently, models are proposed to represent documents in a semantic concept space using lexical ontologies, i.e. WordNet or Wikipedia [7,8]. However, the lexical ontologies are difficult to be constructed and their coverage can be limited. In contrast, topic models are used as an alternative for discovering latent semantic space in corpora based on the per topic word distribution. LDA [1] as a classic topic model identifies topics of documents by evaluating word co-occurrences. Some work attempt to integrate word semantics from lexical resources into topic models [3,4]. Alternatively, our models are fully unsupervised and do not rely on any external semantic resources, which will be extremely applicable for resource poor languages and domains.

2.2   Word Sense Disambiguation and Induction

Word sense disambiguation, which identifies the correct word sense from a set of pre-defined sense candidates, has been proved to benefit various NLP tasks [9]. However, manually-compiled large lexical resources such as WordNet are often required.Instead, Word Sense Induction (WSI) can learn word senses from corpora in an unsuper-vised manner. With respect to the Bayesian approaches, Brody and Lapata [10] used an extended LDA model to induce word senses which provide the state-of-the-art performance in SemEval-2007 evaluation[11] . Yao and Durme [12] used Hierarchical Dirichlet Process (HDP) [5] to induce word senses and empirically verified its advantage over LDA. WSI is also applied in other tasks like information retrieval [2], where word senses for query words are induced. To the best of our knowledge, no work has been reported to exploit WSI in document topic modeling as we do in this paper.

3   Topic Models Incorporating Statistical Word Senses

As shown in Fig. 3, the classic LDA assigns each word in the document a topic and considers the surface words as the basic granularity for a document. Alternatively, our model emits a sense for each surface word and assigns each sense a topic. Therefore, the basic granularity for our model is the word sense (Fig. 3). To address this motivation, we introduce an additional latent variable of word sense to LDA and induce it from the observed surface words.

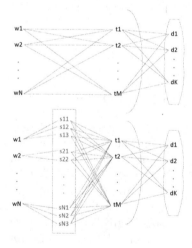

Fig. 3. Illustration of the classic LDA model (above) and the word sense extended LDA models (below). The values in the dot rectangle are assigned to the latent variable (i.e., word sense).

We design several models to implement this purpose as follows:

– Standalone SLDA (SA-SLDA): We isolate the Word Sense Induction (WSI) process as a standalone step. With the induced word senses in hand, we perform the word sense based LDA. .

– Collaborative SLDA (CO-SLDA): We identify the generative story as two iteratively interchangeable steps. Given an observed topic, we generate the word sense from the topic. Given an observed word sense, we generate the topic for each word sense, where the word sense is a point estimate from the mode of the distribution.

3.1   Standalone SLDA Model (SA-SLDA)

In the SA-SLDA model, WSI and documet representation (DR) are considered as standalone modules, where DR takes the output (i.e., word senses) of WSI as input(see Fig.5).

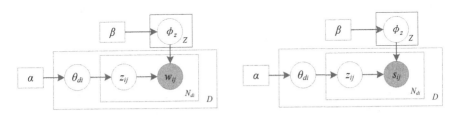

Fig. 4. Illustration of the standard LDA model

Fig. 5. Illustration of the SA-SLDA model

We follow [12] to employ Hierarchical Dirichlet Processes (HDP) for word sense induction.In this paper, we perform HDP on each word. We define a word on which the WSI algorithm is performed as a target word. We also define the words in the context of a target word as context words of the target word. After WSI, we simply take the mode sense in the distribution as the sense of the target word.

As shown in Fig. 5, we replace surface word with word sense in the gray plate. Given D documents and W word types, the formal procedure with Z topics of document representation in SA-SLDA is given as follows:

1. For each topic z:
   (a) choose $\phi_z \sim Dir(\beta)$.
2. For each document d_i:
   (a) choose $\theta_{d_i} \sim Dir(\alpha)$.
   (b) for each word w_{ij} in document d_i:
       i. choose topic $z_{ij} \sim Mult(\theta_{d_i})$.
       ii. choose sense $s_{ij} \sim Mult(\phi_{z_{ij}})$.

where d_i refers to i-th document in the corpus; w_{ij} refers to j-th word in document d_i; z_{ij} refers to the topic that word w_{ij} is assigned; s_{ij} refers to the sense that word w_{ij} is assigned from WSI; α, β are hyper-parameters of the model; $\phi_{z_{ij}}$ and θ_{d_i} are per topic sense distribution and per document topic distribution respectively which are drawn from Dirichlet distributions.

We use Collapse Gibbs Sampling to do inference for SA-SLDA [13]. Compared with LDA, we replace the surface words with the induced word senses. Therefore, the topic inference is similar to the classic LDA, where the condition probability $P(z_{ij} = z|\boldsymbol{z_{-ij}}, \boldsymbol{s})$ is evaluated by

$$P(z_{ij} = z|\boldsymbol{z_{-ij}}, \boldsymbol{s}) \propto \frac{n^{d_i}_{-ij,z} + \alpha}{n^{d_i}_{-ij} + Z\alpha} \times \frac{n^{s}_{-ij,z} + \beta}{n_{-ij,z} + S\beta} \qquad (1)$$

In Eq.1,$n^{d_i}_{-ij,z}$ is the number of words that are assigned topic z in document d_i $n^{s}_{-ij,z}$ is the number of senses with sense s that are assigned topic z , $n^{d_i}_{-ij}$ is the total number of words in document d_i; $n_{-ij,z}$ is the total number of words assigned topic z; S is the number of senses for the data set.$-ij$ in all the above variables refers to excluding the count for word w_{ij}. Further details are similar to the classic LDA [13].

3.2   Collaborative SLDA (CO-SLDA)

Alternatively, we induce word senses and the document topics simultaneously in a joint model (see Fig.6). We are interested in whether the topic assigned to a word has a positive feedback on WSI, which then can be used to refine the topic distribution. Inspired by this motivation, we propose a Collaborative SLDA model which takes the topics of senses from SLDA as the pseudo feedback for WSI and iteratively infers both topics and word senses. Specifically, we achieve a point estimate for the target word in WSI and feed this estimated sense to DR.

In this model, a three-level HDP algorithm is used to capture the relationship between word senses and topics of a target word w (see Fig. 6). In the

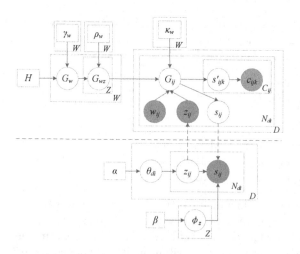

Fig. 6. Illustration of the CO-SLDA model

three-level HDP, for each word type w , we choose for each topic a probability measure G_{wz} which is drawn from Dirichlet Process $DP(\rho_w, G_w)$. For each word w_{ij} in document d_i , given topic $z_{ij} = z$, we use G_{wz} as the base probability measure for the context of w_{ij} and draw its own G_{ij} from Dirichlet process $G_{ij} \sim DP(\kappa_{wz}, G_{wz})$. This means that word w may have different sense distributions in different topics. For each context v_{ij} of the target word w, the sense s_{ijk} for each word c_{ijk} in v_{ij} has a nonparametric prior G_{ij}. H is a Dirichlet distribution with hyper-parameter ϵ. The context word distribution η_s given a sense s is generated from H:$\eta_s \sim H$. Hyper-parameters γ_w, ρ_w and κ_w are the concentration parameters for DP, controlling the variability of the distributions of G_w,G_{wz} and G_{ij} respectively.

We show the graphical presentation for CO-SLDA in Fig. 6. C_{ij} refers to the number of words in the context window v_{ij} for word w_{ij} in document d_i . The above dotted line shows the WSI process while the below shows the DR process. Given observed topics $\{z_{ij}\}$, word senses $\{s_{ij}\}$ are inferred in WSI. Given observed senses $\{s_{ij}\}$, topics $\{z_{ij}\}$ are inferred in DR. The two processes are interchangeably performed. We provide the dashed arrows in Fig. 6 to connect $\{s_{ij}\}$ and $\{z_{ij}\}$ that will change from hidden to observed during the alternation of two processes.

The word sense induction process is as follows:

1. For each word type w:
   (a) choose $G_w \sim DP(\gamma_w, H)$.
   (b) For each topic z:
       i. choose $G_{wz} \sim DP(\rho_w, G_w)$.
2. For each document d_i:
   (a) For each context v_{ij} of word w_{ij}:
       i. choose $G_{ij} \sim DP(\kappa_{wz}, G_{wz})$.
       ii. For each context word c_{ijk} of target word w_{ij}:
           A. choose $s_{ijk} \sim G_{ij}$.
           B. choose $c_{ijk} \sim Mult(\eta_{s_{ijk}})$.
       iii. set $s_{ij} = \arg\max_s P(s|G_{ij})$.

The document representation process is just like SA-SLDA.

For inference, we interchangeably infer two groups of hidden variables in CO-SLDA,

1. Given that the topic for each word sense z_{ij} is observed, we infer the sense distribution G_{ij} in the context window around a target word. This is achieved through the same scheme as [5]. Then we estimate s_{ij} for the target word as sense with the highest probability in G_{ij}.
2. Given that the word sense s_{ij} is observed, we infer the topic z_{ij} for each word sense. This can be achieved using the same inference scheme as SA-SLDA.

As the iteratively process can refine both topics and word senses based on each other's prediction, intuitively, the CO-SLDA model should be advantageous over the SA-SLDA model, which only provides single round estimation of the variables.

4   Evaluation

In the experiments, we first evaluate the latent topics in document clustering task and then analyze the averaged per sense topic distribution and averaged per document topic distribution of the proposed models.

4.1   Document Clustering

In this section, we apply the proposed models on the document clustering task and evaluate the performance against the baselines of LDA and K-means algorithms.

4.1.1   Setup

Data Set: Three data sets are used in our experiments.

1. **TDT4**:Following [14], we use the English documents from TDT2002 and TDT2003, i.e., TDT41 and TDT42. .
2. **Reuters**: Documents are extracted from Reuters-21578[15] with the most frequent 20 categories, i.e., Reuters20.

In the experiments, only nouns and verbs are used as target words for word sense induction and topic inference. We use per sentence as the context window for each target word. TreeTagger[16] is used to for Part-of-speech labeling.

Evaluation Metrics: In the experiments, we intend to evaluate the proposed topic models in document clustering task. Each topic in the test dataset is considered as a cluster and each document is clustered into the topic with the highest probability. We adopt the evaluation criteria proposed by [17]. The calculation starts from maximum F-measure of each cluster. The general F-measure of a system is the micro-average of all the F-measures of the system-generated clusters.

4.1.2   Experiment 1.1: Different Word Sense Induction Approaches

In this experiment, we aim to investigate how well the different word sense induction approaches contribute to the task of document clustering. We compare the performance of two different Bayesian models, i.e. LDA vs. HDP, in our SA-SLDA model.

System Parameters: As we isolate the WSI process from the document representation process in SA-SLDA, we present the parameters accordingly. (1) In the WSI step, for HDP, we set the hyper-parameters γ_w , ρ_w , ϵ for every word type to be $\gamma_w \sim Gamma(1, 0.001)$, $\rho_w \sim Gamma(0.01, 0.028)$, $\epsilon = 0.1$; for LDA, we set $\alpha = 0.2$, $\beta = 0.1$ and set the sense numbers for all words to be 4. (2) In the Document representation step, we set $\alpha = 1.5$ and $\beta = 0.1$. All hyper-parameters are tuned in the TDT42 dataset. The number of topics is set to be equal to the number of clusters in each dataset.

In all experiments, we let the Gibbs sampler burn in for 2000 iterations and subsequently take samples 20 iterations apart for another 200 iterations.

Table 1. Results of SA-SLDA with different WSI approaches (i.e., LDA and HDP)

Method	TDT41	TDT42	Reuters20
SA-SLDA(LDA)	0.787	0.842	0.490
SA-SLDA(HDP)	0.792	0.870	0.512

Experimental results are presented in Table 1.

Discussions: From Table 1, we can find that WSI with HDP outperforms WSI with LDA in all data sets when integrated into the SA-SLDA model. This is because LDA is a parametric model which requires user's explicit setting of the parameters. Alternatively, HDP, as a non-parametric model, can automatically infer the number of senses for each word. This provides reasonable interpretation for word sense modeling and additional flexibility for document representation. This advantage of HDP also provides our series of SLDA models with better interpretation. As a result, we employ HDP as a non-parametric prior for all proposed models.

4.1.3   Experiment 1.2: Different Extended LDA Models

In this experiment, we aim to verify the effectiveness of the proposed models in document clustering. Other than the proposed models, i.e., SA-SLDA and CO-SLDA , we also present K-means and LDA as our baselines. Specifically, we implement the Bisecting K-Means algorithm [17] which computes the cosine similarity between documents based on the TF-IDF features.

System Parameters In the WSI step we set the hyper-parameters γ_w, ρ_w , ϵ for every word type to be $\gamma_w \sim Gamma(8, 0.1)$, $\rho_w \sim Gamma(5, 1)$, $\kappa_w \sim Gamma(0.1, 0.028)$, $\epsilon = 0.1$; (2) in the DR step, we set $\alpha = 1.5$ and $\beta = 0.1$. In LDA, we set $\alpha = 1.5$, $\beta = 0.1$. The number of topics is set to be equal to the number of clusters in each dataset. In K-Means, we set K to be equal to the number of clusters in each dataset.

Experimental results are presented in Table 2.

Discussions: From Table 2, we can find that: First, SLDAs outperform the two baselines in all data sets. This indicates that using word senses other than surface words improves the document clustering results, which is due to the fact that SLDAs are facilitated with more fine-grained features of word sense induced from the context.

Table 2. Results of the proposed models and baselines

Method	TDT41	TDT42	Reuters20
K-Means	0.727	0.843	0.501
LDA	0.744	0.867	0.496
SA-SLDA	0.792	0.870	0.512
CO-SLDA	0.825	0.874	0.597

Second, CO-SLDA outperforms SA-SLDA in all data sets. This indicates that the joint inference process for topics of words and word senses provides positive impact to refine the results. Two reasons are worthy of noting: (1) In common sense, instances of the same word type in different topics may have different senses while instances in the same topic often refer to the same thing. Since CO-SLDA can jointly infer topics and word senses, instances of the same word in the same topic are more likely to be assigned the same sense while instances in different topics are likely to be assigned differently. As a result, word senses will be better identified. (2) Using topics as a pseudo feedback will facilitate the target words with topic-specific senses. For example, the word *election* only has one sense in general cases. However, in the TDT42 data set, topics are labeled in a more fine-grained perspective. For example, the following two sentences are labeled to be from two different topics as the countries of elections are different: z_1: *Ilyescu Wins Romanian Elections*, z_2: *Ghana Gets New Democratically Elected President*. With the joint inference of topic and sense, we can induce the word 'election' with two senses, i.e., *election#1* and *election#2*, related to the electing processes in Romania and Ghana respectively. By incorporating these topic-specific senses, *election* with context word *Romania* is identified as *election#1* and more likely to be assigned topic z_1 while *election* with context word *Ghana* is identified as *election#2* and more likely to be assigned z_2 .

4.2   Distribution Analysis

In this study, we aim to analyze the averaged per sense topic distribution and averaged per document topic distribution of the proposed models.

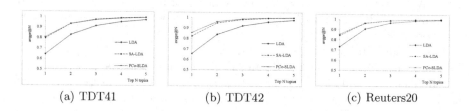

(a) TDT41   (b) TDT42   (c) Reuters20

Fig. 7. Averaged per word (sense) topic distribution on the top-5 topics

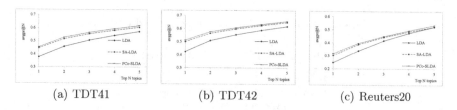

(a) TDT41   (b) TDT42   (c) Reuters20

Fig. 8. Averaged per document topic distribution on the top-5 topics

In Fig. 7, we present the $argpr@N$ values of averaged per word topic distribution for all words in LDA and averaged per sense topic distribution for all senses in SLDAs on the three datasets. Specifically, for each word, the topics that a word w is associated are ranked according to the probability $p(z|w)$[1] . $argpr@N$ is calculated by averaging the probabilities $p(z|w)$ of all words on the top N topics. For each word sense, we calculate $argpr@N$ based on $p(z|s)$. We use the senses in the last iteration of SLDA models(e.g. 2200) and the topics inferred by these word senses. For each data set, we draw the cumulative curve over the top N topics. Furthermore, we measure $avgpr@N$ of document-topic distribution by averaging the probabilities $p(z|d)$ for all documents in LDA and SLDA models on three datasets are presented in Fig. 8.

Discussions: From Fig. 7, we can observed that: First, SLDAs are all above LDA in the cumulative curves. This indicates word senses provide better discriminative capabilities for topic models than surface words. Second, the cumulative curves of CO-SLDA are above SA-SLDA. This benefit comes from that fact that CO-SLDA induces topic-specific senses by using topics as a pseudo feedback. The topic-specific senses are more discriminative than common senses.

From Fig. 8, we can observe that: First, the cumulative curves of SLDAs are all above LDA. This indicates that in SLDAs, documents concentrate on fewer topics which makes topics from sense-based topic models more discriminative. Second, the cumulative curves of CO-SLDA are above SA-SLDA. This suggests that the iteratively refined topics and words senses provide reinforcement of the posterior estimation of topics for documents.

5   Conclusion

In this paper, we propose to represent topics with distributions over word senses. In order to achieve this purpose in a fully unsupervised manner without relying on any external resources, we model the word sense as a latent variable and induce it from corpora via WSI. We design several models for this purpose. Distributions analysis of average sense-topic distribution and the average document-topic distribution shows a sharper distribution of topics in SLDAs which suggests that the proposed models provide more confidence on the posterior estimation. Empirical results verify that the word senses induced from corpora can facilitate the LDA model in document clustering. Specifically, we find the joint inference model (CO-SLDA) outperforms the standalone model (SA-SLDA) as the estimation of sense and topic can be collaboratively improved.

Acknowledgments. This work is supported by NSFC China (No. 61272233). We thank the reviewers for the valuable comments.

---

[1] $p(z|w)$ can be calculated with $p(z|w) \propto p(w|z) \Sigma p(z|d)p(d)$ where $p(w|z)$ and $p(z|d)$ are parameters of the model thus can be estimated while we estimate $p(d)$ to be the proportion of d's document length to the length of the entire document collection.

References

1. Blei, D.M., Ng, A.Y., Jordan, M.I.: Latent dirichlet allocation. J. Mach. Learn. Res. 3, 993–1022 (2003)
2. Navigli, R., Crisafulli, G.: Inducing word senses to improve web search result clustering. In: Proceedings of the 2010 Conference on Empirical Methods in Natural Language Processing, EMNLP 2010, Stroudsburg, PA, USA, pp. 116–126. Association for Computational Linguistics (2010)
3. Boyd-Graber, J.L., Blei, D.M., Zhu, X.: A topic model for word sense disambiguation. In: EMNLP-CoNLL, pp. 1024–1033 (2007)
4. Guo, W., Diab, M.: Semantic topic models: Combining word distributional statistics and dictionary definitions. In: Proceedings of the Conference on Empirical Methods in Natural Language Processing, EMNLP 2011, pp. 552–561. Association for Computational Linguistics, Stroudsburg (2011)
5. Teh, Y.W., Jordan, M.I., Beal, M.J., Blei, D.M.: Hierarchical dirichlet processes. Journal of the American Statistical Association 101 (2004)
6. Salton, G., Wong, A., Yang, C.S.: A vector space model for automatic indexing. Commun. ACM 18(11), 613–620 (1975)
7. Hotho, A., Staab, S., Stumme, G.: Wordnet improves text document clustering. In: Proc. of the SIGIR 2003 Semantic Web Workshop, pp. 541–544 (2003)
8. Gabrilovich, E., Markovitch, S.: Computing semantic relatedness using wikipedia-based explicit semantic analysis. In: Proceedings of the 20th International Joint Conference on Artifical Intelligence, IJCAI 2007, San Francisco, CA, USA, pp. 1606–1611. Morgan Kaufmann Publishers Inc. (2007)
9. Tufiş, D., Koeva, S.: Ontology-supported text classification based on cross-lingual word sense disambiguation. In: Masulli, F., Mitra, S., Pasi, G. (eds.) WILF 2007. LNCS (LNAI), vol. 4578, pp. 447–455. Springer, Heidelberg (2007)
10. Brody, S., Lapata, M.: Bayesian word sense induction. In: Proceedings of the 12th Conference of the European Chapter of the Association for Computational Linguistics, EACL 2009, pp. 103–111. Association for Computational Linguistics, Stroudsburg (2009)
11. Agirre, E., Soroa, A.: Semeval-2007 task 02: Evaluating word sense induction and discrimination systems. In: Proceedings of the 4th International Workshop on Semantic Evaluations, SemEval 2007, pp. 7–12. Association for Computational Linguistics, Stroudsburg (2007)
12. Yao, X., Van Durme, B.: Nonparametric bayesian word sense induction. In: Proceedings of TextGraphs-6: Graph-based Methods for Natural Language Processing, pp. 10–14. Association for Computational Linguistics (2011)
13. Griffiths, T.L., Steyvers, M.: Finding scientific topics. PNAS 101(suppl. 1), 5228–5235 (2004)
14. Kong, J., Graff, D.: Tdt4 multilingual broadcast news speech corpus. Linguistic Data Consortium (2005), http://www. ldc. upenn. edu/Catalog/CatalogEntry. jsp
15. Lewis, D.D.: Reuters-21578 text categorization test collection, distribution 1.0 (1997), http://www.research.att.com/~{}lewis/reuters21578.html
16. Schmid, H.: Probabilistic part-of-speech tagging using decision trees. In: Proceedings of International Conference on New Methods in Language Processing, Manchester, UK, vol. 12, pp. 44–49 (1994)
17. Steinbach, M., Karypis, G., Kumar, V.: A comparison of document clustering techniques. In: KDD Workshop on Text Mining (2000)

How Preprocessing Affects
Unsupervised Keyphrase Extraction

Rui Wang, Wei Liu, and Chris McDonald

School of Computer Science & Software Engineering
The University of Western Australia, Australia
21224938@student.uwa.edu.au, {wei.liu,chris.mcdonald}@uwa.edu.au

Abstract. Unsupervised keyphrase extraction techniques generally consist of candidate phrase selection and ranking techniques. Previous studies treat the candidate phrase selection and ranking as a whole, while the effectiveness of identifying candidate phrases and the impact on ranking algorithms have remained undiscovered. This paper surveys common candidate selection techniques and analyses the effect on the performance of ranking algorithms from different candidate selection approaches. Our evaluation shows that candidate selection approaches with better coverage and accuracy can boost the performance of the ranking algorithms.

Keywords: Unsupervised Keyphrase Extraction, Text Preprocessing.

1 Introduction

Keyphrases provide a high level abstraction of a document's content, which play an important role in many areas of document processing, including document indexing, classification, clustering and summarisation. A keyphrase is a lexical unit or a chain of lexical units that can be a single word, the habitual co-occurrence of two words, or a frequent recurrent uninterrupted string of words [1]. Keyphrase extraction is a task that identifies and extracts meaningful lexical unit chains that can describe a given document.

Unlike other natural language processing tasks, keyphrase extraction technology offers much room for improvement. Most of state of the art systems are still unable to achieve a precision of 50 percent [2,3]. In this paper, we seek a better understanding of the current techniques in keyphrase extraction.

In general, keyphrase extraction can be classified into either supervised and unsupervised techniques. Supervised techniques treat keyphrase extraction as either a classification [4,5] or a learning to rank problem [6,7] requiring annotated training data to build models. In contrast, unsupervised techniques treat keyphrase extraction as a ranking problem, scoring each candidate by considering cues such as word occurrence and co-occurrence frequencies, occurrence position, linguistic features, or information from external semantic networks.

An important step, common to both supervised and unsupervised keyphrase extraction, is candidate phrase selection. The candidate phrase selection process often involves text cleaning, text normalisation, and filtration, such as the

A. Gelbukh (Ed.): CICLing 2014, Part I, LNCS 8403, pp. 163–176, 2014.
© Springer-Verlag Berlin Heidelberg 2014

removal of stop words and punctuation marks, stemming, tokenising, part of speech (POS) tagging, and n-gram filtration.

What constitutes the candidate selection process, and the order of its steps, may have significant impact on the extraction process that follows. For example, stemming will remove a word's suffix, thus changing the word's format and potentially its grammatical tag. Early stemming will affect the accuracy of identifying a word and its grammatical tag. For example, after stemming *classification (noun)* becomes *classif (preposition)*. Much reported research also removes punctuation marks from the text based on the assumption that very few assigned keyphrases would contain punctuation marks. However, the apostrophe and hyphen are not uncommon in assigned keyphrases. Simply removing all punctuation marks will reduce the candidate coverage of potential keyphrases.

As supervised learning involves more features, the effect of candidate selection on the results of keyphrase extraction is not as obvious and significant as with unsupervised learning. Therefore, in this paper we only focus on unsupervised keyphrase extraction techniques.

Specific to unsupervised keyphrase extraction, a ranking algorithm takes the candidate phrases as direct inputs, and outputs a desired number of ranked candidates with associated scores indicating the likelihood of each being a keyphrase. Therefore, the ranking algorithm will not identify any words or phrases not in the candidate list. Thus, the accuracy and coverage of the candidates, as well as the ranking algorithm, will affect the final result.

Despite this, most previous studies discuss the candidate phrase selection and candidate ranking as a single pipeline, with little discussion on how the candidate selection steps are implemented. This leaves difficulty in understanding how the claimed improvements are achieved, let alone identifying whether they are achieved from the candidate selection, the ranking algorithm, or both.

Little research reported of the importance of the candidate phrase select process. Hulth (2003) [8] presents a comparison between three candidate selection techniques: n-grams, POS patterns, and noun-phrase (NP) chunking, on a supervised keyphrase extraction algorithm. More recently, studies show refined candidate selection approaches, but experiments are conducted in conjunction with their ranking algorithms. Kumar and Srinathan (2008) [9] demonstrated how a prepared dictionary of distinct n-grams using *LZ78* data compression could affect the n-gram filtration results. Kim and Kan (2009) [10], and Wang and Li (2010) [11] also focused more on explaining the preprocessing steps during the presentation of their refined NP chunking approach.

In this paper, we carefully examine how the effectiveness of identifying candidate phrases affects the ranking algorithms. We undertake a systematic study of three different candidate selection approaches, and combines each selection approach with four popular ranking algorithms: TF-IDF [12], RAKE [13], KeyGraph [14], and TextRank [15]. The four algorithms are selected to represent the major branches of unsupervised ranking algorithms: statistically-based and graph-based ranking. The results demonstrate that improving the candidate phrase coverage and accuracy on potential keyphrases leads to better performance of the ranking algorithms.

This paper is organised as follows: in Section 2, we provide a detailed review of common candidate phrase selection techniques. In Section 3, we provide an overview of some popular ranking algorithms, and Section 4 presents the implementation details of three candidate selection approaches. The datasets for our evaluation are described in Section 5. Section 6 describes experiments setup and discusses results, and Section 7 presents our conclusions.

2   Candidate Phrase Selection Techniques

Candidate phrase selection cleans and normalises text, and then filters out improper candidates. Cleaning process identifies and optionally removes characters or character chains that carry little or no semantic meaning to the given text. Depending on the observation of the dataset, the semantically meaningless characters may include punctuation marks, stop words, symbols, or mathematical equations. Some studies also apply heuristic rules for cleaning text. For example, Paukkeri and Honkela [16] remove authors' names and addresses, tables, figures, citations, and bibliographic information from scientific articles.

Text normalisation aims to convert a text into a format enabling more efficient filtering. For example, the distinction between 'Cat', 'cat', and 'cats' is ignored after normalisation. Common techniques include converting characters to lowercase, lemmatising, and stemming.

Filtration removes unwanted candidates. Two common techniques are n-gram filtration with heuristic rules, and NP chunking with POS patterns. N-gram filtration requires text segments as inputs to reduce the number of generated grams. A common approach for splitting text uses meaningless characters identified in the cleaning stage, based on the assumption that very few keyphrases would contain these characters. The n-gram generates all possible sequential combinations for each input. For example, the 4-word text segment: 'a b c d', leads to 10 combinations: 'a b c d', 'a b c', 'a b', 'b c d', 'b c', 'c d', 'a','b','c','d'. Rules are then applied to filter improper combinations – for example, selecting longer grams occurring above a frequency threshold [14,17].

NP chunking finds candidate phrases with pre-defined POS patterns, therefore tokenising and POS tagging must be performed before NP chunking. Researchers usually define their own POS patterns based on the analysis of the datasets, but a few basic patterns are shared by many studies – for example, an adjective followed by a number of nouns [18,19].

As opposed to ranking algorithms, the candidate phrase selection process receives much less attention. Previous studies do not explicitly mention the exact techniques used for candidate selection, leaving a lot of ambiguity. We have carefully selected 12 papers which provide relatively clearer descriptions of their selection approaches. We list all the techniques that are explicitly described in these papers, and summarise the processing sequence in Table 1.

Table 1. Common candidate phrase selection techniques

	Processing Sequence	1	2	3	4	5	6	7	8
Ohsawa et al. (1998)[14]	3, 4, 5			✓	✓	✓			
Mihalcea and Tarau (2004)[15]	1, 2, 6, 8	✓	✓				✓		✓
Matsuo and Ishizuka (2004)[20]	4, 5				✓	✓			
Bracewell et al. (2005)[21]	1, 2, 4, 6	✓	✓		✓		✓		
Krapivin et al. (2008)[22]	1, 3, 5, 4	✓		✓	✓	✓			
Liu et al. (2010)[18]	1, 2, 6	✓	✓				✓		
Ortiz et al. (2010)*[17]	3, 4, 5			✓	✓	✓			
Bordea and Buitelaar (2010)*[23]	1, 2, 6, 7	✓	✓				✓	✓	
El-Beltagy and Rafea(2010)*[24]	3, 7, 4, 5			✓	✓	✓		✓	
Paukkeri and Honkela(2010)*[16]	7, 4, 8				✓			✓	✓
Zervanou (2010)*[19]	7, 2, 6		✓				✓	✓	
Rose et al. (2010) [13]	3			✓					
Dostal and Jazek (2011)[25]	1, 3, 2, 6	✓	✓	✓			✓		
Total:		6	6	6	7	5	6	4	2

1: Tokenising 2: POS Tagging 3: Splitting text by meaningless words or characters and/or removing them 4: Stemming 5: n-gram filtering with heuristic rules 6: Phrase chunking with POS patterns 7: other heuristic rules 8: Phrase formation after ranking
∗ Although some of the processing may not be explicitly mentioned in the paper, it participated SemEval2010 workshop shared task 5 and have been reviewed by Kim et al. [3] where more implementation details are provided.

3   Unsupervised Keyphrase Ranking Techniques

Unsupervised keyphrase ranking techniques can be classified into two groups: statistically-based and graph-based. Statistically-based techniques process text into matrices, and assign each candidate a score by applying statistical techniques to the data. Graph-based techniques represent text as graphs, where vertices usually correspond to lexical units, and edges are lexical relations between the two lexical units. Two vertices are connected if they have the specified lexical relationship and the edge is weighted by the strength of that relationship, calculated by a weighting algorithm. This section briefly reviews four ranking algorithms that we use in our experiments.

Term Frequency - Inverse Document Frequency. (TF-IDF) [12] is a weighting scheme that statistically analyses how important a term is to an individual document in a corpus. TF-IDF is calculated as the product of two statistics: a term's TF weight and its IDF weight. The TF scheme analyses the importance of a term against a document, thus a term with higher frequency is assigned a higher TF weight. While the IDF weighting scheme analyses the importance of a term against the entire corpus, a term occurring frequently in a large number of documents gains a lower IDF score. The classic TF-IDF weighting scheme [12] assigns weight to term t_i in document d as:

$$tfidf(t_i) = tf_i \times idf_i = tf_i \times log\frac{|D|}{|\{d \in D : t_i \in d\}|} \qquad (1)$$

where tf_i is the number of times term t_i occurs in d, $|D|$ is the total number of documents in corpus D, and $|\{d \in D : t_i \in d\}|$ is the number of documents in which term t_i occurs.

RAKE. [13] reports word co-occurrence information among candidate phrases, based on the degree and the frequency of a candidate. Frequency is denoted as $freq(w)$. The degree is the sum of the frequency of this word (w) and its co-occurrence frequencies with other words in the candidate phrase list, denoted as $deg(w)$. For a single word phrase, the $Score(w)$ is defined as the ratio of degree to frequency, namely:

$$Score(w) = \frac{deg(w)}{freq(w)}. \qquad (2)$$

For a multi-word phrase (W), the score is calculated as the sum of the ratio of degree to frequency of each word $w \in W$:

$$Score(W) = \sum_{w \in W} \frac{deg(w)}{freq(w)} \qquad (3)$$

KeyGraph. [14] is a graph-based algorithm that uses term frequency and co-occurrence information as the evidence for identifying keyphrases from a single document. In KeyGraph, vertices represent terms and edges represent co-occurrence relations. Weak edges (ones with lower scores) are considered to be the appropriate ones for segmenting the document into clusters. Each cluster then can be seen as a subgraph, within which nodes are strongly connected. A cluster therefore is regarded as a group of supporting terms around certain keyphrases. Consequently, keyphrases are the terms that tie and hold clusters together, representing the document's key points.

TextRank. [15] uses an algorithm derived from Google's PageRank [26] to rank the importance of a vertex within a graph based on both local vertex-specific and global information that is recursively computed from the entire graph. TextRank implements the idea of 'voting'. The link from a vertex V_i to another vertex V_j is treated as that V_i casts a vote for V_j, then the higher the number of votes V_j receives, the more important V_j is. Moreover, the importance of the vote itself is also considered by the algorithms: the more important the voter V_i is, the more important the vote itself is. The votes a vertex received is the local vertex-specific information. The importance of a voter which is recursively drawn from the entire graph is the global information. In TextRank, the importance of a vertex within a graph is not only determined by local vertex-specific information but also global information.

4   Implementation

For our evaluation, we have re-implemented the four ranking algorithms described in Section 3. We also implemented three de-coupled candidate selection approaches, which we have named *Basic Splitter*, *Basic n-gram Filter*, and *Basic Chunker*. We use Python and the *Natural Language Toolkit (NLTK)* [27] package. Figure 1 presents an illustration of the pipelines implemented.

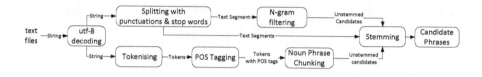

Fig. 1. Decoupled Keyphrase Extraction Pipelines

Basic Splitter. implements a common candidate selection approach, used by many studies [13,17] as a basic text splitting technique. Since keyphrases rarely contains semantically meaningless characters or character chains, such as stop words and punctuation marks, these can be the indicators of the boundary of meaningful phrases. Our basic splitter uses the stop word and punctuation mark lists provided by *NLTK*. We also reduced standard punctuation mark list by removing the hyphen, because it frequently appears in assigned keyphrases. The processing sequence is as follows:

1. convert input text to lowercase, and represent as unicode.
2. split text into segments, delimited by words from the standard stop word list, and characters from our reduced punctuation mark list.
3. stem using Porter's algorithm [28].

Basic n-gram Filter. implements the candidate selection process described in KeyGraph [14]. The basic n-gram filter is built on top of our basic splitter, taking output from the splitter as input. The n-gram filter first generates all possible sequential combinations greater or equal to two tokens, then saves them into n-gram list L. For example, phrase '*a b c*' will generate '*a b c*', '*a b*', '*b c*'. Inputs having fewer than three tokens are saved to the list L directly. After generating all n-grams for all the text segments, a heuristic rule is applied to remove unwanted grams. The processing sequence is as follows:

1. use the basic splitter to get text segments and input to the n-gram filter.
2. generate n-grams for all segments, $2 < n \leq length\ of\ segment$.
3. sort the n-gram list L by frequency. $\forall$ gram $g \in L$, if any gram G in the list where $g \subset G$ and $freq(g) \leq freq(G)$, remove g, otherwise remove all G.

Basic Chunker. discards tokens not fitting into the predefined POS pattern. We choose a simple but widely used pattern – find each phrase that begins with a determiner or personal pronoun (optional), followed by a number of Adjectives or Verb Past Participles or both (optional), and ending with a number of Nouns. This is because most content bearing phrases are noun phrases. The processing sequence is as follows:

1. convert input text to lowercase, and represent as unicode.
2. tokenise the text using the *NLTK tokeniser*.
3. add POS tags using the *NLTK POS tagger*.
4. filter the labelled tokens with the defined pattern.
5. stem identified candidates using Porter's algorithm [28].

5   Datasets

There are a few publicly available datasets for evaluating the keyphrase extraction task[1]. We select 2 datasets for our evaluation: Hulth2003 and SemEval2010. Both Hulth2003 and SemEval2010 consist of training and test sets for supervised extraction evaluations. Since no training data is required for unsupervised techniques, the training sets can also be used as our test sets.

Each dataset article pairs with keyphrases assigned by authors, readers, or both. We use the combination of both authors' and readers' keyphrases as the ground truth for our evaluation. However, many assigned keyphrases do not appear in the actual content of the article. This affects evaluation results, because neither the candidate selection techniques nor the ranking algorithms can extract a keyphrase not appearing in the text. We undertook further investigation of articles not containing their assigned keyphrases, and found that the distribution is random.

For a fair evaluation, we merged the three sets in Hulth2003: training, validation and test, then removed texts not containing all the assigned keyphrases. For SemEval2010, due to the small number of the articles, we have defined a *keyphrases coverage ratio r* as the number of assigned keyphrases appearing in the article, divided by the total number of assigned keyphrases, and

1. removed articles with low keyphrases coverage ($r \leq 0.8$), and
2. for the remaining articles, removed absent assigned keyphrases from the keyphrase list.

After the refinement, SemEval2010 contains 132 full length journal articles, consisting of 32 in distributed systems, 33 in information search and retrieval, 32 in artificial intelligence and multi-agent system, and 35 in social and behavioural sciences - economics. Hulth2003 contains 459 abstracts of journal articles from Computer Science. All assigned keyphrases appear in the articles' actual content.

---

[1] http://github.com/snkim/AutomaticKeyphraseExtraction

Table 2. Refined Datasets

	Total Article	Type	Assigned Keyphrase	mean
SemEval2010	132	Full length journal article	1910	14.5
Hulth2003	459	Abstract of journal article	3154	6.87

6   Evaluation

Two experiments are conducted in this study. We first evaluate how coverage on the assigned keyphrases differs in the various selection approaches described in Section 4. We then examine how candidate selection techniques affect the performance of the ranking algorithms. We test the performance of each ranking algorithm by delivering the outputs from each selection approach.

We stem the assigned keyphrases using Porter Stemmer [28]. An assigned keyphrase matches an extracted phrase when they correspond to the same stem sequence. For example, *fuzzy cats* matches *fuzzy cat*, but not *cat* or *cats fuzzy*.

We employ the *Precision*, *Recall*, and *F-measure* for evaluating the ranking algorithm. The *Precision* is defined as:

$$precision = \frac{the\ number\ of\ correctly\ matched}{total\ number\ of\ extracted} = \frac{TP}{TP + FP} \qquad (4)$$

Recall is defined as:

$$recall = \frac{the\ number\ of\ correctly\ matched}{total\ number\ of\ assigned} = \frac{TP}{TP + FN} \qquad (5)$$

F-measure is defined as:

$$F = 2 \times \frac{precision \times recall}{precision + recall} \qquad (6)$$

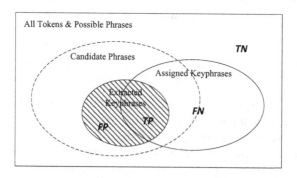

Fig. 2. Venn Diagram illustrating the set relationships

The dashed ellipse in Figure 2 indicates the potential candidate phrases selected by the candidate selection algorithms. The intersection between the candidate phrase set and the assigned keyphrase set is the coverage. The shaded ellipse is the final extracted keyphrases.

6.1 Experiment 1: Evaluation of Candidate Coverage

In this experiment, we evaluate the candidates' coverage on the assigned keyphrases. Again, none of the ranking algorithms can extract a keyphrase not appearing in the candidate list. Thus, if a candidate selection technique produces 70% coverage on assigned keyphrases, we have lost 30% true positive before running any ranking algorithm. Table 3 shows the experiment results.

Table 3. Candidate selection approaches coverage on assigned keyphrases

	Hulth2003 Dataset			SemEval2010 Dataset		
	Assigned	Match	Coverage %	Assigned	Match	Coverage %
BasicSplitter	3153	2223	70.5	1910	1604	84.0
N − gramFilter	3153	2053	65.1	1910	1371	71.8
Basicchunk	3153	2259	71.6	1910	1613	84.5

6.2 Experiment 2: Evaluation of Ranking Algorithm Performance

In this experiment, we deliver the output of each candidate selection approach to the four different ranking algorithms. We prepared a refined candidate list using the basic splitter for candidate selection, but with reference to the assigned keyphrases. In this way, we are able to gain candidate lists with 97% coverage on Hulth2003, and 99% on SemEval2010. This minimises the impact of candidate selection approaches since the ranking algorithms run with a nearly ideal candidate coverage.

For *KeyGraph*, the number of high frequent words is set to 30. The co-occurrence scores (edges) are calculated at the sentence level. For *TextRank*, we choose a window size of 10 for co-occurrence identification and the initial size of 10 for co-occurrence relation identification and the initial value of each node is set to 1, damping factor 0.85, iterations 30, and threshold of breaking 0.0001. *RAKE* and *TF-IDF* do not require any special setting.

Finally, the top 10 ranked candidates are selected from each result set. The results are shown in Table 4.

6.3 Discussion

Candidate Coverage. Of the three candidate selection approaches, the Basic Chunker produced the best coverage on assigned keyphrases. The Basic Splitter produced very close results, which we did not expect. After further analysis we found that because the Basic Splitter uses stop words and punctuation marks as delimiters, it has a higher probability of selecting different combinations of words, thus producing a better coverage. However, the Basic Splitter is unable to identify important phrases as the Chunker does – this is demonstrated in Experiments 2, where the same ranking algorithm has a lower performance when coupled with the Basic Splitter.

Table 4. Performance of four ranking algorithms couple with 3 selection approaches and the refined candidate list

	Hulth2003 Dataset				SemEval2010 Dataset			
	Coverage	Precision	Recall	F-score	Coverage	Precision	Recall	F-score
$TF - IDF$								
rl	97%	35.82	52.14	42.46	99%	34.02	23.51	27.80
pc	71.6%	28.71	41.80	34.04	84.5%	17.12	11.83	13.99
bs	70.5%	24.84	36.16	29.45	84%	15.23	10.52	12.45
nf	65.1%	24.23	35.27	28.72	71.8%	13.64	9.42	11.15
$KeyGraph$								
rl	97%	22.64	32.95	26.84	99%	21.82	15.08	17.83
pc	71.6%	22.05	32.10	26.14	84.5%	13.11	9.06	10.71
bs	70.5%	17.43	25.37	20.66	84%	9.55	6.60	7.80
nf	65.1%	16.51	24.04	19.58	71.8%	11.29	7.80	9.23
$RAKE$								
rl	97%	38.17	55.57	45.25	99%	1.06	0.73	0.87
pc	71.6%	32.35	47.10	38.36	84.5%	0.76	0.52	0.62
bs	70.5%	26.08	37.96	30.92	84%	0.30	0.21	0.25
nf	65.1%	25.08	36.50	29.73	71.8%	0.30	0.21	0.25
$TextRank$								
rl	97%	27.86	40.56	33.04	99%	28.94	20.00	23.65
pc	71.6%	24.12	35.11	28.59	84.5%	13.64	9.42	11.15
bs	70.5	19.63	28.58	23.27	84%	12.42	8.59	10.15
nf	65.1%	19.32	28.13	22.91	71.8%	17.27	11.94	14.12

rl: refined candidate list bs: basic splitter nf: n-gram filter pc: phrase chunker

Changing processing sequences on the selection approaches, described in Section 4, usually results in a lower coverage. For example, early stemming introduced a loss between 0.5% to 10%, depending on the selection approach.

The majority of loss occurs when a candidate is incorrectly split from a text that can be either a substring or a superstring of a assigned keyphrase. The most common stop words occur in the assigned keyphrases are *of, and, on, until, by, with, for, from*. The most common punctuation mark is the apostrophe, following by '.' and '+' which appear in '.net' and 'C++' in many articles in the Computer Science and Information Technology fields. A summary of different types of loss is presented in Table 5.

Table 5. Loss investigation

	Match	Lost	Error1	Error2	Error3	Error4	Error5
Basic Filter	3827	1236	1075	27	40	93	1
n-gram	3424	1639	835	666	40	93	5
Np chunk	3872	1191	516	468	40	93	74

Error 1: candidate identified is too long, being superstring of the assigned phrase
Error 2: candidate identified is too short, being substring of the assigned phrase
Error 3: assigned phrase contains invalid char such as punctuation marks
Error 4: assigned phrase contains stop words **Error 5**: others

Ranking Algorithm Performance. Although the n-gram filter performs worst on the candidate coverage test (65.1% on Hulth2003, and 71.8% SemEval2010), it produces almost identical or slightly better results than the Basic Splitter when coupled with ranking algorithms on both datasets. In contrast, the Basic Splitter produced a relatively impressive result in the candidate coverage evaluation (70.5% on Hulth2003 and 84%, on SemEval2010), it did not boost the performance of ranking algorithms. While the Basic Splitter has a better probability of selecting different combinations of words, it fails to correctly count the phrase occurrence frequency from other noisy candidates. For example, *dogs hate cats* is counted as one phrase by the Basic Splitter, while the correct counts are *dogs* and *cats*. The n-gram filter, however, counts the overall occurrence frequency and selects the most frequent phrase with longer grams. The Basic Chunker produces the best candidate coverage and boosts the performance ranking algorithms in comparison with other two selection approaches. The Basic Chunker is able to identify more correct phrases than others because the POS pattern does not take verbs into account.

In general, the performance of the ranking algorithms is increased when the candidate phrases have a better coverage on assigned keyphrases. As shown in Figure 3, there is a clear impact on the performance of ranking algorithms while the candidate coverage increases on Hulth2003. On SemEval2010, the impact is

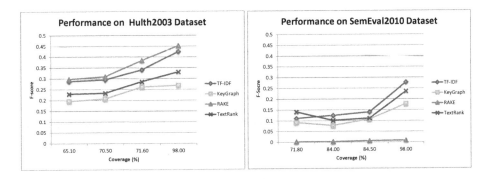

Fig. 3. Ranking algorithm performance

obscure. Further analysis shows that the articles in Hulth2003 have less noisy data than the ones in SemEval2010 set. SemEval2010 consists of full length scientific articles that contain a large number of mathematic equations, figures, numbers, and programming code. The performance of ranking algorithms are heavily affected by these data as none of the selection approaches we implemented has a noisy data cleaning process.

It is worth noting that the evaluation results may not conform with what the authors original claimed due to the different evaluation environments and settings. For example, Mihalcea and Tarau [15] use keyphrase formation approach after ranking.

In short, increasing the candidate coverage on the potential keyphrases leads to a better performance on ranking algorithms, but should be performed in conjunction with differentiating the potential keyphrases from noisy data.

7   Conclusions

In this paper, we have conducted a systematic evaluation of three common candidate phrase selection approaches coupled with four major unsupervised ranking algorithms. The evaluation shows that increasing the candidates' coverage on potential keyphrases results in better performance on the same ranking algorithm. In conclusion, the keyphrase extraction technology can be improved from both candidate selection techniques and the ranking algorithms. In future work, we will focus on developing an improved candidate selection approach with higher coverage and accuracy.

Interested readers are invited to contact the authors to obtain a copy of our software and its documentation.

Acknowledgement. This research was funded by the Australian Postgraduate Awards Scholarship, and a Safety Top-Up Scholarship by The University of Western Australia.

References

1. Daudaravicius, V.: Automatic identification of lexical units. Informatica: An International Journal of Computing and Informatics 34(1), 85–91 (2010)
2. Hasan, K.S., Ng, V.: Conundrums in unsupervised keyphrase extraction: Making sense of the state-of-the-art. In: Proceedings of the 23rd International Conference on Computational Linguistics: Posters, pp. 365–373. Association for Computational Linguistics (2010)
3. Kim, S.N., Medelyan, O., Kan, M.Y., Baldwin, T.: Semeval-2010 task 5: Automatic keyphrase extraction from scientific articles. In: Proceedings of the 5th International Workshop on Semantic Evaluation, pp. 21–26. Association for Computational Linguistics (2010)
4. Frank, E., Paynter, G.W., Witten, I.H., Gutwin, C., et al.: Domain-specific keyphrase extraction. In: Proc. Sixteenth International Joint Conference on Artificial Intelligence, pp. 668–673. Morgan Kaufmann Publishers (1999)

5. Turney, P.D.: Learning algorithms for keyphrase extraction. Information Retrieval 2(4), 303–336 (2000)
6. Jean-Louis, L., Gagnon, M., Charton, E.: A knowledge-base oriented approach for automatic keyword extraction. Computación y Sistemas 17(2), 187–196 (2013)
7. Jiang, X., Hu, Y., Li, H.: A ranking approach to keyphrase extraction. In: Proceedings of the 32nd International ACM SIGIR Conference on Research and Development in Information Retrieval, pp. 756–757. ACM (2009)
8. Hulth, A.: Improved automatic keyword extraction given more linguistic knowledge. In: Proceedings of the 2003 Conference on Empirical Methods in Natural Language Processing, pp. 216–223. Association for Computational Linguistics (2003)
9. Kumar, N., Srinathan, K.: Automatic keyphrase extraction from scientific documents using n-gram filtration technique. In: Proceedings of the Eighth ACM Symposium on Document Engineering, pp. 199–208. ACM (2008)
10. Kim, S.N., Kan, M.Y.: Re-examining automatic keyphrase extraction approaches in scientific articles. In: Proceedings of the Workshop on Multiword Expressions: Identification, Interpretation, Disambiguation and Applications, pp. 9–16. Association for Computational Linguistics (2009)
11. Wang, L., Li, F.: Sjtultlab: Chunk based method for keyphrase extraction. In: Proceedings of the 5th International Workshop on Semantic Evaluation, pp. 158–161. Association for Computational Linguistics (2010)
12. Jones, K.S.: A statistical interpretation of term specificity and its application in retrieval. Journal of Documentation 28(1), 11–21 (1972)
13. Rose, S., Engel, D., Cramer, N., Cowley, W.: Automatic keyword extraction from individual documents. Text Mining, 1–20 (2010)
14. Ohsawa, Y., Benson, N.E., Yachida, M.: Keygraph: Automatic indexing by co-occurrence graph based on building construction metaphor. In: Proceedings of the IEEE International Forum on Research and Technology Advances in Digital Libraries, ADL 1998, pp. 12–18. IEEE (1998)
15. Mihalcea, R., Tarau, P.: Textrank: Bringing order into texts. In: Lin, D., Wu, D. (eds.) Proceedings of EMNLP 2004, pp. 404–411. Association for Computational Linguistics, Barcelona (July 2004)
16. Paukkeri, M.S., Honkela, T.: Likey: Unsupervised language-independent keyphrase extraction. In: Proceedings of the 5th International Workshop on Semantic Evaluation, pp. 162–165. Association for Computational Linguistics (2010)
17. Ortiz, R., Pinto, D., Tovar, M., Jiménez-Salazar, H.: Buap: An unsupervised approach to automatic keyphrase extraction from scientific articles. In: Proceedings of the 5th International Workshop on Semantic Evaluation, pp. 174–177. Association for Computational Linguistics (2010)
18. Liu, Z., Huang, W., Zheng, Y., Sun, M.: Automatic keyphrase extraction via topic decomposition. In: Proc. of the 2010 Conference on Empirical Methods in Natural Language Processing, pp. 366–376. Assoc. for Computational Linguistics (2010)
19. Zervanou, K.: Uvt: The uvt term extraction system in the keyphrase extraction task. In: Proceedings of the 5th International Workshop on Semantic Evaluation, pp. 194–197. Association for Computational Linguistics (2010)
20. Matsuo, Y., Ishizuka, M.: Keyword extraction from a single document using word co-occurrence statistical information. International Journal on Artificial Intelligence Tools 13(01), 157–169 (2004)
21. Bracewell, D.B., Ren, F., Kuriowa, S.: Multilingual single document keyword extraction for information retrieval. In: Proceedings of 2005 IEEE International Conference on Natural Language Processing and Knowledge Engineering, IEEE NLP-KE 2005, pp. 517–522. IEEE (2005)

22. Krapivin, M., Marchese, M., Yadrantsau, A., Liang, Y.: Unsupervised key-phrases extraction from scientific papers using domain and linguistic knowledge. In: Third International Conference on Digital Information Management, ICDIM 2008, pp. 105–112. IEEE (2008)

23. Bordea, G., Buitelaar, P.: Deriunlp: A context based approach to automatic keyphrase extraction. In: Proceedings of the 5th International Workshop on Semantic Evaluation, pp. 146–149. Association for Computational Linguistics (2010)

24. El-Beltagy, S.R., Rafea, A.: Kp-miner: Participation in semeval-2. In: Proceedings of the 5th International Workshop on Semantic Evaluation, pp. 190–193. Association for Computational Linguistics (2010)

25. Dostál, M., Jezek, K.: Automatic keyphrase extraction based on nlp and statistical methods. In: DATESO, pp. 140–145 (2011)

26. Brin, S., Page, L.: The anatomy of a large-scale hypertextual web search engine. Computer Networks and ISDN Systems 30(1), 107–117 (1998)

27. Bird, S., Klein, E., Loper, E.: Natural language processing with Python. O'Reilly (2009)

28. Porter, M.F.: An algorithm for suffix stripping. Program: Electronic Library and Information Systems 14(3), 130–137 (1980)

Methods and Algorithms for Unsupervised Learning of Morphology

Burcu Can and Suresh Manandhar

Department of Computer Science,
University of York, Heslington,
York, YO10 5GH, UK
{burcu,suresh@cs.york.ac.uk

Abstract. This paper is a survey of methods and algorithms for unsupervised learning of morphology. We provide a description of the methods and algorithms used for morphological segmentation from a computational linguistics point of view. We survey morphological segmentation methods covering methods based on MDL (minimum description length), MLE (maximum likelihood estimation), MAP (maximum a posteriori), parametric and non-parametric Bayesian approaches. A review of the evaluation schemes for unsupervised morphological segmentation is also provided along with a summary of evaluation results on the Morpho Challenge evaluations.

Keywords: unsupervised learning, probabilistic models, morphological segmentation, machine learning of morphology.

1 Introduction

Morphology is the study of the internal structure of words. The term *'morphology'* was first introduced by the German linguist August Schleicher in 1859 [67]. Morphology refers to the study of how various sub-word units combine together to form new words through a sequence of rule applications. The sub-word units, called *morphemes*, are the smallest meaning bearing units in a word. For example, the word *interestingly* is made up the morphemes *interest, ing*, and *ly*.

Morphological segmentation is the process of analysing a word by identifying its constituent morphemes. As a computational problem, morphological segmentation has been treated both as a supervised and unsupervised machine learning problem. In this paper, we provide a survey of existing approaches to unsupervised learning of morphology. Unsupervised learning of morphology is attractive for several reasons: 1. it is able to accommodate changes in the language and 2. it does not require manually annotated data which makes it particularly suitable for resource-poor languages.

Morphological segmentation and morphological analysis are essential preprocessing tasks for many NLP applications. *Speech recognition* is one such application that benefits from morphological segmentation as using whole word dictionary becomes problematic especially for morphologically rich languages

A. Gelbukh (Ed.): CICLing 2014, Part I, LNCS 8403, pp. 177–205, 2014.
© Springer-Verlag Berlin Heidelberg 2014

and use of morphemes (or other sub-word unit) sequences rather than word sequences provides better coverage [21,2,47,6,52,65]. *Machine translation* is another field that uses morphological segmentation. Machine translation models either use morphological information within the pre-processing step [13,34,25] in order to prepare the text for the translation, or morphological segmentation is employed as a post-processing step to generate the inflected morphological forms of words [57,48]. *Information retrieval* also benefit from morphological segmentation due to the ambiguity and OOV (out-of-vocabulary) words. Within information retrieval, simple morphological approaches like truncation, stemming, stem generation, or lemmatization are often employed [39,49,42,46]. *Question answering* is another application that benefits from morphological segmentation. In a question answering system, morphological analysis is usually required for extracting questions, as well as for the answers that are retrieved. Similar approaches (i.e. stemming, lemmatization, etc.) to the ones used in information retrieval are adopted in order to obtain morphological information in question answering [7,3].

In this paper, we categorise unsupervised morphology learning methods into the following types:

- *Letter Successor Variety models*: Harris [40], Hafer and Weiss [37], Dejean [26], Bordag [9,10]
- *Minimum Description Length based models*: Brent et al. [12], Goldsmith's Linguistica [31,32], Morfessor Baseline MDL [22], Argamon et al. [1], Kazakov & Manandhar [43,44], Gelbukh et al. [30]
- *Other deterministic approaches*: Bernhard [5], Neuvel and Fulow [61], Keshava and Pitler [45], Monson et al. [58], Lignos et al. [55], Can and Manandhar [14]
- *Maximum likelihood models*: Morfessor Baseline ML [22], Morfessor Categories ML [23], Probabilistic ParaMor [59]
- *Maximum A-Posteriori models*: Morfessor Categories MAP [24]
- *Bayesian parametric models*: Creutz [19], Poon et al. [63]
- *Bayesian non-parametric models*: Goldwater et al. [33], Can and Manandhar [15], Sirts and Alumäe [68], Dreyer and Eisner [28], Snyder and Barzilay [70]

2 Related Work

Hammarström [38] is a survey of the work in morphology learning covering a wide range of work between 1955 and 2006. Hammarström provides a synopsis of the field by categorising the studies into four groups: border and frequency methods that detect the segment boundaries either by investigating the substrings that occur frequently with other adjacent substrings or by using the compression of the frequent long substrings; group and abstract methods that analyse morphologically related words in groups (e.g. paradigms); feature-based methods that see words as consisting of various features; and phonology-based methods that analyse words based on their vowels and consonants. Some prominent examples

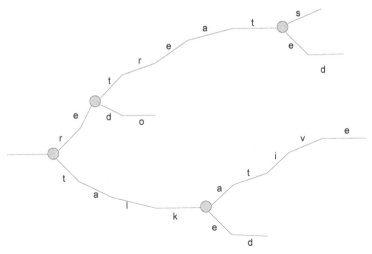

Fig. 1. Word split points in a LSV model

are given for each category. However, the paper does not contain a description of the methods and algorithms employed. It primarily describes the languages that they are tested on, whether the algorithms require any thresholds or parameters to be set by humans, and what the algorithm learns (analysis, paradigms, transducers etc).

Here, our aim is not to survey the same work reviewed by Hammarström from the same perspective. Instead, we aim to focus on the methods and algorithms that have been used for unsupervised morphological segmentation from 1955 till 2013. For this reason, we mainly focus on the methods and algorithms and provide a mathematical overview of the methods from a computational linguistics point of view.

3   Letter Successor Variety (LSV) Models

Harris [40] was the first to introduce the distributional properties of letters within a word and to devise the earliest class of deterministic algorithms for word segmentation. In this model, the potential segmentation points within a word can be characterised by the sharp changes in the number of successors of a letter within a word. For example, a given corpus contains the words *walnut, wall, walks, walked, walking, walk*. The number of letter successors of the prefix *wal-* equals 3, namely, *n, l* and *k*. However, the number of letter successors of *walk* is 4, namely, *s, e, i* and *$* (denoting the word boundary). Harris calls the number of letters that can follow each letter in a word as *successor variety*. Similarly, the letters that precede other letters is called *predecessor variety*.

To determine potential split points, a letter tree (i.e. a trie) is constructed. An example of a letter successor tree is given in Fig. 1. In this example, *re-* is a potential prefix whereas *-s, -ed* and *-ing* are potential suffixes on the tree. Harris chooses a cutoff value manually. However, the cutoff value must be chosen carefully. If it is too small, then words are oversegmented. In contrast, if the cutoff is too big, then most true segments are missed.

The successor counts are applied to all words in the corpus to find morpheme boundaries. For example, the procedure may choose *-ing* as a morpheme. Subsequently, all words that precede *-ing* are considered as stems. However, this is problematics since this will cause the model to segment that do not contain *-ing* as a morpheme such as *sing, string, spring, cling, etc.*

Despite this, many researchers have followed the idea of using statistical properties of letter successors and predecessors to identify potential split points. Hafer and Weiss [37] improve the original idea by using the entropy of the successors and predecessors instead of using raw counts. The letter successor entropy (LSE) of a prefix w is defined as follows:

$$LSE(w) = \sum_{c \in \Sigma} -\frac{f(w_c)}{f(w)} \log_2 \frac{f(w_c)}{f(w)} \tag{1}$$

where Σ is the alphabet, $f(w_c)$ is the number of word entries in the corpus that have prefix w followed by the letter c, and $f(w)$ is the total number of the word entries that begin with w and can be followed with any letter.

Mopheme boundaries typically have high LSE and using it improves detection of real morpheme boundaries from non-boundaries that have lower entropies even though both may have the same letter successor counts.

Dejean [26] improves upon Harris's method by dividing the process into 3 different phases. In the first phase, a morpheme dictionary is constructed by using the letter successor variety technique and choosing only the high frequency morphemes. In the second phase, the words in the corpus are segmented using the morpheme dictionary to generate more morphemes. In the final phase, the corpus is analysed by using the morpheme dictionary. For example, given the words *lights, lighting, lighted, lightly, lightness, lightest, lighten.* In the first phase, the most frequent morphemes are selected such that *-s, -ing, -ed, -ly* that have a higher LSV frequency than a given threshold value. In the second phase *-ness, -est,* and *-en* are captured by segmenting the words *lightness, lightest,* and *lighten.* Finally, the entire corpus is morphologically analysed using the combined morpheme dictionary *-s, -ing, -ed, -ly, -ness, -est, -en.*

For example, the words *started, startled, startling* are segmented as *start+ed, start+led, start+ling* in Harris's approach, whereas in Dejean's approach once the morphemes *-ed* and *-ing* are captured, the words are correctly segmented giving *start+ed, startl+ed, startl+ing.*

Bordag [9] does not use any global LSV cutoff value to segment all the words according to the same threshold. Instead, a local LSV value to segment words that are contextually similar is used. The contextual similarity is intended to group words that are syntactically similar. Thus, the idea is to identify syntactically similar words such as subclasses of *adjectives, verbs* etc. and choose a different *local* LSV cutoff value for each subclass. With this method, orthographically similar words such as *early* and *clearly* are analysed independently since they tend to be contextually different.

Table 1. Local LSV scores of the word *early* [9]

input word:	e	a	r	l	y
final score:		1.0	0.1	1.0	2.0

Table 2. Local LSV scores of the word *clearly* [9]

input word:	c	l	e	a	r	l	y
final score:		0.4	1.2	0.1	0.4	13.4	4.6

Bordag uses the combination of local LSV weights, the inverse bigram weights, in addition to the original LSV score to obtain a combined score. A cutoff threshold is chosen for the combined score. The scores for *ear-ly* (1.2), *clear-ly* (13.4) permit distinguishing the two cases easily (see Table 1 and Table 2).

Bordag [9,10] uses the segmentations produced by the local LSV method to train a classifier. Bordag places the morpheme segmentations on a Patricia trie [60] classifier with their frequencies in order to generalise the results for novel words. An example Patricia trie trained by Bordag [10] is given in Figure 2. If a novel word is to be analysed, the *trie* is searched from the root until the correct branch in the trie is found which gives a split for the word. For example, for the novel word *strong*, the trie gives 0.4 by looking at the *root* node only. However, for the novel word *strongly*, the trie gives 0.66 by looking at the *earl* node. Using *tries* helps to handle exceptions as well. For example, a trie with the words *clear+ly*, *strong+ly* and *early* can classify hundreds of words ending with *-ly*, but still remembers one exception which is *early*.

4 Minimum Description Length (MDL) Based Models

According to the MDL principle, the best description of data or the best hypothesis is the one that leads to the best compression of the data. In order to find the best compression of data, the regularities in data need to be captured, as stated by Grünwald [35]:

> "[The MDL Principle] is based on the following insight: any regularity in a given set of data can be used to compress the data, i.e. to describe it using fewer symbols than needed to describe the data literally."
> Grünwald [35]

From a Bayesian perspective, MDL can be viewed as a prior on the model M:

$$
\begin{aligned}
\arg\max_{M} p(M|D) &= \arg\max_{M} \log_2 p(M|D) \\
&= \arg\min_{M} [-\log_2 p(M|D)] \\
&= \arg\min_{M} -log_2 \frac{p(D|M)p(M)}{p(D)} \\
&\propto \arg\min_{M} -\log_2 [p(D|M)p(M)]
\end{aligned}
$$

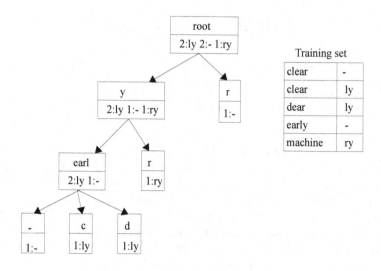

Fig. 2. A sample Patricia trie trained on the training set that contains *clear, clearly, dearly, early,* and *machinery* [10]

Table 3. Input Words   **Table 4.** Stem Table   **Table 5.** Suffix Table

walk	referral
walks	refer
walked	refers
walking	dump
referred	dumps
referring	preferential

stem	code
walk	1
referr	2
refer	3
dump	4
preferenti	5

suffix	code
ϵ	1
s	2
ed	3
ing	4
al	5

Table 6. Encoded Words

stem	suffix	stem	suffix
00	00	01	110
00	01	100	00
00	100	100	01
00	101	101	00
01	100	101	01
01	101	1100	110

Hence, maximising the posterior probability of a model M given data D is equivalent to minimising the description length of the model times the model likelihood. Equivalently, MDL can be thought as an information theoretic regularisation prior within a MAP estimation model.

Brent et al. [12] encodes the stems and suffixes as binary codes and the encodings are kept in tables (see Tables 3, 4, and 5). The most frequent stems and suffixes are encoded with shorter encodings. The Shannon-Fano (SF) coding [12] is used in order to find the optimal length of each code word. The description length (DL) in bits for the SF coding for a morpheme, m, can be approximated with the negative binary logarithm of its relative frequency:

$$DL(m) = -log_2(freq(m)) \tag{2}$$

$$p(M) = \sum_{m \in M} DL(m) \tag{3}$$

A key problem with the approach is that searching through all possible models is not practical. For example, the number of the possible splits of a given text is equal to the product of the lengths of all words in the text. Instead of searching all possible splits of a given text, some heuristics such as first finding the suffix table and then searching for the stem table are employed in Brent's approach.

Linguistica [31,32] is another system that is based on MDL. In addition to using stem and affix codebooks, Linguistica employs *signatures* to encode the data. A *signature* represents the inner structure of a list of words that have similar inflective morphology. Thus their model consists of: a stem list, an affix list, and a signature list (see Figure 3).

The signature list contains only pointers to stems and affixes[31] and can be thought as an optimal encoding of the signature list. The probability of a segmentation $w = t + f$ is given by:

$$p(w = t + f | \sigma) = p(\sigma)p(t|\sigma)p(f|\sigma) \tag{4}$$

where $p(\sigma)$ is the empirical frequency of the signature σ (normalised); and $p(t|\sigma)$, $p(f|\sigma)$ are the empirical stem, and suffix frequencies (normalised) given the signature σ.

In terms of description length, the size of a word becomes the sum of the size of the pointer to its signature, stem, and affix. For the size, inverse logarithm is used as given in Equation 2. The description length of a corpus is computed through all words in the corpus.

In order to compute the length of the model, the lengths of all lists are added up:

$$DL(M) = DL(T) + DL(F) + DL(\Sigma) \tag{5}$$

where T is the stem list, F is the suffix list, and Σ is the signature list in the model. Here, the length of each list is the length of each item in the list plus the number of occurrences of each item in the list. Therefore, the description length of a stem list becomes:

$$DL(T) = log_2(|T|) + \sum_{t \in T} len(t) \tag{6}$$

where $log_2(|T|)$ computes the information needed for the number of items in the stem list and $len(t)$ is the number of bits needed for the stem t, which is computed as follows:

$$len(t) = |t| \ log_2 26 \tag{7}$$

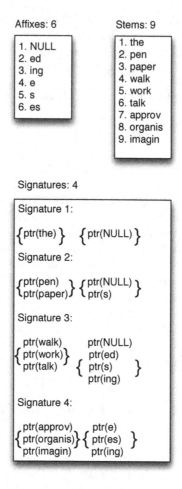

Fig. 3. A sample morphology from Linguistica, that can generate the words: *the, pen, pens, paper, papers, walk, walked walking, walks, work, worked, working, works, talk, talked, talking, talks, approve, approves, approved, organise, organises, organised, imagine, imagines, imagined*

where $|t|$ is the number of letters in the stem t by considering a language with 26 letters. The length of a list of affixes is calculated analogously.

In order to calculate the length of a signature list, the length of a pointer has to be determined since the signatures only keep the pointers to the stems, affixes, and other signatures. The length of a pointer to a stem t, suffix f, and signature σ are computed as follows respectively:

$$\log \frac{|W|}{freq(t)}, \ \log \frac{|W|}{freq(f)}, \ \log \frac{|W|}{freq(\sigma)}$$

where $|W|$ is the number of words in the corpus and $freq()$ gives the number of occurrences of the given segment in the corpus.

Goldsmith also defines a recursive segmentation procedure that segments words with multiple split points. A flag for each stem is placed in the stem list to determine whether the stem is a simple stem or a complex stem with a triple pointer to a signature, stem, and affix. This modification in the definition of a stem enables the analysis of words such as *[organis-ation]-s* where the stem *organis-ation* is decoded as a complex stem that consists of a pointer to a signature which includes the stem *organis* and the affix *-ation*.

Morfessor Baseline defines the total cost as follows:

$$Cost = DL(D) + DL(M)$$
$$= \sum_{i \in D} -\log p(m_i) + \sum_{j \in M} len(m_j) \tag{8}$$

where m_i denotes the morphemes and $p(m_i)$ denotes the maximum likelihood estimate of the morpheme m_i. The maximum likelihood estimate of a morpheme m_i is the number of token count for m_i divided by the total number of token counts in the corpus. Here the corpus is generated by morphemes in the model. Hence, the length of a corpus is computed by the maximum likelihoods of the morphemes. Morfessor Baseline deploys a recursive segmentation where each

Table 7. First-order (Prolog) decision-list rules learnt in Kazakov and Manandhar [44]. Exceptions are towards the top and generic rules are towards the bottom.

1. seg(A,B) :-	append([b,l,e,s,s], B, A), !.
2. seg(A,[a,i]) :-	append(_,[a,i],A), !.
3. seg(A,B) :-	append([c,o,m,t,e], B, A),
	append(C,[e,z],A), !.
4. seg(A,B) :-	append([o,r,g,a,n,i,s], B, A),
	append([o,r,g,a,n,i,s, a], C, A), !.
5. seg(A,[a]) :-	append(_, [a], A),
	append(_, [i,r,a], A), !.

discovered morpheme is analysed recursively as long as it improves the cost. The method does not make use of signatures like Linguistica, instead a single *codebook* is used. A similar approach for recursive segmentation has also been used by Argamon et al. [1].

Kazakov & Manandhar [44] develop a hybrid combination of genetic algorithms and inductive logic programming (ILP). A MDL bias is employed within a genetic algorithm by choosing a suitable fitness function that favours codebooks with shorter description length. The genetic algorithm generates an initial segmentation. In the following step, segmentation rules are learned from the initial segmentations by employing a first-order decision list learner [56]. The decision-list is able to generalise by learning rules for the segmentation of unseen words.

The use of first-order decision lists has two advantages. Firstly, the decision lists easily capture regular expression patterns over which a given segmentation rule applies. Secondly, decision-lists provide a natural mechanism for capturing exceptions since decision-lists are ordered (in terms of priority). Some examples of rules learnt are given in Table 7.

5   Other Deterministic Approaches

We review deterministic methods that do not fall into the categories covered in the previous sections.

Neuvel and Fulow [61] propose an algorithm based on the word-based theory of morphology [29]. In this approach, instead of inducing the morphemes, morphological relations between the words are defined to learn new word forms.

Keshava and Pitler [45] describe an algorithm, RePortS, that is based on using a trie. A forward trie is used for the suffixes, whereas a backward trie is used for the prefixes. Keshava and Pitler define heuristic criteria based on the strings' conditional probabilities on the trie, to identify the suffixes and prefixes by giving them scores. These heuristics are improved by Demberg [27] for handling complex morphology. Lavallée and Langlais [53] use formal analogies to find the relation between 4 word forms, such as {walking, speaker, walks, speaks}. However, due to the large search space, such methods can be considered impractical for large lexicons.

Bernhard [5] uses features that combine the length and frequency of morphemes. Stems are generally longer and less frequent than suffixes, whereas suffixes are shorter and more frequent than stems. In order to extract the prefixes and suffixes, the transitional probabilities between substrings are used. First, for each position of the word k the following function is computed:

$$f(k) = \frac{\sum_{i=0}^{k-1} \sum_{j=k+1}^{n} max[p(s_{i,k}|s_{k,j}), p(s_{k,j}|s_{i,k})]}{k(n-k)} \qquad (9)$$

which gives the mean of the maximum transition probabilities for the position k. Here the transition probability $p(s_{i,k}|s_{k,j})$ is estimated as follows:

$$p(s_{i,k}|s_{k,j}) = \frac{f(s_{i,j})}{f(s_{k,j})} \qquad (10)$$

where $f(s_{i,j})$ is the frequency of the substring $s_{i,j}$ and the transition probability $p(s_{i,k}|s_{k,j})$ is estimated as follows:

$$p(s_{k,j}|s_{i,k}) = \frac{f(s_{i,j})}{f(s_{i,k})} \qquad (11)$$

Local minima of the values of $f(k)$ in a given word correspond to potential morpheme boundaries. Once the morpheme boundaries are found, the longer and less frequent morphemes are chosen as stems and the rest chosen as either

Table 8. A sample subgroup of words that contains the stem *hous* and starts with the empty prefix [5]

Words	Suffixes	Potential stems	New suffixes
housekeeping		-ekeeping	
housing	-ing		
household		-ehold	
house's			-e's
house	-e		
housed			

prefix or suffix depending on its position. Different words sharing the same stem are compared to find other segments.

ParaMor is a system developed by Monson et al. [58] that discovers candidate suffixes and stems to build paradigms. In their approach, candidate suffixes are any final substrings of words that are found iteratively. Once partial paradigms are built, they are merged by clustering. Eventually, words are segmented by stripping off suffixes that occur in these paradigms. The system is rule based and does not involve a confidence measure. Moreover, the authors combine the results of the ParaMor with Morfessor [20] (named as P+M model). The joint P+M model outperforms other ParaMor variants in several Morpho Challenge evaluations (see Section 11) in terms of f-score.

Lignos et al. [55] employ Base and Transforms model [16] that is based on the discovery of the base and derived forms of words. The discovery is performed through transforms, which are orthographic modifications that are applied on a word to derive another form of the same word. A transform given by (s_1, s_2) removes the suffix s_1 from the word and adds another suffix s_2 to derive another form of the word. Lignos [54] develops an inference procedure that can learn the base form of a word when it is absent in the corpus. The new model handles compounding by decomposing a word into its component words by choosing the highest geometric mean of the component frequencies.

Can and Manandhar [14] propose a deterministic model that makes use of syntactic categories. Syntactic information and morphology are strongly connected to each other. For example, words ending with *-ly* are generally adverbs, words ending with *-ed* are generally verbs, etc. Syntactic categories are induced using context distribution clustering [17]. Potential suffixes in each syntactic category are ranked by their conditional probability $p(m|c)$ where m denotes the suffix and c denotes the syntactic category. The definition of a morphological paradigm is somewhat different to that of others. Each paradigm consists of a list of morpheme/cluster pairs, $m_i/c_{i,}$, and a list of stems, s_i. A paradigm, P, has the form:

$$P = < \{m_1/c_1, \ldots, m_r/c_r\}\{s_1, \ldots, s_k\} >$$

For example, two sample paradigms are:
$P_1 = < \{s/2, ing/1\}\{walk, fight, repeat, play\} >,$

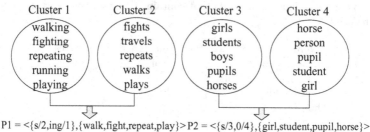

Fig. 4. Paradigm capturing across syntactic categories in the deterministic approach by Can and Manandhar [14]

$P_1 =< \{s/3, 0/4\}\{girl, student, pupil, horse\} >$ (see Figure 4).

Suffix pairs that have the maximum number of common stems across two different syntactic categories are merged and a new paradigm is created (see Figure 4). Once the initial morphological paradigms are learnt, they are merged based on their *accuracy* (*Acc*) as defined below:

$$Acc_1 = \frac{S}{S+N_1}, \ Acc_2 = \frac{S}{S+N_2}, \ Acc = \frac{Acc_1+Acc_2}{2} \tag{12}$$

where S is the number of common stems between the two paradigms, N_1 is the number of stems that are present in the first paradigm, but absent in the second paradigm (and vice versa for N_2). Higher values of N_1 and N_2 will result in smaller *Acc* scores and correspondingly lower possibility of merging. Similarly, higher values of S will be preferred for merging. The merging process creates increasingly more general paradigms. The results clearly demonstrate that using syntactic information can help morphological segmentation:

6   Methods Based on Maximum Likelihood (ML)

Within Bayesian statistics, Maximum Likelihood (ML) estimation provides, conceptually, the simplest inference procedure for learning models that generalise from data. In morphological segmentation, typically, the model is a probability assignment to possible morphemes. In ML estimation, there is no prior bias towards any model, and the model M that maximises the likelihood function is chosen:

$$M_{ML} = \arg\max_i p(D|M_i) = \arg\max_i log(p(D|M_i)) \tag{13}$$

In Morfessor Baseline ML [22], a model M_i gives a probability distribution over a collection of morphemes. Given such a model, a corpus can be split into its constituent morphemes:

$$log(p(D|M_i)) = \sum_{m\in D} log\, p(m|M_i) \tag{14}$$

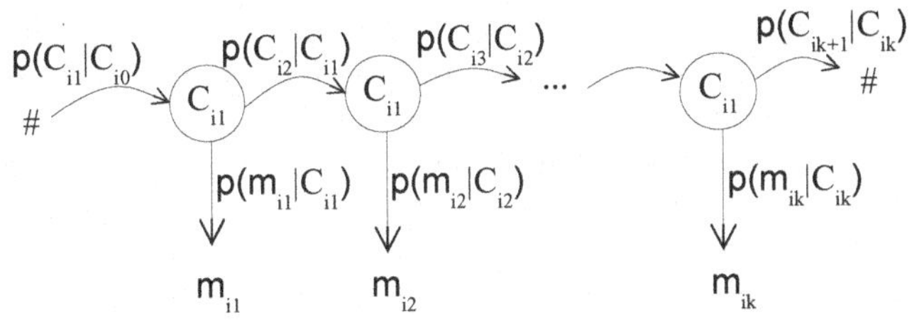

Fig. 5. Transition and emission probabilities of a word w according to Equation 15

As this is ML estimation, the model prior is not involved. Initially, words are split with the suffix length drawn from a Poisson distribution. The algorithm employs two *hard* conditions that always reject rare morphemes and single letter morphemes. In that case, word is re-segmented randomly. Otherwise, the segmentation is accepted. An Expectation Maximization (EM) algorithm is employed to find increasingly better segmentations. The inference involves a number of iterations in which 1. the morpheme probabilities are estimated for a given segmentation 2. the text is re-segmented by using the Viterbi algorithm in order to find the segmentation with the lowest cost for each word 3. the segmentation of the word is either accepted or rejected.

The results show that ML approach tends to overspilt when compared to the MDL approach [22]. For example, the word *affectionate* is split as *affecti+on+at+e* in ML approach, where as it is split as *affect+ion+ate* in MDL approach.

Morfessor Categories ML [23] is a Morfessor variant that is also based on ML estimation. In contrast to Morfessor Baseline ML, a hidden Markov model (HMM) is used to assign probabilities to each possible split of a word form. In the model, each morph is emitted from a hidden state that can be interpreted as either prefix, suffix, stem etc. Within a bigram model, the probability of a segmentation of a word w into the morphemes $m_1, m_2, \ldots, m_k$ is computed as follows:

$$p(m_1, m_2, \ldots, m_k | w) = [\prod_{i=1}^{k} p(C_i | C_{i-1}) p(m_i | C_i)] p(C_{k+1} | C_k) \qquad (15)$$

To learn the HMM transition probabilities, $p(C_i | C_{i-1})$, and the emission probabilities, $p(m_i | C_i)$ (see Figure 5), words are initially segmented by applying the Morfessor Baseline ML [19]. Category membership probabilities $p(C_i | m_i)$ are estimated using a *perplexity* measure. The perplexity measure expresses the predictability of the preceding and following words of a given word. EM is employed to estimate the probabilities in each iteration after re-tagging the words using the Viterbi algorithm.

Morfessor Categories ML improves upon the Morfessor Baseline for English. Alhough, the Baseline performs slightly better precision, the recall of the Categories ML model is a lot better than the baseline model. In Finnish, for smaller datasets Morfessor Categories ML and Baseline perform on a similar level, however for bigger datasets Morfessor Categories ML performs far better. This work shows that the dependencies between the morphemes play an important role in morphology learning.

Probabilistic ParaMor [59] extends the original ParaMor [58] algorithm by training a finite-stage tagger that will mimic the results of the original ParaMor. The statistical model learns whether each character in a word is the beginning of a new stem or a suffix. The surrounding characters and morpheme-tags (i.e. stem vs suffix) are used as features in the tagger. For the surrounding characters, character unigram, bigram, and trigram morpheme tags are used. For example, in the word *strongly*, the character features for the letter *'o'* consist of *'stro'*, *'tro'*, *'ro'*, *'o'*, *'on*, *'ong'*, *and 'ongly'*. Monson et al. [58] use the averaged perceptron algorithm [18] to train the finite-state tagger. Viterbi search is used for the decoding process. Eventually, the tagger tags each split point within a word as a morpheme boundary or as a continuation of a morpheme. Therefore, the segmentation process is akin to a part-of-speech tagging process.

The probabilistic ParaMor has a higher accuracy compared to the baseline ParaMor. Moreover, the authors combine the results of the baseline ParaMor with Morfessor [20] to train the tagger (named as P+M Mimic model).

7   Methods Based on Maximum A-Posteriori (MAP) Estimation

In contrast to ML estimation, the *maximum a-posteriori estimation* (MAP) approach allows specifying model prior, $p(M_i)$.

$$M_{MAP} = \arg \max_i p(D|M_i)p(M_i) \tag{16}$$

The MDL models described in Section 6 can be viewed as MAP estimation models with description length (DL) as the the model prior. In this section, we focus on model priors other than those based on DL.

Morfessor Categories MAP [24] employs a first-order HMM in order to model the internal word syntax as given in Figure 5. Morfessor Categories MAP defines a prior for each morpheme using two parameters: meaning and form. The *form* of a morpheme refers to the substructure of the morpheme (made of letters or made of two sub-morphemes). The *meaning* of a morpheme consists of the features such that length, frequency and right/left perplexity of the morpheme. Therefore, the prior probability of a model, M, becomes the combination of the meaning and the form of each morpheme, m_i:

$$p(M) = |M|! \prod_{i=1}^{M} M[p(meaning(m_i))p(form(m_i))] \tag{17}$$

The term $|M|!$ accounts for the $|M|!$ possible orderings of the morphs in the model. Thus, the prior favours smaller number of morphemes.

In order to find the model and the corpus segmentation with the minimum cost, a greedy search algorithm is used in Morfessor Categories MAP. In each

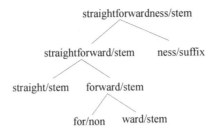

Fig. 6. The hierarchical segmentation of the English word 'straightforwardness' by the Morfessor Categories MAP [24]

step, different segmentations for each word are suggested and the one with the maximum probability is chosen. The segmentation of each word is kept in a binary splitting tree. Figure 6 provides an example.

The results for Morfessor Categories MAP are below that of the Morfessor Categories ML. However, the effects of different types features within the prior in MAP models is yet to be explored.

8   Bayesian Parametric Models

Bayesian modelling employs the full form of Bayes' theorem that defines a posterior probability distribution over the parameters in terms of the likelihood $p(D|M)$ and the prior model probability $p(M)$:

$$p(M|D) = \frac{p(D|M)p(M)}{p(D)} \qquad (18)$$

Both ML and MAP estimates are point estimates that correspond to the modes of the above distribution. Bayesian modelling introduces a different perspective by representing the estimate in the form of a probability distribution rather than a single point estimate.

One common way to estimate the parameters is to draw random samples from the posterior distribution. Markov Chain Monte Carlo (MCMC) methods are the most common methods employed for sampling from the underlying posterior probability distribution. Samples drawn from the posterior distribution form a Markov chain such that each state is dependent only on the previous state:

$$p(X_{n+1} = x|X_1 = x_1, \ldots, X_n = x_n) = p(X_{n+1} = x|X_n = x_n) \qquad (19)$$

The Markov chain converges to a distribution over states, called an equilibrium. Gibbs sampling and Metropolis-Hastings algorithm are the two common MCMC algorithms used for learning segmentation.

Creutz [19] proposes a generative probabilistic model that is intended to overcome the over-segmentation problem in Baseline Morfessor. The proposed model uses prior information on the morpheme lengths and morpheme frequencies, within a generative probabilistic model framework. The model is based on the probabilistic model by Brent [11].

The generative story can be told as follows. The total number of morphemes n is sampled with a uniform distribution. Morpheme lengths l_{m_i} are then drawn from a gamma distribution:

$$p(l_{m_i}) = \frac{1}{\gamma(\alpha)\beta^\alpha} l_{m_i}^{\alpha-1} e^{-l_{m_i}/\beta} \tag{20}$$

where α and β are constants, and γ is the gamma function. Once the lengths are drawn, the letters that each morpheme consists of are drawn according to the maximum likelihood of each letter c_j:

$$p(c_j) = \frac{n_{c_j}}{\sum_k n_{c_k}} \tag{21}$$

where n_{c_j} is the frequency of the letter c_j in the corpus, and $\sum_k n_{c_k}$ is the total number of letters in the corpus. Finally, the model/lexicon is created with these morphemes regardless of the order they are created:

$$p(M) = p(n)\, n! \prod_{i=1}^{n} p(m_i) \tag{22}$$

$$p(m_i) = p(l_m) \prod_{j=1}^{l_{m_i}} p(c_j) \tag{23}$$

where n is the number of morphemes in the model, l_m is the length of each morpheme and c_j denotes the letters within morphemes.

Once the model is created, corpus requires to be built by using the morphemes in the model. First, morpheme frequencies are determined by Mandelbrot's correction of Zipf's formula (see Baayen [4]). Finally, the corpus is created according to a particular order by using the inverse of the multinomial:

$$p(Corpus) = \left(\frac{(\sum_{i=1}^{n} f_{m_i})!}{\prod_{i=1}^{n} f_{m_i}!} \right)^{-1} \tag{24}$$

where the numerator is the summation of the morpheme frequencies in the model and the denominator is the product of the factorial of the frequency of each morpheme in the model. The optimal model is searched following a similar recursive search algorithm which is used in the Baseline Morfessor [22]. Results show that the usage of prior information increases the accuracy of the algorithm.

Poon et al. [63] develop a log-linear model where the joint probability between the corpus and all possible segmentations is defined. Since it is not possible to derive all the pairs belonging to the joint probability, a normalisation constant Z is estimated to normalise the joint probability. A few techniques are suggested earlier to compute the normalisation constant. Smith and Eisner [69] apply contrastive estimation by searching around the neighbourhood of the data, whereas

Rosenfeld [66] and Poon et al. [64] use sampling to compute the normalisation constant. Poon et al. use both contrastive estimation and sampling to compute the normalisation constant. The neighbourhood is searched by transposing pairs of letters to create invalid words. Gibbs sampling is used to find the optimum segmentation. In the model, also a prior information that is inspired by the MDL model which controls the number of morpheme types in the lexicon and the morpheme tokens in the corpus is used.

9   Bayesian Non-parametric Models

Bayesian non-parametric models potentially permit an infinite number of parameters to be learnt. In other words, in a non-parametric model, the number of parameters can grow with data. Typically, for example, within morphological segmentation, the number of morpheme classes is not known in advance. Thus, rather than fixing the number of classes in advance, non-parametric models provide a more realistic and flexible framework to capture the irregularities in data by permitting flexibility in the parameter space.

A well-known approach in Bayesian non-parametric modelling is the *Dirichlet Process*. A Dirichlet process defines a probability distribution over an infinite number of objects [62].

Given data points $x = \{x_1, \ldots, x_N\}$ generated from a Dirichlet process $DP(\alpha, H)$ with a concentration parameter α and a base distribution H (see Figure 7 for the plate diagram):

$$x_i \sim G$$
$$G \sim DP(\alpha, H)$$

$$\tag{25}$$

the probability of a future observation $x_{N+1} = j$ is given by [8]:

$$p(x_{N+1} = j | x, \alpha, H) = \frac{1}{N+\alpha} \sum_{i=1}^{N} I(x_i = j) + \frac{\alpha}{N+\alpha} H(j)$$
$$= \frac{n_j + \alpha H(j)}{N + \alpha}$$

$$\tag{26}$$

Here I is an indicator function that outputs 1, if $x_i = j$, otherwise it outputs 0.

This formulation of the Dirichlet process leads to the Chinese Restaurant Process (CRP). Imagine a restaurant that consists of an infinite number of tables with an infinite number of seats at each table where each customer chooses a table and sits down(see Figure 8). At each table, a (possibly) different type of meal is served. The customer chooses an occupied table with a probability which is proportional to the number of customers who are already sitting at the table,

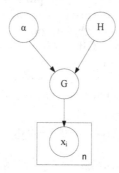

Fig. 7. Plate diagram of a Dirichlet process: $DP(\alpha, H)$ that produces x_i for n times by using the concentration parameter α and the base distribution H

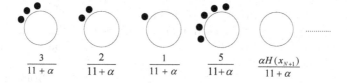

$$\frac{3}{11+\alpha} \qquad \frac{2}{11+\alpha} \qquad \frac{1}{11+\alpha} \qquad \frac{5}{11+\alpha} \qquad \frac{\alpha H(x_{N+1})}{11+\alpha}$$

Fig. 8. An illustration of the Chinese Restaurant Process. The new customer x_{N+1} sits at a table which is already occupied with a probability proportional to the number of customers sitting at the table; which is $\frac{3}{11+\alpha}$, $\frac{2}{11+\alpha}$, $\frac{1}{11+\alpha}$, and $\frac{5}{11+\alpha}$ respectively. The customer sits at a table which is empty with a probability proportional with the concentration parameter; which is $\frac{\alpha H(x_{N+1})}{11+\alpha}$.

whereas she chooses an empty table with a probability proportional to a defined constant α. Therefore, tables which have a great number of customers attract more customers according to the rich-get-richer principle.

Goldwater et al. [33] introduce a two stage model that extends the Chinese restaurant metaphor, where each table is labelled with a word from a corpus. In their model, initially these labels are generated by a generator component that draws the labels from a multinomial distribution:

$$p(l_k = w) = \sum_{c,t,f} I(w = t + f)p(c_k = c)p(t_k = t|c_k = c)p(f_k = f|c_k = c) \quad (27)$$

where c denotes the class label (which involves a distribution over stems and suffixes), t denotes the stem, and f denotes the suffix that belongs to word w having the label l_k. According to the generative story, first the class label, c_k, is drawn, then the stem, t_k, and suffix, f_k, of the word are drawn conditionally with the class label. Each of these are drawn from multinomial distributions with symmetric Dirichlet priors as follows:

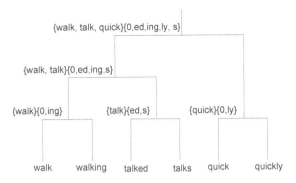

Fig. 9. A sample tree structure

$$x_k \sim Multinomial(\theta)$$
$$\theta \sim Dirichlet(\beta)$$

$$(28)$$

In the second stage, the actual sequence of words is generated by estimating the frequencies of the words in order to create a power-law distribution. Goldwater et al. [33] use Pitman-Yor process [41][1] for generating the ith word conditioned on all previous words:

$$p(w_i = w | \boldsymbol{w_{-i}}, \boldsymbol{z_{-i}}, \theta) = \sum_{k=1}^{K(\boldsymbol{z}_{-i})} \frac{n_k^{(\boldsymbol{z}_{-i})} - a}{i - 1 + b} I(l_k = w) + \frac{K(\boldsymbol{z}_{-i})a + b}{i - 1 + b} \theta_w \quad (29)$$

where z_i denotes the class that generates the ith word, l_k denotes the multinomial distribution over words that belong to the class k, $\boldsymbol{w_{-i}}$ represent the previously generated words, $\boldsymbol{z_{-i}}$ denotes the current seating arrangement, a and b are the parameters of the process, and $K(\boldsymbol{z}_{-i})$ is the number of tables that are occupied. The approach allows different analyses for different tokens of the same word, however only one split point is generated for each word.

Can and Manandhar [15] propose a Dirichlet Process based approach that learns morphological paradigms (see Figure 9). In their approach, morphological paradigms are learned in a hierarchical structure where each node corresponds to a morphological paradigm. The likelihood of data under any subtree is defined recursively by:

$$p(D_k | T_k) = p(D_k)p(D_l | T_l)p(D_r | T_r) \quad (30)$$

where the probability is defined in terms of left T_l and right T_r subtrees. Thus, the likelihood is decomposed recursively until the leaf nodes are reached. The marginal probability is used as prior information since it bears the probability

---

[1] The Pitman-Yor process is a generalisation of the Dirichlet process (see [41]).

of having the data from the left and right subtrees within a single cluster. The marginal likelihood of words in the node k is defined such that:

$$p(D_k) = p(S_k)p(M_k)$$
$$= p(s_1, s_2, \ldots, s_n)p(m_1, m_2, \ldots, m_n)$$

where $s_1, s_2, \ldots, s_n$ are the stems and $m_1, m_2, \ldots, m_n$ are the suffixes in the node/paradigm k.

Can and Manandhar define two Dirichlet processes to generate stems and suffixes in each node on the hierarchical structure independently:

$$G_s|\beta_s, P_s \sim DP(\beta_s, P_s)$$
$$G_m|\beta_m, P_m \sim DP(\beta_m, P_m)$$
$$s|G_s \sim G_s$$
$$m|G_m \sim G_m$$

where $DP(\beta_s, P_s)$ denotes a Dirichlet process that generates stems and $DP(\beta_m, P_m)$ denotes a Dirichlet process that generates suffixes, where β_s and β_m are the concentration parameters that determine the number of stem/suffix types in the model. P_s and P_m are the base distributions on the letters that each morpheme consists of, where letters are assumed to be distributed uniformly. Therefore, morphemes having shorter lengths are favoured.

Sirts and Alumäe [68] introduce a non-parametric Bayesian approach for jointly learning morphological segmentation of words along with their part-of-speech tags. Sirts and Alumäe employ a trigram hidden Markov model (HMM) for the part-of-speech tags. The trigram transitions are modelled by hierarchical Dirichlet process (HDP):

$$G^U \sim DP(\alpha^U, H) \tag{31}$$
$$G_j^B \sim DP(\alpha^B, G^U) \tag{32}$$
$$G_{jk}^T \sim DP(\alpha^T, G_j^B) \tag{33}$$

where G_U, G_j^B, and G_{jk}^T are unigram, bigram, and trigram DP's. Unigram DP is used as a base distribution for the bigram DP, where bigram DP is used as a base distribution in the trigram DP. This forms an HDP model. The emission probabilities are modelled with a simple Multinomial-Dirichlet conjugacy. Finally, the segmentations are also modelled as a HDP:

$$G^S \sim DP(\alpha^S, S) \tag{34}$$
$$G_j^{TS} \sim DP(\alpha^{TS}, G^S) \tag{35}$$

where G^S is the common base distribution that is used as a base distribution for the tag-specific DP G_j^{TS} defined for the morphological segments. Here, S is the general base distribution and consists of two components: a geometric distribution over the segment lengths and collapsed Dirichlet-multinomial over character unigrams. Sirts and Alumäe sample tags and morphological segments jointly in their inference algorithm. The results show that learning morphological segments jointly with the part-of-speech tags improve the segmentation. When the tags are fixed and only the morphological segments are learned, it gives lower scores, whereas when both are learned jointly, the results are comparably higher.

Dreyer and Eisner [28] propose an infinite Diriclet mixture model for learning the part-of-speech tag, inflection, lexeme, morphological paradigm of each word in the corpus. For example, *learned* belongs to a verb part-of-speech class, it has past participle inflection, belongs to the lexeme *learn*, and belongs to a morphological paradigm that consists of words *learn, learns, learned, learning*. Dreyer and Eisner construct morphological paradigms via an infinite Dirichlet process mixture model, where each paradigm corresponds to a mixture component having the forms of the same lexeme and word tokens are generated from each paradigm.

10   Evaluation of Morphology Segmentation Algorithms

The evaluation of morphological segmentation requires a gold standard to compare with the suggested analyses, common with most natural language processing tasks. The evaluation process, at first glance, appears straighforward as system generated segmentation need to match the split points in the gold standard. However, especially in morphologically complex languages, additional features such as morphological ambiguity, morphophonology, stem changes etc. can be present. Taking these into consideration, obtaining a gold standard in a range of languages is a demanding task.

Spiegler et al. [71] define the features of a good evaluation metric as:

- Correlating well with other NLP tasks.
- Being computationally easy.
- Being robust.
- Being informative about the strengths and weaknesses of the system.
- Being able to account for the linguistic structure of the language, such as morphophonology, allomorphy, syncretism, and ambiguity.

The evaluation methods for morphological segmentation can be investigated using two categories: intrinsic methods based on a comparison against a gold standard, and, extrinsic methods based on evaluating how the segmentation improves the performance of a NLP task.

Evaluation Using a Gold Standard Segmentation. For morphological segmentation, precision, recall and f-score are predominantly used evaluation measures, as in many NLP tasks. F-score is computed as the harmonic mean of precision and recall scores:

$$F\text{-}score = \frac{1}{1/Precision + 1/Recall} \tag{36}$$

Some researchers have used a gold standard consisting of segmentation of all words in a corpus [33,63]. For this purpose, either a highly accurate morphological analyser is used (for Arabic such as the one by Habash and Rambow [36], or some heuristics are used for the construction of a gold standard (for English, see Goldwater et al. [33].

Instead of using the full corpus for evaluation, Morpho Challenge [51] uses a sampled set of gold standard words for evaluation. In both cases, the gold standard consists of words with their segmentations plus additional morphological information.

For example, given below is example segmentation data from Morpho Challenge. Here morpheme labels represent inflection classes; i.e. plural, past tense form, participle etc.:

ablatives ablative:ablative_A s:+PL
abounded abound:abound_V ed:+PAST
carriages carri:carry_V age:age_s s:+PL
detraction detract:detract_from_V ion:ion_s
entitling entitl:entitle_V ing:+PCP1

To measure precision, from the system generated segmentation of the test words. For each morpheme in the list, another word is found that includes the same morpheme. This will create a word pair list. Finally, word pairs are checked in the gold standard to see whether the pairs share a common morpheme. For each true guess, one point is given. The score is computed by dividing the total number of received points by the number of sampled words. Recall is measured analogously to precision, where the word pairs are sampled from the gold standard, and comparisons are made through the resulting segmentations.

Spiegler and Monson [71] propose a novel evaluation metric called EMMA that does not perform a one-to-one comparison with the gold standard data, but instead finds the maximum match between the suggested segmentations and the gold standard segmentations using an optimal maximum matching (in a bipartite graph).

Evaluation via Other Tasks. Another way of evaluating the results of a morphological segmentation is to embed the suggested segmentations into a real world NLP task which utilises the analysed words. In addition to the traditional evaluation metric which is described earlier, Morpho Challenge [50] performs information retrieval and machine translation tasks. In both tasks, words are replaced with the word segmentations. In information retrieval, queries are replaced with their segmentations, whereas in the machine translation task the source language is replaced with its segmentations. Finally, the tasks are evaluated using average precision and *BLEU* score respectively.

11   Evaluation Results in Morpho Challenge

We summarise the Morpho Challenge results for 2007, 2008, and 2009 here to give a better comparison between the models in terms of their accuracy. A wide

Table 9. Comparison of methods competed in Morpho Challenge between years 2007 and 2009 for the English language

Method	2007			2008			2009		
	P	R	F	P	R	F	P	R	F
Bernhard 1 [5]	72.05	52.47	60.72	-	-	-	75.61	57.87	65.56
Bernhard 2 [5]	61.63	60.01	**60.81**	-	-	-	67.42	65.11	66.24
Bordag 5 [9]	59.80	31.50	41.27	-	-	-	-	-	-
Bordag 5a [9]	59.69	32.12	41.77	-	-	-	-	-	-
Can & Manandhar [14]	-	-	-	-	-	-	58.52	44.82	50.76
Lignos [55]	-	-	-	-	-	-	83.49	45.00	58.48
Monson ParaMor [58]	48.46	52.95	50.61	58.50	48.10	52.79	63.32	51.96	57.08
Monson P+M [58]	41.58	65.08	50.74	50.64	63.30	**56.26**	70.09	67.38	**68.71**
Monson P+M Mimic [59]	-	-	-	-	-	-	54.80	60.17	57.36
Morfessor CatMap. [24]	82.17	33.08	47.17	82.17	33.08	47.17	84.75	35.97	50.50
Morfessor Baseline. [22]	-	-	-	71.93	43.27	54.04	74.93	49.81	59.84

Table 10. Comparison of methods competed in Morpho Challenge between years 2007 and 2009 for the German language

Method	2007			2008			2009		
	P	R	F	P	R	F	P	R	F
Bernhard 1 [5]	63.20	37.69	47.22	-	-	-	66.82	42.48	51.94
Bernhard 2 [5]	49.08	57.35	52.89	-	-	-	54.02	60.77	57.20
Bordag 5 [9]	60.71	40.58	48.64	-	-	-	-	-	-
Bordag 5a [9]	60.45	41.57	49.27	-	-	-	-	-	-
Can & Manandhar [14]	-	-	-	-	-	-	57.67	42.67	49.05
Monson ParaMor [58]	59.05	32.81	42.19	53.42	38.15	44.51	56.98	42.10	48.42
Monson P+M [58]	51.45	55.55	**53.42**	49.53	59.51	**54.06**	64.06	61.52	**62.76**
Monson P+M Mimic [59]	-	-	-	-	-	-	51.07	57.79	54.22
Morfessor CatMap. [24]	67.56	36.92	47.75	67.56	36.92	47.75	84.75	35.97	50.50
Morfessor Baseline. [22]	-	-	-	80.23	19.22	31.01	81.70	22.98	35.87

range of approaches have competed in Morpho Challenge. These have been based on using - Bayesian and frequentist statistics, information theory and heuristics. Depending on the approach taken, the Morpho Challenge evaluations show that some are better in some languages, whereas others are better in other languages.

For English, Bernhard 2 [5] outperforms the other systems in 2007. ParaMor-Morfessor (P+M) [58] outperforms the other systems in 2008. ParaMor-Morfessor (P+M) still outperforms other systems in 2009. For German, ParaMor-Morfessor (P+M) [58] outperforms the other systems in all years. For Turkish Morfessor CatMap. [24] outperforms other systems in 2007. However, Monson ParaMor-Morfessor [59] outperforms others in 2008, and Monson ParaMor-Morfessor [59] Mimic outperforms other systems in 2009.

Table 11. Comparison of methods competed in Morpho Challenge between years 2007 and 2009 for the Turkish language

Method	2007			2008			2009		
	P	R	F	P	R	F	P	R	F
Bernhard 1 [5]	78.22	10.93	19.18	-	-	-	-		-
Bernhard 2 [5]	73.69	14.80	24.65	-	-	-	-	-	-
Bordag 5 [9]	81.44	17.45	28.75	-	-	-	81.19	23.44	36.38
Bordag 5a [9]	81.31	17.58	28.91	-	-	-	81.06	23.51	36.45
Can & Manandhar. [14]	-	-	-	-	-	-	41.39	38.13	39.70
Monson ParaMor [58]	-	-	-	56.67	39.42	46.50	57.35	45.75	50.90
Monson P+M [58]	-	-	-	51.88	52.10	**51.99**	66.78	57.97	62.07
Monson P+M Mimic [59]	-	-	-	-	-	-	48.07	60.39	**53.53**
Morfessor CatMap. [24]	76.36	24.50	**37.10**	-	-	-	79.38	31.88	45.49
Morfessor Baseline. [22]	-	-	-	-	-	-	89.68	17.78	29.67

The results show that hybrid approaches that implement system combinations such as ParaMor-Morfessor (P+M) perform well and there is a still a long way to go to for unsupervised systems.

12   Conclusions

Morphological analysis has a very long history in natural language processing. Modern work in unsupervised morphological segmentation dates back to the work of Harris in the 1950s.

The primary goal of this paper is to survey the methods and algorithms used for unsupervised morphological segmentation with the goal of robust morphological segmentation without using any tagged data. A wide range of methods have been used for unsupervised morphological segmentation. All current methods approach the problem from slightly different perspectives. Some methods employ a form of clustering to segment morphologically similar words cooperatively, while other methods model the internal word syntax for example by using a sequence model such as a HMM. And, some methods benefit from employing syntactic/PoS classes. A wide range of mathematical and algorithmic methods have been employed including Bayesian, frequentist, heuristic and information theoretic methods.

As shown in this review, the literature is rather broad and there has been wide range of approaches adopted. Despite the use of current machine learning algorithms, morphological segmentation and more generally unsupervised morphological analysis remains a challenging unsolved problem. Future research could address non-concatenative morphology, stem alternation, morpheme clustering and morphological transformation rule induction.

References

1. Argamon, S., Akiva, N., Amir, A., Kapah, O.: Efficient unsupervised recursive word segmentation using minimum description length. In: Proceedings of the 20th International Conference on Computational Linguistics, COLING 2004, pp. 1058–1064. Association for Computational Linguistics, Stroudsburg (2004)
2. Arısoy, E., Dutağacı, H., Arslan, L.M.: A unified language model for large vocabulary continuous speech recognition of Turkish. Signal Process. 86, 2844–2862 (2006)
3. Aunimo, L., Heinonen, O., Kuuskoski, R., Makkonen, J., Petit, R., Virtanen, O.: Question answering system for incomplete and noisy data. In: Sebastiani, F. (ed.) ECIR 2003. LNCS, vol. 2633, pp. 193–206. Springer, Heidelberg (2003)
4. Baayen, R.: Word Frequency Distributions. Kluwer Academic Publishers (2001)
5. Bernhard, D.: Unsupervised morphological segmentation based on segment predictability and word segments alignment. In: PASCAL Challenge Workshop on Unsupervised Segmentation of Words into Morphemes (2006)
6. Berton, A., Fetter, P., Regel-Brietzmann, P.: Compound words in large-vocabulary German speech recognition systems. In: Proceedings of the Fourth International Conference on Spoken Language, ICSLP 1996, vol. 2, pp. 1165–1168 (October 1996)
7. Bilotti, M.W., Katz, B., Lin, J.: What works better for question answering: Stemming or morphological query expansion? In: Proceedings of the Information Retrieval for Question Answering (IR4QA) Workshop at SIGIR (2004)
8. Blackwell, D., MacQueen, J.B.: Ferguson distributions via polya urn schemes. The Annals of Statistics 1, 353–355 (1973)
9. Bordag, S.: Two-step approach to unsupervised morpheme segmentation. In: Proceedings of 2nd Pascal Challenges Workshop, pp. 25–29 (2006)
10. Bordag, S.: Unsupervised and Knowledge-Free Morpheme Segmentation and Analysis. In: Peters, C., Jijkoun, V., Mandl, T., Müller, H., Oard, D.W., Peñas, A., Petras, V., Santos, D. (eds.) CLEF 2007. LNCS, vol. 5152, pp. 881–891. Springer, Heidelberg (2008)
11. Brent, M.R.: An efficient, probabilistically sound algorithm for segmentation and word discovery. Machine Learning 34, 71–105 (1999)
12. Brent, M.R., Murthy, S.K., Lundberg, A.: Discovering morphemic suffixes a case study in mdl induction. In: Fifth International Workshop on AI and Statistics, Ft., pp. 264–271 (1995)
13. Brown, P.F., Della Pietra, V.J., Della Pietra, S.A., Mercer, R.L.: The mathematics of statistical machine translation: Parameter estimation. Comput. Linguist. 19(2), 263–311 (1993)
14. Can, B., Manandhar, S.: Clustering morphological paradigms using syntactic categories. In: Peters, C., Di Nunzio, G.M., Kurimo, M., Mandl, T., Mostefa, D., Peñas, A., Roda, G. (eds.) CLEF 2009. LNCS, vol. 6241, pp. 641–648. Springer, Heidelberg (2010)
15. Can, B., Manandhar, S.: Probabilistic hierarchical clustering of morphological paradigms. In: EACL, pp. 654–663 (2012)
16. Chan, E.: Structures and distributions in morphology learning. PhD thesis, University of Pennsylvania (2008)
17. Clark, A.S.: Inducing syntactic categories by context distribution clustering. In: Proceedings of CoNLL 2000 and LLL 2000, pp. 91–94 (2000)

18. Collins, M.: Discriminative training methods for hidden markov models: Theory and experiments with perceptron algorithms. In: Proceedings of the ACL 2002 Conference on Empirical Methods in Natural Language Processing, EMNLP 2002, vol. 10, pp. 1–8. Association for Computational Linguistics, Stroudsburg (2002)

19. Creutz, M.: Unsupervised segmentation of words using prior distributions of morph length and frequency. In: Proceedings of the 41st Annual Meeting on Association for Computational Linguistics, ACL 2003, vol. 1, pp. 280–287. Association for Computational Linguistics, Stroudsburg (2003)

20. Creutz, M.: Induction of the Morphology of Natural Language: Unsupervised Morpheme Segmentation with Application to Automatic Speech Recognition. PhD thesis, Computer and Information Science, University of Technology, Espoo, Finland (2006)

21. Creutz, M., Hirsimäki, T., Kurimo, M., Puurula, A., Pylkkönen, J., Siivola, V., Varjokallio, M., Arisoy, E., Saraçlar, M., Stolcke, A.: Morph-based speech recognition and modeling of out-of-vocabulary words across languages. ACM Trans. Speech Lang. Process. 5, 1–29 (2007)

22. Creutz, M., Lagus, K.: Unsupervised discovery of morphemes. In: Proceedings of the ACL 2002 Workshop on Morphological and Phonological Learning, MPL 2002, vol. 6, pp. 21–30. Association for Computational Linguistics, Stroudsburg (2002)

23. Creutz, M., Lagus, K.: Induction of a simple morphology for highly-inflecting languages. In: Proceedings of the 7th Meeting of the ACL Special Interest Group in Computational Phonology: Current Themes in Computational Phonology and Morphology, SIGMorPhon 2004, pp. 43–51. Association for Computational Linguistics, Stroudsburg (2004)

24. Creutz, M., Lagus, K.: Inducing the morphological lexicon of a natural language from unannotated text. In: Proceedings of the International and Interdisciplinary Conference on Adaptive Knowledge Representation and Reasoning, pp. 106–113 (2005)

25. de Gispert, A., Mariño, J.: On the impact of morphology in English to Spanish statistical mt. Speech Communication 50, 1034–1046 (2008)

26. Déjean, H.: Morphemes as necessary concept for structures discovery from untagged corpora. In: Proceedings of the Joint Conferences on New Methods in Language Processing and Computational Natural Language Learning, NeMLaP3/CoNLL 1998, pp. 295–298. Association for Computational Linguistics, Stroudsburg (1998)

27. Demberg, V.: A language-independent unsupervised model for morphological segmentation. In: Proceedings of the 45th Annual Meeting of the Association of Computational Linguistics, pp. 680–685 (2007)

28. Dreyer, M., Eisner, J.: Discovering morphological paradigms from plain text using a dirichlet process mixture model. In: Proceedings of the 2011 Conference on Empirical Methods in Natural Language Processing, pp. 616–627. Association for Computational Linguistics, Edinburgh (July 2011)

29. Ford, A., Singh, R., Martohardjono, G.: Pace Panini. Peter Lang (1967)

30. Gelbukh, A., Alexandrov, M., Han, S.-Y.: Detecting inflection patterns in natural language by minimization of morphological model. In: Sanfeliu, A., Martínez Trinidad, J.F., Carrasco Ochoa, J.A. (eds.) CIARP 2004. LNCS, vol. 3287, pp. 432–438. Springer, Heidelberg (2004)

31. Goldsmith, J.: Unsupervised learning of the morphology of a natural language. Computational Linguistics 27(2), 153–198 (2001)

32. Goldsmith, J.: An algorithm for the unsupervised learning of morphology. In: Natural Language Engineering, vol. 12, pp. 353–371 (2006)

33. Goldwater, S., Griffiths, T.L., Johnson, M.: Interpolating between types and tokens by estimating power-law generators. In: Advances in Neural Information Processing Systems, vol. 18. MIT Press, Cambridge (2006)
34. Goldwater, S., McClosky, D.: Improving statistical mt through morphological analysis. In: Proceedings of the Conference on Human Language Technology and Empirical Methods in Natural Language Processing, HLT 2005, pp. 676–683. Association for Computational Linguistics, Stroudsburg (2005)
35. Grünwald, P.: A tutorial introduction to the minimum description length principle. In: Advances in Minimum Description Length: Theory and Applications. MIT Press (2005)
36. Habash, N., Rambow, O.: Arabic tokenization, part-of-speech tagging and morphological disambiguation in one fell swoop. In: Proceedings of the 43rd Annual Meeting on Association for Computational Linguistics, ACL 2005, pp. 573–580. Association for Computational Linguistics, Stroudsburg (2005)
37. Hafer, M.A., Weiss, S.F.: Word segmentation by letter successor varieties. Information Storage and Retrieval 10(11-12), 371–385 (1974)
38. Hammarstrm, H.: A survey and classification of methods for (mostly) unsupervised learning of morphology. In: The 16th Nordic Conference of Computational Linguistics, NODALIDA 2007, Tartu, Estonia, May 25-26. NEALT (2007)
39. Harman, D.: How effective is suffixing. Journal of the American Society for Information Science 42(1), 7–15 (1991)
40. Harris, Z.S.: From phoneme to morpheme. Language 31(2), 190–222 (1955)
41. Ishwaran, H., James, L.F.: Generalized weighted chinese restaurant processes for species sampling mixture models. Statistica Sinica 13 (2003)
42. Järvelin, K., Pirkola, A.: Morphological processing in mono- and cross-lingual information retrieval. In: Arppe, A., Carlson, L., Lindén, K., Piitulainen, J., Suominen, M., Vainio, M., Westerlund, H., Yli-Jyrä, A. (eds.) Inquiries into Words, Constraints and Contexts. Festschrift for Kimmo Koskenniemi on his 60th Birthday, pp. 214–226. CSLI Publications, Stanford (2005)
43. Kazakov, D.: Unsupervised learning of naive morphology with genetic algorithms. In: ECML/Mlnet Workshop on Empirical Learning of Natural Language Processing Tasks, Prague, pp. 105–112 (1997)
44. Kazakov, D., Manandhar, S.: Unsupervised learning of word segmentation rules with genetic algorithms and inductive logic programming. In: Machine Learning, pp. 43–121 (2001)
45. Keshava, S., Pitler, E.: A simpler, intuitive approach to morpheme induction. In: PASCAL Challenge Workshop on Unsupervised Segmentation of Words into Morphemes, pp. 31–35 (2006)
46. Kettunen, K., Kunttu, T., Järvelin, K.: To stem or lemmatize a highly inflectional language in a probabilistic ir environment? Journal of Documentation 61(4), 476–496 (2005)
47. Kirchhoff, K., Vergyri, D., Bilmes, J., Duh, K., Stolcke, A.: Morphology-based language modeling for conversational Arabic speech recognition. Computer Speech & Language 20(4), 589–608 (2006)
48. Toutanova, K., Suzuki, H., Ruopp, A.: Applying morphology generation models to machine translation. In: Proceedings of the 46th Annual Meeting of the Association for Computational Linguistics: Human Language Technologies, pp. 514–522. Association for Computational Linguistics, Columbus (2008)
49. Krovetz, R.: Viewing morphology as an inference process. In: Proceedings of the 16th Annual International ACM SIGIR Conference on Research and Development in Information Retrieval, SIGIR 1993, pp. 191–202. ACM, New York (1993)

50. Kurimo, M., Lagus, K., Virpioja, S., Turunen, V.: Morpho challenge 2010 (June 2011), http://research.ics.tkk.fi/events/morphochallenge2010/

51. Kurimo, M., Virpioja, S., Turunen, V.: Proceedings of the morpho challenge 2010 workshop. In: Proceedings of the 11th Meeting of the ACL Special Interest Group on Computational Morphology and Phonology, SIGMORPHON 2010, pp. 87–95. Association for Computational Linguistics, Stroudsburg (2010)

52. Larson, M., Willett, D., Khler, J., Rigoll, G.: Compound splitting and lexical unit recombination for improved performance of a speech recognition system for German parliamentary speeches. In: International Conference on Spoken Language Processing, pp. 945–948 (2000)

53. Lavallée, J.F., Langlais, P.: Morphological acquisition by formal analogy. In: Working Notes for the CLEF 2009 Workshop (September 2009)

54. Lignos, C.: Learning from unseen data. In: Kurimo, M., Virpioja, S., Turunen, V., Lagus, K. (eds.) Proceedings of the Morpho Challenge 2010 Workshop, Aalto University, Espoo, Finland, pp. 35–38 (2010)

55. Lignos, C., Chan, E., Marcus, M.P., Yang, C.: A rule-based unsupervised morphology learning framework. In: Working Notes for the CLEF 2009 Workshop (September 2009)

56. Manandhar, S., Deroski, S., Erjavec, T.: Learning multilingual morphology with clog. In: Page, D. (ed.) ILP 1998. LNCS, vol. 1446, pp. 135–144. Springer, Heidelberg (1998)

57. Minkov, E., Toutanova, K., Suzuki, H.: Generating complex morphology for machine translation. In: Proceedings of the 45th Annual Meeting of the Association of Computational Linguistics, pp. 128–135. Association for Computational Linguistics, Prague (2007)

58. Monson, C., Carbonell, J.G., Lavie, A., Levin, L.: Paramor: Finding paradigms across morphology. In: Peters, C., Jijkoun, V., Mandl, T., Müller, H., Oard, D.W., Peñas, A., Petras, V., Santos, D. (eds.) CLEF 2007. LNCS, vol. 5152, pp. 900–907. Springer, Heidelberg (2008)

59. Monson, C., Hollingshead, K., Roark, B.: Probabilistic ParaMor. In: Proceedings of the 10th CLEF Conference on Multilingual Information Access Evaluation: Text Retrieval Experiments, CLEF 2009 (September 2009)

60. Morrison, D.R.: Patricia - practical algorithm to retrieve information coded in alphanumeric. Journal of the ACM 15, 514–534 (1968)

61. Neuvel, S., Fulop, S.A.: Unsupervised learning of morphology without morphemes. In: Proceedings of the ACL 2002 Workshop on Morphological and Phonological Learning, MPL 2002, vol. 6, pp. 31–40. Association for Computational Linguistics, Stroudsburg (2002)

62. Orbanz, P., Teh, Y.W.: Bayesian nonparametric models. In: Encyclopedia of Machine Learning, pp. 81–89. Springer (2010)

63. Poon, H., Cherry, C., Toutanova, K.: Unsupervised morphological segmentation with log-linear models. In: Proceedings of Human Language Technologies: The 2009 Annual Conference of the North American Chapter of the Association for Computational Linguistics, NAACL 2009, pp. 209–217. Association for Computational Linguistics, Stroudsburg (2009)

64. Poon, H., Domingos, P.: Joint unsupervised coreference resolution with Markov logic. In: Proceedings of the Conference on Empirical Methods in Natural Language Processing, EMNLP 2008, pp. 650–659. Association for Computational Linguistics, Stroudsburg (2008)

65. Roeland Ordelman, A.V.H., Jong, F.D.: Compound decomposition in Dutch large vocabulary speech recognition. In: Proceedings of Eurospeech 2003, pp. 225–228 (2003)
66. Rosenfeld, R.: A whole sentence maximum entropy language model. In: Proceedings of the IEEE Workshop on Speech Recognition and Understanding (1997)
67. Schleicher, A.: Zur Morphologie der Spreche, St. Pétersburg. moires de l'Académie Impériale des Sciences de St. Pétersburg Series VII, vol. 1(7) (1859)
68. Sirts, K., Alumäe, T.: A hierarchical dirichlet process model for joint part-of-speech and morphology induction. In: Proceedings of the 2012 Conference of the North American Chapter of the Association for Computational Linguistics: Human Language Technologies, NAACL HLT 2012, pp. 407–416. Association for Computational Linguistics, Stroudsburg (2012)
69. Smith, N.A., Eisner, J.: Contrastive estimation: training log-linear models on unlabeled data. In: Proceedings of the 43rd Annual Meeting on Association for Computational Linguistics, ACL 2005, pp. 354–362. Association for Computational Linguistics, Stroudsburg (2005)
70. Snyder, B., Barzilay, R.: Unsupervised multilingual learning for morphological segmentation. In: Proceedings of ACL 2008: HLT, pp. 737–745. Association for Computational Linguistics, Columbus (June 2008)
71. Spiegler, S., Monson, C.: Emma: A novel evaluation metric for morphological analysis. In: Proceedings of the 23rd International Conference on Computational Linguistics, COLING (August 2010)

Morphological Analysis of the Bishnupriya Manipuri Language Using Finite State Transducers

Nayan Jyoti Kalita, Navanath Saharia, and Smriti Kumar Sinha

Department of Computer Science and Engineering, Tezpur University, India
nayan.jk.123@gmail.com, {nava_tu,smriti}@tezu.ernet.in

Abstract. In this work we present a morphological analysis of Bishnupriya Manipuri language, an Indo-Aryan language spoken in the north eastern India. As of now, there is no computational work available for the language. Finite state morphology is one of the successful approaches applied in a wide variety of languages over the year. Therefore we adapted the finite state approach to analyse morphology of the Bishnupriya Manipuri language.

1 Introduction

Morphology involves the study of inner structures of words and their different forms and is an essential step to any natural language processing task. The input to a morphological analyser is a word and it gives the lexical form of the input words, consisting of the root of the word and other morphological information the word has as output. There are two challenges in the morphological analysis; the morphotactics and the orthography. The morphotactics show how the morphemes combine together to form a word. The orthography is the model of how spelling changes when the morphemes combine together. One of the most efficient approaches to morphological analysis is the finite state transducer (FST) approach which determines whether an input string of morphemes makes up a word or not. [1] successfully implemented the finite state approach to morphology. He combined both the morphotactics and the orthography into a single lexical transducer. There are number of tools available for the construction of FST based morphological analyser among which XFST[1], SFST[2] and OFST[3] are popular among researchers. Our experiment in this work is based on XFST tool. In this work, we report the development of a morphological analyser using finite state transducers on Bishnupriya Manipuri. The Bishnupriya Manipuri language is an Indo-Aryan language spoken in the boarder of Tibeto-Burman language

---

[1] Xerox Finite State Transducer;
http://www.fsmbook.com
[2] Stuttgart Finite State Transducer;
http://www.cis.uni-muenchen.de/~schmid/tools/SFST
[3] OpenFST;
http://www.openfst.org/twiki/bin/view/FST/webhome

A. Gelbukh (Ed.): CICLing 2014, Part I, LNCS 8403, pp. 206–213, 2014.

family. We feel, being a boarder language and a resource scare language this work is significant.

The organization of the paper is as follows. Section 2 gives an outline on the Bishnupriya Manipuri language. Section 3 describes the related work on the morphological analysis. Section 4 describes the corpus development. Section 5 gives the architecture of the experiment. Section 6 gives the results obtained from our experiment. Section 7 concludes the report.

2   The Bishnupriya Manipuri Language

The Bishnupriya Manipuri language is of Indo-Aryan origin and is a kin to Assamese, Bengali and Oriya and neighbour of the Meitei language[4]. Being a boarder language of another language family, the Bishnupriya Manipuri language has around 4,000 borrowed root words originated from Meitei. Though the speaker of the language was originally confined only to the surroundings of the lake Loktak in Manipur of India, it is practically dead in its place of origin [2] . However, the language is retained by its speakers in diaspora mostly in Assam, Tripura, a few in Manipur of India and Bangladesh According to Ethnologue[5], total number of speakers of this language is approx. 117500, where 77,500 speakers are from India (2001 census). There are two dialects in the Bishnupriya Manipuri language, namely, the *Madai Gang* or the dialect of the village of the queen and the *Rajar Gang* or the dialect of the village of the king. Being an agglutinative language, Bishnupriya Manipuri has the capability of generating hundreds of words from a single noun and verb root. For example, the root word মানু (**manu** : *man*) may form different inflected words.

1. মানুয়ে (manuje) → মানু (manu) + য়ে (je)

   → Noun + NCM
2. মানুহাবি (manuhabi) → মানু (manu) + হাবি (habi)

   → Noun + Pl
3. মানুরে (manuɹe) → মানু (manu) + রে (ɹe)

   → Noun + ACM
4. মানুগয়ে (manugɔje) → মানু (manu) + গ (gɔ) + য়ে (je)

   → Noun + DM + NCM
5. মানুহাবিয়েহে (manuhabijehe) → মানু (manu) + হাবি (habi) + য়ে (je) + হে (he)

   → Noun + Pl + NCM + EM

A noun root can take case marker (locative, genitive, ablative, nominative, accusative, vocative, instrumental, dative), emphatic marker, number (singular

---

[4] A language of the Kuki-Chin branch of the Tibeto-Burman language family
[5] http://www.ethnologue.com/language/bpy

and plural) marker and definitive marker in the form of sequence of suffix. Likewise, a verb root may take tense marker (simple present, simple past, simple future), aspect marker, person (1st, 2nd, 3rd) marker and mood (indicative, imperative, subjunctive) marker as suffix. Table 1 shows some examples of inflections in the Bishnupriya Manipuri language. Table 2 shows the verbal inflection of root কর (kɔɹ : *to do*)

During the study we found a peculiar characteristic of this language. The Bishnupriya Manipuri language is an Indo-Aryan language. Since it was developed in the context of Tibeto-Burman languages, it has borrowed a number of Tibeto-Burman noun and verb roots, mainly from Meitei. Interestingly, the Bishnupriya Manipuri suffixes cannot be added directly with the Meitei verb roots. An Indo-Aryan verb root is to be added first with the Meitei verb root forming this basis for affixation. For example-

হংকরানি (hɔŋkɔɹani : *to make*)→ হং (*Meitei root*)+ কর (Sanskrit/Indo-Aryan root)+ আনি (*suffix*)

Table 1. Examples of inflections in Bishnupriya Manipuri language. PDM→Plural Definitive Marker; Pl→Plural Marker; LCM→Locative Case Marker; PsPSg→Past-Perfect-Singular; PsPPl→Past-Perfect-Plural Marker.

Words (IPA)	Output
দাদাগাছি (dadagaʃi)	দাদা_Noun+Pl
মাছগি (maʃgi)	মাছ_Noun+PDM
গুরুমাহেই (guɹumahei)	গুরু_Noun+Pl
মাছগুলি (maʃguli)	মাছ_Noun+Pl
মানুহাবি (manuhabi)	মানু_Noun+Pl
ঘরে (gʰɔɹe)	ঘরে_Noun+Sg+LCM
পাছিলে (paʃile)	পা_Verb+PsPSg
পাছিলায় (paʃilai)	পা_Verb+PsPPl

Table 2. Inflectional forms of the verb কর (kɔɹ : *to do*) with tense and person

	1st person	2nd person	3rd person
Present	করুরি (kɔɹuɹi)	করর (kɔɹɔɹ)	করের (kɔɹeɹ)
Past	করলু (kɔɹlu)	করলে (kɔɹle)	করল (kɔɹlɔ)
Future	করতৌ (kɔɹtou)	করতেই (kɔɹtei)	করতই (kɔɹtɔi)

3   Related Works

Number of researches has been done for developing morphological analyser in various languages based on finite state transducers. [3] showed how the morphotactics and the variation rules of Arabic have been described using only finite state operation and implemented a significant morphological analyser using this approach. [4] designed a morphological analyser and generator for Aymara language using Xerox finite-state machine. [5] described some of the challenges encountered in a computational morphological analysis of Persian and developed a morphological analyser for Persian using Xerox finite-state tools.

In Indian language context, [6] developed a morphological analyser for Hindi. They used SFST tool for generating the finite states. A morphological generator was developed for Kannada language using FST [7]. [8] described a finite state morphological analyser for the nouns in Oriya. He specified the co-occurrence restrictions of the morphemes in a nominal form and used the FSA to determine whether an input string of morphemes makes up an Oriya noun or not. [9], reported a paradigm-based finite state morphological analyser for Marathi language. Among languages spoken in north eastern India, we do not find any work on morphological analysis using FST.

4   Bishnupriya Manipuri Corpus

For the morphological analysis of the Bishnupriya Manipuri language, we have developed a Bishnupriya Manipuri Corpus. For this corpus, we have collected the texts from the blogs, websites. We have also collected the texts from the Bishnupriya Manipuri version of Wikipedia. But, there exists a few numbers of resources available on the Internet. Moreover, most of the resources present in the Internet are in image format, we do not have the sophisticated Optical Character Recognition (OCR) tool to extract text format.

Further we have collected large number of text written in Smriti Font, which is an ASCII Font. We have developed an ASCII-to-UNICODE Converter to convert Smriti font texts to UTF-8 texts. Presently, this corpus contains approx. one lakh words, with 10,196 sentences.

5   Experiments

We have developed the morphological analyser using XFST tool developed by Xerox. It supports UTF-8 character coding which is important for the implementation of Bishnupriya Manipuri computational morphologies. The tool is based on a lexicon and a set of rules for root and morphemes. This lexicon contains the list of root words and its category separated by a tab. The analyser fails on giving a complex word as an input and the corresponding root word does not exist in the lexicon file. We have developed the Bishnupriya Manipuri lexicon and the rules file required for analysis. The lexicon is designed to reflect the word categories in the Bishnupriya Manipuri language.

The lexicon contains different states for each of the root words, starting with the declaration of the tags.

```
Multichar_Symbols
```

+Pl +Sg +CM +DM

The declaration of tags is followed by the different LEXICONS. For instance, the states for noun can be encoded as-

```
LEXICON Nouns
    মামা N;
LEXICON N
    %+Noun:0 NPL;
    %+Noun:0 PostPositional;
LEXICON NPL
    +Pl:গাছি NCM;
    +Pl:গুলি NCM;
    +Sg:0 NCM;
LEXICON NCM
    +CM:CO NEM;
    +CM:য NEM;
    +CM:রে NEM;
LEXICON NEM
    +EM:হে #;
    +EM:ক #;
```

The root words and its category are separated by a semicolon. The left side of the colon represents the upper side or the analysis form of the transducer, and the right side shows the lower side or the surface form. The hash symbol at the end of a row indicates the end of the transition, and therefore, that state is the final state. The analyser takes the surface form as input and produces the result as the grammatical structure of the word or the lexicon form.

In the word formation, the morphemes can only appear in certain order and in certain combinations. For example, the initial ই (i) of the suffix ইলু (ilu) is dropped when combine with the roots ending in consonants. Therefore, the following root word কর (kɔɹ : *to do*) changes:

করলু (kɔɹlu) → কর (kɔɹ) + ইলু (ilu)

Again, the ending আ or ○া (a) in পা (pa: *to get*) changes to এ or CO (e) when combine with ইলে (ile).

পেইলে (peile) → পা (pa) + ইলে (ile)

To implement these changes in xfst, we have to include the phonological rules in a separate file with an extension *.regex*. Regular expressions are generally used to manipulate the phonological rules. In the case of পেইলে (peile), the phonological rule will be:

[আ | ○া → এ | CO || _ই]

which translates as: আ or ○া becomes এ or CO iff ই follows আ or ○া.

Figure 1 describes the architecture of used approach.

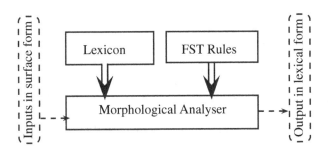

Fig. 1. Pictorial work-flow of our approach

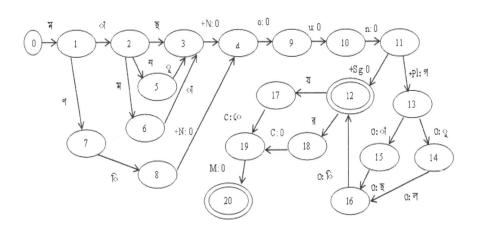

Fig. 2. Diagrammatic representation of example words

6   Result and Analysis

We have tested the developed rules with 1000, 2000 and 3000 texts separately. Some of the rules for the noun are given below:

1. Noun + Pl
2. Noun + Pl + CM + EM
3. Noun + PDM
4. Noun + PDM + CM + EM
5. Noun + Sg + SDM + CM + EM

where Pl→Plural Marker, Sg→Singular, CM→Case Marker, EM→Emphatic Marker, PDM→Plural Definitive Marker and SDM→Singular Definitive Marker.

The texts used in this work are part of Bishnupriya Manipuri corpus consisting of web blogs, web sites, Wikipedia and some typed text. We have 150 noun and 50 verb roots as lexicon, in the rule file. After manual verification on the output of the tests, table 3 gives the obtained results with different test data. The experiment was evaluated on the basis of precision and recall values.

Precision: For the morphological analysis, the Precision value is defined as the ratio of the number of words correctly generated to the total number of inputs that should have correctly generated.

Recall: The Recall value is defined as the ratio of the number of words generated to the total number of input words.

F-Score: The F-Score value is define as F= ((2*Precision*Recall)/ (Precision + Recall))

Table 3. Obtained results

Test data size	Precision	Recall	F-Score
1000	93.24	41.24	56.91
2000	91.13	38.01	53.64
3000	91.03	39.38	54.98

Figure 2 illustrates the transitions of five words, viz. মামাগাছি (mamagaʃi : *maternal uncles*), মানুয়ে (manuje), মানুর (manuɹ), মাছগুলি (maʃɔguli : *a few fishes*), মণি (mɔni : *jewel*). The output of these five words is as below:

মামাগাছি→মামা+Noun+Pl
মানুয়ে→মানু+Noun+Sg+CM
মানুর→মানু+Noun+Sg+CM
মাছগুলি→মাছ+Noun+Pl
মণি→মণি+Noun+Sg

The automation for these five words is represented as a finite state transducer: a finite set of states together with a set of arcs between pairs of states. States are represented as circles and arcs are represented by arrows from one state to another state. The final states are represented by two concentric circles. The morphological analyser starts at the initial state and goes through a sequence of states by computing the morphemes. If it matches the symbol on an arc leaving the present state, then it moves to the next state through that arc and moves to the next symbol of the input word. Thus it successfully recognizes all the morphemes in an input string.

7 Conclusion

This paper has described the morphological analyser for the Bishnupriya Manipuri language based on the finite state transducer. The designed rules are successfully tested with the text collected from the Bishnupriya Manipuri Corpus. Our main motivation is to develop the resources of linguistics work on this language and being the first work of this kind for the Bishnupriya Manipuri language, we obtained a good precision for our work.

References

1. Karttunen, L.: Finite-state lexicon compiler. Xerox Corporation (1993)
2. Sinha, K.P.: The Bishnurpiya Manipuri Language. Firma KLM Prvt Ltd. (1981)
3. Beesley, K.R.: Arabic morphology using only finite-state operations. In: Proceedings of the Workshop on Computational Approaches to Semitic Languages, pp. 50–57 (1998)
4. Beesley, K.R.: Finite-state morphological analysis and generation for Aymara. In: Proceedings of the Workshop of Finite-State Methods in Natural Language Processing, pp. 19–26 (2003)
5. Megerdoomian, K.: Finite-state morphological analysis of Persian. In: Proceedings of the Workshop on Computational Approaches to Arabic Script-based Languages, pp. 35–41 (2004)
6. Kumar, D., Singh, M., Shukla, S.: Fst based morphological analyzer for Hindi language. International Journal of Computer Science Issues 9(4) (2012)
7. Melinamath, B.C., Mallikarjunmath, A.G.: A morphological generator for Kannada based on finite state transducers. In: 3rd International Conference on Electronics Computer Technology, vol. 1, pp. 312–316 (2011)
8. Sahoo, K.: Oriya nominal forms: a finite state processing. In: Conference on Convergent Technologies for Asia-Pacific Region, pp. 730–734 (2003)
9. Bapat, M., Gune, H., Bhattacharyya, P.: A paradigm-based finite state morphological analyzer for Marathi. In: Proceedings of the 1st Workshop on South and Southeast Asian Natural Language Processing, pp. 26–34 (2010)

A Hybrid Approach to the Development
of Part-of-Speech Tagger for Kafi-noonoo Text

Zelalem Mekuria[1] and Yaregal Assabie[2]

[1] Department of Computer Science and Information Technology, Aksum University
zelalem.mgd@gmail.com
[2] Department of Computer Science, Addis Ababa University
yaregal.assabie@aau.edu.et

Abstract. Although natural language processing (NLP) is now a popular area of research and development, less-resourced languages are not receiving much attention from developers. One of such under-resourced languages is Kafi-noonoo which is spoken in the south-western regions of Ethiopia. This paper presents the development of part-of-speech tagger for Kafi-noonoo. In order to develop the tagger, we employed a hybrid of two systems: statistical and rule-based taggers. The lexical and transitional probabilities of word classes are modeled using HMM. However, due to the limitation of corpus for the language, a set of transformation rules are applied to improve the result. The system was tested with test corpus and, with 90% of the corpus used for training, the hybrid tagger yielded an accuracy of 80.47%.

Keywords: Kafi-noonoo NLP, Part-of-Speech Tagging, Hybrid Systems.

1   Introduction

Part-of-speech (POS) tagging, also called grammatical tagging or word category disambiguation, is the process of labeling a word in a text with a particular word class, based on both its definition as well as its context i.e. relationship with adjacent and related words within a phrase, sentence or paragraph [3], [7]. In other words, POS tagger reads text in given language and assigns parts-of-speech such as noun, verb, adjective, etc. to each word within the text. It is an important component of high level natural language processing applications and plays an important role in parsing, machine translation, grammar checking, speech synthesis, information retrieval, word sense disambiguation, etc. Most tagging algorithms fall in to one of two classes [7]: rule-based and statistical taggers. Rule-based taggers use hand coded rules to determine the lexical categories of a word [1]. Words are tagged based on the contextual information around a word that is going to be tagged. Part-of-speech distribution and statistics for each word can be derived from annotated corpora dictionaries. The dictionary provides a list of words with their lexical meanings. In the dictionaries there are many citations that describe a word in different context. These contextual citations provide information that is used as a clue to develop a rule and determine lexical categories of the word [8]. On the other hand, statistical methods assign tag for a word

A. Gelbukh (Ed.): CICLing 2014, Part I, LNCS 8403, pp. 214–224, 2014.
© Springer-Verlag Berlin Heidelberg 2014

by calculating the most likely tag in the context of the word and its immediate neighbor [5]. The main idea behind all statistical taggers is a simple generalization and picks the most-likely tag for the word [8]. A statistical approach includes most frequent tag, n-gram and hidden Markov model (HMM). Nowadays, part-of-speech tagger is developed for various languages around the world and it remains an intensive area of research and development for other languages [2], [5], [8], [10]. Although Kafi-noonoo is one of the major languages commonly used in the south-western regions of Ethiopia, to our best knowledge, there is no POS tagger or other NLP applications developed for the language.

This paper presents part-of-speech tagger for Kafi-noonoo text. The remaining part is organized as follows. Section 2 presents the characteristics of Kafi-noonoo language with emphasis to its POS tagsets. The proposed system is discussed in Section 3. Experimental results are presented in Section 4, and conclusion and future works are highlighted in Section 5. References are provided at the end.

2   Linguistic Characteristics of Kafi-noonoo

2.1   Kafi-noonoo Language

Kafi-noonoo is a language that is spoken by about 3 million people in south western part of Ethiopia. It belongs to the Afro-Asiatic language super family of the North-Omotic Southern Gonga sub-group [6]. Kafi-noonoo uses latin script for writing. It has 22 consonant phonemes. Out of these, six of them are both long and short consonants. Among the 22 consonants, five of them are borrowed from English and Amharic languages. In addition to the consonants, it has five long and short vowels. The long vowels and consonants can be obtained by doubling the corresponding short vowels and consonants, respectively. The difference in length of both vowels and consonants induces difference in meaning. For example, the Kafi-noonoo word "baro" means *corn* while "baaro" means *forehead*. In Kafi-noonoo, tone has a semantic and grammatical function. For example, "kemo" (with high and low tones) can mean *buy* and the same word (with both high tones) can also mean *sell*.

2.2   Kafi-noonoo Tagsets

Words in Kafi-noonoo can be divided into two broad categories: closed class types and open class types [2]. Closed classes are those that have relatively fixed members while open classes are those that continually changed or borrowed from other languages. Since Kafi-noonoo is under-resourced language, to our best knowledge, there were no defined tagsets and/or tagged corpus available for research and development. Thus, in consultation with linguists, we identified a total of 34 tagsets for the language. The tagsets are defined according to the hierarchies of word classes and subclasses of nouns, verbs, adjectives, pronouns, adverbs, prepositions. In addition to these, conjunction, interjections, numerals and punctuations are also included as basic classes of Kafi-noonoo language.

Nouns: In this class, we identify proper noun, proper noun with conjunction, noun conjunction, noun preposition, and noun possession as sub-classes.

- Nouns that represent the name of a person, place, thing, organization, etc. are considered to be proper noun and tagged as NP.
- Proper noun can be attached with conjunction. This type of proper noun classified under proper noun with conjunction sub-class is tagged as NPC (e.g., "yeeri_n_").
- Nouns affixed with conjunction are considered as noun conjunction and tagged as NC (e.g., "Johni_naa_ Deevid"/*John and David*).
- Nouns suffixed with preposition are considered to be noun preposition sub-class and tagged as NPERP (e.g., "kaameloo_na_"/*by car*).
- Noun can be attached with possession and is tagged as NPOSS (e.g., "ashi_ch_").
- Other forms of nouns that cannot be classified under the above classes such as collective nouns, abstract nouns, and common nouns are tagged as N.

Pronouns: In this class, we identify demonstrative pronoun, pronoun conjunction, interrogative pronoun, and pronoun preposition as sub-classes.

- Pronouns that point or identify noun or pronoun are classified under demonstrative pronoun and tagged as PROND (e.g., "ebi"/*this*).
- Pronouns attached with conjunction are classified under pronoun conjunction and tagged as PRONC (e.g., "biin_naa_ biinna/*him and her*).
- Pronouns that can be used to ask questions are classified as interrogative pronoun and tagged as PRONI (e.g., "koni"/*who*).
- Pronouns attached with preposition are classified under pronoun preposition sub-class and tagged as PRONPREP ("noo_ch_"/*for us*).
- All pronouns that cannot be classified under the above sub-classes are tagged as PRON.

Verbs: In this class, we identify sub-classes of verbs presented as follows.

- Verbs attached with conjunction are categorized under verb conjunction sub-class and tagged as VC (e.g., "maaoh_iu_chiye"/*ate and drank*)
- Verbs that show infinitive are tagged as VI (e.g., "dic_hoo_"/*to develop*).
- An infinitive verb may be attached with conjunction and is tagged as VIC ("xiishiy_oonaa_/*to assure and*).
- Auxiliary verbs are tagged as AUX.
- Verbs that cannot be classified under the above sub-classes are tagged as V.

Adjectives: In this class, we identify sub-classes that are presented as follows.

- Adjectives attached with conjunction are tagged by ADJC (e.g., "maccoo_naa_ ikkonoomee"/*social and economic*)
- Adjectives attached with possession are tagged as ADJPOSS.
- Adjectives attached with preposition are tagged as ADJPREP (e.g., "digooy_ich_"/*for peaceful*).
- Adjectives that cannot be classified under the above sub-classes are tagged as ADJ.

Adverbs: In this class, we identify subclasses presented as follows.

- Adverbs attached with conjunction are tagged as ADVC (e.g., "shatiyoo_naa_"/*warning and*).

- Adverbs attached with preposition are tagged as ADVPREP (e.g., "ame ya-woona/*in what way*).
- Adverbs that cannot be classified under the above sub-classes are tagged as ADV.

Numerals: Kafi-noonoo numerals, like that of English, can be cardinal or ordinal. Moreover, Kafi-noonoo numerals can be attached with conjunction or preposition.

- Cardinal numerals are tagged as CARDN.
- Ordinal numerals are tagged as ORDN.
- Cardinals attached with conjunctions or prepositions are tagged as CARDNC or CARDNPREP, respectively.
- Ordinals attached with conjunctions or prepositions are tagged as ORDNC or ORDNPREP, respectively.

Prepositions: In Kafi-noonoo, prepositions can be attached with other basic classes like noun, adjective, adverb, etc. In this case, they are tagged as PREPC (e.g., "daggeexo"/*at the middle*). Otherwise, they are tagged as PREP.

Conjunctions: In Kafi-noonoo, conjunctions can be attached with other classes as discussed above. However, they can also appear independently as separate words. In such cases they are tagged as C.

Interjections: Interjections are are tagged as INT.

Punctuations: Punctuation marks are tagged by PUNC.

3   The Proposed Kafi-noonoo Part-of-Speech Tagger

3.1   General Architecture

The proposed Kafi-noonoo POS tagger is a hybrid of statistical and rule-based methods. The system has three main components: statistical tagger, output analyzer and rule-based tagger. The statistical tagger computes the lexical and contextual probabilities using HMM. The output analyzer takes the results of the statistical component and analyzes if the lexical tag probabilities for a given sentence (confidence measure) are above a threshold value. If the confidence is below the threshold, the sentence will be passed to the rule-based component for further analysis. The overall architecture of the system is shown in Figure 1.

3.2   Statistical POS Tagger

The statistical component of the tagger is developed using HMM. During the training phase, we used supervised learning approach where POS tagged text is used for training. The lexical POS tag probability of each word in the training set is computed from the corpus itself. A bi-gram model, which is recommended for large number of tagsets or small training data [9], was used to compute the contextual POS tag probability. The lexical and contextual probabilities provide information about the POS tag probability of each word and contextual POS tag probabilities for a given word, respectively. During the test phase, the HMM decodes the optimal sequence of POS tags in a given sentence using the Viterbi algorithm.

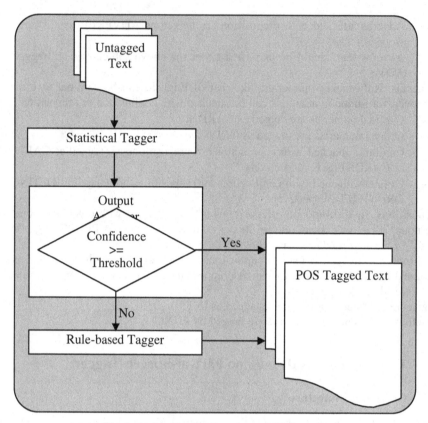

Fig. 1. Architecture of the proposed POS tagger

3.3   Output Analyzer

The purpose of the output analyzer is to compute sentence level POS tag probability and decide if the statistical component of the system needs to be complemented by the rule-based method. The sentence level POS tag probability measures the confidence on POS tags at sentence level. It is computed by adding the probabilities of POS tags for each word in a given sentence and normalized by the number of words. The sentence level POS tag *SPOS* can be expressed mathematically as:

$$SPOS = \frac{\sum_{i=0}^{n} P(W_i / T_i)}{n} \tag{1}$$

where n is the number of words in the sentence and $P(W_i / T_i)$ the lexical POS tag probability for a given word. If the *SPOS* value (confidence measure) for a given sentence is greater than or equal to a pre-defined threshold value, then the HMM result will be accepted. Otherwise, the rule-based tagger will be employed for further analysis.

3.4   Rule-Based POS Tagger

As Kafi-noonoo is one of the least under-resourced languages, it was difficult to get large corpus to be used for training. Thus, the statistical approach was implemented using limited amount of corpus. This situation induces a necessity to supplement the statistical tagger with rule-based tagger. The rule-based tagger will be employed when the output analyzer decides that no sufficient confidence exists in the outputs of the statistical tagger. To develop the rule-based POS tagger, we adopted transformation-based error-driven learning (TEL) approach [4] with modifications on the learners' templates to fit with Kafi-noonoo language features such as prefixes, the maximum length of characters that can be deleted from the beginning of the words so as to predict the tag of unknown words, the length of words/tags that can be allowed before and/or after a given word to find the tag of the word based on contextual information, etc. The rule-based POS tagger has four sub-components: *Initial-State Tagger, Lexical Rule Learner, Contextual Rule Learner,* and *Rule.* The general structure of the rule-based POS tagger is illustrated in Figure 2.

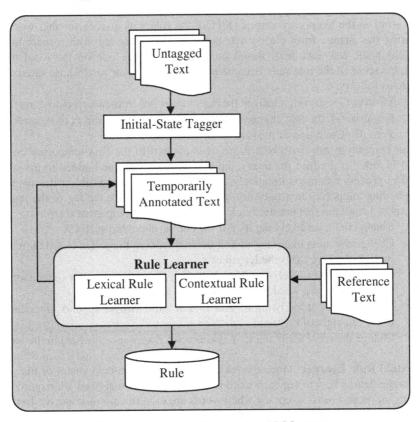

Fig. 2. Architecture of the rule-based POS tagger

Initial-State Tagger: The initial-state tagger sub-component takes untagged Kafi-noonoo text and tags with their most likely POS tag. Different initial-state taggers that range from stochastic n-gram taggers to default taggers (e.g. that label all words as nouns) can be employed. In this work, we used the results of the statistical POS tagger as initial-state tagger.

Lexical Rule Learner: The lexical rule learner is used to derive lexicon and the rules which assign the most likely tag for a given word that may or may not be seen in the training corpus. The lexicon is computed using a statistical method and it contains every word within the training corpus associated with its most frequent tag. It is used to tag untagged words that are seen at least once during the training phase. In order to generate the lexical rules, the lexical rule learner takes untagged Kafi-noonoo text and passes it through the initial-stage tagger to produce a Kafi-noonoo temporary corpus called KTC_0. Following this, based on the condition which is defined in the lexical rule learner template, it finds the rule which gets the best permissible score when applied to KTC_0. A best score for a rule means a rule that gives better resemblance with a reference text when applied to KTC_0. It can be computed as follows: for each tagged word in the temporary corpus (KTC_0), the rule gets a score for that word by comparing the change from the current tag to the resulting tag with respect to the word within the reference text. Based on the effect of the rule on the word to be tagged, the score of the rule may become positive, negative or zero whose interpretations are as follows:

-   Positive (+): the rule changes the tag of the word from incorrect to correct.
-   Negative (-): the rule changes the tag of the word from correct to incorrect.
-   Zero (0): Condition of the rule is not satisfied.

After computing rule with best score, it is applied to the first temporary corpus (KTC_0) in order to produce the next temporary corpus KTC_1 and added to the set of rules. The process continues iteratively to produce all the permissible rules with the corresponding temporary text until no rule can further improve the tag of the temporary corpus. Templates that are used in lexical rule learner component are:

-   Change the most likely tag to Y if the current word has suffix X
-   Change the most likely tag to Y if deleting/adding the suffix X, |X|<3, results in a word where |X| is the length of x.
-   Change the most likely tag from X to Y if deleting the prefix (character) X, |X|<3, results in a word where |X| is length of x.
-   Change the most likely tag from X to Y if word W ever appears immediately to the left/right of the word.
-   Change the most likely tag to Y if character Z appears anywhere in the word.

Contextual Rule Learner: Once the lexical rule learner sub-component of the rule-based tagger learns how to tag each word in a corpus, contextual rules are required for disambiguation and better accuracy when words are used in various contexts. In order to make accurate prediction of tags for words, the contextual rule learner finds rule on the basis of context of the word. To generate contextual rule, the learner initially accepts both temporary and reference texts as inputs. Then, the learner generates all possible rules from the predefined contextual rule template whenever there is a trigger. A trigger is a set of predefined conditions in the form of templates that must be satisfied (become true) to generate the contextual rule. Triggers that are used in the contextual rule learner template are:

- The preceding/following word is tagged as X
- One of the two preceding/following words is tagged as X
- One of the three preceding/following word is tagged as X
- The preceding word is tagged as X and the following word is tagged as Y
- The preceding/following two words are tagged as X and Y
- The two words before/after are tagged as X

After generating all possible rules, the learner computes the score of each rule for a particular word. Based on the score, the learner picks rules with highest score and stores it in the sub-component of rule-based tagger called rules. For each word W in the temporary corpus, the learner computes the score. This can be achieved by comparing the tag of words in the temporary corpus after applying the rule with that of the reference text. If the rule is applied to the word and corrects an error, the score of the rule is +1; while the rule introduces an error, the score of the rule is -1; and otherwise, the score of the rule is 0. The total score of each rule is computed by adding scores of the rule when it is applied to each word within the temporary corpus. Following this, the learner takes the rules that are stored in the rule component of the tagger and applies them on the temporary corpus to generate another temporary corpus. The process continues in the same fashion until no rule exists that makes the temporary corpus resemble with that of the reference text.

Rule: The rule sub-component has two main parts, namely triggers (condition or current tag) and rewrite (resulting tag). The rules take the form of "if *trigger*, then *change tag X to Y*". The "if" part represents the "trigger" component and the "then" part represents the "rewrite" component of the rules. After a set of lexical and contextual rules are learned through lexical and contextual rule learner sub-components, the rules are stored in a file. In addition, the rule-based component builds a lexicon to tag words that are seen at least once during training phase. All rules and entries in the lexicon are stored permanently. As a result, the rule-based component of the system becomes a trained model that is able to tag the untagged texts of Kafi-noonoo language using the stored set of rules.

4   Experiment

4.1   Corpus Preparation

To our best knowledge, there is no POS tagged Kafi-noonoo corpus that can be used as input for training. Thus, we prepared a tagged corpus from a flat file using an incremental corpus preparation approach. Incremental corpus preparation is a semi-automatic approach that involves three main stages: manual, automatic and correction stages. In the manual stage, the text collected from different sources is passed to the annotators for manual annotation and the output of the manual annotation is used to train the POS tagger. The trained POS tagger will then tag new text in the automatic stage. In the correction stage, the output of the trained tagger is passed to the annotator for manual correction. The output obtained from the correction stage is added to the training set to contain the approved text which is later used for training the tagger. Starting from the automatic stage, the process is repeated until all the raw text is tagged. In this way, a corpus size of 354 sentences was tagged. A total of 34 POS tags were identified and the tagset indicates only word classes rather than gender, number, tense, etc.

4.2   Implementation

The proposed Kafi-noonoo POS tagger was implemented using Python programming language. We used the open source toolkit NLTK which contains Python modules, linguistic data and documentation for research and development in the area of natural language processing [3].

4.3   Test Results

To test the performance of the proposed POS tagger, experiments were conducted for the statistical tagger, rule-based tagger and hybrid tagger. The tests were made by using various sizes of the corpus as training set. Figure 3 shows the performance curve analysis of statistical and rule-based taggers. As it can be seen in the figure, the statistical tagger outperformed the rule-based tagger especially when higher percentage of corpus is used for training. However, given the limited corpus resource available for the language, the statistical tagger alone could not be reliable. Thus, the rule-based tagger was used to complement the statistical tagger by analyzing its output. The output of the statistical tagger is analyzed by taking the confidence measure of the individual tags of words in a sentence as discussed in Section 3.3. When the confidence measure falls below a specific threshold value, results from the rule-based tagger will be selected. Otherwise, results from the statistical tagger will be selected.

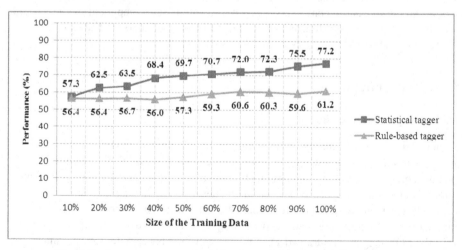

Fig. 3. Performance curve analysis for the statistical and rule-based taggers

The performance of the hybrid tagger was evaluated empirically for various threshold values and it yielded better results than the individual statistical and rule-based taggers. Figure 4 depicts the performance of the hybrid tagger for various threshold values when 90% of the corpus was used for training. We can see from the graph that the best tag results are obtained for threshold values close to 0.5.

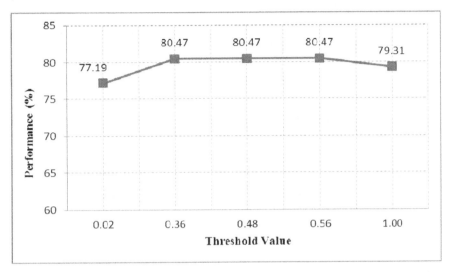

Fig. 4. Performance of the hysbrid tagger for various threshold values

5   Conclusion and Future Work

Kafi-noonoo is one of the most under-resourced languages. Resources used to develop NLP applications for the language are very limited. Although part-of-speech tagger is an essential component required in many high level NLP applications, the development of Kafi-noonoo POS tagger is itself deterred by unavailability of sufficient corpus. Noting these problems, we developed a hybrid system that takes advantage of the synergy effect of statistical and rule-based taggers. Although statistical approaches are known to be among the most successful methods to develop POS taggers, the limited corpus available for under-resourced languages hinders to use the full potential of statistical approaches. Thus, in the proposed hybrid system, the statistical tagger is complemented by a rule-based tagger. The selection of the statistical or rule-based tagger results is made by an output analyzer that measures the confidence of the statistical tagger at sentence level. Test results showed that the hybrid of statistical and rule-based taggers performed better than individual components. To train and test the developed system, we collected a Kafi-noonoo corpus of 354 sentences and identified 34 tagsets for the language. The corpus was annotated using an incremental corpus preparation approach. Future work is recommended to be focused on refining the transformational rules in the rule-based component and training the statistical component with better corpus size.

References

1. Allen, J.: Natural language Understanding. The Benjamin/Cummings Publishing Company, Redwood (1995)
2. Altunyurt, L., Orhan, Z., Güngör, T.: A Composite Approach for Part of Speech Tagging in Turkish. In: Proceeding of International Scientific Conference on Computer Science, Istanbul, Turkey (2006)
3. Bird, S., Klein, E., Loper, E.: Natural Language processing with python: Analyzing text with the natural language toolkit. O'Reilly Media, Cambridge (2009)
4. Brill, E.: Transformation-Based Error-Driven Learning and Natural Language Processing: A Case Study in Part-of-Speech Tagging. Computational Linguistics 21(4), 543–565 (1995)
5. Dand, S., Sarkar, S., Basu, A.: Automatic Part-of-Speech Tagging for Bengali: An Approach for Morphologically Rich Languages in a Poor Resource Scenario. In: Department of Computer Science and Engineering, Kharagpur, India, Indian Institute of Technology (2007)
6. Harold, F.: The non-Semitic languages of Ethiopia. Michigan State University, Michigan (1976)
7. Jurafsky, D., Martin, J.: Speech and Language Processing: An Introduction to Natural Speech Recognition. Prentice-Hall, New Jersey (2000)
8. Mamo, G., Meshesha, M.: Part-of-Speech Tagging for Afaan Oromo Language. Inter. Journal of Advanced Computer Science and Applications 1(3), 1–5 (2011)
9. Nivre, J.: Sparse data and smoothing in statistical part-of-speech tagging. Journal of Quantitative Linguistics, 1–17 (2000)
10. Zin, K.: Hidden Markov Model with Rule Based Approach for Part of Speech Tagging of Myanmar Language. In: Proceedings of the 3rd International Conference on Communications and Information Technology, Florida, pp. 123–128 (2009)

Modified Differential Evolution for Biochemical Name Recognizer

Utpal Kumar Sikdar, Asif Ekbal, and Sriparna Saha

Computer Science and Engineering
Indian Institute of Technology patna
Patna, India
{utpal.sikdar,asif,sriparna}@iitp.ac.in

Abstract. In this paper we propose a modified differential evolution (MDE) based feature selection and ensemble learning algorithms for bio-chemical entity recognizer. Identification and classification of chemical entities are relatively more complex and challenging compared to the other related tasks. As chemical entities we focus on IUPAC and IU-PAC related entities. The algorithm performs feature selection within the framework of a robust machine learning algorithm, namely Conditional Random Field. Features are identified and implemented mostly without using any domain specific knowledge and/or resources. In this paper we modify traditional differential evolution to perform two tasks, *viz.* determining relevant set of features as well as determining proper voting weights for constructing an ensemble. The feature selection technique produces a set of potential solutions on the final population. We develop many models of CRF using these feature combinations. In order to further improve the performance the outputs of these classifiers are combined together using a classifier ensemble technique based on modified DE. Our experiments with the benchmark datasets yield the recall, precision and F-measure values of 82.34%, 88.26% and 85.20%, respectively.

Keywords: Modified Differential Evolution (MDE), Conditional Random Field (CRF), Feature Selection, Ensemble, Biochemical Named Entity.

1 Introduction

In recent times, information extraction has drawn huge attention to the practitioners and researchers. A large amount of online information is unorganized and a large number of data documents are added to it daily, so organizing, finding and extracting relevant information from such a huge amount of data is an important challenge in our day-to-day life. In life science publications and patents, chemical compounds like small signal molecules or other biological active chemical substances are the important entity classes. There exist many representations and nomenclatures for chemical names. Some examples are SMILES, InChI and IUPAC, out of which the first two allow a direct structure search, but IUPAC

A. Gelbukh (Ed.): CICLing 2014, Part I, LNCS 8403, pp. 225–236, 2014.

like names are more frequent in biochemical texts. Trivial chemical names can be easily found using a dictionary-based approach and can be subsequently mapped to their corresponding structures. In contrast it is not feasible to enumerate all IUPAC like names. Automatic identification of mentions of chemical compounds in text is of interest for a variety of reasons. This has potential application to the different text mining tasks that include but not limited to the predictions of drug-drug/protein-protein interactions, finding relations to adverse reactions of chemical compounds and their associations to toxicological endpoints or the extraction of pathway and metabolic reaction relations. It helps in semantic search by enabling the search engine to return documents containing elements of the entity class.

The performance of any classification technique depends on the features of training and test data sets and parameters of the model. Feature selection [8,7], also termed as variable selection, attribute selection or variable subset selection, is a commonly used technique in pattern recognition and machine learning domains. In order to solve any problems related to these domains multiple features or attributes are extracted. All such features are not effective to solve the target problem. Efficient feature selection, is thus, plays an important role. In this paper, we propose single objective optimization based feature selection and classifier ensemble techniques where we optimize F-measure metric. As an optimization technique we use *modified differential evolution* algorithm. The modified differential evolution is quite similar to the conventional differential evolution [10] except the mutation operator. In the first stage, we perform feature selection within the framework of a robust machine learning algorithm, namely Conditional Random Field (CRF)[6] using a modified differential evolution algorithm. Thereafter in the second step we combine the solutions, obtained in the first step, using a modified differential algorithm based ensemble technique. For ensemble we determine the appropriate voting weights for each class in a classifier.

In recent past there has been some efforts [2,4,1] for building evolutionary algorithm based feature selection and ensemble techniques, particularly focussing on text processing domains. A single objective optimization (SOO) based classifier ensemble technique was proposed in [3]. This was evaluated for named entity extraction in multiple natural language texts. In addition a genetic algorithm (GA) based feature selection technique was also introduced. This optimization technique is based on genetic algorithm [5] which is a randomized search and optimization technique guided by the principles of evolution and genetics, having a large amount of implicit parallelism. In [1], a GA based classifier ensemble selection technique was developed. This approach determines only a subset of classifiers that can form the final classifier ensemble. In [4] a multiobjective optimization based ensemble technique was developed for classifier ensemble. Along with feature selection exhaustive evaluation was also carried out. In [9], a differential algorithm based feature selection and ensemble technique was developed. This algorithm was based on single objective optimization.

The work reported in this paper deals with the problems of information extraction, especially entity extraction in biochemical domain, which is more difficult and challenging. The inherent structures of the these entities pose a big challenge for their identification. Our current paper focusses on developing single objective feature selection and ensemble learning techniques. Unlike the previous ones [2,4,1,3], here differential evolution is employed as an optimization technique. Differential evolution has a different perspective compared to genetic algorithm. The technique proposed in [9] was evaluated for named entity (NE) extraction in three different Indian languages, namely Bengali, Hindi and Telugu. Here in our present work we develop an algorithm based on the modified differential evolution. Therefore the algorithms proposed in [9] are different to what we propose here. In addition the proposed algorithm is evaluated for a more complex problem, i.e. entity extraction in biochemical domain. Evaluation on a benchmark dataset yields the recall, precision and F-measure values of 82.34%, 88.26% and 85.20%, respectively.

2   Overview of Modified Differential Evolution

Differential Evolution (DE) [10] is a parallel direct search method which performs search in large, complex and multi-modal landscapes, and provides near-optimal solutions of an optimization problem. In DE, within the search space parameters are encoded to form a chromosome. A set of such type of chromosomes is called a population denoted by NP. It is a collection of NP number of D-dimensional parameter vectors $x_{i,G} = [x_{1,i,G}, x_{2,i,G}, \ldots, x_{D,i,G}]$, $i = 1, 2, \ldots, NP$ for each generation G. The value of D represents the number of real parameters on which optimization or fitness function depends. During the optimization process the size of the population, NP does not change. The initialization of the first generation population is chosen randomly and it represents different search points and should cover the entire parameter space. A fitness or an objective function is associated with each chromosome that indicates the goodness of the chromosome. In DE, new parameter vectors are generated by adding the weighted difference between two chromosomes to a third chromosome in a population. This operation is called mutation. In this paper we modify *the mutation operation* of traditional DE. In case of new mutation operator, at first best vector(i.e chromosome) with respect to the objective function is identified from the current population, and the weighted difference between two randomly chosen population vectors are added to the best vector. The parameters of the predetermined vector(also called target vector)and mutated vector are mixed and it produces a new vector, called the trial vector. Such type of parameter mixing is often called to as "crossover". If the fitness value of the trail vector is better than the target vector, then the target vector is replaced with the trail vector in the next generation. This last operation is called "selection". The selection, crossover and mutation processes run for the maximum number of generations. The pseudo code for single objective modified differential evolution is shown in Algorithm 1.

Algorithm 1. Pseudo Code for Single Objective Modified Differential Evolution

1: G=0
2: /* Initialization*/
3: Create a random initial population $X_{i,G}, \forall i, i = 1, \ldots, NP$
4: Select best vector, rb from the initial population $X_{i,G}, \forall i, i = 1, \ldots, NP$
5: **for** G=1 to MAX_GEN **do**
6:    **for** i=1 to NP **do**
7:       Select randomly two different chromosomes r1 and r2
8:       /*generate a random integer value from 1 to D */
9:       $j_{rand} = \text{randomInt}(1,D)$
10:      **for** j=1 to D **do**
11:         /*Mutation and Crossover*/
12:         **if** $rand_j[0,1]$ ¡ CR or j=j_{rand} **then**
13:            $u_{i,j,G+1} = x_{rb,j,G} + F \times (x_{r1,j,G} - x_{r2,j,G})$
14:         **else**
15:            $u_{i,j,G+1} = x_{i,j,G}$
16:         **end if**
17:      **end for**
18:      Evaluate $f(U_{i,G+1})$ and $f(X_{i,G})$
19:      /* Selection*/
20:      **if** $f(U_{i,G+1}) > f(X_{i,G})$ **then**
21:         $X_{i,G+1} = U_{i,G+1}$
22:      **else**
23:         $X_{i,G+1} = X_{i,G}$
24:      **end if**
25:   **end for**
26:   Select the best vector rb from the next generation population $X_{i,G+1}, \forall i, i = 1, \ldots, NP$
27: **end for**
end

3   Proposed Method for Feature Selection

In this section we describe the problem of relevant feature selection within the framework of modified differential evolution. Suppose, the D number of available features for a given classifier are denoted by $F_1, \ldots, F_D$. Let, $\mathcal{A} = \{F_i : i = 1; D\}$. The feature selection method is then stated as follows:

Find the appropriate subset of features $\mathcal{A}' \subseteq \mathcal{A}$ such that the classifier, CRF trained using these subset of features should have optimized some classification quality measure, which is in this case F-measure.

3.1   Chromosome Representation and Population Initialization

If a chromosome contains D number of features, then the length of the chromosome is D. Each bit of the chromosome denotes the presence or absence of the corresponding feature. The chromosomes in a population are randomly initialized to either 0 or 1. Here, if the parameter value of the ith position is 1 then the ith feature participates in constructing the CRF based classifier. Else, if the ith feature parameter value is 0 then the feature does not participate in constructing the CRF based classifier. If the size of population is NP then all the NP number of chromosomes are initialized in the above way.

3.2   Fitness Computation

For the objective function or fitness function computation, the following steps are executed.

1. Suppose, in a particular chromosome K number of features are present (i.e., there are total K number of 1's present in the chromosome).
2. Using these K number of features, construct the classifier.
3. We perform three-fold cross validation. Initially we divide the training data into three parts. Two subsets are used for training and the remaining part is used for testing. The classifier is trained using the features encoded in the chromosome.
4. Compute the recall, precision and F-measure values of this classifier on the test data.
5. Steps 2-3 are repeated three times to perform three-fold cross validation. The average F-measure value is used as the objective function, and this is maximized using the search capability of the modified differential evolution algorithm.

3.3   New Mutation Operator

For each target vector $x_{i,G}$; $i = 1, 2, 3, \ldots, NP$, a mutant vector/donor vector is generated according to

$$v_{i,G+1} = x_{rb,G} + F(x_{r1,G} - x_{r2,G}), \tag{1}$$

where rb is the best vector with respect to the objective function value within the current population (vector with maximum F-measure value) and $r1$ and $r2$ are the random indices and belong to $\{1, 2, \ldots, NP\}$. The value of $r1$ and $r2$ are mutually different and $F > 0$. The value of $r1$ and $r2$ which are chosen randomly are different from the running index rb and i, so that NP must be greater or equal to four(three in case when i and rb arc the same vectors). The value of F is a real constant factor which belongs $[0, 1]$. Here the value of F is 0.5. It controls the amplification of the differential variation $(x_{r1,G} - x_{r2,G})$.

3.4   Crossover

Crossover is needed because of increasing the diversity of the target vectors. This is well-known as the process of recombination (or crossover). To this end, the trial vector:

$$u_{i,G+1} = (u_{1i,G+1}, u_{2i,G+1}, \ldots, u_{Di,G+1}) \tag{2}$$

is formed, where

$$u_{j,i,G+1} = v_{j,i,G+1} \quad \text{if} \quad (randb(j) \leq CR) \quad \text{or} \quad j = rnbr(i) \tag{3}$$
$$= x_{j,i,G} \quad \text{if} \quad (randb(j) > CR) \quad \text{and} \quad j \neq rnbr(i) \tag{4}$$

for $j = 1, 2, \ldots, D$,

In Equation 3, the value of $randb(j)$ which is chosen randomly, belongs to $[0, 1]$. CR is the crossover constant which has to be determined by the user. It can take any value between $[0, 1]$. In our case we keep the value of CR equal 0.5. $rnbr(i)$ is a randomly chosen index x that belongs to $\{1, 2, \ldots, D\}$ which ensures that trial vector $u_{i,G+1}$ gets at least one parameter from mutant vector $v_{i,G+1}$.

3.5   Selection

In selection process, the fitness value of the target vector $x_{i,G}$ is compared to the fitness value of trial vector $u_{i,G+1}$ using the greedy criterion. If target vector $x_{i,G}$ yields a higher cost function value than the trial vector $u_{i,G+1}$, then the next generation target vector $x_{i,G+1}$ is assigned to the trial vector $u_{i,G+1}$; otherwise, the old value of the target vector $x_{i,G}$ is retained to the next generation target vector $x_{i,G+1}$.

3.6   Termination Condition

The processes of mutation, crossover (or, recombination), fitness computation and selection are executed for a maximum number of generations and the last generation provides the best subset of features.

4   Method for Classifier Ensemble

The first step of the algorithm yields a set of solutions on the final best population. The solutions that are generated during feature selection are equally important and we generate several different classifiers using these feature combinations. We combine all these solutions using a single objective optimization based ensemble technique.

The weighted vote based classifier ensemble problem   [3] is stated below. Suppose, there are N number of classifiers that are denoted by $C_1, \ldots, C_N$. Let, $A = \{C_n : n = 1; N\}$ and M output classes. We form the classifier ensemble problem as follows:

Find out the voting weights V per classifier which will optimize fitness function $F(V)$ using the search capability of the modified differential evolution. The size of the V is $N \times M$ and it represents a real array. $V(n, m)$ represents the voting weights of the nth classifier for the mth class. These weights are very important, and can vary from one generation to the other. The algorithm will find the appropriate values while combining the outputs of the classifiers.

The ensemble problem under single objective optimization can be formulated as: For each classifier, find the weights of votes V per classifier such that, $maximize$ $[F(V)]$, where $F \in \{\text{recall}, \text{precision}, \text{F-measure}\}$. We optimize $F=$ F-measure as the objective function.

4.1 Chromosome Representation and Population Initialization:

If N is the total number of classifiers and M is the total number of output classes, then the chromosome length is $D = N \times M$. Weights of votes are assigned to each classifier for the possible M classes.

As an example, the chromosome representation is shown in Figure 1. Here, if $N = 3$ and $M = 3$ then total nine($3 \times 3 = 9$) votes can be possible. The chromosome represents the following ensemble:

Suppose, the voting weights of output classes are 0.59, 0.12 and 0.56, respectively for classifier 1. Similarly, for classifier 2, weights of votes for 3 different output classes are 0.09, 0.91 and 0.02, respectively and 0.76, 0.5 and 0.21, respectively for classifier 3.

We apply real encoding, and all the chromosomes in the entire population are randomly initialized to a real value (r) which belongs to $[0, 1]$. Here, $r = \frac{rand()}{RAND\_MAX+1}$. If the population size is NP then all the NP number of chromosomes are initialized in the above way.

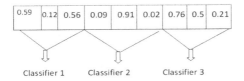

Fig. 1. Chromosome representation for the ensemble selection

4.2 Objective Functions Computation

We execute the following steps to compute the objective function values.

1. Suppose, there are total N number of classifiers. Let, the overall F-measure values of these N classifiers be F_n, $n = 1 \ldots N$.
2. We have M classes (each from a different classifier) for each token. Now for the ensemble classifier, the output class for each token is determined using the weighted voting of these N classifiers' outputs. The weight of the output class provided by the nth classifier is equal to F_n. The final weight of a particular class for a particular token t is:

$$f(o_m) = \sum F_n \times C(n, m),$$

$$\forall n = 1 \text{ to } N \text{ and } op(t, n) = o_m$$

Here, $C(n, m)$ corresponds to the nth classifier and m^{th} class; and $op(t, n)$ denotes the output class provided by the classifier n for the token t. Hence the token receives that particular class label that gets maximum voting weight.
3. We compute F-measure value which is optimized using the modified differential evolution algorithm.

Operators: Other operators of the modified differential evolution are similar to those of the feature selection approach as described above.

5   Features for Chemical Entity Extraction

We implement a diverse set of features for constructing the classifier. Though we apply these feature for the chemical domain these can be useful for any other related domains such as entity extraction in biomedical texts.

Context Words: Local contextual information carries effective information for identification of biochemical names. Here we consider the contextual information of the previous three and next three tokens.

Word Prefix and Suffix. These are the word prefix and suffix character sequences of length up to n. The sequences are stripped from the leftmost (prefix) and rightmost (suffix) positions of the words. We experiment with $n=3$ (i.e., 6 features) and 4 (i.e., 8 features) both.

Word Length. We define a binary-valued feature that fires if the length of w_i is greater than a pre-defined threshold. Here, the threshold value is set to 5. This feature captures the fact that short words are likely not to be the chemical names.

Infrequent Word. A list is compiled from the training data by considering the words that appear less frequently than a predetermined threshold, i.e. 10 in our current experiment. Now, a feature is defined that fires if w_i occurs in the compiled list. This is based on the observation that more frequently occurring words are rarely the chemical names.

Part-of-Speech (PoS) Information: PoS information is a critical feature for entity extraction. In this work, we use PoS information of the current and/or the surrounding token(s) as the features. This information is obtained using GENIA tagger V2.0.2 [1]. The PoS class for each word is assigned by GENIA tagger.

Chunk Information: Shallow parsing information such as phrase helps to identify the boundaries of entities properly. This was extracted from the GENIA tagger.

Lemma Information: We use the stems or root forms of the surface-level wordforms as the feature. This was again extracted from the GENIA tagger.

Unknown Token Feature: We define a feature that is set to high for the unknown token, i.e. for the token that appears in the test set but does not appear in the training set. For the training set the values to this feature were set at random. **Word Normalization:** Word shapes refer to the mapping of each word to their equivalence classes. Here each capitalized character of the word is replaced by 'A', small characters are replaced by 'a' and all consecutive digits are replaced by '0'. For example, 'IL-88' is normalized to 'AA-00'. This feature will group similar names into the same chemical class.

Orthographic Features: We define a number of orthographic features depending upon the contents of the wordforms. These features are: initial capital,

---

[1] http://www-tsujii.is.s.u-tokyo.ac.jp/GENIA/tagger

all capital, capital in inner, initial capital then mix, only digit, digit with special character, initial digit then alphabetic, digit in inner. The presence of some special characters like (',','-','.',')','(' etc.) is very much helpful to detect chemical names. For example, many biochemical names have '-' (hyphen) in their construction. We also use the features that check the presence of ATGC sequence and stop words. In total, we define 24 features based on the orthographic constructs.

Informative Words: Most frequently occurring words in the surrounding contexts of the chemical names provide useful evidences for chemical name identification. By extracting the frequent such context words from the training data, we prepare two lists, one for the words that appear preceding chemical name and the other for the words that follow the chemical name.

Chemical Prefix and Suffix: From the IUPAC and IUPAC like names present in the training data, we extract the most frequently occurring prefixes and suffixes of length two. Thereafter two binary valued features are defined. The values of these features are set to 1 if the sequences of two characters stripped either from the beginning or from the end positions of words do have matches with the strings stored in the lists. **PubChem Prefix and Suffix:** Most frequently occurring prefixes and suffixes of length two are extracted from the IUPAC chemical names of PubChem database [2]. A binary valued feature is then defined that fires if and only if any of these inflections matches with the character sequences stripped either from the beginning or from the end positions of words.

Dynamic NE Information: This is the output label(s) of the previous token(s). The value of this feature is determined dynamically at run time.

6   Dataset

Biochemical names can be presented in various forms, one standardized representation is the International Union of Pure and Applied Chemistry (IUPAC). It is a systematic way of naming convention that map their chemical structures. The description of datasets includes IUPAC and IUPAC like names. We experiment with the benchmark datasets, available at this site [3]. The training and test datasets were taken from the Medline database and collection of patent documents, respectively. The test dataset contains seven classes, namely IUPAC(e.g. N-methyl), PARTIUPAC(partial chemical names such as 3H-Testosterone, here "3H" is an IUPAC name), TRIVIAL(trade, common or generic names of compounds such as paracetamol, aspirin etc.), MODIFIER, SUM(molecular formula such as C9H8O4), ABBREVIATION(abbreviations and acronyms of chemicals compounds and drugs such as DMSO) and FAMILY(chemical names associated to some chemical structure like terpenoids). However, the training dataset has only the instances of IUPAC, PARTIUPAC and MODIFIER classes. Thus, in the test data we convert the classes that are not available in the training data

---

[2] http://pubchem.ncbi.nlm.nih.gov/

[3] http://www.scai.fraunhofer.de/chem-corpora.html

Table 1. Statistics of datasets: #abstracts (Total number of abstracts), #sentences (Total number of sentences), #tokens (Total number of tokens/words) and #IUPAC (IUPAC and IUPAC like names)

Dataset	#abstracts	#sentences	#tokens	#IUPAC
Training dataset	463	3,700	1,61,591	3,712
Test dataset(Patent)	27	160	4,417	471

to the other-than-chemical-name denoted by "O" class. Statistics of the training and test datasets are presented in Table 1.

7   Experiments

The parameters of the proposed algorithm are selected by conducting a thorough sensitivity analysis on the development data. We use a part of training as the development set. The parameters of MDE for feature selection are set as follows: population size = 30, CR (probability of crossover) = 0.5, number of generations = 20 and F (mutation factor) = 0.5. Here we generate different feature combination using the proposed feature selection technique, where we use both the modified differential evolution and simple differential evolution. We identify the fourteen promising classifiers from the feature selection technique with respect to the values of precision and recall, and combine them using the ensemble technique. The parameters of MDE for ensemble are set as follows: population size:60; number of generations:300; other parameters are kept same as that of the feature selection.

We define the baseline by training CRF with the local contexts of previous one and next one tokens along with all the features listed in section 5. Based on proposed techniques, we develop the following three models:

- *Model 1*: In this model, simple differential evolution based feature selection technique is applied on the test data. Here we select the best solution with respect to F-measure value from the final population.
- *Model 2*: This model corresponds to the modified differential evolution based feature selection technique. Here we select the best solution with respect to the F-measure value from the final population.
- *Model 3*: In this model, simple differential evolution(DE) based classifier ensemble technique is used to combine the fourteen classifiers selected from the first stage of the proposed approach.

Results of the baseline and three different models are shown in Table 2. The baseline which is constructed by including all the features in CRF model yields the recall, precision and F-measure values of 90.22%, 72.91% and 80.65%, respectively. The first model where the simple differential evolution is used demonstrates an increment of 2.56 point F-measure over the baseline. Hence this is evident that careful feature selection is important. The second model that makes use

Table 2. Overall evaluation results

Methods	recall	precision	F-measure
Baseline	90.22	72.91	80.65
Model 1	78.80	88.11	83.19
Model 2	79.21	88.76	83.71
Model 3	81.02	87.38	84.08
Our Proposed Method	82.34	88.26	85.20

of the modified DE yields the recall, precision and F-measure values of 79.21%, 88.76% and 83.71%, respectively. The third model based on conventional DE achieves further performance improvement. Finally our proposed algorithm obtains the recall, precision and F-measure values of 88.26%, 82.34%, and 85.20%, respectively.

8   Conclusion

In this paper we present our work on feature selection and classifier ensemble for biochemical entity extraction. Our proposed methods were based on the modified differential evolution, where we changed in the mutation operator. Feature selection was performed for Conditional Random Field (CRF) using both simple DE and the modified DE. The classifier is trained using a diverse feature set. One important characteristic of the features is that though these have been applied on the biochemical domain, these can also be applied for the other related domains. We select some promising solutions from the final population, and combine the outputs using a modified DE based ensemble technique. The proposed technique is evaluated on a benchmark dataset of chemical domain. In future we plan to carry out experiments on the recent datasets, made publicly available through the BioCreative campaigns.

References

1. Ekbal, A., Saha, S.: Classifier ensemble selection using genetic algorithm for named entity recognition. Research on Language and Computation 8, 73–99 (2010)
2. Ekbal, A., Saha, S.: Weighted vote based classifier ensemble selection using genetic algorithm for named entity recognition. In: Proceedings of the Natural Language Processing and Information Systems, NLDB 2010, pp. 256–267 (2010)
3. Ekbal, A., Saha, S.: Weighted vote-based classifier ensemble for named entity recognition: A genetic algorithm-based approach. ACM Trans. Asian Lang. Inf. Process. 10(2) (2011)
4. Ekbal, A., Saha, S.: Multiobjective optimization for classifier ensemble and feature selection: an application to named entity recognition. IJDAR 15(2), 143–166 (2012)
5. Goldberg, D.E.: Genetic Algorithms in Search, Optimization and Machine Learning. Addison-Wesley, New York (1989)

6. Lafferty, J.D., McCallum, A., Pereira, F.C.N.: Conditional Random Fields: Probabilistic Models for Segmenting and Labeling Sequence Data. In: ICML, pp. 282–289 (2001)
7. Liu, H., Motoda, H.: Feature Selection for Knowledge Discovery and Data Mining. Kluwer Academic Publishers, Norwell (1998)
8. Liu, H., Yu, L.: Toward integrating feature selection algorithms for classification and clustering. IEEE Trans. on Knowl. and Data Eng. 17(4), 491–502 (2005)
9. Sikdar, U.K., Ekbal, A., Saha, S.: Differential evolution based feature selection and classifier ensemble for named entity recognition. In: COLING, pp. 2475–2490 (2012)
10. Storn, R., Price, K.: Differential evolution a simple and efficient heuristic for global optimization over continuous spaces. J. of Global Optimization 11(4), 341–359 (1997)

Extended CFG Formalism for Grammar Checker and Parser Development

Daiga Deksne, Inguna Skadiņa, and Raivis Skadiņš

Tilde, Riga, Latvia
{daiga.deksne,inguna.skadina,raivis.skadins}@tilde.lv

Abstract. This paper reports on the implementation of grammar checkers and parsers for highly inflected and under-resourced languages. As classical context free grammar (CFG) formalism performs poorly on languages with a rich morphological feature system, we have extended the CFG formalism by adding syntactic roles, lexical constraints, and constraints on morpho-syntactic feature values. The formalism also allows to assign morpho-syntactic feature values to phrases and to specify optional constituents. The paper also describes how we are implementing the grammar checker by using two sets of rules – rules describing correct sentences and rules describing grammar errors. The same engine with a different rule set can be used for the different purposes – to parse the text or to find the grammar errors. The paper also describes the implementation of Latvian and Lithuanian parsers and grammar checkers and the quality measurement methods used for the quality assessment.

Keywords: parsing, grammar checking, inflected languages.

1 Introduction

Proofing tools have been in development for a rather long time. Tools for checking spelling are available for many languages in different word processing applications as well as other natural language applications. However, a more complicated task for computers is grammar checking. Due to the high ambiguity of languages, grammar checking tools are available for a rather small number of languages. Moreover, even grammar checking tools for the English language only allow for the correction of certain types of errors.

The problem becomes even more complicated when it concerns highly inflected languages that have a rather free word order. Only a few grammar checkers have been developed for such languages (e.g., there are several grammar checkers for the Russian language).

In this paper, we present a framework for grammar checking that is derived from a context-free grammar (CFG) formalism. A classical CFG performs poorly on inflected languages, e.g., large numbers of non-terminals are necessary for representation of morpho-syntactic features, as well as parser output usually consists of many parse trees. Thus different syntactic formalisms derived from CFG (e.g., Generalized Phrase

A. Gelbukh (Ed.): CICLing 2014, Part I, LNCS 8403, pp. 237–249, 2014.
© Springer-Verlag Berlin Heidelberg 2014

Structure Grammar introduces mechanism for feature passing [10], Definite Clause Grammar expresses grammar as clauses of first-order predicate logic [19]) have been developed. In addition many syntactic formalisms that adopt phrase structure are proposed: for instance, Head-Driven Phrase Structure Grammar adopts the basic phrase structure syntax through unification of feature structures [20], Lexical Functional Grammar use constituent structure together with feature structure for syntax representation [14], Augmented transition network formalism [4] realizes unification through recursive transition network.

In this paper we propose to extend CFG by adding morpho-syntactic features and syntactic roles, and by introducing two operators - constraint checking operator and assignment operator. Additionally, rules for grammar checking are divided into two sets – one rule set for parsing and recognizing correct patterns and another rule set for error detection and correction. Our grammar checking system allows to correct 23 types of errors, including syntactic errors, style errors, and capitalization errors. For inflected languages, the most important groups of errors are word agreement errors and errors that are related to punctuation in specific constructions.

The developed framework is used to implement grammar checkers for two languages of the Baltic language group - Latvian and Lithuanian. The evaluation results for these languages are presented and discussed in this paper. In addition, we also demonstrate how the developed framework can be used for grammar checking of other inflected languages, e.g., the Slavic language group.

2   Related Work

The grammar checking problem has been actual since the 1970s, when language technologies obtained their intelligence. The first grammar checkers checked punctuation and style inconsistencies. In the early 80s, grammar checkers were released for personal computers, and, soon afterwards, grammar checkers that could detect writing errors beyond simple style errors were developed. Among grammar checkers we would like to mention the grammar checker in Word97 for English [11], the rule-based system for Dutch [27], ReGra for Brazilian Portuguese [16], first grammar checker for Latvian [7], grammar checkers for Swedish [1], [8], [22], German [23], and Arabic [24].

Different approaches have been used for grammar checking; the most popular being rule-based (e.g., constraint grammar, context-free grammar), statistical [2], [13], [25], and hybrid [8], [9], [28].

The rule-based approach usually uses manually created rules that can be easily modified, added, or removed. Such rules are linguistically motivated. However, it is not easy to maintain larger systems. One popular approach is Constraint Grammar (CG) formalism, which was originally designed by Fred Karlsson [15] for grammar-based parsing. However, CG parser can be used not only to tag a sentence with surface syntactic functions, but also to mark possible grammar errors. It has been used for the detection of syntactic errors in Swedish [1], [5] and Norwegian [12].

LanguageTool grammar checker[1] [17] is a rule-based open source grammar check-ing system that is a plug-in for *OpenOffice.org*. Currently, it supports 29 languages. However, support significantly varies from language to language.

Despite a long history of development of grammar checkers, there is a lot of space for improvements where it concerns language coverage and algorithms as it is demon-strated for English [18].

3   Extended Context-Free Grammar Formalism

Extended context-free grammar formalism is derived from CFG. Similar to CFG, our formalism contains a description of phrase structure. However, it usually also has a rule body consisting of constraints, lexical restrictions, and value assignment/inheritance statements.

3.1   Structure of the Rules

Context-free grammar, formalized independently by Chomsky [6] and Backus [3], is defined as a 4-tuple (P, N, T, S) with the following components:

— **P** is a set of grammar rules or productions (i.e., items of the form $X \to a$, where X is a non-terminal symbol, and a is a string of terminal and non-terminal symbols)
— **N** is the set of non-terminal symbols (i.e., grammatical or phrasal categories)
— **T** is the set of terminal symbols (i.e., words of the language)
— **S** is a designated start symbol, normally interpreted as representing a full sentence

Classical CFG is powerful and efficient for describing sentence structures of ana-lytic languages that convey grammatical relationships without the use of inflectional morphemes. However, it is not so efficient in describing the sentence structure of synthetic languages that use inflectional morphemes and have a quite free word order. It is especially difficult to describe agreement (or disagreement) between words or phrases. Therefore, we have introduced morpho-syntactic properties to CFG, have allowed non-terminals to inherit these properties from their constituents, and have used these properties to restrict rules.

We use terminals and non-terminals in the same way as CFG does. However, on the right side of the production rule, syntactic roles are added to each constituent as shown in (1) for noun phrase NP:

$$NP \to attr:NP\ main:N \qquad (1)$$

In each rule, the syntactic role *main* is mandatory for the head constituent, and other possible roles include *subj, obj, mod,* etc.

The body of the rule usually contains some constraints and some assignments. The constraints are used to restrict the application area for the rule and to avoid over-generation. They are realized through morpho-syntactic properties of terminals and

---

[1] https://www.languagetool.org/

non-terminals. The set of properties and property values is language specific. For Latvian and Lithuanian, we use 26 properties, such as, number, case, gender, etc. Table 1 provides a summary of comparison and assignment operators.

Table 1. Comparison and assignment operators

Operation	Sample	Explanation
Strict comparison with a constant	`attr:P.Case==dative`	The case of the pronoun (P) must be dative
Comparison with a constant when the property's value is defined	`attr:P.Case===dative`	If the case of the pronoun is defined, it must be dative
Comparison with a constant for inequality	`main:VP.Person!=III`	The person of the verb phrase (VP) must not be 3rd
Strict comparison of property values for two right side constituents	`mod:P.Case==main:NP.Case`	The case values of pronoun (P) and noun phrase (NP) must be equal
Comparison of property values for two right side constituents when their values are defined	`mod:P.Case===main:NP.Case`	The case values of pronoun (P) and noun phrase (NP) must be equal, if the case values are defined
Comparison for inequality of two right side constituents	`mod:P.Case!=main:NP.Case`	The case values of pronoun (P) and noun phrase (NP) must be different
Assignment of constant	`NP.Person=III`	The 3rd person is assigned to the left side noun phrase
Inheritance/ assignment of property values of right side constituents	`NP.Case=main:NP.Case`	The case of the right side noun phrase is assigned to the left-side noun phrase

Two functions are introduced that allow to check agreement between constituents with a single statement. Function *Agree(item1, item2, property-1, property-2, …, property-n)* allows to check whether values of *property-1, …, property-n* are equal for constituents *item1, item2*. For instance, (2) checks the agreement of noun (N) and adjective (A) in case, number, and gender.

$$\text{Agree(attr:A, main:N, Case, Number, Gender)} \qquad (2)$$

Similarly, the *Disagree(item1, item2, property-1, property-2, ..., property-n)* function checks whether at least one of the properties *property-1, property-2, ..., property-n* differs between *item1* and *item2*. This is especially useful for grammar checking. For instance, in the case of error in a noun phrase, there could be a disagreement between noun and adjective in gender, case, or number (3).

$$\text{Disagree(attr:A, main:N, Gender, Number, Case)} \qquad (3)$$

With a LEX statement the terminal symbol lexical values – base forms – can be specified. The number of words in a LEX statement must be equal to the number of right side constituents in the rule. The '*' symbol is used to allow any value for the constituent. There can be several LEX statements in a rule, as shown in Fig. 1, where the adverbial phrase (ADVP) consists of two adverbs (R), and the first adverb could only be either *pavisam* ('entirely') or *īpaši* ('especially').

```
ADVP -> ad:R main:R
  LEX pavisam *
  LEX  paši *
```

Fig. 1. Sample of rule with lexical constraints

3.2   Rule Specifics for Grammar Checking

For grammar checking, we also introduce error rules. In error rules, the left side non-terminal is in the form 'ERROR-id'. All rules that describe the same type of error have the same id. Error rules do not contain value assignment operators which assign values to the left side non-terminal; they contain only constraint expressions (Fig. 2).

```
ERROR-1 -> attr:A main:N
  Disagree(attr:A,main:N, Case, Number, Gender)
```

Fig. 2. Sample of error rule describing disagreement between noun N and adjective A in case, number, or gender

The error description is followed by correction suggestion part of the rule. It starts with the label "GRAMCHECK" and is followed by the markup operator that tags an error in the phrase. Operator *MarkAll* tags the whole phrase, while operator *Mark(some right side constituent[+other right side constituent]*)* tags the part of the phrase represented by the right side constituent(s). Property assignment statements allow for the changing of the properties of the right side items and are used for generating suggestions.

Finally, the statement SUGGEST is used to form correct output (concatenated with '+'). An example of the grammar checking rule is shown in Fig. 3.

The error rules may contain phrases that are created with parsing rules (e.g., noun phrase, adjective phrase, etc.), and there usually are some agreement or disagreement statements (between properties of several phrases that are correct within themselves) in the body of the error rule.

```
ERROR-1 -> attr:AP main:NP
   Disagree(attr:AP,main:NP, Case, Number, Gender)
GRAMMCHECK MarkAll
   attr:AP.Case=main:NP.Case
   attr:AP.Number=main:NP.Number
   attr:AP.Gender=main:NP.Gender
SUGGEST(attr:AP+main:NP)
```

Fig. 3. Grammar checking rule that corrects disagreement between noun phrase (NP) and adjective phrase (AP) in case, number, or gender

Error rules are often coupled with rules describing correct grammar, i.e., there is a correct grammar rule with the same right side constituents as some error rule, and only the constraint operators differ (see Fig. 3 and Fig. 4).

```
NP -> attr:CAP main:NP
  Agree(attr:CAP, main:NP, Case, Number, Gender)
```

Fig. 4. Parsing rule that contains the same constituents as the error rule in Fig. 3, but differs in constraints

If all comparison operators in the error rule are true, it does not guarantee that this error will be in a final parse tree. For the error rule to succeed, the phrase it covers must be bigger than the phrase for which the parsing rule works. In Fig. 5, the error rule is applied for the three subsequent words, while the parsing rule covers the phrase with five words which include the shorter phrase. Thus, there is no error.

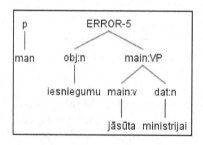
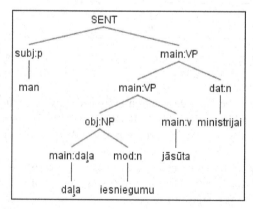

Fig. 5. Parsing example of a grammatically incorrect phrase 'man iesniegumu jāsūta ministrijai' ('I application must send to the ministry') on the left and a parse tree for a correct phrase 'man daļa iesniegumu jāsūta ministrijai' ('I part of applications must send to the ministry') containing three subsequent words from the left side phrase

4   Parsing with the Extended CFG Rules

We use the Cocke-Younger-Kasami (CYK) algorithm [26] for parsing. It allows partial parsing which is important for ungrammatical sentences. This algorithm requires grammar in the Chomsky Normal Form. During compilation, our rule compiler expands the rules with optional parts, inserts the unary rules, and transforms the rules into binary form.

Our parser generates parse trees in two formats – either constituency parse trees or dependency parse trees. Every constituency parse tree can be converted to a dependency parse tree by traversing the parse tree from the root node to the child with a syntactic role "main" first and moving it to the parent position. See Fig. 6, for an example of constituency and dependency trees for the same sentence.

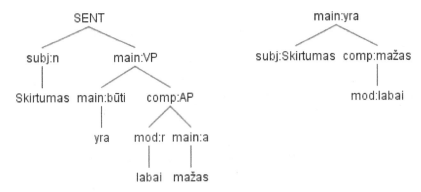

Fig. 6. Constituency (on the left) and dependency (on the right) parse trees for the sentence 'Skirtumas yra labai mažas.' ('The difference is very small.')

5   Evaluation

We have developed several sets of rules for parsing and for grammar checking that are used in different combinations. Correct syntax rules are used to determine the syntactic structure of the sentence. Correct syntax rules, together with error rules, are used to find syntactic errors in a text. Another error rule set is used to find errors in subsequent words in an incorrect text.

Table 2. Rule set statistics

Rule type	Latvian	Lithuanian
Correct syntax rules	580	179
Error rules which depend on phrases described by correct syntax rules	263	72
Error rules which contain only terminal symbols	239	560
Total	**1082**	**811**

We use several data sets to evaluate the quality of the grammar checkers.

The **Latvian Balanced corpus** contains 9,358 sentences. Sentences are taken from different types of texts – news, travel information, student papers, legal texts, blogs, e-mails, non-edited marketing materials, project drafts, etc. They represent the diversity of texts that the potential grammar checker user might check.

The **Lithuanian Balanced corpus** contains 10,000 sentences that are split in two similar parts – each part contains 5,000 sentences. The content is similar to the Latvian balanced corpus.

The **Corpus of Latvian Student papers** contains texts from student essays and abstracts of scientific papers. Intentionally, low quality texts with many grammatical errors are included in this corpus. This corpus is split in two similar parts (development and test) – each part contains 5,157 sentences.

The grammar checker can find a wide variety of errors. All errors that can be flagged by our grammar checking system are divided into 23 groups. This division is based on the theory of language syntax, theoretical literature about common error types in language, and analysis of real texts from different domains.

Simple punctuation errors include errors related to incorrect usage of whitespace characters and punctuation marks for general, language-independent cases (e.g., number of brackets). They are located using search with regular expressions.

Capitalization errors are related to incorrect usage of upper/lower case letters in named entities.

Style errors include different errors for cases where some words are misused, overused, used ungrammatically, or the word sequence is borrowed directly from another language. If the style error rule describes the misusage of individual words, lexical statements must be added to the error rule. If the style error rule describes a phrase with ungrammatically ordered sub-phrases, a set of correct grammar rules together with error rules must be used as in the case of syntax errors.

Syntax errors are related to different agreement errors, punctuation errors in sub-clauses, wrong mood for a verb, word or sentence part sequence errors, errors in address, punctuation errors in grouping, comma errors (between equal parts of sentence, in insertions, etc.), and other syntax errors. To locate the syntax errors, full parsing of the sentence must be done. A set of correct grammar rules is applied together with the rules describing the errors.

At first, we manually annotated the above mentioned evaluation corpora to create a Gold Standard. During evaluation, the Gold Standard was updated with previously unknown cases and incorrect error detection samples from the output of the grammar checker.

The record in the Gold Standard has four TAB separated fields: sentence number in corpus, error type or symbol '0' for a correct sentence, correctness tag ('COR' - correct or 'INCOR'- incorrect) and suggested correction (Fig. 7).

```
1   ERROR-1 COR 'Vakar susitikau su geru draugu.'
2   0
```

Fig. 7. Records in the Gold Standard added by annotator. The first sentence has error of type ERROR-1 and suggestion for error correction, the second sentence is correct.

For negative samples, i.e., if there is no error of type x in the sentence, the '!' symbol appears before the error type, and a phrase for which there should not be this error appears after the INCOR mark.

Fig. 8 shows sentences from the Gold Standard which were previously annotated, and afterwards the information was updated. For the first sentence, the grammar checker detects ERROR-1, but generates the wrong corrections. In the second sentence, the grammar checker incorrectly detects ERROR-1.

```
1   ERROR-1 INCOR 'Vakar susitikau su geru draugai.'
2   !ERROR-1  INCOR 'aš skai iau'
```

Fig. 8. Records appended to the Gold Standard after running grammar checker

In order to evaluate the quality of the grammar checker in general and for certain error types specifically, we calculate recall, precision, and f-measure [21]. Evaluation results are summarized in Table 3.

Table 3. Evaluation results for all error types and for the two most common error types

Corpus	Error type	Precision	Recall	F-measure
Lithuanian	all error types	0.898	0.412	0.564
Balanced	vocabulary errors	0.956	0.535	0.686
	incorrect usage of cases	0.734	0.259	0.383
Latvian	all error types	0.780	0.455	0.575
Balanced	punctuation in sub-clauses	0.757	0.643	0.695
	punctuation in participle clauses	0.617	0.671	0.643
Latvian	All error types	0.652	0.231	0.341
Student	punctuation in sub-clauses	0.706	0.586	0.641
papers (dev)	punctuation in participle clauses	0.656	0.560	0.604
Latvian Stu-	all error types	0.753	0.203	0.320
dent papers	punctuation in sub-clauses	0.773	0.588	0.668
(test)	punctuation in participle clauses	0.766	0.685	0.723

We also performed a human evaluation of the grammar checker on 150 sentences (divided into five files containing 30 sentences each) from the Corpus of Latvian student papers. Five human annotators were involved; each file was evaluated by two annotators.

The evaluation was done in two steps – without help from the grammar checker and with help from the grammar checker. At first annotators were asked to find and correct grammar errors in a file. The next day, the files with the same sentences where given to human annotators for correction, but this time files also contained information about how the grammar checker would correct these sentences.

Table 4. Human evaluation results

Situation	Cases %	Hypothesis
First step - annotators agree	62.50	Annotators often disagree. The sentences are not simple, and annotators have different language skills.
Second step - annotators agree	85.83	Agreement is higher, and the grammar checker helps.
Second step – annotator corrects the previously unnoticed error, if the grammar checker suggests it	51.43	The grammar checker helps, but annotators do not blindly accept all suggestions made by the grammar checker.
Second step – annotator does not correct the previously corrected error, if the grammar checker does not suggest it	70.97	Annotators do not read sentences as carefully as before, and they rely on the grammar checker.
First step - sentences which annotator corrects	37.04	
Second step - sentences which annotator corrects	27.78	
Sentences which grammar checker corrects	27.33	

The errors which the human annotators did not notice before, but fixed after seeing the grammar checker's suggestions are: date formatting errors, punctuation errors in participle clauses and sub-clauses, wrong forms of similarly written words, and writing style errors.

Although our main task was to evaluate the grammar checkers, we also did an initial evaluation of the Lithuanian parser. For evaluation, we created a Gold Standard containing 115 correct dependency parse trees. As the syntactic rules that have been developed so far do not cover all of the syntactic constructions used in the Lithuanian language, we included sentences in the Gold Standard from news texts which do not have a very complex structure, but still represent the main syntactic constructions of the Lithuanian language. We compared dependency trees from the Gold Standard with dependency trees generated by the parser. As it is hard to compare two dependency trees, we first converted them into triplets: <parent>:<parent start position in sentence>, < syntactic role>, <child>:<child start position in sentence> (see Fig. 9).

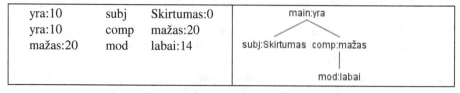

Fig. 9. Triplets for the sentence 'Skirtumas yra labai mažas' ('The difference is very small')

The quality of the parser is calculated by measuring the precision and the recall of triplets. For the initial Gold Standard, we obtained precision - 0.935 and recall - 0.922.

6   Conclusion and Future Work

In this paper, we introduced extended CFG formalism for grammar checking of inflected languages, allowing powerful grammar checkers to be built for a practical application. The proposed grammar checking formalism has been implemented and tested for Latvian and Lithuanian. The obtained precision and recall numbers (precision over 0.78 and recall over 0.41) allow us to conclude that it can be used in commercial applications.

Our investigations also show that it can also be used for grammar checking of other inflectional languages. We have investigated its possible application to the Polish language. The Polish language belongs to the West-Slavonic group of the Indo-European family of languages. The main verbal morpho-syntactic features (tense, person, aspect, mode, and voice) and nominal features (case, number, and gender) as well as the syntactic structure of the sentence (main parts - subject and predicate; secondary parts - attribute, adverbial modifier, and complement) are similar to the Baltic language group. Our proposed formalism allows the describing of such structures. Similar error types that are common for the Baltic languages are also common in Polish and other Slavic languages: agreement between words, wrong noun case usage, punctuation errors in subclauses, errors in negation, subject and predicate agreement errors, etc.

Our proposed formalism can also be used for named entity recognition and information extraction, and it can be incorporated into hybrid machine translation systems.

The next steps are to add the possibility to specify the weights or probabilities of the rules in the formalism and to implement the CYK algorithm for parsing weighted CFG grammar.

Acknowledgements. The research leading to these results has received funding from the research project "Information and Communication Technology Competence Center" of EU Structural funds, contract nr. L-KC-11-0003 signed between ICT Competence Centre and Investment and Development Agency of Latvia, Research No. 2.8 "Research of automatic methods for text structural analysis".

References

1. Arppe, A.: Developing a grammar checker for Swedish. In: 12th Nordic Conference in Computational Linguistics (Nodalida 1999), pp. 13–27. Trondheim (2000)
2. Atwell, E.S.: How to detect grammatical errors in a text without parsing it. In: 3rd Conference of the European Chapter of the Association for Computational Linguistics, pp. 38–45. Association for Computational Linguistics, Copenhagen (1987)
3. Backus, J.W.: The syntax and semantics of the proposed international algebraic language of the Zurich ACM-GAMM Conference. In: International Conference on Information Processing, pp. 125–132. UNESCO (1959)

4. Bates, M.: The theory and practice of augmented transition networks. In: Bolc, L. (ed.) Natural Language Communication with Computers. LNCS, vol. 63, pp. 191–254. Springer, Heidelberg (1978)

5. Birn, J.: Detecting grammar errors with Lingsoft's Swedish grammar checker. In: 12th Nordic Conference in Computational Linguistics (Nodalida 1999), pp. 28–40. Trondheim (2000)

6. Chomsky, N.: Syntactic structures. Mouton, The Hague (1957)

7. Deksne, D., Skadiņš, R.: CFG Based Grammar Checker for Latvian. In: 18th Nordic Conference in Computational Linguistics (NODALIDA 2011), pp. 275–278. Riga (2011)

8. Domeij, R., Knutsson, O., Carlberger, J., Kann, V.: Granska: An efficient hybrid system for Swedish grammar checking. In: 12th Nordic Conference in Computational Linguistics (Nodalida 1999), pp. 49–56. Trondheim (2000)

9. Ehsan, N., Faili, H.: Grammatical and context-sensitive error correction using a statistical machine translation framework. Software: Practice and Experience 43(2), 187–206 (2013)

10. Gazdar, G.: Generalized Phrase Structure Grammar. Harvard University Press (1985)

11. Heidorn, G.E.: Intelligent writing assistance. In: Dale, R., Moisl, H., Somers, H. (eds.) Handbook of Natural Language Processing, ch. 8, pp. 181–207. Marcel Dekker, New York (2000)

12. Hagen, K., Johannessen, J. B., Lane, P.: Some problems related to the development of a grammar checker. Paper presented at NODALIDA 2001, the 2001 Nordic Conference in Computational Linguistics, May 21–22, 2001 (2001)

13. Izumi, E., Uchimoto, K., Saiga, T., Supnithi, T., Isahara, H.: Automatic error detection in the Japanese learners English spoken data. In: 41st Annual Meeting of the Association for Computational Linguistics (ACL 2003), Sapporo, pp. 145–148 (2003)

14. Kaplan, R.M., Bresnan, J.: Lexical-Functional Grammar: A formal system for grammatical representation. In: Bresnan, J. (ed.) The Mental Representation of Grammatical Relations, pp. 173–281. The MIT Press, Cambridge (1982)

15. Karlsson, F.: Constraint Grammar as a framework for parsing running text. In: 13th International Conference on Computational Linguistics (COLING 1990), Helsinki, vol. 3, pp. 168–173 (1990)

16. Martins, R.T., Hasegawa, R., Das Gracas VolpeNunes, M., Montilha, G., De Oliveira, O.N.: Linguistic issues in the development of ReGra: A grammar checker for Brazilian Portuguese. Natural Language Engineering 4(4), 287–307 (1998)

17. Naber, D.: A rule-based style and grammar checker. Master's thesis, University of Bielefeld (2003)

18. Ng, H.T., Wu, S.M., Wu, Y., Hadiwinoto, C., Tetreault, J.: The CoNLL-2013 Shared Task on Grammatical Error Correction. In: 17th Conference on Computational Natural Language Learning (CoNLL 2013), pp. 1–12. Association for Computational Linguistics (2013)

19. Pereira, F., Warren, D.: Definite clause grammars for language analysis–A survey of the formalism and a comparison with augmented transition networks. In: Artificial Intelligence, vol. 13(3), pp. 231–278. (1980)

20. Pollard, C., Sag, I.A.: Head-driven phrase structure grammar. University of Chicago Press, Chicago (1994)

21. Van Rijsbergen, C.J.: Evaluation. In: Information Retrieval, 2nd edn. Butterworth, Newton (1979)

22. Sågvall-Hein, A.: A Chart-Based Framework for Grammar Checking. In: 11th Nordic Conference in Computational Linguistics (Nodalida 1998), pp. 68–80 (1998)

23. Schmidt-Wigger, A.: Grammar and Style Checking in German. In: 2nd International Workshop on Controlled Language Applications (CLAW 1998). Language Technologies Institute, Carnegie Mellon University, Pittsburgh (1998)
24. Shaalan, K.: Arabic Gramcheck: A Grammar Checker for Arabic. Software: Practice and Experience 35(7), 643–665 (2005)
25. Sjöbergh, J., Knutsson, O.: Faking errors to avoid making errors: Very weakly supervised learning for error detection in writing. In: Recent Advances in Natural Language Processing IV (RANLP 2005), Borovets, pp. 506–512 (2004)
26. Younger, D.: Recognition and parsing of context-free languages in time n3. Information and Control 10(2), 189–208 (1967)
27. Vosse, T.: The Word Connection. Grammar-Based Spelling Error Correction in Dutch. Neslia Paniculata, Enschede (1994)
28. Xing, J., Wang, L., Wong, D.F., Chao, S., Zeng, X.: UM-Checker: A Hybrid System for English Grammatical Error Correction. In: 17th Conference on Computational Natural Language Learning (CoNLL-2013), vol. 34 (2013)

Dealing with Function Words
in Unsupervised Dependency Parsing

David Mareček and Zdeněk Žabokrtský

Charles University in Prague
Faculty of Mathematics and Physics
Institute of Formal and Applied Linguistics
{marecek,zabokrtsky}@ufal.mff.cuni.cz

Abstract. In this paper, we show some properties of function words in dependency trees. Function words are grammatical words, such as articles, prepositions, pronouns, conjunctions, or auxiliary verbs. These words are often short and very frequent in texts and therefore many of them can be easily recognized. We formulate a hypothesis that function words tend to have a fixed number of dependents and we prove this hypothesis on treebanks. Using this hypothesis, we are able to improve unsupervised dependency parsing and outperform previously published state-of-the-art results for many languages.

1 Introduction

Function words (also known as grammatical words) are words which have no or very little lexical meaning, in contrast to content words (lexical words), which have some meaningful content. Function words are articles, pronouns, prepositions, conjunctions, particles or auxiliary verbs and all belong to closed-class words. They are used to express grammatical attributes of content words or grammatical relationships between two or more content words.

In some representations of linguistic structure, function words are treated differently from content words. Tesnière [1] introduces the notion of empty words (function words) and argues that they cannot occupy alone a position in the dependency structure. Functional Generative Description [2] uses so called tectogrammatical representation, in which only content words are represented by nodes and function words are there in forms of their attributes. In Logical Forms [3], some of the function words become labels of edges connecting content words. Another example where the function words are excluded from a sentence structure is the Abstract Meaning Representation [4]. Nevertheless, even within a chosen formalism, the boundaries between function and content words are not entirely straightforward and they are often very fuzzy.

This paper is organized as follows: In Section 2, we describe how to recognize function words in a language, if either only raw texts are available or words in the corpus are labeled with POS tags. Properties of function words in dependency structures are discussed in Section 3. Section 4 explains how these properties can be used in unsupervised dependency parsing task. Experiments are shown in Section 5 and Section 6 concludes.

A. Gelbukh (Ed.): CICLing 2014, Part I, LNCS 8403, pp. 250–261, 2014.

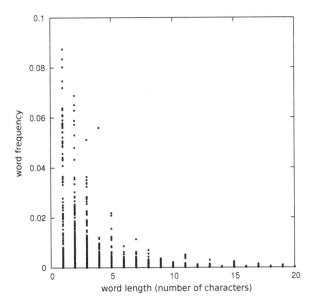

Fig. 1. Relationship between length (number of characters) of a word and its relative frequency. Each point in the graph corresponds to one word type. The words were taken from 20 different treebanks.

2   Recognition of Function Words

We introduce two simple function words properties that can be used for recognizing them in an unsupervised way (with no manual effort). The first one is their frequency – function words are more frequent in language than content words. The second one is their length – function words are relatively short. The well-known relationship between length of a word and its frequency is caused by the *economy* of language [5]; we show this relationship in Figure 1.

It is apparent that the frequent words are mostly short, however it is not true that the short words are frequent. Many short words are abbreviations or numbers, which are definitely not function words. Therefore, we decided to recognize function words using their relative frequency in corpus and not using their length. The relative frequency of word is computed simply by dividing the number of its occurrences by the number of all the words in the corpus:

$$F_W(w) = \frac{count(w)}{total}.$$

The more frequent a word is in a corpus, the more likely it is a function word.

If we have a corpus which is manually or automatically tagged with part-of-speech (POS) tags, we can compute the aggregated word frequency for individual POS tags. The aggregated frequency is computed by averaging the relative frequency of all tokens in the corpus labeled by that POS tag.

$$F_T(t) = \sum_{w; tag(w)=t} \frac{count(w,t)}{count(t)} \cdot \frac{count(w,t)}{total}$$

The formula above expresses the weighted average over the relative frequencies of word types, which is equal to uniform average over the tokens. The more frequent a word is, the higher influence on the aggregated frequency it has. Such weighted average of word frequencies seems to be able to to sufficiently separate the function words POS tags.

3   Properties of Function Words in Dependency Trees

3.1   Types of Function Words

For the purposes of this work, we divide function words into several groups:

- **Function words as content word modifiers** – They express attributes of content words. For example in majority of treebank annotations, *articles* modify nouns. They determine the definiteness and, in some languages, grammatical categories, e.g. the noun case in German. Another example may be *negative particles* that modify verbs and determine their negation. Such function words are mostly annotated as leaves in dependency treebanks.

- **Function words as grammatical relations between content words** – In the second group, there are function words connecting two content words, for example *prepositions* or *postpositions*. They usually connect a noun with another noun or verb and express the type of such connection. In the level of sentences, we have *subordinating conjunctions*, which have a similar role in connecting clauses. Such function words have often just one dependent – a content word which is in the relation with its grandparent through the parent function word.

- **Other function words** – The third group is for the rest of function words, which usually have more than one dependents. The forms of the verbs *'to be'*, *'to do'*, *'to have'*, etc. and their equivalents in other languages are the typical examples. These verbs should be treated as function words, nevertheless, they can be in a role of the main finite verb, in which they can have many dependents. The verb *'was'* in the sentence *'He was not yesterday in the bar with that girl.'* has five dependents and cannot be included into any of the previous two groups. Other *auxiliary* and *modal verbs* could be considered to belong to this group as well, since the content verbs are attached below them in many treebanks.

Note that the assignment of a function word to one of the proposed groups as well as the boundary between function and content words can differ across different linguistic theories. We do not want to argue about the correct annotation of function words.[1] We only show that function words can be easily grouped according to the number of their dependents.

---

[1] Differences in annotations over different treebanks are discussed e.g. in [6].

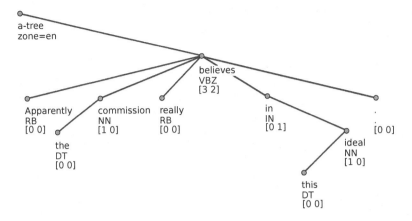

Fig. 2. An example of an English dependency tree. Fertility patterns are given in square brackets.

3.2   Fertility patterns

We use the term *fertility pattern* to express how many of left and right dependents (children) a word has in the dependency structure. We define it as a pair $[l, r]$, where l is the number of children preceding this word, and l is the number of children following this word. An example of word fertility patterns in an English sentence is given in Figure 2. The fertility patterns are shown in square brackets.

From the analysis given in Section 3.1, we can say that the fertility pattern of function words is often fixed. Majority of function words have either no dependents (pattern $[0\ 0]$, e.g. articles, pronouns, auxiliary verbs) or just one dependent to the right (fertility pattern $[0\ 1]$, e.g. prepositions or conjunctions) or to the left (fertility pattern $[1\ 0]$, e.g. postpositions).

To support this hypothesis, we perform the following experiments that are run across 20 testing treebanks from CoNLL shared tasks 2006 [7] and 2007 [8].

3.3   Most Frequent Fertility Patterns for Word Forms

The first experiment explores the relation between the most frequent fertility pattern of a given word[2] and its relative frequency.

1. For each word in a treebank, we go through all its occurrences in the treebank and collect counts of its fertility patterns.
2. We find the most frequent fertility pattern for each word and denote its relative frequency as $HF(word)$. This score says how much the fertility pattern is stable (fixed) for the particular word.

---

[2] We choose relative frequency of the most frequent fertility pattern as a stability rate. Similarly, we could use entropy or other information theory quantities; this is left for further research.

Table 1. Statistics of fertility patterns for selected English words from the example in Figure 2

rank	the		commission		believes		in		this	
1st $(HF(word))$	[0 0]	1.00	[0 0]	0.33	[1 1]	0.37	[0 1]	0.96	[0 0]	0.98
2nd	–	–	[3 0]	0.19	[1 2]	0.24	[0 0]	0.02	[1 0]	0.01
3rd	–	–	[2 0]	0.14	[2 1]	0.16	[1 1]	0.01	[0 2]	0.00
4th	–	–	[1 0]	0.12	[3 2]	0.05	[0 2]	0.00	[0 1]	0.00
5th	–	–	[1 1]	0.05	[4 1]	0.05	[0 1]	0.00	[0 3]	0.00

3. We compute the word relative frequencies $F_W(word)$ for each word in the treebank.
4. We plot the points representing individual words into the graph. Each word is parametrized by $HF(word)$ and $F_W(word)$.

An example is given in Table 1, in which the five most frequent fertility patterns are listed for selected words from the example in Figure 2. The most frequent pattern $HF(word)$ is the one in the first row. As it was expected, the function words (*the, in, this*) have one dominant fertility pattern (their HF is 1.00, 0.96, and 0.98 respectively), whereas the content words (*commission, believes*) have much more options.

The graph showing the relation between $HF(word)$ scores and $F_W(word)$ frequencies generally is depicted in Figure 3. There are all the words from all the 20 testing treebanks plotted in one graph. We can see that the very frequent words (with $F_W(word) > 0.02$) have often very stable fertility patterns ($HF(word) > 0.7$), which supports our previous hypothesis.

3.4 Most Frequent Fertility for Part-of-Speech Tags

We compute analogous statistics for part-of-speech (POS) tags.

1. For each POS tag in a treebank, we go through all its occurrences in the treebank and collect counts of its fertility patterns.
2. We find the most frequent fertility pattern for each POS tag and denote its relative frequency as $HF(tag)$. This score says how much the fertility pattern is stable for that POS tag.
3. We compute $F_T(tag)$, the aggregated relative frequency of words labeled by that POS tag as defined in Section 2, for all the POS tags in the treebank.
4. We plot the points representing individual POS tags into one graph. Each POS tag is parametrized by $HF(tag)$ and $F_T(tag)$. Moreover, we can express frequencies of individual POS tags by different sizes of the points. It is worth here since the POS tag relative frequency differs from the $F_T(tag)$.

The generated plot over all 20 treebanks is shown in Figure 4. We can observe a similar shape as for word forms in Figure 3. Almost all the tags with the

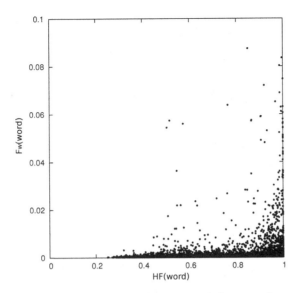

Fig. 3. Relationship between relative frequency of the most frequent fertility pattern of a word ($HF(word)$ as defined above) and the word relative frequency. There are all word types from 20 testing treebanks plotted into one graph.

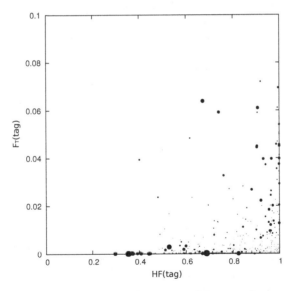

Fig. 4. Relationship between relative frequency of the most frequent fertility pattern of a POS tag ($HF(tag)$ as defined above) and the aggregated relative frequency of words labeled by that POS tag ($F_T(tag)$ as defined in Section 2). There are all POS tags from 20 testing treebanks plotted into one graph. The sizes of points shows relative frequencies of individual POS tags.

aggregated word frequency F_T higher than 0.02 have their most frequent fertility pattern $HF(tag)$ higher than 0.7. Therefore, we can say that our hypothesis holds for individual part-of-speech tags as well and across different treebanks.

4   Applying Function Word Properties in Unsupervised Parsing

In this section, we employ our previously described properties of function words in the task of unsupervised dependency parsing.

4.1   Introduction to the Unsupervised Dependency Parsing System

We use the unsupervised dependency parsing system described by Mareček and Straka in [9]. The software is freely available at http://ufal.mff.cuni.cz/udp/. The unsupervised parser is based on Dependency Model with Valence, which was introduced by Klein and Manning [10] and further improved by Headden et al. [11] and Spitkovsky et al. [12,13]. Inference procedure is based on blocked Gibbs sampling technique [14,15].

The Dependency Model with Valence is a generative model based on two probabilities. The first one, $P_{choose}(t_d|t_g, dir)$, expresses the probability of POS tag t_d of the dependent given the POS tag t_g of its governing word and the direction dir, which represents the *left* or *right* attachment.

The other one, $P_{stop}(x|t_g, dir, adj)$, expresses the probability that a word labeled by POS tag t_g has ($x = STOP$) or has not ($x = CONT$) children in the direction dir. The adjacency parameter adj determines whether we predict the first child in a given direction or a next child after one that already exists. For example, $P_{stop}(STOP|NN, left, 1)$ gives the probability that a noun (tag NN) has no children in the left direction; $P_{stop}(CONT|NN, left, 0)$ gives the probability that the noun that already has children in the left direction will have one more child in that direction.

Another important feature, from which the unsupervised dependency parser by Mareček and Straka [9] benefits, is so called reducibility. The idea is that if a word, or a short sequence of words $w_1, \ldots, w_n$, can be removed from a sentence without damaging its correctness, nothing else in the sentence can depend on any of the words $w_1, \ldots, w_n$. In other words, such a sequence of words forms a subtree or more adjacent subtrees in the respective dependency structure. By computing such statistics on large corpora, the prior probabilities for the P_{stop} model can be estimated and used to force the inference procedure to tend to better solutions.

4.2   Predicting P_{stop} Probabilities for Functional Words

As shown above, we hypothesize that if a word is frequent, it should have a stable fertility pattern, i.e. fixed number of left and right children.

To predict the number of children, we use the reducibility principle, which we described in [16]. A phrase (sequence of words) is reducible if the rest of the sentence after removing this phrase exists elsewhere in the corpus as a sentence. For example, the phrase *'on Monday'* is reducible from the sentence *'He arrived to London on Monday.'*, if the rest of the sentence *'He arrived to London.'* exists elsewhere in the corpus as well. The phrase *'arrived to'* is probably not reducible, since the sentence *'He London on Monday.'* can hardly be found in the corpus. It is evident that we can find only very few reducible sequences with this procedure. However, even if it leads to very sparse statistics on words, it is already sufficient for recognizing prototypical properties of individual part-of-speech tags. We search for reducible sequences on large collections of Wikipedia articles provided by [17] containing between 10 and 80 million words for each language.

By this simple procedure, we can search for reducible sequences. For our purposes, we need to compute the following counts:

- $red(t)$ – number of times a single word labeled by POS tag t can be removed from a sentence,
- $red_l(t)$ – number of times a two- or three-word phrase beginning with a word labeled by POS tag t can be removed,
- $red_r(t)$ – number of times a two or three word phrase ending with a word labeled by POS tag t can be removed.

Only the reducible sequences (phrases) consisting of at most three words are taken into account, since longer reducible sequences do not reflect grammatical properties and introduce a significant noise into the counts.

In the following experiments, we will use three rules designed to recognize fertility patterns of function words POS tags using the precomputed reducibility statistics $red(t)$, $red_l(t)$, and $red_r(t)$.

1. **function words with one right dependent** (fertility pattern $[0\ 1]$) are words for which $red_l(t) > 3\,red(t)$ and $red_l(t) > 3\,red_r(t)$,
2. **function words with one left dependent** (fertility pattern $[1\ 0]$) are words for which $red_r(t) > 3\,red(t)$ and $red_r(t) > 3\,red_l(t)$,
3. **function words with no dependents** (fertility pattern $[0\ 0]$) are words for which $red(t) > 10$ and $red(t) > red_l(t)$ and $red(w) > red_r(t)$.

We set the threshold for function words POS tags frequency to $F_T(t) = 0.005$. If the conditions of one of the three rules are fulfilled for a particular POS tag with $F_T(t) \geq 0.005$, we set its left and right P_{stop} prior probabilities to 1.0 or 0.0 according to the predicted fertility pattern. All the constants included in the proposed rules were set manually. Automatic optimization procedure was left for future research.

5   Experiments and Results

We follow the same experimental settings as Mareček and Straka [9]. To compute the reducibility scores of individual POS tags, we use Wikipedia articles

Table 2. Unlabeled attachment scores (in %) of unsupervised dependency parsing measured on 20 treebanks from CoNLL 2006 and 2007 shared tasks. The average across all the treebanks for each experiments is shown in the last row.

CoNLL		this work				previous works
language	year	baseline (mar13)	rules 1,2	rule 3	rules 1,2,3	spi13
Arabic	06	**35.3**	34.9	**35.3**	35.2	9.3
Arabic	07	38.2	43.0	38.6	**43.3**	26.8
Basque	07	**35.5**	35.3	**35.5**	35.3	24.4
Bulgarian	06	54.9	56.6	54.6	56.8	**63.4**
Catalan	07	67.0	43.8	65.0	43.9	**68.0**
Czech	06	52.4	**55.9**	53.7	55.6	44.0
Czech	07	51.9	**54.6**	54.2	54.3	34.3
Danish	06	41.6	**45.9**	42.5	45.5	21.4
Dutch	06	47.5	30.8	51.3	**54.8**	48.0
English	07	55.4	56.1	57.9	**58.5**	58.2
German	06	52.4	45.2	54.0	45.0	**56.2**
Greek	07	26.3	33.7	26.2	34.5	**45.4**
Hungarian	07	34.0	34.3	34.0	34.1	**58.3**
Italian	07	39.4	**51.0**	39.7	**51.0**	34.9
Japanese	06	61.2	61.0	62.2	61.9	**63.0**
Portuguese	06	69.6	46.2	72.0	**75.1**	74.5
Slovenian	06	35.7	47.4	35.9	47.3	**50.9**
Spanish	06	61.1	61.1	61.2	**61.7**	61.4
Swedish	06	54.5	54.2	55.6	**55.6**	49.7
Turkish	07	56.9	57.0	56.6	**57.0**	37.9
Average:		48.5	47.4	49.3	**50.3**	46.5

collection by [17], which was tagged using the TnT tagger [18]. On the same data, we compute the $red(t)$, $red_l(t)$, and $red_r(t)$ counts for the function words fertility pattern predictions. As the testing treebanks, we use 20 dependency treebanks from CoNLL shared tasks 2006 [7] and 2007 [8], which comprise 18 different languages.

Table 2 shows our results. We evaluate four different configurations – the baseline p_{stop} priors (results from our last work (mar13) [9]), p_{stop} priors with updated values for POS tags with fertility patterns [0 1] and [1 0] (rules 1 and 2), p_{stop} priors with updated values for POS tags with fertility pattern [0 0] (rule 3), and p_{stop} priors with values updated by all the three rules. We compare our results with the state-of-the-art results reported by Spitkovsky et al. [19] (spi13).

We can see that the configuration, in which all the three rules were applied, has the highest average attachment score (50.3%) over our 20 testing treebanks. It is also important that this configuration improved the scores for almost all the treebanks, compared to the baseline configuration. It worsened the attachment score significantly only for German and Catalan.

Table 3 provides a list of POS tags affected by our rules for all testing treebanks. Interestingly, the rule 2 was not applied at all. The reason may be the fact

Table 3. POS tags affected by the rules 1 and 3. Rule 2 did not affect any POS tag. Basic types of POS tags are in brackets: *prep* are prepositions, *punc* is punctuation, *det* are determiners, *pron* are pronouns, and *conj* are conjunctions. The last column (*err/all*) shows the number of POS tags, for which the pattern prediction was not correct, and the total number of POS tags affected.

CoNLL		pattern [0 1] (rule 1)	pattern [0 0] (rule 3)	err/all
Arabic	06	–	–	0/0
Arabic	07	P-*(prep)*	G-*(punc)*	0/2
Basque	07	–	PUNC*(punc)* KOMA*(punc)*	0/2
Bulgarian	06	R*(prep)*	–	0/1
Catalan	07	da*(det)* pr*(pron)*	Fc*(punc)*, Fp*(punc)*	**2/4**
Czech	06	R*(prep)*	:*(punc)*	0/2
Czech	07	R*(prep)*	:*(punc)*	0/2
Danish	06	SP*(prep)*	XP*(punc)*	0/2
Dutch	06	Prep*(prep)*	Punc*(punc)*, Art*(det)*	0/3
English	07	IN*(prep)*	,*(punc)* .*(punc)* DT*(det)*	0/4
German	06	APPR*(prep)* ART*(det)*	$(*(punc)* $,*(punc)* $.*(punc)*	**1/5**
Greek	07	AtDf*(det)* AsPpSp*(prep)* AsPpPa*(prep)*	PUNCT*(punc)*	**1/5**
Hungarian	07	Tf*(det)*	WP*(punc)* SP*(punc)* Cc*(conj)*	**1/4**
Italian	07	RD*(det)* E*(prep)*	PU*(punc)* C*(conj)*	**1/4**
Japanese	06	–	–	0/0
Portuguese	06	prp*(prep)*	punc*(punc)* art*(det)*	0/3
Slovenian	06	ad-pr*(prep)*	PUNC*(punc)*	0/2
Spanish	06	sp*(prep)*	Fc*(punc)* Fe*(punc)* Fp*(punc)* di*(det)* da*(det)*	0/6
Swedish	06	–	IK*(punc)* PO*(pron)* IP*(punc)*	0/3
Turkish	07	–	–	

that function words are often attached to (or govern) the following content word, at least for the languages we experiment with. Therefore the fertility pattern [0 1] is much more probable than the pattern [1 0]. The correct fertility pattern was predicted for 48 POS tags out of 54 POS tags for which the prediction was made (see the last column in Table 3).

We can also find out why the parsing accuracy of German and Catalan was worsened by our predictions. In both cases, the patterns for articles (German *ART* and Catalan *da*) was wrongly predicted as [0 1] instead of [0 0], probably because they are obligatory in that languages in majority of cases. This caused that the following nouns were more forced to become their dependents, which is not in accordance with the treebanks' annotation rules. However, note that choosing articles as the noun governors is not entirely an error. See the debate about the DP-hypothesis in [20]. The fertility pattern of Hungarian articles (*Tf*) was wrongly predicted as well, however, it does not affect the attachment score, since the same problem occurred in the baseline dependency trees.

6   Conclusions

We described the properties of function words and we proposed methods how to recognize them and how to predict whether they tend to be leaves in the dependency trees or they tend to have left or right dependents. We employed such methods in unsupervised dependency parsing system and show substantial improvement in attachment scores when testing on 20 different dependency tree-banks from CoNLL shared tasks. To our knowledge, the achieved performance constitutes a new state of the art for about a half of the languages under study.

Acknowledgments. This research has been supported by the grant no. GPP406/14/06548P of the Grant Agency of the Czech Republic and by the European Union Seventh Framework Programme under grant agreement FP7-ICT-2013-10-610516 (QTLeap). This work has been using language resources developed and/or stored and/or distributed by the LINDAT/CLARIN project of the Ministry of Education of the Czech Republic (project LM2010013).

References

1. Tesnière, L.: Eléments de syntaxe structurale. Editions Klincksieck, Paris (1959)
2. Sgall, P., Hajičová, E., Panevová, J.: The Meaning of the Sentence in Its Semantic and Pragmatic Aspects. Reidel, Dordrecht (1986)
3. Menezes, A., Richardson, S.D.: A best-first alignment algorithm for automatic extraction of transfer mappings from bilingual corpora. In: Proceedings of the Workshop on Data-driven Methods in Machine Translation, vol. 14, pp. 1–8 (2001)
4. Banarescu, L., Bonial, C., Cai, S., Georgescu, M., Griffitt, K., Hermjakob, U., Knight, K., Koehn, P., Palmer, M., Schneider, N.: Abstract meaning representation for sembanking. In: Proceedings of the 7th Linguistic Annotation Workshop and Interoperability with Discourse, pp. 178–186. Association for Computational Linguistics, Sofia (August 2013)
5. Zipf, G.K.: The Psychobiology of Language. Houghton Mifflin, Boston (1935)
6. Zeman, D., Mareček, D., Popel, M., Ramasamy, L., Štěpánek, J., Žabokrtský, Z., Hajič, J.: HamleDT: To Parse or Not to Parse? In: Proceedings of the Eight International Conference on Language Resources and Evaluation (LREC 2012). European Language Resources Association (ELRA), Istanbul (2012)
7. Buchholz, S., Marsi, E.: CoNLL-X shared task on multilingual dependency parsing. In: Proceedings of the Tenth Conference on Computational Natural Language Learning, CoNLL-X 2006, pp. 149–164. Association for Computational Linguistics, Stroudsburg (2006)
8. Nivre, J., Hall, J., Kübler, S., McDonald, R., Nilsson, J., Riedel, S., Yuret, D.: The CoNLL 2007 Shared Task on Dependency Parsing. In: Proceedings of the CoNLL Shared Task Session of EMNLP-CoNLL 2007, pp. 915–932. Association for Computational Linguistics, Prague (June 2007)
9. Mareček, D., Straka, M.: Stop-probability estimates computed on a large corpus improve Unsupervised Dependency Parsing. In: Proceedings of the 51st Annual Meeting of the Association for Computational Linguistics, vol. 1: Long Papers, pp. 281–290. Association for Computational Linguistics, Sofia (August 2013)

10. Klein, D., Manning, C.D.: Corpus-based induction of syntactic structure: models of dependency and constituency. In: Proceedings of the 42nd Annual Meeting on Association for Computational Linguistics, ACL 2004. Association for Computational Linguistics, Stroudsburg (2004)
11. Headden III, W.P., Johnson, M., McClosky, D.: Improving unsupervised dependency parsing with richer contexts and smoothing. In: Proceedings of Human Language Technologies: The 2009 Annual Conference of the North American Chapter of the Association for Computational Linguistics, NAACL 2009, pp. 101–109. Association for Computational Linguistics, Stroudsburg (2009)
12. Spitkovsky, V.I., Alshawi, H., Jurafsky, D.: Punctuation: Making a point in unsupervised dependency parsing. In: Proceedings of the Fifteenth Conference on Computational Natural Language Learning, CoNLL 2011 (2011)
13. Spitkovsky, V.I., Alshawi, H., Jurafsky, D.: Three Dependency-and-Boundary Models for Grammar Induction. In: Proceedings of the 2012 Conference on Empirical Methods in Natural Language Processing and Computational Natural Language Learning, EMNLP CoNLL 2012 (2012)
14. Gilks, W.R., Richardson, S., Spiegelhalter, D.J.: Markov chain Monte Carlo in practice. Interdisciplinary statistics. Chapman & Hall (1996)
15. Mareček, D., Žabokrtský, Z.: Gibbs Sampling with Treeness constraint in Unsupervised Dependency Parsing. In: Proceedings of RANLP Workshop on Robust Unsupervised and Semisupervised Methods in Natural Language Processing, Hissar, Bulgaria, pp. 1–8 (2011)
16. Mareček, D., Žabokrtský, Z.: Exploiting reducibility in unsupervised dependency parsing. In: Proceedings of the 2012 Joint Conference on Empirical Methods in Natural Language Processing and Computational Natural Language Learning, EMNLP-CoNLL 2012, pp. 297–307. Association for Computational Linguistics, Stroudsburg (2012)
17. Majliš, M., Žabokrtský, Z.: Language richness of the web. In: Proceedings of the Eight International Conference on Language Resources and Evaluation (LREC 2012). European Language Resources Association (ELRA), Istanbul (May 2012)
18. Brants, T.: TnT - A Statistical Part-of-Speech Tagger. In: Proceedings of the Sixth Conference on Applied Natural Language Processing, pp. 224–231 (2000)
19. Spitkovsky, V.I., Alshawi, H., Jurafsky, D.: Breaking out of local optima with count transforms and model recombination: A study in grammar induction. In: Proceedings of the 2013 Conference on Empirical Methods in Natural Language Processing, pp. 1983–1995. Association for Computational Linguistics, Seattle (October 1995)
20. Abney, S.P.: The English Noun Phrase In Its Sentential Aspect. PhD thesis. MIT (1987)

When Rules Meet Bigrams*

Eric Wehrli and Luka Nerima

Laboratoire d'Analyse et de Technologie du Langage - CUI
University of Geneva
Battelle - 7 route de Drize, CH-1227 Carouge
{Eric.Wehrli,Luka.Nerima}@unige.ch

Abstract. This paper discusses an on-going project aiming at improving the quality and the efficiency of a rule-based parser by the addition of a statistical component. The proposed technique relies on bigrams of pairs (word+category) selected from the homographs contained in our lexical database and computed over a large section of the Hansard corpus, previously tagged. The bigram table is used by the parser to rank and prune the set of alternatives. To evaluate the gains obtained by the hybrid system, we conducted two manual evaluations. One over a small subset of the Hansard corpus, the other one with a corpus of about 50 articles taken from the magazine *The Economist*. In both cases, we compare analyses obtained by the parser with and without the statistical component, focusing only on one important source of mistakes, the confusion between nominal and verbal readings for ambiguous words such as *announce, sets, costs, labour*, etc.

1   Introduction

Can statistical data help improving the efficiency of a rule-based parser? That is the question that we would like to address in this paper, reporting on an on-going research project. The question is by no means new, since it was already the topic of Klavans and Resnik (1996) and numerous subsequent papers, a majority of them either using statistics at the pre-syntactic level of POS-tagging by means of a statistical tokenizer (see, for instance, Adolphs et al. 2008; Blache & Rauzy, 2013), or building statistical systems on the basis of linguistically richer treebanks (cf. Nivre et al. 2007).

The research described in this paper concerns a different type of hybridization, attempting to integrate a statistical component as a ranking device within the syntactic parser. To the best of our knowledge, this brand of hybridization has been much less pursued, with the notable exception of G. Schneider's hybrid dependency parser (cf. Schneider, 2008; Sennrich et al. 2009).

Overgeneration is a well-known (and arguably unavoidable) feature of rule-based parsers. This is due to the compounding effect of lexical ambiguity and of

---

* Thanks to Meghdad Farahmand and Yves Scherrer for useful comments and contributions to this paper.

the number of rules that can apply in any particular configuration. The result is a fast-growing number of alternatives which may bring the parser to its knees, if unchecked. To avoid such dire straits and more generally to limit the number of alternatives in order to guarantee a reasonable level of efficiency for demanding applications such as machine translation, pruning techniques based on mostly ad hoc ranking schemes must be activated. In the case of the Fips parser, ranking is established by means of scores assigned to constituents either on the basis of psycholinguistic criteria (e.g. local attachments are preferred to non-local ones), congruence with lexical selectional properties, as well as purely ad hoc penalties to discourage some particular attachments. The goal of our current research is to investigate whether statistical data could also be used to rank alternative analyses. As a first step, we will consider bigrams of adjacent words.

2   Bigram Acquisition

Since bigram probabilities are used to help the parser disambiguate lexical items, bigrams will only be computed for pairs of words where at least one of the words is ambiguous, that is, belongs to more than one lexical category (e.g. *break*, *labour*, *set*, which can all be nouns and verbs). The list of ambiguous words is drawn from our lexical database. Furthermore, to distinguish between the probabilities of the different readings of an ambiguous word, the bigrams consist of a pair of [orthographic word, POS-tag] as illustrated in (1).

(1) [labour-NOUN, costs-NOUN]

2.1   Corpus

To overcome the problem of scarce bigrams, it is necessary to compute them on large corpora. We considered several candidates: an excerpt of the English Wikipedia (≈ 800 Mo), a corpus of about 10,000 articles from the magazine *The Economist* (≈ 60 Mo), a part of the Hansard Senate Debates (≈ 80 Mo) and of the Hansard House Debates (≈ 400 Mo). Given the fact that the corpus must first be POS-tagged, the best results were obtained with well-written corpora, such as the Hansard or *The Economist*. The Wikipedia corpus is often noisy, due mainly to remaining html tags, tables, figures, etc.

2.2   Part-of-Speech Tagging

We first considered using the Stanford Parser (cf. Klein & Manning, 2003) in order to perform the part-of-speech tagging of the corpora. There were several good reasons for this choice: (i) we would exploit linguistic knowledge external of our system, (ii) the reported performance is very good, and (iii) the Stanford Parser is fast, enabling us to parse large corpora. However, the overall results were disappointing, for two main reasons, which are independent from the performance of the Stanford Parser. The first reason is that the word segmentation is in many cases not equivalent to that of our parser: every single word is considered and tagged separately by the Stanford Parser. Our lexicon contains many compound words, nouns like *Prime Minister, playing card, high school*, complex prepositions like *according to, just like*, complex conjunctions like *instead of, rather than*, adverbials *by the way, more or less*, etc. In such cases, the bigrams obtained could not match the lexical units retrieved by our parser, and were therefore useless, or that not match those searched by our parser which are therefore useless, and, even worse, had a negative effect. The second reason is that the tagset used by the Stanford Parser (those defined in the Penn Treebank project, see Santorini 1990) are different from those used by our parser. For instance, the Penn Treebank POS-tag **IN** meaning "Preposition or subordinating conjunction" is too vague to help solving ambiguities.

For those reasons, we decided to use our Fips parser (Wehrli, 2007; Wehrli & Nerima, 2009) to POS-tag the corpora, which can produce various and variably fine-grained POS-tags. For this task, we used a limited tagset, close to the Universal Tagset (Petrov et al., 2012), with the following tags: NOUN, VERB, ADJ, PREP (preposition), DET (determiner), ADV, CONJ (conjunction), INTERJ (interjection), POSS (possessive marker), PART (particle), PREFIX, PONC (punctuation) and OTHER.

2.3   Calculating Bigram Probability

As we have seen, a bigram is constituted by a pair of words, each one described by the tuple [orthographic string, POS tag]. Two probabilities are assigned to the bigram:

- given the first word of the bigram and its POS tag, the probability that the second word bear a particular POS tag
- given the second word of the bigram and its POS tag, the probability that the first word bear a particular POS tag

The sum of the probabilities for a given word and a given POS tag as the first constituent of the bigrams is 1. Symmetrically, the sum of the probabilities for a given word and a given POS tag as the second constituent of the bigrams is 1.

As an illustration, consider the bigrams where *labour* is the first word and *costs* the second. Based on the Hansard corpus we computed the following probabilities:

(2) *P(costs* NOUN | *labour* NOUN*)* = 0.94
     P(costs VERB | *labour* NOUN*)* = 0.06

Table 1 shows the number of bigrams extracted from each corpus.

For the evaluation reported below the bigrams extracted from the Hansard House Debates corpus will be used.

Table 1. Numbers of extracted bigrams for each corpus

Corpus	Corpus size	Number of bigrams
Wikipedia	800 Mo	3,252,7570
The Economist	60 Mo	908,545
Hansard Senate Debates	80 Mo	729,996
Hansard House Debates	400 Mo	2,187,882

3   Statistical Data to Rank Alternatives

As already mentioned, rule-based parsers overgenerate and must be complemented with some pruning mechanism to maintain the proliferation of alternatives within reasonable limits. In the case of Fips, the pruning mechanism takes place after the attachment of each new incoming word.

3.1   Overview of the Parsing Mechanism

In a nutshell, the parsing algorithm works as follows: the parser keeps a list of items, each one of them corresponding to a potential analysis of the input data up to the current position. An item keeps a non-empty stack of (active) constituents, call it the active stack. If the active stack contains just one constituent, this means that a complete analysis has been achieved for the current section of the input data, otherwise the active stack will contain more than one constituents.

Upon reading a new incoming word, the parser tries to combine it with the top constituent of the active stack of each one of the items. Combining means either left attachment, right attachment or even no attachment. In each case, the incoming word is first projected to a constituent level (eg. a noun becomes a noun phrase, a verb a verb phrase, and so on). Left attachment means that the constituent projected from the incoming word takes the constituent on the active stack as its left subconstituent. Right attachment means that the incoming constituent can be attached as a right subconstituent of the constituent on the active stack. Finally, no attachment means simply that the incoming constituent is added on top of the active stack. Left and right attachments are governed by rules and conditions (eg. agreement, selectional restrictions, etc.), see Wehrli (2007) for more details.

3.2   Using Bigrams

As explained in the previous section, our goal focuses on the lexical ambiguity problem. To this effect, the bigram table only takes into account pairs of (orthographical) words where at least one the two terms is lexically ambiguous, ie. associated to more than one category in the lexical database. The probabilities associated to bigrams are used after each attachment action. In other words, probabilities do not affect attachments, they do not interfere with grammar rules. Bigram probabilities are only used to rank partial and complete analyses. Consider a concrete example, as illustrated in (3):

(3) Unit labour **costs** measure the average cost of labour per unit of output.

Once the parser has treated the word *costs* (noun or verb), consider, among the many analyses produced, the one with *labour* as a noun along with *costs* as either a noun or a verb. At that stage, the bigram table will be consulted with the following pairs:

(4) a. [labour-NOUN, costs-NOUN]
    b. [labour-NOUN, costs-VERB]

The probability associated with the bigram (4a) is much higher than the one associated with bigram (4b) (≈ 0.94 vs 0.06), and hence will be ranked higher.

Notice that, although ranked lower, the second alternative is not ruled out and could resurface given an (unlikely) appropriate right context (eg. *labour costs more in certain countries*).

4   Evaluation

To evaluate the impact of the statistics-based ranking scheme on the parser results, we ran the parser with and without the statistical component in a series of tests. To simplify the evaluation, we selected the POS-tagging mode of the parser, and further restricted the output to the triple (word, POS-tag, position). For the POS tagset, we opted for the universal tagset (cf. Petrov et al. 2012). Both output files could then easily be manually compared using a specific user interface as illustrated in Figure 1 below, where differences are put in red.

Notice that in order to facilitate the manual evaluation, we only took into account differences involving the NOUN and VERB tags for ambiguous words such as *announce, sets, costs, labour*, etc. Not only is the confusion between the nominal reading and the verbal reading of such words easy to spot for a human evaluator (cf. Figure 1), it also corresponds to an important mistake, one that would have significant consequences in NLP applications (eg. translation).

In the screenshot the two result files (from a small extract of the Hansard corpus) are displayed, on the left the results obtained with the bigram table, on the right the results obtained without the bigram table. For each file, one line contains the input lexical item (simple word or compound), its tag, and its position with respect to the beginning of the sentence. Differences (restricted here to NOUN vs VERB tags) between the two files are indicated in red. For each difference, the user selects the best choice, using the *Better left* or *Better right* button or the *Skip* button if the difference is irrelevant. After each choice, the next difference is immediately displayed.

Two distinct corpora were used for the evaluation. The first corpus is a small fraction of the Hansard corpus (from which the bigram table was computed)

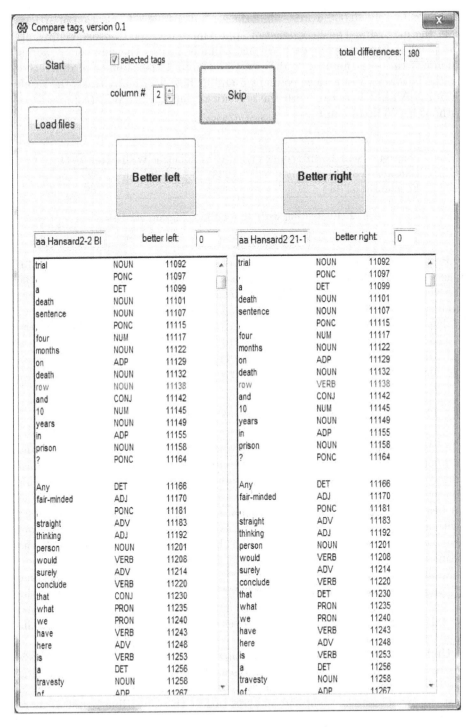

Fig. 1. Manual evaluation user interface

of approximately 40,000 words, while the second corpus is a selection of articles from the magazine *The Economist* of approximately 52,000 words. The results are given in Figure 2. For each corpus, column 2 gives the number of words, column 3 the total number of differences, column 4 the number of differences for the NOUN vs VERB tags, columns 5 and 6 show how many times the result (NOUN / VERB) is better with the statistical component [WS] (column 5) or without it [WO](column 6).

corpus	words	differences	noun vs verb	better WS	better WO
Hansard	37314	1126	228	146	82
Economist	52514	1857	460	292	168

Fig. 2. Evaluation with and without bigram statistics

Results show a clear improvement in terms of precision for the NOUN vs VERB tags. For the Hansard corpus, of the 228 such differences, 146 were better in the WS version versus 82 for the WO one. Results from the second corpus are very similar with 292 better choices in the WS version versus 168 for the WO out of 460 differences. Given that the bigrams were computed on the basis of the Hansard corpus, this second result comes as a good surprise. It seems to indicate that the choice of the corpus used for the bigram extraction has little impact on the quality of the results and that the extracted bigrams have generic properties.

5   Discussion and Concluding Remarks

The results achieved by this first attempt to add to our linguistic parser a ranking component based on bigram probabilities are positive in some respect, as illustrated above. In other respects, though, they do not meet our expectations. This concerns the number of complete analyses achieved by the parser for a particular beam size. The table in Figure 3 below shows results obtained by the parser (without statistical component) for a corpus of approximately 75,000 words from the magazine *The Economist* with various beam-width. Column 1 indicates the beam-width (i.e. the maximal number of alternatives the parser can pursue), column 2 gives the treatment time (word/second), and column 3 the percentage of complete analyses. The results displayed in Figure 3 clearly demonstrate the strong impact of the beam-width on the number of complete analyses achieved by the parser, and unsurprisingly the even more obvious impact it has on the speed of parser.

Under the (reasonable) assumption that the number of complete analyses correlates with the quality of the analyses, those results show that increasing the beam-width leads to better analyses. Such qualitative improvements have a cost, as an increase of complete analyses of less than 4% (line 2 and 3) more than doubles treatment time (93 vs 198 w/s). Finally, as the beam-width is completely independent of the grammar rules, the reduction of the number of

beam-width	word/second	complete analyses
120	59	77.98%
80	93	75.45%
40	198	71.80%
10	579	59.40%
5	793	50.94%

Fig. 3. Number of complete analyses given a particular beam-width

complete analyses (and hence of the quality) triggered by reduction of the beam is due to the pruning mechanism, which eliminates analyses that should have been kept.

For those reasons, it was expected that using a better ranking scheme would lead to a better pruning of alternatives, which in turn would lead to a higher number of complete analyses, or to the same number of complete analyses with a smaller beam. Unfortunately, those expectations have not yet been met, and the parser with or without the statistical component basically reaches the same number of complete analyses for a give beam-width. This last point will be further investigated in future work, as well as the possibility to use bigrams of dependencies rather than bigrams of adjacent words.

Finally, in order to improve the precision of the bigram probabilities, we will restrict the acquisition of bigrams to the sentences for which the parser returns a complete analysis. We observed that when the parser fails to compute a full analysis the risk of inaccurate POS-tag increases significantly.

References

Adolphs, P., Oepen, S., Callmeier, U., Crysmann, B., Flickinger, D., Kiefer, B.: Some Fine Points of Hybrid Natural Language Parsing. In: Proceedings of LREC 2008, Marrakesh, Morocco (2008)

Blache, P., Rauzy, S.: Probabiliser les Grammaires de Propriétés. In: Proceedings of the TALN-Mixeur Workshop, TALN 2013, Sables d'Olonne, pp. 108–111 (2013)

Klavans, J., Resnik, P. (eds.): The Balancing Act: Combining Symbolic and Statistical Approaches to Language. MIT Press (1996)

Klein, D., Manning, C.D.: Accurate Unlexicalized Parsing. In: Proceedings of the 41st Meeting of the Association for Computational Linguistics, pp. 423–430 (2003)

Nivre, J., Hall, J., Nilsson, J., Chanev, A., Eryigit, G., Kübler, S., Marinov, S., Marsi, E.: MaltParser: A Language-independent System for Data-driven Dependency Parsing. Natural Language Engineering 13(2), 95–135 (2007)

Petrov, S., Dipanjan, D., McDonald, R.: A Universal Part-of-Speech Tagset. In: Proceedings of the LREC 2012, Istanbul, Turkey (2012)

Santorini, B.: Part-of-Speech Tagging Guidelines for the Penn Treebank Project (3rd Revision, 2nd printing) (1990), http://www.cis.upenn.edu/treebank

Schneider, G.: Hybrid Long-Distance Functional Dependency Parsing. Ph.D. dissertation, Institute of Computational Linguistics, University of Zurich (2008)

Sennrich, R., Schneider, G., Volk, M., Warin, M.: A New Hybrid Dependency Parser for German. In: Proceedings of GSCL-Conference (2009)

Wehrli, E.: Fips, a 'Deep' Linguistic Multilingual Parser. In: Proceedings of the ACL 2007 Workshop on Deep Linguistic Processing, Prague, Czech Republic, pp. 120–127 (2007)

Wehrli, E., Nerima, L.: L'Analyseur Syntaxique Fips. In: IWPT 2009, Workshop on French Parsers, Paris (2009), http://alpage.inria.fr/iwpt09/atala/fips.pdf

Methodology for Connecting Nouns to Their Modifying Adjectives

Nir Ofek[1], Lior Rokach[1], and Prasenjit Mitra[2]

[1]Ben-Gurion University of the Negev, Israel
[2]The Pennsylvania State University, US
{nirofek,liorrk}@bgu.ac.il, pmitra@ist.psu.edu

Abstract. Adjectives are words that describe or modify other elements in a sentence. As such, they are frequently used to convey facts and opinions about the nouns they modify. Connecting nouns to the corresponding adjectives becomes vital for intelligent tasks such as aspect-level sentiment analysis or interpretation of complex queries (e.g., "*small hotel with large rooms*") for fine-grained information retrieval. To respond to the need, we propose a methodology that identifies dependencies of nouns and adjectives by looking at syntactic clues related to part-of-speech sequences that help recognize such relationships. These sequences are generalized into patterns that are used to train a binary classifier using machine learning methods. The capabilities of the new method are demonstrated in two, syntactically different languages: English, the leading language of international discourse, and Hebrew, whose rich morphology poses additional challenges for parsing. In each language we compare our method with a designated, state-of-the-art parser and show that it performs similarly in terms of accuracy while: (a) our method uses a simple and relatively small training set; (b) it does not require a language specific adaptation, and (c) it is robust across a variety of writing styles.

Keywords: Parsing, Relation Extraction, Information Retrieval.

1 Introduction

Given a sentence, we seek to identify adjectives and the nouns they modify. Connecting nouns to their corresponding modifying adjectives is an important task for properly understanding the meaning of a sentence. For example, in the excerpt "*you cannot compare Brazil, which is vast, with Greece – a small country on the other side of the globe,*" a fact (or opinion) is conveyed by an adjective *vast* that modifies the proper noun *Brazil*. Because natural languages are versatile and do not always comply with simple rules, connecting nouns to their modifying adjectives is not a simple task. In the previous example, applying a rule that connects adjectives with the nearest noun, or within a window of two tokens, will identify *Greece* as being *vast*. Additionally, a single adjective can modify more than a single noun, as in "*the room and service were both great*", or no nouns at all ("*the building looks good*").

A. Gelbukh (Ed.): CICLing 2014, Part I, LNCS 8403, pp. 271–284, 2014.

The general problem we present is important to various applications. The following are examples of three tasks that can benefit from connecting adjectives to nouns:

Sentiment Analysis. People use adjectives to express sentiment. Hatzivassiloglou and Wiebe [17] indicate that adjectives are good indicators of subjectivity in sentences. Blair-Goldensohn, et al. [8] generate a sentiment lexicon where nearly 90% of the terms reported are adjectives. Thus, not only computing adjectives' sentiment score is important to identify expressed sentiments, but connecting them with corresponding nouns becomes crucial for aspect-level sentiment analysis. Consider the following review: "*The staff was not friendly, but the food was excellent and the open air bar quaint and relaxing.*" The Stanford Parser[1] indicates that *quaint* is connected with *staff*, what might lead to error sentiment prediction. Our method can successfully connect *quaint* with *bar*.

Information Retrieval. Identifying adjectives that modify nouns is important in order to parse queries such as *rechargeable battery*; or to provide shopping agents with the ability to mine textual content that includes specifications or properties of products. For example, if a user is looking for a hotel with large rooms, it is crucial to know which adjectives are modifying the noun *room* in a set of relevant reviews.

Spell-Checker. Identifying relationships between nouns and adjectives can boost the performance of text-correction applications. That is true, especially in languages such as Hebrew where adjectives can be plural or singular according to the associated noun.

Dependency-based parsers are widely used for establishing binary relations between words, and as such, they can also address the noun-adjective dependency problem. However, they require a relatively large amount of training data [9] and tend to work well mainly in the language, or the language types, they were specifically designed for. Other studies report that parsers perform less well with content types that differ from their training set. For example, Foster et al. [12] report a drastic drop in performance when moving from the Wall Street Journal (WSJ) domain (training set) to the Twitter dataset (used for evaluation). Consider the following tweet: "*big day tomorrow!*" While the Stanford Parser does not connect *big* with *day*, our method is able to do so.

Despite the prevalence of methods that identify different types of relations such as complex noun [7], neither of the studies has addressed the noun-adjective relationship using machine learning methods. The goal of this research is to develop and examine a machine learning approach for connecting adjectives to their corresponding nouns.

We propose a supervised machine learning method where each instance is defined as a sentence and a pair consisting of a noun and its potentially associated adjective. It begins with the step of tokenizing and POS tagging of input set of sentences and continues to the extraction of all noun-adjective pairs. The goal is to classify each pair with one of the two labels: Modifying or Non-Modifying.

---

[1]   http://nlp.stanford.edu:8080/parser/index.jsp

Our algorithm learns indicative part-of-speech (POS) patterns in the text in order to perform the classification process. We consider the following three sequence position types: POS tags sequences that (1) appear before either word (i.e., either the noun or the adjective, depending upon which comes first), (2) appear between the two words (i.e., in between the noun and adjective), and (3) appear after the last word (i.e., either the noun or the adjective, depending upon which comes last). Fig. 1 illustrates an example of extracted sequences. We cluster sequences by the abovementioned types into three groups; each group's sequences are generalized to patterns, and these patterns are used as features in the binary (Modifying or Non-Modifying) classification task. This approach ensures that the feature set is not tied to a specific language and can be determined from the training set itself.

Since our problem is a sub-problem of the dependency parsing task, we apply a simpler technique (a classifier instead of a parser). Therefore, the training instances contain only the information about adjective-noun relations. This information can be obtained from any treebank, but is not dependent on its availability; and labeling instances specific to the sub-problem is much easier than entire trees.

To summarize, our method has the following properties: (1) it can be applied with languages and content types for which treebanks are not available, since it is trained with a simple and small dataset, (2) it avoids language-specific tuning, and (3) it remain robust across content types, such as user-generated content (UGC).

The capabilities of the new method are demonstrated in two languages: English, arguably the most widely used language in the world, especially as a lingua franca, and Hebrew. Hebrew was selected, because its syntax is fundamentally different than the syntax of English and it is more complex (for example, Hebrew requires a by-gender by-singular/plural adjective form). In addition, Hebrew, a Semitic language, has a rich morphology property which imposes additional complexities [2, 28], as having an average of 2.7 possible morphological analyses per token – compared to 1.4 in English [1].

Our experiments show that the method achieves comparable accuracy compared to parsers trained with many more input trees.

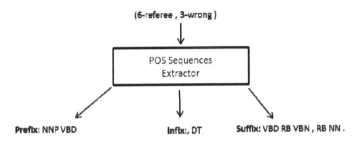

Fig. 1. Three POS sequences. The adjective *wrong* is a modifier candidate for the noun *referee*.

2   Related Work

The goal of the parsing process is to build a dependency structure, consisting of words connected by binary dependency relations. However, the main drawbacks with respect to connecting nouns with their modifying adjectives are:

1. Training Data. For training parsers, annotated corpora (treebanks) of thousands to tens of thousands of sentences are necessary [9]. However, there is a dearth of annotated trees in some languages [28]. Since our method is designed to solve a subproblem, it can work well under a very small training set, particularly due to the generalization process used.
2. Language – Dependent. Since most parsers are dedicated to a specific language, adapting to other languages can impose additional challenges, such as whether the morphological information can be encoded on dependency relations. More challenges related to the parsing of morphologically rich languages, such as using Morphological CASE, are discussed by [28]. Moreover, De Marneffe et al., [10] use a general method for producing dependencies which requires appropriate rules for each language. The proposed method is designed to perform a more specific task and extracts features on-the-fly, i.e., the method is adjusted to adapt attributes of the trained language at learning time.
3. Vulnerability to Content Type. Well-known parsers are trained on the Wall Street Journal (WSJ) domain. Moving from the WSJ domain test set to the Twitter dataset resulted in a drastic drop in performance [12]. For example, the Stanford Parser does not connect trip with good in the statement "*the trip wasnt good*" due to the missing apostrophe.

More drawbacks of parsers are reported by [10]; for example, Minipar [20], it is reported as being confused by punctuation and conjunctions since most of these tags are generalized and others are ignored.

The information extraction approaches are another machine learning techniques that can be used to solve our problem. These methods are aimed at extracting indicative semantic relations between the two entities (or phrases). Relation extraction systems are designed to automatically populate knowledge bases by learning relations between entities. These approaches are useful for handling tasks such as entity disambiguation, information retrieval and question answering [5]; however, most intelligent systems rely upon existing NLP tools such as parsers, as indicated by the recent work of Xu et al. [30].

Some information extraction studies focus on specific problems, such as complex noun phrases [7, 18]. Agichtein and Gravano [4] propose a snowball method, but unlike our system, their approach expects to be provided with a set of general regular expression that the entities must match. However, in the case of our problem, the use of the snowball method could lead to poor recall. One of the most well-known systems is Open IE [11], a system that relies heavily on a very large unlabeled training set. Since this approach is unsupervised, it might result in contradictory information, as reported by Minack, Gianluca and Wolfgang [22].

3   Generating Part-Of-Speech Patterns

The proposed method extracts POS sequences that are used as a binary feature in a classification process. In order to avoid the curse of dimensionality and potential overfitting, we employ a generalization process that reduces the feature space from sequences to patterns. For example, an experiment on the Hebrew treebank initially extracted a few hundred sequences and the generalization process reduced them to seven patterns. Our method can be used in any language that has a POS tagger and a component that can tokenize words.

The process begins with a set of input labeled sentences and outputs a set of generalized POS patterns. It is comprised of four steps. The implementation is available in the online software[2] that can be used to learn other relationships as well.

Step 1: Tokenization and POS Tagging. We begin with a set of input sentences. Each input sentence is then tokenized into word tokens and is tagged with its POS tags; thus, it can be applied to any language in which a POS tagger and the tokenization component are available.

In English, we use the easy-to-implement and popular segmentation and tagging components available in nltk[3]. Similar to Goldberg and Elhadad [15], in Hebrew, we use the tokenizer and POS tagger components developed in a continuous work by Adler, Goldberg and Elhadad [1, 3, 13].

Step 2: Preparing the Training Repository. Our method is supervised and needs to be provided with a set of sentences and a Modifying or Non-Modifying label for each noun-adjective pair. This information can be easily obtained from an annotated treebank, but it is not limited to situations only in which treebanks are available. For each sentence we first collect all noun-adjective pairs, i.e., each noun with each adjective constitute an instance. Each instance will be labeled manually with either the Modifying label (indicating existing dependency) or the Non-Modifying label (reflecting that they are not dependent).

If annotated treebanks are not available, instances should be given with one of the two labels of our problem. Otherwise, extracting the label from the treebank is mainly a technical issue and involves the following tasks: First, we convert constituency trees to dependency trees using the Stanford Parser's convertor[4]. We considered the following dependency relation types as Modifying: 'amod', 'nsubj', 'dep', i.e., a noun-adjective pair is considered connected if it has one of those dependency labels. In case that the treebank scheme does not follow the Penn Treebank scheme, (different dependency labels are used), we expect a list of the equivalent dependency labels.

Step 3: Extracting POS Sequences. For each instance, we extract three types of sequences that follow the three difference positions: (1) prefix – the POS sequence that appears before either word (i.e., either the noun or the adjective, depending upon which comes first), (2) infix – the POS sequence that appears between the two words (i.e., in between the noun and adjective), and (3) suffix – the POS sequence that

---

[2] http://www.ise.bgu.ac.il/faculty/liorr/ofek_cicling2014.zip
[3] http://nltk.org/api/nltk.tag.html
[4] http://nlp.stanford.edu/software/dependencies_manual.pdf

appears after the last word (i.e., either the noun or the adjective, depending upon which comes last). Fig. 1 provides an example.

The extracted sequences are later used as features by the classifier. However, the feature space dimensionality is high since each sequence position type consists of hundreds or thousands of instances. The next step is used for feature space reduction.

Step 4: Generalizing Sequences to Patterns. At this step we want to reduce the feature space by generalizing POS sequences into patterns in order to avoid the curse of dimensionality and the subsequent issue of overfitting.

In each of the three sequence position types (prefix, infix, suffix), we group sequences into one of two sets, according to the class it was extracted from. Thus far, we have six groups of POS sequences. Next, we generalize each of the six groups' sequences, independently, to a six groups of patterns. This ensures that the result patterns are generalized over sequences from the same position and class.

In our evaluation we chose to use the Teiresias algorithm designed by Rigoutsos and Floratos [25] to perform generalization, since it was found useful in similar problems [26]; however, the generalization process is not tied to a specific algorithm. The Teiresias algorithm searches for patterns that satisfy certain density constraints, limiting the number of wild-cards occurring in any stretch of pattern. More specifically, Teiresias looks for maximal <L,W> patterns with the support of at least K, where a pattern P is called an <L,W> pattern if every sub-pattern of P with a length of at least W words (combination of specific words and wild-cards) contains at least L specific words. In all of our experiments, for both evaluated languages, we found the following values as efficient: W=5, L=1. For detailed information and usage, refer to [26].

After generalization, the two sets of patterns (correspond with the two class labels - Modifying and Non-Modifying) at each of the three positions types are merged into a single patterns set. A classification process (described below) takes place in which patterns are used as features. This approach ensures that the feature set is not tied to a specific language since features are determined from the training set itself.

4   Classification

The classification phase aims to train a model to discriminate between Modifying and Non-Modifying instances. Each POS pattern generated in the previous phase is used as a binary feature to indicate whether the pattern matches with the instance or not, in the position type (prefix, infix, and suffix). Thus, each pattern corresponds to only one of the three positions it was originally extracted from, and each pattern has a position property to indicate its position type. For example, the pattern Prefix_{NN * JJ} is translated to a feature that indicates whether the prefix of the instance matches with the pattern NN * JJ.

Table 1 illustrates instances derived from a single sentence. Each row corresponds to one noun-adjective instance. Each column corresponds to one pattern (feature). The binary values in the table indicate if the pattern exists in the sentence ('1') or not ('0'). In addition, we add a binary feature to indicate whether the noun precedes the adjective or not. In fact, we used this feature to split the training set and train two models, but that did not improve performance.

Table 1. Representation of the six instances, derived from the sentence: *"The room had a wide window with an old frame."*

Instance	Pattern 1	Pattern 2	...	Pattern n	Class
room,wide	0	1		0	Non-Modifying
room,old	1	1		1	Non-Modifying
window,wide	0	0		1	Modifying
window, old	0	1		0	Non-Modifying
frame,wide	0	1		0	Non-Modifying
frame, old	0	0		1	Modifying

The algorithm we used is J48 [29], which is a Java implementation for Quinlan's C4.5 [24] algorithm. Other algorithms were evaluated, but since none outperformed J48, we opted to use this model for its comprehensiveness.

Recall the sentence from Twitter *"big day tomorrow!"* that poses difficulties for existing parsers. Fig. 2 illustrates a decision sub-tree model trained on a sample of 2,000 trees from Penn TreeBank in this case. We can see that in node '1' the instance (*day, big*) travels left since there are not any POS tags in between the noun-adjective pair. In node '2' the instance travels right for the decision leaf Modifying since the noun does not precede the adjective. Node '2' in the figure illustrates how the last feature (which comes first in the sentence – the adjective or the noun) can be used in the classification process.

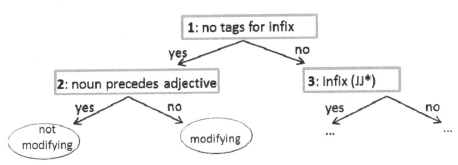

Fig. 2. Illustration of a decision sub-tree model; elliptic-shaped nodes are leaves

5   Evaluation

To evaluate our method, we experimented with both English and Hebrew. Since English is well studied in the literature, it will be the primary focus of this section. The results are compared with state-of-the-art parsers for both languages. In addition, a simple baseline was used to understand the complexity of the problem. It is important to mention that only the training set was used to generate patterns.

5.1   Evaluation Measures: AUC

The evaluated datasets are imbalanced; namely, there are many more Non-Modifying instances than Modifying. For example, in the Penn TreeBank, the imbalance rate is 1:10. Standard machine learning techniques might be 'overwhelmed' by the majority class and ignores by the minority class. A classifier can achieve a high accuracy (90% in this case) by always predicting the majority class. The ROC (Receiver Operating Characteristic) curve is a graph produced by plotting the fraction of true positives versus the fraction of false positives for a binary classifier as its discrimination threshold varies. In addition to standard measures, we will use the Area Under the ROC Curve (AUC) measure, in our evaluation. As indicated by Oommen, Baise, and Vogel [23], "AUC is more robust over other measures and is not influenced by class imbalance or sampling bias". The AUC value of the best possible classifier will be equal to 1, which means that we can find a discrimination threshold under which the classifier will obtain 0% false positives and 100% true positives. A higher AUC value means that the ROC graph is closer to the optimal threshold and is a single measure for our task.

The evaluated parsers output only a deterministic answer i.e., Modifying or not, without the probabilities that are essential to compute the true-positive rate (TPR) and false-positive rate (FPR) across all alternative acceptance points. To plot the ROC for parsers and compute their AUC, we plotted the single TPR and FPR point (calculated based on the parser's configuration) and connected it with the minimum (0,0) and maximum (1,1) values of the TPR and FPR.

5.2   English

In English, we used the nltk components for tokenization and POS tagging.

Experiment Setup. The Penn TreeBank (PTB), whose source is the WSJ domain [21], is a well-known manually annotated treebank. The training and test sets from this source were randomly sampled, and, to allow repeatability, a whole section of the treebank was sampled. In order to perform cross-learning evaluations for UGC and non-UGC types, we also used the Web2.0 dataset denoted as the UGC. It is a small treebank of 1000 syntactically annotated sentences, 519 of which were taken from Twitter and 481 from bbc.co.uk (football discussion board). More information is provided by Foster et al. [12].

As a baseline, we used the Stanford Parser [10] version 2.0.5, a state-of-the-art parser that has been used by many researchers, as well as in a number of recent studies [27, 30]. Following the documentation, we considered adjective-noun pairs as Modifying if they have one of the following dependency labels: 'amod', 'nsubj', 'dep'.

Results. Table 2 summarizes the experiments in the English datasets. Our method was trained on a randomly sampled subset from the PTB. Both methods were tested on the UGC and on a subset of PTB. Despite the fact that Stanford Parser's API allows for training on different set of trees, it had too many errors for misidentifying the correct POS tag. Since for the best of our knowledge there is no way to train the POS tagger and the parser on different sets, the comparison with our method could not be fairly conducted with lower number of trees. Therefore, Stanford Parser was used with its default settings and trained model.

Table 2. Performance of our method and the Stanford Parser tested on the UGC dataset. The size of the training trees is indicated in parentheses.

Method	Training	Testing	Accuracy	F-score	AUC
Stanford	PTB (tens of thousands)	UGC	0.957	0.955	0.869
POS_seq	PTB sec_00 (1,921)	UGC	0.957	0.958	0.935

Fig. 3 illustrates the ROC curves for the experiment of Table 2. As mentioned above, the parser's curve is estimated since it outputs a deterministic result without probabilities.

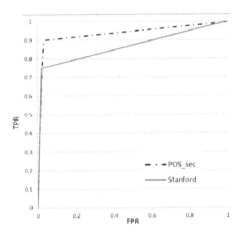

Fig. 3. ROC curves of the two methods for the UGC dataset

We evaluated our method on various training set sizes from the UGC. Fig. 4 shows the trend. The experiment type is cross-validation, where the folds are determined by the size of the training set.

Evaluation of our method continued and also included additional experiments, such as cross-learning evaluation, presented in Table 3.

5.3   Hebrew

Hebrew is a Semitic, morphologically rich language.

We used the most recent tokenizer and POS tagger components proposed by Adler, Goldberg and Elhadad [2, 3, 13], and this was used by the Hebrew parser as well.

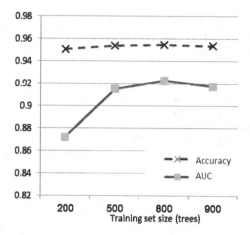

Fig. 4. Performance of our method in various training set sizes for the UGC

Table 3. Performance of our method on several datasets. The size of the training trees is indicated in parentheses.

Training	Testing	Accuracy	F- score	AUC
UGC - 10 Folds c.v. (900)		0.953	0.954	0.917
PTB sec_00 (1,921)	PTB sec_15	0.975	0.976	0.967
UGC (1000)	PTB sec_00	0.975	0.976	0.955

Experiment Setup. The Hebrew Treebank [16] consists of two different tree sets. One tree set containing 5258 annotated trees is used for training (HTB_train), and another gold-set containing 474 trees is used for testing (HTB_test). Goldberg and Elhadad [14] provide more details about the Hebrew Treebank.

As a baseline, we used the Easy-First parser for Modern Hebrew [15], which is considered to be the only parser for Modern Hebrew. In order to evaluate the performance of the parser, identified labels represent noun-adjective dependency in the parser's output format. The set of relevant dependency labels was provided by the linguist that was involved with the annotation of the training trees for the Easy-First parser. Thus 'ADJ', 'MOD' and 'dep' dependency types were considered as Modifying.

Results. The 474 trees in HTB_test contain 5997 noun-adjective pairs. There are nearly 12.7 candidate pairs in each sentence. We evaluated our method on training trees of various sizes but found no trend, i.e., the performance remains similar at 300 trees and up to 5258. Results of the Easy-First parser and our POS_seq method are presented in Table 4.

Table 4. Performance of our method and the Easy-First Parser. Both were trained on the HTB_train and tested on the HTB_test.

Method	Accuracy	F- score	AUC
Easy-First	0.970	0.969	0.865
POS_seq	0.968	0.969	0.912

5.4   Simple Baselines

The results of evaluation of the method on the PTB provide an indication of the complexity of the problem. The following rules were employed, resulting in the performance indicated. For each adjective (a) always connect with the nearest noun (53% accuracy), (b) always connect with the nearest noun on the left (1.38% accuracy) and (c) on the right (80.8% accuracy).The performance is relatively poor for all of the rules. That is, since the problem is more complex due to versatility of languages, simple heuristics could not hold. The complexity can be demonstrated by the two examples: (1) "*the mother and her son are old*", (2) "*the boy looks sad*". In the first sentence a single adjective modifies two nouns and in the second the adjective does not modify any noun.

6   Discussion

Table 2 demonstrates that our method achieves similar results compared with state-of-the-art parser. The comparable results are achieved while using a relatively smaller training set. This observation is further supported by Fig. 4, in which our method still performs well, even when using far fewer training instances. It can be explained by the fact that our method is designed to solve a specific problem in which a relatively small number of training trees is needed. Moreover, the generalization process is an important factor in enabling efficient performance while using small number of training trees. Since every tree provides on average 11 instances for our task, a small number of trees still provides a satisfactory number of instances. The ROC curves on Fig. 3 demonstrate that our method was found superior in terms of AUC – which is important for imbalanced datasets since it is not influenced by the majority class (non-Modifying). However, as mentioned in section 5.1, the parser's curves are estimated since they provide only deterministic results.

Table 4 demonstrates that our method achieves comparable results with a state-of-the-art parser in Hebrew. This is a promising finding since our method could be used with other languages without additional tuning, as long as they have POS tagger and tokenization components available, and as long as sequential information is meaningful. There are some free-order languages (e.g., Russian, Polish, and Greek) where the POS functionality can be interpreted by the suffix of the word while the order of the words is not always determined by their syntactic role. However, since there is still a typical word order, our sequence-based approach could be beneficial to some of those as well. The success of the method seems to critically depend on the granularity of POS scheme used, and thus it might not work well in some languages. For example, in some morphologically rich languages like Czech, many POS schemes include information such as case.

It is shown by the shared task from 2006 on multilingual dependency parsing [9] that parsers' performances vary across languages. For example, by averaging the labeled attachment score (LAS) for all parsers, we can see that the score for Arabic, a Semitic language like Hebrew, is 59.9 whereas the score for Japanese is 85.9. The standard deviation over 13 evaluated languages, on average across all participating parsers, is 8.8. A drop in performance is observed also by Kübler [19], who shows that there is a drop in performance when parsing German language text with respect to English. Based on the two evaluated languages, our method was found robust by providing similar results for the two.

Our method tends to be susceptible to larger margins of error when used with complex sentences where the modifying adjective is positioned far from the noun. This occurs because the variance of longer POS sequences is relatively high; thus, these sequences are considered as outliers in the generalization process. For example, in the following review, taken from TripAdvisor.com: *"Breakfast at the hotel is very good and the service of the staff here excellent,"* our model fails to connect *excellent* with *service* unlike the Stanford Parser, but it does not repeat Stanford Parser's mistake of connecting *excellent* with *hotel.*

7   Conclusion and Future Work

Given a natural language sentence, this study addresses the problem of finding semantically connected noun-adjective pairs. The supervised approach uses a small number of training instances and ensures that the feature set is not tied to a specific language since they are determined from the training set itself. Therefore it is suitable to scenarios where training instances are not widely available or derived from a specific content type to be used in other content types.

The method should not be limited to identifying noun-adjective relations. In the future, we intend to extract other relation types where the design should remain the same: a binary classification problem consisting of positive [presence of relation] and negative [absence of relation], as long as sequential information is important. Every specific relation will still demand a relevant annotated corpus. One relation we are interested in investigating is noun-noun relations, i.e., whether one noun semantically contains another noun as its property or aspect. For example consider the two sentences: (1) *"the room had a nice view"* and (2) *"I was waiting in the room while she was enjoying the view."* In the first, the *view* is a property of the *room*, while in the second it is not the case.

References

1. Adler, M.: Hebrew Morphological Disambiguation: An Unsupervised Stochastic Word-based Approach. Ph.D. thesis, Ben-Gurion University of the Negev, Beer-Sheva, Israel (2007)
2. Adler, M., DahanNetzer, Y., Goldberg, Y., Gabay, D., Elhadad, M.: Tagging a Hebrew Corpus: The Case of Participles. In: LREC (2008)
3. Adler, M., Elhadad, M.: An Unsupervised Morpheme-Based HMM for Hebrew Morphological Disambiguation. In: Proceedings of COLING-ACL 2006 (2006)
4. Agichtein, E., Gravano, L.: Snowball: Extracting relations from large plain-text collections. In: Proceedings of the Fifth ACM Conference on Digital Libraries. ACM (2000)

5. Alfonseca, E., Filippova, K., Delort, J.-Y., Garrido, G.: Pattern learning for relation extraction with a hierarchical topic model. In: Proceedings of the 50th Annual Meeting of the Association for Computational Linguistics: Short Papers, vol. 2, pp. 54–59. Association for Computational Linguistics (2012)

6. Banko, M., Etzioni, O., Center, T.: The Tradeoffs Between Open and Traditional Relation Extraction. In: ACL, vol. 8, pp. 28–36 (2008)

7. Barker, K., Szpakowicz, S.: Semi-automatic recognition of noun modifier relationships. In: Proceedings of the 17th International Conference on Computational Linguistics, vol. 1. Association for Computational Linguistics (1998)

8. Blair-Goldensohn, S., Hannan, K., McDonald, R., Neylon, T., Reis, G.A., Reynar, J.: Building a sentiment summarizer for local service reviews. In: WWW Workshop on NLP in the Information Explosion Era (NLPIX), New York, NY, USA. ACM (2008)

9. Buchholz, S., Marsi, E.: CoNLL-X Shared Task on Multilingual Dependency Parsing. In: Proceedings of the Tenth Conference on Computational Natural Language Learning (CoNLL-X), New York, NY (2006)

10. De Marneffe, M.-C., MacCartney, B., Manning, C.D.: Generating typed dependency parses from phrase structure parses. In: Proceedings of LREC, vol. 6 (2006)

11. Etzioni, O., Banko, M., Soderland, S., Weld, D.S.: Open information extraction from the web. Communications of the ACM 51(12), 68–74 (2008)

12. Foster, J., Çetinoglu, Ö., Wagner, J., Le Roux, J., Hogan, S., Nivre, J., Hogan, D., Van Genabith, J.: # hard-to-parse: POS Tagging and Parsing the Twitterverse. In: Proceedings of the Workshop on Analyzing Microtext (AAAI 2011), pp. 20–25 (2011)

13. Goldberg, Y., Adler, M., Elhadad, M.: EM Can Find Pretty Good HMM POS-Taggers (When Given a Good Start). ACL (2008)

14. Yoav, G., Elhadad, M.: Hebrew Dependency Parsing: Initial Results. In: Proceedings of IWPT (2009)

15. Goldberg, Y., Elhadad, M.: Easy first dependency parsing of modern Hebrew. In: Proceedings of the NAACL HLT 2010 First Workshop on Statistical Parsing of Morphologically-Rich Languages. Association for Computational Linguistics (2010)

16. Guthmann, N., Krymolowski, Y., Milea, A., Winter, Y.: Automatic annotation of morpho-syntactic dependencies in a modern Hebrew treebank. In: Proceedings of TLT (2009)

17. Hatzivassiloglou, V., Wiebe, J.M.: Effects of adjective orientation and gradability on sentence subjectivity. In: Proceedings of the 18th Conference on Computational Linguistics, vol. 1. Association for Computational Linguistics (2000)

18. Kim, S.N., Nakov, P.: Large-scale noun compound interpretation using bootstrapping and the Web as a corpus. In: Proceedings of the Conference on Empirical Methods in Natural Language Processing. Association for Computational Linguistics (2011)

19. Kübler, S.: The PaGe 2008 shared task on parsing German. In: Proceedings of the Workshop on Parsing German, pp. 55–63. Association for Computational Linguistics (2008)

20. Lin, D.: MINIPAR: a minimalist parser. Maryland linguistics colloquium (1999)

21. Marcus, M.P., Marcinkiewicz, M.A., Santorini, B.: Building a large annotated corpus of English: The Penn Treebank. Computational Linguistics 19(2), 313–330 (1993)

22. Minack, E., Demartini, G., Nejdl, W.: Current approaches to search result di-versification. In: First International Workshop on Living Web: Making Web Diversity a True Asset, Washington DC (2009)

23. Oommen, T., Baise, L.G., Vogel, R.M.: Sampling bias and class imbalance in maximum-likelihood logistic regression. Mathematical Geosciences 43(1), 99–120 (2011)

24. Quinlan, J.R.: Induction of decision trees. Machine Learning 1, 81–106 (1986)

25. Rigoutsos, I., Floratos, A.: Combinatorial pattern discovery in biological sequences: The TEIRESIAS algorithm. Bioinformatics 14(2), 229 (1998)

26. Rokach, Romano, Maimon: Negation recognition in medical narrative reports. Information Retrieval 11(6), 499–538 (2008)

27. Swanson, B.: Exploring Syntactic Representations for Native Language Identification. In: NAACL/HLT 2013, p. 146 (2013)

28. Tsarfaty, R., Seddah, D., Goldberg, Y., Kübler, S., Candito, M., Foster, J., Versley, Y., Rehbein, I., Tounsi, L.: Statistical parsing of morphologically rich languages (SPMRL): what, how and whither. In: Proceedings of the NAACL HLT 2010 First Workshop on Statistical Parsing of Morphologically-Rich Languages, pp. 1–12. Association for Computational Linguistics (2010)

29. Witten, I.H., Frank, E.: Data Mining: Practical machine learning tools and techniques. Morgan Kaufmann (2005)

30. Xu, Y., Kim, M.-Y., Quinn, K., Goebel, R., Barbosa, D.: Open Information Extraction with Tree Kernels. In: Proceedings of NAACL-HLT, pp. 868–877 (2013)

Constituency Parsing of Complex Noun Sequences in Hindi

Arpita Batra[1], Soma Paul[1], and Amba Kulkarni[2]

[1] Language Technologies Research Centre, International Institute of Information Technology, Hyderabad, India
`arpita.batra@students.iiit.ac.in, soma@iiit.ac.in`
[2] Department of Sanskrit Studies, University of Hyderabad, India
`apksh@uohyd.ernet.in`

Abstract. A complex noun sequence is one in which a head noun is recursively modified by one or more bare nouns and/or genitives[1] Constituency analysis of complex noun sequence is a prerequisite for finding dependency relation (semantic relation) between components of the sequence. Identification of dependency relation is useful for various applications such as question answering, information extraction, textual entailment, paraphrasing.

In Hindi, syntactic agreement rules can handle to a large extent the parsing of recursive genitives (Sharma, 2012)[12].This paper implements frequency based corpus driven approaches for parsing recursive genitive structures that syntactic rules cannot handle as well as recursive compound nouns and combination of gentive and compound noun sequences. Using syntactic rules and dependency global algorithm, an accuracy of 92.85% is obtained.

Keywords: constituency parsing, bracketing, complex noun sequence, compound noun, genitives.

1 Introduction

A noun can have various pre-modifiers such as adjective, adjectival phrase, bare noun (henceforth, compound noun), genitive noun. The case becomes complex when a head noun[2] is modified recursively as in

1. (*ladake*   *kA*   (*mittI*   *kA*   *ghar*))
   "boy" genitive-marker "mud" genitive-marker "house"
2. (*jilA*   (*nirvAchan*   *adhikArI*))
   "district" "election" "officer"

Or a head noun is modified by a complex modifier. Example:

3. ((*AdamI*   *ke*   *bete*)   *kA*   *ghar*)
   "man" genitive-marker "son" genitive-marker "house"

---

[1] Genitive markers in Hindi are kA, and its allomorphic variations ke and kI.
[2] Hindi is a head final language.

A. Gelbukh (Ed.): CICLing 2014, Part I, LNCS 8403, pp. 285–296, 2014.

4. ((*krishi*   *prasanskaraNa*)   udyog)
   "agriculture"   "processing"   "industry"

Complex noun sequence is a sequence having multiple nouns. Nouns may or may not be separated by genitive markers. When no genitive marker is present in between the nouns, then such sequence is known as compound noun. A noun sequence can be represented as the following regular expression[1]:

(*noun+* [3]   *genitive-marker*)* [4]   *noun+*

Binary constituency parsing of noun with complex modifier is an important requirement for determining the semantic relation between noun and its modifier. (Sharma, 2012)[12] uses agreement rules for parsing nouns having recursive genitive modifiers and reports an accuracy of 80%. Syntactic agreement rules alone fail to determine the constituents when the allomorphic forms of genitive are same as in:

5. (*ladake*   kA   (*patthar*   kA   *ghar*))
   "boy"   genitive-marker   "stone"   genitive-marker   "house"
6. ((*vimAnoM*   kI   *kharId*)   kI   *yojanA*)
   "aircraft"   genitive-marker   "purchases"   genitive-marker   "plan"

Syntactic rules also fail for bare noun sequences. For handling cases where rules cannot be applied, we use the frequency based method. Detail approach and results are discussed in Section 3 and 4 respectively.

2   Related Work

Corpus driven approaches are prevalent for handling constituent analysis of compound nouns. (Pustejovsky et. al., 1993)[11] has used bigram frequency for bracketing compound nouns having three noun constituents. For compound noun a-b-c, where "a", "b" and "c" are noun constituents, if frequency of a-b is greater than frequency of b-c, then the result is ((a-b)-c), else the result is (a-(b-c)). (Lauer, 1995)[9] has used conceptual association instead of lexical association for finding the better pair. This helps in handling sparsity. Conceptual association is found in between noun categories, obtained from Roget's thesaurus. In this paper, the concept of adjacency and dependency pairs was used for the first time. In adjacency model, semantically better adjacent pairs, a-b or b-c is found. If a-b is semantically better than b-c, then the output is ((a-b)-c), otherwise the output is (a-(b-c)). For dependency model, we find whether a modifies b or a modifies c. If a-b is semantically better than a-c, then the result is left bracketed ((a-b)-c), otherwise the result is right bracketed (a-(b-c)). (Keller and Lapata, 2005)[8] has also used both the adjacency and dependency based models. But

---

[3] x+ matches any sequence that contains atleast one x.

[4] x* matches any sequence that contains zero or more occurrences of x.

association is compared using frequencies of the adjacency and dependency pairs. These frequecies are obtained from interpolation of frequency obtained from web and corpus. (Girju et. al., 2005)[3] has used the conceptual association to find the parse. Noun classes from WordNet are used for determining the output.

(Kulkarni and Kumar, 2011)[6] has used conditional probability to decide the compatibility between the two words. For noun compound a-b-c, it compares the probability of occurrence of a-b and b-c. If occurrence of a-b is more probable, then the ouput is ((a-b)-c), else the ouptut is (a-(b-c)). This approach is also used in (Kulkarni et. al., 2012)[7] for bracketing compounds of Indian languages like Sanskrit, Hindi, Marathi. (Kavuluru and Harris, 2012)[5] has used both non-greedy and greedy based approach for bracketing compound nouns having four constituents. These compound nouns are from biomedical domain. Greedy approach is based on comparing frequencies of the possible components which can be joined. Initially, a and b, b and c, c and d are the possible components which can be joined. After this combination, second best combination is found. For greedy approach, we have used lexical association[10], jaccard index[4] to find the best pair. This helps in normalizing the frequency. In non-greedy approach, cohesion measure is calculated. Cohesion values are found for all possible trees. This is obtained by summing the relatedness of each node and relatedness is found by using jaccard index. Then, we choose the output with the highest cohesion value.

3   Approaches

In this paper, we handle complex noun constructions where the modifier can be a series of bare nouns or a combination of genitive and bare nouns. As a first step, we locally group bare sequences of noun within a complex sequence. Such group is called here local group. The head of the local group can either modify a noun outside the group (as in 7) or it can itself be modified by a noun outside the group (as in 8):

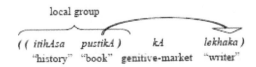

7.

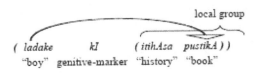

8.

After the formation of local group (if there is any), the syntactic rules (Sharma, 2012)[12] are applied for genitives. Finally, frequency based approach is applied for the cases where syntactic rules fail. Same approach is also followed for further constituent analysis of local groups. The detail of this step is discussed below.

Frequencies for both adjacency and dependency pairs are used to parse the noun sequences. Adjacency pairs refer to the two sequences of noun adjacent to each other in a complex sequence. Dependency pairs are the two sequences of noun which are not necessary to be adjacent to each other but their relative positions are maintained. Let a-b-c-d be the sequence with a, b, c, d as the constituent nouns or sequence of nouns. Adjacency pairs in this compound are: a-b, b-c and c-d. Dependency pairs in this compound are: a-b, a-c, a-d, b-c, b-d and c-d.

We have used both greedy and global approaches for finding the parsed output. Greedy approach refers to the concept in which we find the best pair from all possible pairs and then we find the second best pair. If a wrong pair is chosen, then the error percolates up in the whole parse. Consider an example with a-b-c-d as the sequence. Let the expected output be (a((bc)d)). If a-b is found to be the best pair initially, then possible generated output are ((ab)(cd)) and (((ab)c)d). Due to wrong selection of (ab), output (a((bc)d)) cannot be obtained. To avoid this, global approach has been used. Global approach refers to the concept in which we consider all possible bracketing of a compound. Sequence a-b-c-d have five possible bracketings which are (((ab)c)d), ((ab)(cd)), ((a(bc))d), (a((bc)d)), (a(b(cd))[5]. For all five cases, cohesion value is calculated. We choose the output which results in maximum cohesion value [5].

In total, four approaches have been used: adjacency global, adjacency greedy, dependency global and dependency greedy for parsing sequences where rules do not apply. For all approaches, if jaccard index[4] measured for the pair is found to be zero, then we take left bracketed result as default because it has been observed (see Table 1) that left bracketing is more frequent than right bracketing in Hindi. Bare nouns in a corpus are found to be very less. Therefore, to get better frequencies of the sequence, we have considered both bare as well as genitive nouns because of their same nature. Ex: *"kavi sammelan"* is equivalent to*"kavi kA sammelan"* (meeting of poets). This helps in dealing with sparsity to an extent.

3.1   Adjacency Greedy Approach

It is the approach in which we have considered bracketing adjacent pairs greedily. Lexical association[10] for adjacent pairs is found using jaccard index[4]. In this approach lexical association of all pairs are compared and the best pair is chosen. After this step, another best pair is chosen till the whole parse tree is formed. Algorithm in detail is described below:

---

[5] For sequence having 'n+1' number of noun constituents, number of possible bracketing is equal to catalan number, C_n[6].

Table 1. Distribution of left bracketed and right bracketed sequences having three noun constituents. (*a, b, c in below table denotes noun and gen denotes genitive marker.*)

Type of Complex Noun Sequence	Left Bracketed	Right Bracketed
Compound Noun	159	78
(a b c)	((a b) c)	(a (b c))
Nouns separated by one genitive-marker	683	306
(a gen b c) or (a b gen c)	((a b) gen c)	(a gen (b c))
Nouns separated by two genitive-markers	502	42
(a gen b gen c)	((a gen b) gen c)	(a gen (b gen c))

Algorithm:

(a) Let there be 'n' noun components: W_i, i ranging from 1 to n.
(b) Compute the lexical association for each adjacent pair of constituents.
(c) Select the pair with the highest association. Let the pair be W_{k-1} and W_k.
(d) Join W_{k-1} and W_k nodes.
(e) Remove W_{k-1} and W_k from the constituent list.
(f) Insert W_{k-1}-W_k as the single constituent in the list.
(g) Repeat steps "a" to "f" till all the nodes are joined.

$$Association(p, q) = \frac{freq(pq)}{freq(p) + freq(q) - freq(pq)} \tag{1}$$

where, "p" and "q" are the elements of constituent list.

3.2 Adjacency Global Approach

It is the approach in which cohesion value[5] for all possible bracketing for adjacent pair is considered and one with the highest cohesion value is chosen. Cohesion for a tree is measured by summing the association[10] corresponding each node of a tree. Association for a node in a tree is obtained by using jaccard index[4]:

$$Association(i, j) = \frac{freq(ij)}{freq(i) + freq(j) - freq(ij)} \tag{2}$$

where, "i" and "j" are the nodes which are combined to form a single node.

Consider (a((bc)d)) as an example. To find the cohesion of this tree, lexical association for b-c, bc-d and a-bcd are added. For more clarification, association for each node corresponding (a((bc))d)) tree is shown below. Tree representation for (a((bc)d)) is shown in Fig. 1.

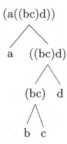

Fig. 1. Tree Representation for bracketed (a((bc)d) sequence

$$Association(b, c) = \frac{freq(bc)}{freq(b) + freq(c) - freq(bc)} \tag{3}$$

$$Association(bc, d) = \frac{freq(bcd)}{freq(bc) + freq(d) - freq(bcd)} \tag{4}$$

$$Association(a, bcd) = \frac{freq(abcd)}{freq(a) + freq(bcd) - freq(abcd)} \tag{5}$$

$$Cohesion([a[[bc]d]]) = Association(b, c) + Association(bc, d) + Association(a, bcd) \tag{6}$$

3.3   Dependency Greedy Approach

It is the approach in which we have considered bracketing dependency pairs greedily. Lexical association[10] for dependency pairs is found using jaccard index (see Formula 1)[4]. Lexical association of all pairs are compared and the best pair is chosen. After this step, another best pair is chosen till the whole parse tree is formed. For approaches which use dependency pairs, the concept of head percolating up has been used. Since, Hindi is a head final language, therefore the last noun is percolated up. This approach is better than adjacency greedy because it is difficult to find the sequence having more than two nouns in a corpus. However, it is relatively easier to find the frequency of the sequence having two nouns from the corpus.

Algorithm:

(a) Let there be 'n' noun components: W_i, i ranging from 1 to n.
(b) Compute the lexical association for each dependency pair of constituents.
(c) Select the pair with the highest association. Let the pair be W_{k-1} and W_k.
(d) Join W_{k-1} and W_k nodes.
(e) Remove W_{k-1} from the constituent list. (head percolation)
(f) Repeat steps "a" to "e" till all the nodes are joined.

3.4 Dependency Global Approach

It is the approach in which cohesion value[5] for all possible bracketing is measured and one with the highest is chosen. In this approach, lexical association for a node depends on the head. It does not depend on all the constituents. Formula for calculating association value for each node is shown below:

$$Association(i,j) = \frac{freq(head(i)head(j))}{freq(head(i)) + freq(head(j)) - freq(head(i)head(j))} \tag{7}$$

Where, "i" and "j" are the nodes being joined. head(i) is the function which returns the head of a node. Since, Hindi is a head final language, therefore the last noun is returned.

Consider (a((bc)d)), cohesion is measured by adding jaccard index value for b-c, c-d and a-d. In adjacency model, jaccard index for b-c, bc-d and a-bcd were measured. Here, instead of bc-d, we take c-d, since c is the head of bc which is already been bracketed. Association for each node of (a((bc)d)) is shown below. For more clarification, tree representation is already shown in Fig. 1.

$$Association(b,c) = \frac{freq(bc)}{freq(b) + freq(c) - freq(bc)} \tag{8}$$

$$Association(bc,d) = \frac{freq(cd)}{freq(c) + freq(d) - freq(cd)} \tag{9}$$

$$Association(a,bcd) = \frac{freq(ad)}{freq(a) + freq(d) - freq(ad)} \tag{10}$$

$$Cohesion([a[[bc]d]]) = Association(b,c) + Association(bc,d) + Association(a,bcd) \tag{11}$$

4 Experiments and Results

For experiment, we have extracted complex noun sequences from Hindi Dependency TreeBank that has been released in (Coling, 2012)[2] for the shared task on machine translation and parsing. We have obtained total 2322 noun sequences. Out of them, the number of noun compounds is 258. There are 603 sequences of genitives of the structure [noun1 kA noun2 (kA noun3) +]). Rest of the data contains both genitive as well compound sequences. Table 2 shows the distribution of complex noun sequences in the data used for conducting the experiments. The third column in the table shows the number of cases that cannot be handled by syntactic agreement rules of genitives and therefore required to be parsed by our proposed approach.

Table 2. Data used for the experiments and distribution of various types of complex nouns

Type of Complex NounSequence	Total number of Sequences	Total number of sequences where agreement rules not help
Compound nouns [a-b-c+]	258	258
No compound noun in genitives [a gen b (gen c)+]	603	228
Noun sequence [(a+ gen)* b]	2322	654

The following table presents the result of applying the syntactic agreement rules and the four frequency based experiments conducted in the present work on all kinds of complex nouns.

Table 3. Accuracy for (a) compound noun, (b) no compound noun in genitives and (c) all complex noun sequences

Method	Accuracy for compound nouns [a b (c)+]	Accuracy for the case where no compound noun is in genitives [a gen b (gen c)+]	Accuracy for complex noun sequences [(a+ gen)* b+]
Adjacency greedy	65.89	88.05	89.62
Adjacency global	61.62	89.05	89.92
Dependency greedy	67.82	88.22	89.70
Dependency global	**68.60**	**93.03**	**92.85**

We find from the above table that dependency global approach outperforms the other three approaches for all kinds of complex noun sequences. However the result of the analysis of only compound nouns is the poorest as is evident from the above table.

In Hindi, noun sequence with three noun constituents occur more often than others. Therefore, detail results for sequences containing three noun constituents is shown in Table 4. Accuracy obtained for compound nouns using dependency global approach is 71.30% while for genitives, it is 93.93%. Here also, we can see that dependency global is performing better than adjacency global and adjacency

Table 4. Accuracy for sequence [a b c] and [a gen b gen c] containing three noun constituents

Method	Accuracy for compound nouns [a b c]	Accuracy for the case where no compound noun is in genitives [a gen b gen c]
Adjacency greedy	67.08	89.15
Adjacency global	63.29	89.52
Dependency greedy	69.19	89.15
Dependency global	**71.30**	**93.93**

Table 5. Precision, recall and f-score for three constituent compound nouns [a b c]

Method	((a b) c)			(a (b c))		
	Precision	Recall	f-score	Precision	Recall	f-score
Adjacency greedy	76.12	74.21	75.15	50	52.56	51.25
Adjacency global	72.78	72.32	72.55	44.30	44.87	44.58
Dependency greedy	80.71	71.06	75.58	52.57	64.38	52.28
Dependency global	71.98	93.71	**81.42**	51.21	53.84	**52.5**

Table 6. Precision, recall and f-score for complex sequence containing three nouns separated by two genitive-markers[a gen b gen c]

Method	((a gen b) gen c)			(a gen (b gen c))		
	Precision	Recall	f-score	Precision	Recall	f-score
Adjacency greedy	96.63	91.43	93.96	37.68	61.90	46.84
Adjacency global	96.64	91.83	94.17	38.80	61.90	47.70
Dependency greedy	96.63	91.43	93.96	37.68	61.90	46.84
Dependency global	96.99	96.41	**96.70**	60	64.28	**62.06**

greedy. However, in case of genitives, much difference is not seen in adjacency greedy method and adjacency global method. Their accuracies are 89.15% and 89.52% respectively.

Table 5 shows f-score for left bracketed and right bracketed bare nouns. Table 6 shows f-score for left bracketed and right bracketed sequence containing three noun constituents and two genitive markers. Left bracketed sequence is having better f-score than right bracketed sequence in all the four methods. F-score of 81.42 and 52.5 is obtained for left bracketed and right bracketed nominal

Table 7. Accuracy for compound nouns and genitives containing four noun constituents

Method	Accuracy for compound nouns	Accuracy for genitives
Adjacency greedy	57.89	76.78
Adjacency global	47.36	83.92
Dependency greedy	57.89	78.57
Dependency global	42.10	83.92

compounds respectively. For the case of genitives, f-score is better than that of compound nouns. F-score measured for left bracketed and right bracketed genitives is 96.70 and 62.06 respectively. Accuracy for nominal compounds and genitives with four noun constituents is shown in Table 7.

5 Analysis of Results

We see in Table 3 that accuracy for constituency parsing of recursive compound nouns is less than that of the genitives. This is because syntactic agreement rules resolves the parse for genitives for 71.83% cases. We also observe that approaches based on dependency pairs perform better than approaches based on adjacency pairs. Adjacency pairs even include sequences with more than two nouns. Such sequences are difficult to be found in corpus[6]. For the same reason, not much difference is seen in adjacency greedy and adjacency global methods. But for the models based on dependency pairs, frequency for two noun constituents is needed, which is found relatively easier in the corpus. This helps in improving accuracy to an extent.

Further, we see that global based approaches are better than greedy based approaches. This is because in global based approaches, whole context is considered in deciding the output. While in greedy based approaches, only nearby sequence is considered. Whole context is not seen in this case. Therefore, during computation, some information gets missed.

We see that right bracketed sequence has better precision than left bracketed sequence. This means that there are less number of cases where ((ab)c) is the expected output but the result obtained is (a(bc)). Reason for this is lexical association for b-c is found to be better than association of a-b in adjacency based approach. And for dependency based approach, the association of a-c or b-c is better than association of a-b. This type of error was found in the sequence,

rakshA   shodha   sevA
"defense" "research" "service"

For the above sequence, expected output is ((rakshA shodha) sevA), but the output obtained is (rakshA (shodha sevA)). For this sequence, lexical association

---

[6] corpus was collected by crawling web pages.

of "shodha sevA" and "rakshA sevA" is found to be better than the association value of "rakshA shodha". Therefore, in both adjacency and dependency based approach result is (rakshA (shodha sevA)).

Less precision is observed for the right bracketed sequence because there are large number of cases where (a(bc)) is the expected output and the result obtained is ((ab)c). One of the reasons for this is non-occurrence of a-b, a-c and b-c in the corpus from which frequency is obtained and therefore, left bracket is taken as the default output. Another reason is lexical association for a-b is found to be better than the association of b-c in adjacency based approach. And for dependency based approach, the association of a-b is better than a-c and b-c. This type of error was observed in the sequence,

rAjya   pashupAlana   vibhaAga
"state"   "herding"   "department"

For the above sequence, expected output is (rAjya (pashupAlana vibhAga)). But the result obtained is ((rAjya pashupAlana) vibhAga). Here, association value of "rAjya pashupAlana" is better than association value of "pashupAlana vibhAga" and "rAjya vibhAga". Therefore, left bracket is obtained as the result. There are more such similar cases and hence resulting in less precision of right bracketed sequence.

6   Conclusion

This paper presents four approaches for constituency parsing of complex nouns in Hindi. Dependency models are found more efficient than adjacency models in handling sparsity of noun sequences and therefore performs better parsing. As part of future work, we attempt to perform the global dependency experiment on a bigger corpus. We also aim to use similarity measures to handle data sparsity issues.

References

1. Chapter 9: Regular expressions, the Open Group Base Specifications Issue 6, IEEE Std 1003.1, 2004 Edition. The Open Group (2004)
2. Hindi dependency treebank, workshop on MT and Parsing in Indian Languages. 24th International Conference on Computational Linguistics (2012)
3. Girju, R., Maldovan, D., Tatu, M., Antohe, D.: On the semantics of noun compounds. Computer Speech and Language (2005)
4. Jaccard, P.: The distribution of the flora in the alpine zone. The New Phytologist (1912)
5. Kavuluru, R., Harris, D.: A knowledge-based approach to syntactic disambiguation of biomedical noun compounds. In: Proceedings of COLING (2012)
6. Kulkarni, A., Kumar, A.: Statistical constituency parser for sanskrit compounds. In: Proceedings of ICON (2011)
7. Kulkarni, A., Paul, S., Kulkarni, M., Kumar, A., Surtani, N.: Semantic processing of compounds in indian languages. In: Proceedings of COLING (2012)

8. Lapata, M., Keller, F.: Web-based models for natural language processing. ACM (2005)
9. Lauer, M.: Corpus statistics meet the noun compound: Some empirical results (1995)
10. Pecina, P.: Lexical association measures. Institute of Formal and Applied Linguistics (2009)
11. Pustejovsky, J., Bergler, S., Anick, P.: Lexical semantic techniques for corpus analysis. Association for Computational Linguistics (1993)
12. Sharma, S.: Disambiguating the parsing of hindi recursive genitive constructions, IIIT Hyderabad, India (2012)

Amharic Sentence Parsing Using Base Phrase Chunking

Abeba Ibrahim and Yaregal Assabie

Department of Computer Science, Addis Ababa University
abeba.ibrahim@gmail.com, yaregal.assabie@aau.edu.et

Abstract. Parsing plays a significant role in many natural language processing (NLP) applications as their efficiency relies on having an effective parser. This paper presents Amharic sentence parser developed using base phrase chunker that groups syntactically correlated words at different levels. We use HMM to chunk base phrases where incorrectly chunked phrases are pruned with rules. The task of parsing is then performed by taking chunk results as inputs. Bottom-up approach with transformation algorithm is used to transform the chunker to the parser. Corpus from Amharic news outlets and books was collected for training and testing. The training and testing datasets were prepared using the 10-fold cross validation technique. Test results on the test data showed an average parsing accuracy of 93.75%.

Keywords: Amharic Parsing, Base Phrase Chunking, Bottom-up Parsing.

1   Introduction

To process and understand natural languages, the linguistic structures of texts are required to be organized at different levels. A structured text increases the capability of NLP applications [2], [4]. The syntactic level of linguistic analysis concerns how words are put together to form correct sentences and determines what structural role each word plays in the sentence. Broadly speaking, the syntactic level deals with analyzing a sentence that generally consists of segmenting a sentence into words, grouping these words into a certain syntactic structural units, and recognizing syntactic elements and their relationships within a structure. Syntactic level also indicates how the words are grouped together into phrases, what words modify other words, and what words are of central importance in the sentence [2], [7]. Parsing can be described as a procedure that searches through various ways of combining grammatical rules to find a combination that generates a tree representing the syntactic structure of the input sentence. Parsing uses the syntax of languages to determine the functions of words in a sentence in order to generate a data structure that can help to analyze the meaning of the sentence [7]. In addition to this, parsing deals with a number of sub-problems such as identifying constituents that can fit together. In general, parsing assists to understand how words are put together to form the correct phrases or sentence along with the structural roles of the words, and it plays a significant role in many NLP applications as it helps to reduce the overall structural complexity of sentences [13]. Some of the NLP applications where parser is used as a component are

A. Gelbukh (Ed.): CICLing 2014, Part I, LNCS 8403, pp. 297–306, 2014.
© Springer-Verlag Berlin Heidelberg 2014

semantic analysis, grammar checking, automatic abstracting, text summarization, machine translation, etc.

Over the years, many algorithms have been proposed to deal with parsing and they can be broadly classified in to two as top-down and bottom-up parsing. Top-down parsing starts with the sentence and then applies the grammar rules forward until the symbols at the terminals of the tree correspond to the components of the sentence being parsed. In the top-down approach, a parser tries to derive the given string from the start symbol by rewriting non-terminals one-by-one using production (grammatical) rules. The non-terminal on the left hand side of a production rule is replaced by its right hand side in the string being parsed. On the other hand, bottom-up parsing starts with words in a sentence and applies grammar rules backward to reduce the sequence of symbols until it consists solely of the start symbol. It begins with the sentence to be parsed and applies the grammar rules backward until a single tree whose terminals are the words of the sentence and whose top node is the start symbol has been produced. In the bottom-up approach, a parser tries to reduce the given string to the start symbol step by step using production rules. The right hand side of a production found in the string being parsed is replaced by its left hand side. Among the widely known top-down and bottom-up parsers are Early parser and CYK parser, respectively [5], [8]. Since automatic parsing is a complex task, various techniques have been employed to improve its efficiency. One of such strategies is chunking whose task is dividing a text into syntactically correlated parts of words. These words are non-overlapping which means that a word can only be a member of one chunk and non-exhaustive, i.e., not all words are in chunks [17]. Abney [1] introduced the concept of chunk as an intermediate step providing input to further full parsing stages. In addition to being a component in parsing, chunkers are also used for the development of different NLP applications such as information retrieval, information extraction, named entity recognition, etc. Although the inherent characteristics of grammatical structures vary from one language to another, there are some models and algorithms that are commonly used to develop chunkers for various languages. These include conditional random fields, hidden Markov models, transformation-based learning, maximum entropy principle, etc. These models and algorithms have been used to develop chunkers for various languages around the world such as English, Chinese, Turkish, Vietnamese, etc. [6], [10], [12], [13], [16], [17], [18].

Although Amharic is the working language of Ethiopia with a population of about 90 million at present, it is still one of less-resourced languages with few linguistic tools available for Amharic text processing. Since chunkers are identified as key components in many NLP applications, we have developed Amharic base phrase chunker using a hybrid of HMM and rule-based methods [9]. Thus, in this work, we used the chunker to develop an Amharic parser. The remaining part of this paper is organized as follows. Section 2 presents Amharic language with emphasis to its phrase structure. Amharic base phrase chunking along with error pruning is discussed in Section 3. In Section 4, we present sentence parsing by making use of chunk results as inputs. Experimental results are presented in Section 5, and conclusion and future works are highlighted in Section 6. References are provided at the end.

2 Structures of Amharic Language

2.1 Amharic Language

There are over 80 languages spoken in Ethiopia which has a population of over 90 million at present. Amharic is the working language of the federal government of the country. Amharic is spoken as a mother tongue by a large segment of the population and it is the most commonly learned second language throughout the country. As a result, Amharic is the *lingua franca* of the country in the modern era [11]. The language is believed to be evolved from Geez which has been used over many years as the liturgical language of Ethiopia. Along with dozens of other Ethiopian languages, Amharic is written using Ethiopic script which has a total of a total of 435 characters, with several languages having their own special sets of characters representing the unique sounds of the respective languages. Out of the whole set of Ethiopic characters, Amharic uses 33 consonants (base characters) from which six other orders of characters representing combinations of vowels and consonants are derived for each base character. The Amharic alphabet is conveniently written in a tabular format of seven columns where the first column represents the base characters and others represent their derived vocal sounds. The vowels of the alphabet are not encoded explicitly but appear as modifiers of the base characters. In addition, there are about two scores of labialized characters such as $\Lambda(l^w a)$, $\eta(m^w a)$ $\mathcal{Z}(r^w a)$ $\mathbf{\Delta}(s^w a)$, etc. used by the language for writing. Part of a Amharic alphabet is shown in Table 1.

Table 1. Part of the Amharic alphabet

Base sound	Base character ä (ሽ)	Orders of the base character					
		u (ሁ)	i (ሂ)	a (ህ)	e (ሄ)	ĭ (ህ)	o (ሆ)
1	h	hä (ሀ)	hu (ሁ)	hi (ሂ)	ha (ሃ)	he (ሄ)	hĭ (ህ)	ho (ሆ)
2	l	lä (ለ)	lu (ሉ)	li (ሊ)	la (ላ)	le (ሌ)	lĭ (ል)	lo (ሎ)
3	h	hä (ሐ)	hu (ሑ)	hi (ሒ)	ha (ሓ)	he (ሔ)	hĭ (ሕ)	ho (ሖ)
4	m	mä (መ)	mu (ሙ)	mi (ሚ)	ma (ማ)	me (ሜ)	mĭ (ም)	mo (ሞ)
.	.	.	.	.	.	.	.	.
.	.	.	.	.	.	.	.	.
.	.	.	.	.	.	.	.	.
33	p	pä (ፐ)	pu (ፑ)	pi (ፒ)	pa (ፓ)	pe (ፔ)	pĭ (ፕ)	po (ፖ)

2.2 Grammatical Rules of Amharic

Yimam [19] and Amare [3] classified phrase structures of the Amharic language as: noun phrases (NP), verb phrases (VP), adjectival phrases (AdjP), adverbial phrases (AdvP) and prepositional phrases (PP). These phrases have principal word classes as heads. For example, an Amharic noun phrase has a noun as its head; an Amharic verb phrase has a verb as its head; etc. Amharic phrases, except prepositional phrases, can be made from a single head word or with a combination of other words. Unlike other phrase constructions, prepositions cannot be taken as a phrase. Instead they should be combined with other constituents and the constituents may come either previous to or subsequent to the preposition. If the complements are nouns or NPs, the position of prepositions is in front of the complements whereas if the complements are PPs, the position will shift to the end of the phrase. Examples are: እንደ ሰው (*ĭndä säw*/like a human), ከቤቱ ኣጠገብ (*käbetu aṭägäb*/close to the house), etc. In Amharic phrase construction, the head of the phrase is always found at the end of the phrase except for prepositional phrases.

Amharic language follows subject-object-verb grammatical pattern unlike, for example, English language which has subject-verb-object sequence of words [3], [19]. For instance, the Amharic equivalent of sentence "John killed the lion" is written as "ጆን (*jon*/John) አንበሳውን (*anbäsawn*/the lion) ገደለው (*gädäläw*/killed)". Amharic sentences can be constructed from simple or complex NP and simple or complex VP. Simple sentences are constructed from simple NP followed by simple VP which contains only a single verb. Complex sentences are sentences that contain at least one complex NP or complex VP or both complex NP and complex VP. Complex NPs are phrases that contain at least one embedded sentence in the phrase construction. The embedded sentence can be complements.

3   Base Phrase Chunking

This section discusses about the Amharic base phrase chunker we used as a component to develop the parser. The Amharic chunker system is exposed further in detail in [9]. The output of the system, i.e. the tag of chunks can be noun phrases, verb phrases, adjectival phrases, etc. in line with the natural language construction rule. In order to identify the boundaries of each chunk in sentences, the following boundary types are used [15]: IOB1, IOB2, IOE1, IOE2, IO, "[", and "]". The first four formats are complete chunk representations which can identify the beginning and ending of phrases while the last three are partial chunk representations. All boundary types use "I" tag for words that are inside a phrase and an "O" tag for words that are outside a phrase. They differ in their treatment of chunk-initial and chunk-final words.

- IOB1:   the first word inside a phrase immediately following another phrase receives **B** tag.
- IOB2:   all phrases- initial words receive **B** tag.
- IOE1:   the final word inside a phrase immediately preceding another same phrase type receives **E** tag.
- IOE2:   all phrases- final words receive **E** tag.
- IO: words inside a phrase receive **I** tag, others receive **O** tag.
- "[":   all phrase-initial words receive "[" tag other words receive "." tag.
- "]":   all phrase-final words receive "]" tag and other words receive "." tag.

We considered six different kinds of chunks, namely noun phrase (NP), verb phrase (VP), Adjective phrase (AdjP), Adverb phrase (AdvP), prepositional phrase (PP) and sentence (S). To identify the chunks, it is necessary to find the positions where a chunk can end and a new chunk can begin. The part-of-speech (POS) tag assigned to every token is used to discover these positions. We used the IOB2 tag set to identify the boundaries of each chunk in sentences extracted from chunk tagged text. Using the IOB2 tag set along with the chunk types considered, a total of 13 phrase tags were used in this work. These are: B-NP, I-NP, B-VP, I-VP, B-PP, I-PP, B-ADJP, I-ADJP, B-ADVP, I-ADVP, B-S, I-S and O. For example, the IOB2 chunk representation for the sentence ካሳ የመጣው ትንሽ ልጅ እንደ አባቱ በጣም ታመመ (*kasa yamäṭaw tǐnǐš lǐj ǐndä abatu bäṭam tamämä*/The little boy that Kassa has brought became very sick like his father) is shown in Table 2. Accordingly, the chunk tagged sentence would be "ካሳ N B-S የመጣው VREL I-S ትንሽ ADJ B-NP ልጅ N I-NP እንደ P B-PP አባቱ N I-PP በጣም ADJ B-VP ታመመ V I-VP".

Table 2. IOB2 chunk representation for "ካሳ የመጣው ትንሽ ልጅ እንደ አባቱ በጣም ታመመ"

Words	IOB2 chunk representation
ካሳ (*kasa*/Kassa)	B-S
የመጣው (*yamäṭaw*/that [Kassa] has brought)	I-S
ትንሽ (*tǐnǐš*/little)	B-NP
ልጅ (*lǐj*/boy)	I-NP
እንደ (*ǐndä*/like)	B-PP
አባቱ (*abatu*/his father)	I-PP
በጣም (*bäṭam*/very)	B-VP
ታመመ (*tamämä*/became sick)	I-VP

To implement the chunker component, we use hidden Markov model (HHM) enhanced by a set of rule used to prune errors. In the training phase of HMM, the system first accepts words with POS tags and chunk tags. Then, the HMM is trained with a training set built from sentences where words are tagged with part-of-speeches and chunks. Likewise in the test phase, the system accepts words with POS tags and outputs appropriate chunk tag sequences against each POS tag using HMM model. We use POS tagged sentence as input from which we observe sequences of POS tags. However, we also hypothesize that the corresponding sequences of chunk tags form hidden Markovian properties. Thus, we used a hidden Markov model (HMM) with POS tags serving as states. The HMM model is trained with sequences of POS tags and chunk tags extracted from the training corpus. The HMM model is then used to predict the sequence of chunk tags for a given sequence of POS tag by making use of the Viterbi algorithm. The output of the decoder is the sequence of chunks tags which group words based on syntactical correlations. The output chunk sequence is then analyzed to improve the result by applying linguistic rules derived from the grammar

```
1. If POS(w)=ADJ and POS(w+1)=NPREP, NUMCR ,then chunk
   tag for w is O
2. If POS(w)=ADJ and POS(w-1)!= ADJ and POS(w+1)= AUX,V,
   then chunk tag for w is B-VP
3. If POS(w)=NPREP and POS(w+1)=N ,then chunk tag for w
   is B-NP
4. If POS(w)=NUMCR and POS(w+1)=NPREP, then chunk tag for
   w is O
5. If POS(w)=N and POS(w+1)=VPREP and POS(w-1)=N, ADJ,
   PRON,NPREP, then chunk tag for w is B-VP
6. If POS(w)=ADJ and POS(w+1)=ADJ, then chunk tag for w
   is B-ADJP
```

Algorithm 1. Sample rules used to prune chunk errors

of Amharic. For a given Amharic word w, linguistic rules (from which sample rules are shown in Algorithm 1) were used to correct wrongly chunked words ("w-1" and "w+1" are used to mean the previous and next word, respectively).

4   Sentence Parser

In this work, bottom-up approach is employed for sentence parsing by using the output of the chunker as an input and recursively remove the head words to make new phrases until individual words are reached. The parse tree is then constructed while head words are recursively removed and new phrases are formed. When we obtain no new phrases during the recursive process, it means that we complete the process of parsing. The algorithm that is used for parsing is given in Algorithm 2.

```
1. Take the tagged document
2. Use chunker to identify base phrases
3. If the base phrase is VP, NP, AdjP, or AdvP
        Replace all identified phrases with their head
   Else {if the base phrase is PP or S}
        The current phrase takes the word next to it
        and makes new phrase by taking the new word as
        a head
4. Find base phrases in the new data stream
5. If Step 4 discovered new phrases
        Repeat Steps 3-5
   Else
        Stop
```

Algorithm 2. Algorithm for parsing a sentence

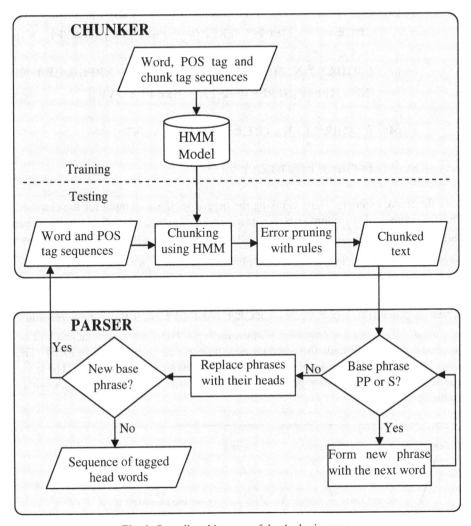

Fig. 1. Overall architecture of the Amharic parser

The Amharic base phrase chunker was integrated in the parser. The overall architecture of the parser including the chunker is shown in Figure 1.

The following example shows how parsing is performed using the proposed algorithm for a given POS tagged sentence: "ወንበዴዎች N በገፈቃደኛች NPREP የገነቡትን VREL ድርጅት N ከጥቅም NPREP ውጭ PREP ኣደረጉት V".

Step1:   ወንበዴዎች N በገፈቃደኛች NPREP የገነቡትን VREL ድርጅት N
ከጥቅም NPREP ውጭ PREP ኣደረጉት V

Step2: [('መንበደዎች', 'N'), ('በገፈቃደኛች NPREP የገኑቡትን VREL', 'S'),
('ድርጅት', 'N'), ('ከጥቅም NPREP ውጭ PREP', 'PP'), ('አደረጉት', 'V')]

Step3: ['መንበደዎች N', ('በገፈቃደኛች NPREP የገኑቡትን VREL ድርጅት N',
'NP'), ('ከጥቅም NPREP ውጭ PREP አደረጉት V', 'VP')]

Step4: [('መንበደዎች', 'N'), ('ድርጅት N አደረጉት V', 'VP')]

Step5: ['መንበደዎች N', 'አደረጉት V']

In the above example, *Step1* is taking the tagged sentence as input for the chunker.
The first output of the chunker is generated in *Step2* which identifies possible base
phrases. In this example, ('በገፈቃደኛች NPREP የገኑቡትን VREL', 'S') and ('ከጥቅም
NPREP ውጭ PREP', 'PP') are the base phrases identified in *Step2* of the algorithm.
In *Step3*, the prepositional phrase and the subordinate clause or sentence are combined with the next word and converted to verb phrase and noun phrase, respectively.
A new sentence [('መንበደዎች', 'N'), ('ድርጅት N አደረጉት V', 'VP')] is now generated
as a result of looking for new base phrases in *Step4*. Here, the new sentence will be
processed recursively until there are no new base phrases discovered in *Step4*. The
parse tree is built in the process by taking base phrases as nodes of the tree. The parse
tree representing the parsing process for the aforementioned example is shown in
Figure 2.

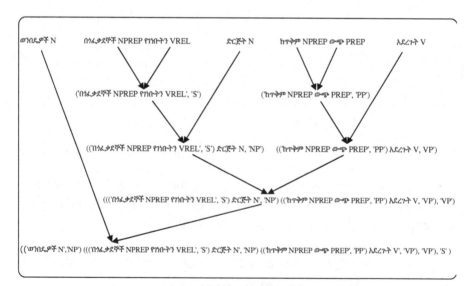

Fig. 2. Parse tree representing the parsing process of an Amharic sentence

5   Experiment

5.1   The Corpus

The major source of the dataset we used for training and testing the system was Walta Information Center (WIC) news corpus which is at present widely used for research on Amharic natural language processing. The corpus contains 8067 sentences where words are annotated with POS tags. Furthermore, we also collected additional text from an Amharic grammar book authored by Yimam [19]. The sentences in the corpus are classified as training data set and testing data set using 10 fold cross validation technique.

5.2   Test Result

In 10-fold cross-validation, the original sample is randomly partitioned into 10 equal size subsamples. Of the 10 subsamples, a single subsample is used as the validation data for testing the model, and the remaining 9 subsamples are used as training data. The cross-validation process is then repeated 10 times, with each of the 10 subsamples used exactly once as the validation data. Accordingly, we obtain 10 results from the folds which can be averaged to produce a single estimation of the model's predictive potential. By taking the average of all the ten results the overall chunking accuracy of the system was 85.31% for the HMM chunking model. However, the result was improved to an accuracy of 93.75% when the HMM was pruned with rules. Test results also show that the parser correctly parses all sentences that are chunked correctly. However, the parser fails to correctly parse sentences that are chunked wrongly. So, the overall accuracy of the parser is the same as that of the hybrid chunker, i.e. 93.75%.

6   Conclusion

Amharic is one of the most morphologically complex and less-resourced languages. This complexity poses difficulty in the development of natural language processing applications for the language. Despite the efforts being undertaken to develop various Amharic NLP applications, only few usable tools are publicly available at present. One of the main reasons frequently cited by researchers is morphological complexity of the language. Amharic text parsing also suffers from this problem. In this work, we tried to overcome this problem by employing chunker. Test results have shown that all sentences that are correctly chunked are parsed correctly which is a promising result. The performance of the parser we developed can be enhanced by improving the effectiveness of the chunking module. It appears that chunking is more manageable problem than parsing because the chunker does not require deeper analysis of texts which will be less affected by the morphological complexity of the language. Thus, future work is recommended to be directed at improving the chunking component of the parser.

References

1. Abney, S.: Parsing by chunks. In: Berwick, R., Abney, S., Tenny, C. (eds.) Principle-Based Parsing. Kluwer Academic Publishers (1991)
2. Abney, S.: Chunks and Dependencies: Bringing Processing Evidence to Bear on Syntax. In: Computational Linguistics and the Foundations of Linguistic Theory. CSLI (1995)
3. Amare, G.: ዘመናዊ የአማርኛ ሰዋስው በቀላል ከቀረረብ (Modern Amharic Grammar in a Simple Approach), Addis Ababa, Ethiopia (2010)
4. Bird, S., Klein, E., Loper, E.: Natural Language Processing with Python. O'Relly Media Inc., Sebastopol (2009)
5. Earley, J.: An efficient context-free parsing algorithm. Communications of the ACM 13(2), 94–102 (1970)
6. Grover, C., Tobin, R.: Rule-based chunking and reusability. In: Proceedings of the Fifth International Conference on Language Resources and Evaluation, LREC 2006 (2006)
7. Jurafsky, D., Martin, H.: Speech and Language Processing: An Introduction to Natural Language Processing, Speech Recognition, and Computational Linguistics, 2nd edn. Prentice-Hall (2009)
8. Hopcroft, J.E., Motwani, R., Ullman, J.D.: Introduction to Automata Theory, Languages, And Computation, ch. 7, pp. 228–302. Addison-Wesley (2001)
9. Ibrahim, A., Assabie, Y.: Hierarchical Amharic Base Phrase Chunking Using HMM With Error Pruning. In: Proceedings of the 6th Conference on Language and Technology, Poznan, Poland, pp. 328–332 (2013)
10. Kutlu, M.: Noun phrase chunker for Turkish using dependency parser. Doctoral dissertation. Bilkent University (2010)
11. Lewis, P., Simons, F., Fennig, D.: Ethnologue: Languages of the World, 17th edn. SIL International, Dallas (2013)
12. Li, S.J.: Chunk parsing with maximum entropy principle. Chinese Journal of Computers: Chinese Edition 26(12), 1722–1727 (2003)
13. Manning, C., Schuetze, H.: Foundations of Statistical Natural Language Processing. MIT Press, Cambridge (1999)
14. Molina, A., Pla, F.: Shallow parsing using specialized HMMs. The Journal of Machine Learning Research 2, 595–613 (2002)
15. Ramshaw, A., Marcus, P.: Text chunking using transformation-based learning. In: Proceedings of the Third ACL Workshop on Very Large Corpora, pp. 82–94 (1995)
16. Thao, H., Thai, P., Minh, N., Thuy, Q.: Vietnamese noun phrase chunking based on conditional random fields. In: International Conference on Knowledge and Systems Engineering (KSE 2009), pp. 172–178. IEEE (2009)
17. Tjong, E.F., Sang, K., Buchholz, S.: Introduction to the CoNLL-2000 shared task: Chunking. In: Proceedings of the 2nd Workshop on Learning Language in Logic and the 4th Conference on Computational Natural Language Learning, vol. 7, pp. 127–132 (2000)
18. Xu, F., Zong, C., Zhao, J.: A Hybrid Approach to Chinese Base Noun Phrase Chunking. In: Proceedings of the Fifth SIGHAN Workshop on Chinese Language Processing, Sydney (2006)
19. Yimam, B.: የአማርኛ ሰዋስው (Amharic Grammar), Addis Ababa, Ethiopia (2000)

A Machine Learning Approach to Pronominal Anaphora Resolution in Dialogue Based Intelligent Tutoring Systems

Nobal B. Niraula and Vasile Rus

Department of Computer Science, The University of Memphis
Memphis, TN, 38152, USA
{nbnraula,vrus}@memphis.edu

Abstract. Anaphora resolution is a central topic in dialogue and discourse that deals with finding the referent of a pronoun. It plays a critical role in conversational Intelligent Tutoring Systems (ITSs) as it can increase the accuracy of assessing students' knowledge level, i.e. mental model, based on their natural language inputs. Although anaphora resolution is one of the most studied problems in Natural Language Processing, there are very few studies that focus on anaphora resolution in dialogue based ITSs. To this end, we present Deep Anaphora Resolution Engine++ (DARE++) that adapts and extends existing machine learning solutions to resolve pronouns in ITS dialogues. Experiments showed that DARE++ achieves a F-measure of 88.93%, proving the potential of the proposed method for resolving pronouns in student-tutor dialogues.

Keywords: Anaphora Resolution, Tutoring System, Machine Learning.

1   Introduction

The task of anaphora resolution is to identify the referent of a pronoun in dialogue and discourse. It is one of the important tasks in many NLP applications such as information extraction, automated essay grading, and summarization. In this paper, we focus on the task of anaphora resolution in a relatively new NLP application, dialogue-based Intelligent Tutoring Systems (ITSs).

ITSs form a category of advanced educational technologies that tailor instruction to each individual student in order to maximize learning for every single student. Indeed, ITSs have already proven to be very effective at inducing learning gains in students [12]. In dialogue based or conversational ITSs, students have a conversation with the system, which helps them solve problems (e.g. Physics problems) through hint-like questions and other types of feedback.

Students' natural language responses to tutors' questions are a major source of information about what a student knows. Incorrect assessment of student responses could lead to incorrect feedback provided by the system which, in turn, could frustrate students sometimes to the point of quitting using the system, an undesired effect. Because student responses often contain pronouns, the accuracy

A. Gelbukh (Ed.): CICLing 2014, Part I, LNCS 8403, pp. 307–318, 2014.

of the inferred student model is directly dependent on resolving anaphors in such student responses.

Consider the real student-tutor interaction below from a state-of-the-art ITS, DeepTutor[12]:

PROBLEM: *A mover pushes a desk with constant velocity V_0 across a carpeted floor. Suddenly, the mover stops pushing. What can you say about the motion of the desk after the mover stops pushing ? Explain why.*
STUDENT ANSWER: *The desk will stop moving because **it** was only moving due to the applied force of the mover pushing on **it**. **It** does not have a constant velocity or acceleration to keep **it** going.*

The student answer in the example above has four pronouns, all referring to *desk*. To fully understand the student response these pronouns must be resolved. A pronoun resolution algorithm such as the one proposed here could help resolve the four pronouns. The need for such an algorithm is further emphasized by the fact that students' use of pronouns while conversing with a computer tutor is quite frequent [7]. The authors reported 5,881 pronouns in 25,945 student turns. Moreover, our analysis shows that about 22% of the total students turns contain at least one pronoun.

Three types of anaphora usage in students' answers can be identified in student-tutor interactions. They include *Intra-turn, Inter-turn intermediate* and *Inter-turn history* anaphora - see Table 1. In the case of *Intra-turn* anaphora, the referents are found within the student's current dialogue turn. In *Inter-turn intermediate* anaphora, the referents lie in the most recent tutor turn [13] and in *Inter-turn history* anaphora, the referents are located in earlier dialogue turns or even the problem description.

Table 1. Use of pronouns in students' responses

(a) Intra-turn :
TUTOR:What does Newton's second law say?
STUDENT:for every force, there is another equal force to counteract *it*
(b) Inter-turn immediate:
TUTOR:What can you say about the acceleration of the piano based on Newton's second law and the fact that the force of gravity acts on the piano?
STUDENT: *It* remains constant.
(c) Inter-turn history:
TUTOR: Since the ball's velocity is upward and its acceleration is downward, what is happening to the ball's velocity?
STUDENT: increasing
TUTOR: Can you please elaborate?
STUDENT: *it* is increasing

While anaphora resolution is a well-studied problem in written texts [4,8,11] and dialogue [15,16,9], there are very limited works which address anaphora resolution in dialogue based ITSs which are more specific systems with different assumptions. Due to the peculiarities of tutorial dialogues, existing solutions for anaphora resolutions must be adapted to get optimal resolutions of anaphors in ITSs dialogues. To this end, we propose Deep Anaphora Resolution Engine++ (DARE++) for resolving pronouns in conversational ITSs. DARE++ is the improved version of DARE [7], a previously developed heuristics-based anaphora resolution engine for dialogue based ITSs. DARE++ is one of the first machine learning techniques proposed for resolving pronouns in ITSs. It is guided by thousands of student-tutor interactions obtained from a state-of-the-art tutoring system, DeepTutor[1].

The rest of the paper is organized as follows. We present related works in Section 2. Data sets and methodology are described in Sections 3 and 4 respectively. Experiments, results and future works are discussed in Section 5, followed by conclusion in Section 6.

2   Related Works

The more general problem of finding coreferents, i.e. words and expressions referring to the same entity or event, is called coreference resolution. Anaphora is the special case of finding referents of pronouns. The literature on anaphora / coreference resolution for written texts is rich [8,11,18,5,10,17]. Similarly, considerable work on resolving pronouns in dialogue can be found in the literature [2,9,15,16,14].

Methodologies for resolving pronouns in dialogue and discourse can be classified into knowledge-poor and classification approaches. Knowledge-poor approaches rely on hand-crafted rules or heuristics. A simple rule based approach proposed for ITSs and closest to our work is by [7]. The authors learned simple rules from few annotated instances and applied them on top of an existing state-of-the-art coreference tool. The limitation of their approach is that learned rules using a few hundred observations is not sufficient for handling all the cases. Moreover, peculiar characteristics of the dialogue based ITSs are underutilized.

Classification approaches, on the other hand, work by means of models acquired through annotated corpora using machine learning algorithms. One such example is by Soon et al. [14] who used a decision tree algorithm for coreference resolution. Michael and Muller [16] proposed a machine learning approach to resolve pronouns in spoken dialogue. They also used decision tree to classify valid antecedent-pronoun pairs using their feature sets. Stent and Bangalore [15] used logistic regression for mention-referent classification. Kernel based methods are also found in the literature to classify the pairs [19].

Anaphora resolution techniques proposed for English written texts need to be adapted when applied to texts in specific domains, genres (e.g. dialogue) and languages (other than English) as anaphora instances exhibit different characteristics than in professionally written texts such as newspaper articles. The technique proposed by [1] is such an example where authors adapted existing

---

[1] http://www.deeptutor.org

anaphora solutions in English to the Basque language. Similarly, Stent and Bangalore [15] adapted solutions to resolve pronouns in a spoken dialogue system by adding spoken dialogue related features to existing solutions. Anaphora resolution in biomedical texts is another example of such adaptation [3].

ITSs have some commonalities with spoken dialogue systems in that both use dialogues in the interactions. It should be noted that we used data from ITSs that interact with students through typed dialogue, i.e. a chatroom-like type of interaction as opposed to spoken dialogue interaction. Furthermore, the dialogues are in the context of science learning while spoken dialogue systems were studied mostly for common tasks such as airline ticket reservations. In both systems, antecedents corresponding to anaphors belong to current or previous utterances. However, there are differences too. First, in spoken dialogue systems, the majority of pronouns are personal and demonstrative pronouns [16]. However, in tutorial dialogues, the pronouns are mostly *it, they, he* and *she* [7]. Second, referents can be VP-antecedents or NP-antecedents in spoken dialogue systems but almost all antecedents in ITSs are NP-antecedents.

Given the above peculiarities of tutorial dialogues compared to written texts and spoken dialogues, existing approaches to pronoun resolution should be adapted in order to maximize accuracy. To this end, we have proposed DARE++ that resolves anaphors in ITS dialogues using machine learning approaches.

3   Data

We extracted and annotated 1,000 pronoun instances from student-tutor interaction logs collected in an experiment involving high-school students interacting with the intelligent tutoring system DeepTutor[12] in the domain of conceptual Physics. We describe the details of the data set creation at [6]. The data is freely available for public usage[2].

A typical collected instance is presented in Table 2. Each instance has a unique id (e.g. 3,624 in the example). The log files are records of the actual dialogue between the computer tutor and students. Student's current response is designated by A (student answer) and the corresponding utterance from the tutor, usually in the form of a guiding question from DeepTutor, is denoted by Q. Previous student responses are denoted with A1, A2, and so on, while previous DeepTutor turns are denoted with Q1, Q2, and so on. The goal is to resolve pronouns in A to their referent, which could be in the same student response A, the previous tutor turn Q, earlier in the dialogue history (and thus part of the common ground built by the two conversation partners), or even the current problem description.

Once the set of 1,000 instances was collected, we annotated the instances following a set of guidelines developed by linguistics experts [6] which also borrowed some ideas from the guidelines used for annotating the data set used in the Message Understanding Conference (MUC-6 [3]). For annotation, we formed

---

[2] http://language.memphis.edu/nobal/AR

[3] http://www.cs.nyu.edu/cs/faculty/grishman/muc6.html

Table 2. A typical instance for anaphora resolution

INSTANCE: 3624
PROBLEM: A stuntman must drop from a helicopter onto a target on the roof of a moving train. The plan is for the helicopter to hover over the train, matching the train's constant speed before the stuntman drops.
Q2: Where should the helicopter be positioned relative to the target? Please begin by briefly answering the above question. After briefly answering the above question, please go on to explain your answer in as much detail as you can.
A2: in front of the target due to wind resistance
Q1: Let me try again. Which principle can be applied when the motion of an object is complex, for instance, it can be thought of as motion in two perpendicular dimensions?
A1: decomposition
Q: What can you say about <p id="3624_2" min="motion">the motion of the stuntman</np> after he jumps?
A: <p id="3624_2" refid="3624_1">it</p> will be parabolic

five pairs of annotators and trained them to annotate the instances. Each annotator in a pair annotated the same 100 instances independently, resolved their differences on the first 100 instances before repeating the annotation for another 100 instances. Average kappa statistic for the annotation was 0.84.

Table 3. Distribution of anaphors

Pronouns	Count	Percentage %
hasRef (e.g. it, he, she)	1003	78.11
first person personal pronouns	170	13.23
pleonastic	32	2.49
communication breakdown (Soft)	32	2.49
communication breakdown (Hard)	27	2.10
others	20	2.49

Once the annotated corpus was ready, we analyzed the annotated instances to first understand pronoun usage in our tutorial dialogues (see Table 3). A student answer can contain more than one pronoun and each pronoun may or may not have a referent (due to pleonastic pronouns, elipsis, etc.). About 78% of the pronouns have referents, clearly demanding a method to resolve them. Students also used first person personal pronouns (e.g. I, we, and my) in their responses. About 13% of the pronouns are pleonastic. About 2.49% of pronouns need some form of inference to correctly identify their referents as the student answer does not precisely refer to an explicitly mentioned referent. We call such designate such case *communication breakdowns - soft*; [6]). About 2.1% of pronouns' are found to be irrelevant to the context such that it is very hard to find their referents even by human experts (*communication breakdown - hard'* [6]).

Table 4. Most common pronouns

Pronoun	Count	Percentage(%)
it	658	53.47
they	153	11.94
its	120	9.37
i	61	4.76
you	55	4.29
her	36	2.81
she	34	2.65
them	21	1.63
he	19	1.48
their	18	1.40
his	17	1.33

Table 4 shows the most used pronouns sorted by their frequency. The pronouns *it, they and its* are the three most frequent pronouns and account for more than 70% of pronoun usage. Since *it* can be pleonastic, identifying and resolving this pronoun is particularly challenging.

We further generated statistics about the locations of the referents corresponding to the students' pronouns and presented the top locations in Table 5. More than 50% of the pronouns refer entities in Q (the immediate tutor question), about 30% of the pronouns have their referents in A (i.e. in the student answer as the pronoun to be resolved), and about 11% of the referents are found in the problem descriptions (Ps). Very few pronouns refer to the entities in the previous tutor questions in the dialogue history (Q_i).

Table 5. Top five locations for antecedents

Location	Count	Percentage(%)
Q	577	53.22
A	342	31.54
P	125	11.53
Q_1	28	2.6
Q_2	5	0.46

4   Methodology

Machine learning based methods are among the most popular approaches to the problem of coreference resolution [15]. The standard coreference pipeline for such methods include identification of *mentions* which are co-referring expressions, extraction of *features* describing these mentions, determining *mention-pair* candidates which are pairs of mentions that corefer, and *clustering* mention-pairs in order to identify mentions that form a chain, i.e. refer to the same entity.

We adopted this coreference pipeline with some modifications. First of all, we do not generate all mention-pairs as our objective is not to generate the complete coreference chain rather just resolve the pronouns in the students answer to the corresponding entity, typically the most recent non-anaphoric reference of the entity. That is, we are interested in finding the referents (if any) of only pronouns that appear in a student answer but not necessarily finding chains of pronouns or other types of referents to the same entity. This is sufficient for our goal of best understanding the current student answer. This simplification significantly reduces the search space of mention-pairs. Additionally, we don't need to cluster the mentions as we need only one referent of a pronoun. Thus, our model generates a limited number of mention-pairs and classifies them as either P(ositive) which means the two mentions (typically a noun and a pronoun mention) corefer or N (egative), otherwise. We present next the major phases in our anaphora pipeline.

4.1   Generation of Mention-Pairs

Our mention-pair construction algorithm works as follows. We use a parser to parse the problem text and tutor-turns and extract noun and noun phrases (we do not consider previous student turns for mention candidates as pronouns in student answers almost never refer to something in a previous student answer/turn). Next, we parse student's answer (i.e. A) and identify pronouns to be resolved. These pronouns are then paired with nouns to get mention-pairs. We exemplify this process for the instance shown in Table 2. We parse the sentences in PROBLEM, Q2, Q1, Q, and A and get the following mention-pairs: (*stuntman,it*), (*helicopter,it*), (*target,it*), (*room,it*), (*train,it*), (*principle,it*), (*motion,it*), etc.

4.2   Feature Selection

In order to use machine learning techniques to automatically induce a classifier, we need to devise a set of features that are useful for classifying the mention-pairs as P or N. This is a crucial step as the accuracy of the induced classifiers relies significantly on these features. We used lexical, syntactic, semantic, and dialog related features which are listed in Table 6.

Lexical Features: Lexical features include lengths of A, Q, A_1, Q_1, A_2, and Q_2, pronoun (P)'s position in A (i.e. the token index), total number of pronouns in student's answer, percentage of tokens before and after a referent candidate (C). We also have boolean features to check whether the candidate referent C is in student's answer A, whether student's answer contains any WH-words and simple negative cue words. We used lists of WH-words and negative cue words, respectively, for this purpose. Type of question is determined by checking first token in Q in this list: (what:1, when:2, where:3, which:4, who:5, whom:6, whose:7, how:8, none of above:-1).

Syntactic Features: To capture the grammatical functions of antecedent candidates, we counted the number of dependency relations and the number of relations with the candidate being a head word (governor). We also computed

Table 6. List of features (P= pronoun, C = a referent candidate)

Type	Features
Lexical	lengths (of A, Q, A_1, Q_1, A_2, and Q_2)(1-6), P's token position in A (7) no. of Ps in A(8), % of tokens before & after C (10-11), is C in A ? (12) has WH-Word in A(13), has negation word in A? (14), question type (15)
Syntactic	dependency relation counts (governer & total) of P (19-20) and C (21-22), present/absent 135 dependency relations (24-158)
Semantic	gender agrees ? (16), number agrees ? (17), person of P (18), is C a proper noun ?(23)
Dialogue	location of C in dialogue stack (9)

these features for pronouns. Moreover, we used binary features for 135 dependency relations each indicating true when the referent candidate is either its governor or dependent.

Semantic Features: We created a dictionary to get the gender of pronouns and of the characters (e.g., John) used in the problem descriptions. Values of *gender agrees* feature can be 1 (matched), 0 (not matched) and 2 (not available). For the *number*(s/p/na), we use simple rules using POS tags. For example, if a noun's POS is NN or NNP, we considered that noun a singular whereas if the POS is NNS or NNPS we consider it as a plural(p). Similarly, a noun is deemed a proper noun if its first character is capitalized (which can also be detected through the NNP or NNPS tags).

Dialogue Features: We used one dialogue feature: *the location of candidate* referent which takes value from 0 to 9 (A:0, Q:1, A_1:2, Q_1:3, A_2:4, Q_2:5, A_3:6, Q_3:7, problem description:8, none of above:9).

4.3   Generation of Training Examples

The machine learning algorithms we experimented with require both positive and negative instances in order to learn the target function, which in our case is a classification function. We generated positive (P) and negative (N) examples of mention-pairs using the annotated data set. Note that an example (training or testing) is a vector containing values corresponding to the feature set. Positive examples are easy to generate as pronouns and their correct referents are marked in the annotated instances. For example, for the instance in Table 2, we generate the following positive mention-pair (*motion,it*).

To generate negative examples, we follow an approach similar to [14]. Following this approach, we generate negative examples by using *(entity, pronoun)* pairs where *entity* refers to any noun between the pronoun and its annotated referent. To achieve this, we start going backwards from the pronoun to be resolved and scan for nouns until we reach its correct referent. We form *(noun, pronoun)* pairs for every identified noun in this span of dialogue. All the pairs

Table 7. Performance Comparison

Method	Acc.	Pre.	Rec.	Fm.	Kappa	RME
Baseline [7]	39.10	38.03	53.96	44.26	-	-
Naive Bayes	82.33	66.9	78.11	72.11	0.59	0.35
SVM	87.78	84.06	71.83	77.47	0.69	0.34
Logistic Regression	88.06	81.24	76.85	79.00	0.70	0.29
Decision Trees (J48)	**93.54**	**89.07**	**88.79**	**88.93**	**0.84**	**0.24**
Multilayer Perceptron	88.82	86.96	72.67	79.17	0.71	0.31

except *(correct-referent,pronoun)* are used to generate negative examples. As an example, we generate the following negative instance of a mention-pair from the annotated instance in Table 2: *(stuntman,it)*. This mention pair is negative because *stuntman* is between the pronoun *"it"* and its referent *"motion"*. If we had other entities like *stuntman* in between *"it"* and *"motion"*, we would have generated other negative examples as well.

4.4   Resolution of Mention-Pairs

To automatically learn how to classify a mention-pair as P or N, we used a number of classifiers which were trained using the positive and negative examples described in Section 4.3. Once induced, the classifier can be used to classify future instances as either P or N. For instance, the referent of a new pronoun would correspond to the referent in the mention-pair classified as P.

5   Experiment Setup and Results

We used the previously mentioned technique to extract the positive and negative examples from the annotated corpus. In total, we obtained 955 positive and 2,312 negative examples. Although our corpus has 1,000 annotated instances, the positive examples are less because not all pronouns in the annotation corpus has a referent (e.g. pleonastic pronouns etc.). We considered the examples corresponding to pronouns without any referents as negative examples as we want our classifier to learn to reject such pairs in the future.

We used ten-fold cross-validation on the 3,267 examples for a number of classifiers as done in [1]. For comparison purpose, we used the DARE system[7] as our baseline. Results are reported in terms of precision, recall, accuracy, F-measure and kappa statistic.

Table 7 shows the results for the baseline, and the best results obtained for DARE++ using Naive Bayes, Support vector machine (SVM), Logistic Regression, Decision trees and Multilayer perceptron classifiers. The results reported were obtained after tuning various parameters of these machine learning algorithms. Note that all the classifiers have large performance gains over the baseline [7] in terms of accuracy, precision, recall and F-measure scores. It is found that poor performance of the baseline system is due to its fairly simple assumptions about the referents' locations which do not cover all the cases. For instance, one simple rule in the baseline algorithm stipulates that referents of pronouns that occur in the middle of a student answer are located in the same student answer [7]. While this seems right for some cases, it is often not true.

Among all classifiers, Logistic regression, Decision Tree (J48) and Multilayer perceptron are the best performing classifiers in terms of F-measure, Kappa-statistic and the root mean squared error (RME). These classifiers have F-measures over 79%. Decision Tree using J48 has the highest accuracy, precision, F-measure and kappa statistics and the lowest root mean squared error.

For tutorial dialogues, false positives are very important because declaring a noun as a referent of a pronoun, when it was actually not, leads to a different interpretation of the student's response. On the other hand, false negatives are less sensitive than false positives as they do not add wrong information during the interpretation process (e.g. suggesting *a pronoun doesn't have a referent when it had one* is not as severe as suggesting *an pronoun has a referent when it didn't have one*). Thus, we paid attention to the false positive counts of the classifiers. We found that the best performing classifiers also have lower false positive counts, satisfying the conditions for tutorial dialogues.

5.1   Feature Analysis

We experimented with adding unigrams, bigrams, and trigrams features for the tokens in A, Q and Q_i and their part-of-speeches as done by Stent et al. [15] for spoken dialogues. However, the performance didn't improve. Thus, the set of features presented in Table 6 is the best for tutorial dialogues.

We further studied the features in order to understand which ones are the most informative in tutorial dialogues. We used information gain and gain ratio to rank the features. The most informative features turned out to be: *the location of referent, prep_about* (dependency relation), *% of tokens after candidate, number agrees?, gender agrees?, det* (dependency relation), *governor relation counts for candidate, is candidate a proper noun, person of pronoun, prep_of* (dependency relation).

It is not surprising to see that the gender, number, and person features are crucial while determining the referents of pronouns in general. Interestingly, *the location of referent* is one of the most informative features for anaphora resolution in tutorial dialogues, which is different compared to the role of this feature in anaphora resolution for written texts. As suggested by the Table 5, more than 80% of the antecedents are located in Q and A alone. *governor relation counts for candidate* is another informative feature in tutorial dialogues. This is the case because words with many governor relations are more likely to be the focus of the tutor question which is typically referred by students in their answers. The dependency relations such as *prep_about* and *prep_of* are found to be other useful features for tutorial dialogues. The tutors typically ask students the following type of questions: *What can you say about XX of the YY ?* Student may reply with: *It equals ZZ.* In such examples, the pronoun *it* in the student answer refers to XX in tutor's question which is involved in a *prep_about* relation.

Due to the relative high frequency of such tutor question - student answer pattern the *prep_about* relation becomes salient.

6 Discussions and Conclusions

In this paper, we presented a solution to the problem of pronoun resolution in tutorial dialogues obtained from dialogue-based ITSs. Although pronoun resolution for written texts and spoken dialogues is well studied, it is not explored much for tutorial dialogues. Our experiments show that DARE++ can achieve a F-measure of 88.93%, showing its robustness in resolving pronouns.

Although the performance of DARE++ is impressive, it can be improved further. Demonstrative pronouns, ellipsis, soft, and hard communication breakdowns (see Table 3) are the major factors limiting its performance. Next important factor is having pronouns without antecedents (e.g. pleonastic pronoun). In addition, we don't consider *cataphora* currently. They are less frequent in tutorial dialogues but should be handled to make the system more robust. Finally, we would like to explore other models that use a reduced set of features based on the feature analysis we presented here or future feature analyses.

Acknowledgment. This research was supported in part by Institute for Education Sciences under awards R305A100875. Any opinions, findings, and conclusions or recommendations expressed in this material are solely the authors.

References

1. Arregi, O., Ceberio, K., Díaz de Illarraza, A., Goenaga, I., Sierra, B., Zelaia, A.: A first machine learning approach to pronominal anaphora resolution in basque. In: Kuri-Morales, A., Simari, G.R. (eds.) IBERAMIA 2010. LNCS (LNAI), vol. 6433, pp. 234–243. Springer, Heidelberg (2010)
2. Ferguson, G., Allen, J., Galescu, L., Quinn, J., Swift, M.: Cardiac: An intelligent conversational assistant for chronic heart failure patient health monitoring. In: AAAI Fall Symposium Series: Virtual Health Care Interaction (VHI 2009), Arlington, VA (2009)
3. Gasperin, C., Briscoe, T.: Statistical anaphora resolution in biomedical texts. In: Proceedings of the 22nd International Conference on Computational Linguistics, vol. 1, pp. 257–264. Association for Computational Linguistics (2008)
4. Mitkov, R.: Anaphora resolution: The state of the art. Tech. rep., University of Wolverhampton (1999)
5. Mitkov, R., Evans, R., Orăsan, C.: A new, fully automatic version of mitkov's knowledge-poor pronoun resolution method. In: Gelbukh, A. (ed.) CICLing 2002. LNCS, vol. 2276, pp. 168–186. Springer, Heidelberg (2002)
6. Niraula, N.B., Rus, V., Banjade, R., Stefanescu, D., Baggett, W., Morgan, B.: The dare corpus: A resource for anaphora resolution in dialogue based intelligent tutoring systems. In: Proceedings of Language Resources and Evaluation, LREC (2014)

7. Niraula, N.B., Rus, V., Stefanescu, D.: Dare: Deep anaphora resolution in dialogue based intelligent tutoring systems. In: Proceedings of the 6th International Conference on Educational Data Mining (EDM 2013), pp. 266–267 (2013)

8. Poesio, M., Kabadjov, M.A.: A general-purpose, off-the-shelf anaphora resolution module: Implementation and preliminary evaluation. In: Proceedings of LREC (2004)

9. Poesio, M., Patel, A., Di Eugenio, B.: Discourse structure and anaphora in tutorial dialogues: An empirical analysis of two theories of the global focus. Research on Language and Computation 4(2-3), 229–257 (2006)

10. Qiu, L., Kan, M.Y., Chua, T.S.: A public reference implementation of the rap anaphora resolution algorithm. In: Proceedings of the Fourth International Conference on Language Resources and Evaluation (2004)

11. Rahman, A., Ng, V.: Supervised models for coreference resolution. In: Proceedings of EMNLP, pp. 968–977. ACL (2009)

12. Rus, V., D'Mello, S., Hu, X., Graesser, A.C.: Recent advances in conversational intelligent tutoring systems. AI Magazine 34(3) (2013)

13. Rus, V., Niraula, N., Lintean, M., Banjade, R., Stefanescu, D., Baggett, W.: Recommendations for the generalized intelligent framework for tutoring based on the development of the deeptutor tutoring service. In: AIED 2013 Workshops Proceedings, vol. 7, p. 116 (2013)

14. Soon, W.M., Ng, H.T., Lim, D.C.Y.: A machine learning approach to coreference resolution of noun phrases. Computational Linguistics 27(4), 521–544 (2001)

15. Stent, A.J., Bangalore, S.: Interaction between dialog structure and coreference resolution. In: 2010 IEEE Spoken Language Technology Workshop (SLT), pp. 342–347. IEEE (2010)

16. Strube, M., Müller, C.: A machine learning approach to pronoun resolution in spoken dialogue. In: Proceedings of the 41st Annual Meeting on Association for Computational Linguistics, vol. 1, pp. 168–175. Association for Computational Linguistics (2003)

17. Su, J., Yang, X., Hong, H., Tateisi, Y., Tsujii, J.: Coreference resolution in biomedical texts: a machine learning approach. Ontologies and Text Mining for Life Sciences 8 (2008)

18. Versley, Y., Ponzetto, S.P., Poesio, M., Eidelman, V., Jern, A., Smith, J., Yang, X., Moschitti, A.: Bart: A modular toolkit for coreference resolution. In: Proceedings of ACL. pp. 9–12. ACL (2008)

19. Yang, X., Su, J., Tan, C.L.: Kernel-based pronoun resolution with structured syntactic knowledge. In: Proceedings of the 21st International Conference on Computational Linguistics and the 44th Annual Meeting of the Association for Computational Linguistics, pp. 41–48. Association for Computational Linguistics (2006)

A Maximum Entropy Based Honorificity Identification for Bengali Pronominal Anaphora Resolution

Apurbalal Senapati and Utpal Garain

Computer Vision and Pattern Recognition Unit
Indian Statistical Institute
203, B.T. Road, Kolkata-700108, India
`apurbalal.senapati@gmail.com, utpal@isical.ac.in`

Abstract. This paper presents a maximum entropy based method for determining honorific identities of personal nouns in Bengali. Later this information is used for pronoun (anaphora) resolution system for Bengali as honorificity plays an important role for pronominal anaphora resolution in Bengali. Experiment has done on a publicly available dataset. Experimental result shows that when the module for honorific identification is added to the existing pronoun resolution system, the accuracy (avg. F1-score) of the system is improved from 0.602 to 0.703 and this improvement is shown to be statistically significant.

Keywords: Pronominal anaphora resolution, Bengali, honorific.

1 Introduction

Little computational linguistics research has been done for Indic languages until recently. One major reason has been the unavailability of annotated datasets for developing basic NLP tools. In the recent past efforts have been taken to bridge this gap and data-sets like tree-banks, annotated dataset for anaphora resolution have been developed for NLP research for some Indic languages like Hindi, Bengali, Tamil, etc. [1, 2, 3].

This paper is focused on a statistical approach for identification of honorific information of Bengali personal nouns. The problem is already identified before and different rule-based approaches were discussed [4, 5]. It is observed that unlike in English gender information has no role for pronoun resolution (in Bengali *he* and *she* represents the same pronoun সে/*se*) in Bengali but similar significant role is played by honorific information of Bengali nouns. Therefore, identification of the honorific information is very crucial for pronoun resolution in Bengali. This paper addresses the problem of identification of honorific information of a person and its effect on anaphora resolution.

Certain honorific addressing terms in Bengali like "বাবু" (*babu*), "ডঃ" (*dr*), "মহাশয়" (*mohasaya*), etc. some-times convey honorific information in Bengali like use of "Mr.", "Mrs.", etc. in English to convey gender information. These honorific's whenever used can appear just before or just after the respective nouns (unlike English where honorific's like "Mr.", "Mrs.", etc. appear only before nouns). Moreover,

A. Gelbukh (Ed.): CICLing 2014, Part I, LNCS 8403, pp. 319–329, 2014.

nature of inflection of the main verb of the containing sentence often convey honorific information of the subject.

As multiple features contribute differently in determining honorificity of Bengali nouns in a sentence, we preferred to use a maximum entropy model [6] based approach rather than the co-occurrence based gender determination approach followed in [7] where only <referent, pronoun> co-occurrences were considered. In fact, in the paper [5], the authors adopted a co-occurrence based approach for honorificity detection where whether a noun co-occurs with any of the honorifics (given in a predefined list) in a sentence was checked. The rest of the paper presents a brief description of honorific information in Bengali, the features for the present experiment, dataset and experimental results.

2   Brief Description of an Existing Anaphora Resolution in Bangla

Here is the brief description of an improved version of existing anaphora resolution approach [4] which has been considered in our evaluation phase. This system consists of five modules; the linguistic resource, rule base, pronoun emission, global discourse knowledge and conflict resolution module. Each module is described below in brief.

The linguistic resource is the classification of all available pronouns based on their compatibility with respect to noun together with some agreements like number, animate/inanimate, honorificity etc. It also contains some other information like honorific context {বাবু/*babu*, ড:/*dr*, মহাশয়/*mahasaya*, ...}, nominal relations {মা/*ma*, বাবা/*baba*, ভাই/*bhai*, বোন/*bon*, দাদা/*dada*, দিদি/*didi*, কাকা/*kaka*, কাকি/*kaki*, ...}, common noun antecedent, etc. The subsequent modules use this information. Table 1 shows a snap of the classification of Bengali pronouns.

Table 1. Classification of Bengali Pronouns

Category	Permissible pronouns
First person	আমি/*aami* (I),আমার/*aamar* (my),আমাকে/*aamake* (me), ...
Second person	তুমি/*tumi*(you), তোমার/*tomar*(your),আপনি /*aapni*(you), ...
Third person	সে/*se*(he), তার/*tar*(his), তারা/*tara*(they), এর/*er*(his), ...
Honorific singular	তিনি/*tini*(he), তাঁর /*tanr*(his/her), তাঁকে /*tanke*(him), ...
Honorific plural	তাঁরা/*tanra*(they), যাঁরা/*janra*(those), আপনারা/*aapnara*(you), ...
Inanimate singular	এটা/*eta*(it), ওটা /*ota*(this), সেটা /*seta*(that), ...
Inanimate plural	এগুলো/*egulo*(these), সেগুলো/*segulo*(those), ...
....	....

The rule base consists of five basic rules. The first one is used for honorific identification of a person. The rule is that if a person co-occurs (left or right) with honorific addressing term (given in linguistic resource) then the person is honourable. Example: ড: অনুপ কর /*dr anup kar* , the person অনুপ কর /*Anup Kar* is honourable, since the honorific addressing term ড:/*dr* co-occurs with the noun. Second rule is for reflexive pronoun. A reflexive pronoun in Bengali co-refers with animate and in a simple sentence the pronoun with its subject. Third rule is for consecutive pronouns. In case of consecutive co-referring pronouns appearing in a sentence, they are person compatible (i.e. either both are in first or second person) and they also follow the honorific agreement. Example, আমি তোমাকে ভালবাসি/*aami tomake bhalobasi*, here the pronoun আমি/*aami* (I), তোমাকে/*tomake* (you) are not co-referring since the pronouns are not person compatible. The fourth rule is for co-occurring pronoun. In Bengali, some pronoun pairs oc-occur as co-referring in the same sentence. In such cases the pronoun pair is co-referring when they appear in the same sentence. The fifth rule is for plural pronoun agreement. An organization or community though they look like singular but may refer to plural pronoun. Example: *The club* has changed *their* requirement policy. Here though *the club* (an organization) is singular is the referent of plural pronoun (*their*).

The pronoun emitting module use to construct antecedent object. A noun emits its permissible pronoun list (available from Table 1) is considered as the antecedent object. It also contains other lexico-syntax information such as sentence number, token (individual word in the text) number, honorific information (if available), co-reference information (if available), etc. In the following text (in Table 2) while found the noun ব্যারেট/*Byarat* (the name of a person) it emits its permissible pronouns {তাঁর/*tanr*, তাঁকে/*tanke*, তিনি/*tini*, তিনিই/*tini-i*, ... } from linguistic resource and become antecedent object looks like ব্যারেট -> {তাঁর/*tanr*, তাঁকে/*tanke*, তিনি/*tini*, তিনিই/*tini-i*, ... } (since from global discourse knowledge it identified the ব্যারেট/*Byarat* is honorable person and hence it emits only honorific pronouns).

Table 2. Sample text from story2.txt

0	ব্যারেট	NNP	B-NP	B-PERSON -
1	বলে	VM	B-VGF	o -
2	উঠলেন	VAUX	I-VGF	o -
3	যে	CC	B-CCP	o -
4	তিনিই	PRP	B-NP	o -
5	বড়	JJ	B-NP	o -
6	হাতীটাকে	NN	I-NP B-LIVTHINGS -
7	গুলি	NN	B-NP	o -
8	করবেন	VM	B-VGF	o -
9	।	SYM	I-VGF	o -

Some knowledge from the entire discourse is also extracted prior to pronoun resolution and has been used in resolution system. The honorific information of a person, hidden person identification and the alias name are considered in the discourse knowledge. The honorific information is identified by the rule (the first rule in rule base described in section 2). In many cases some person appears in the discourse indirectly (defined as hidden person) and they appear as an antecedent. For example, in the story *Lalu.txt* (extended data) the term লালুর বাবা /*lalur baba* (Lalu's father) appears in this story five times as "লালুর বাবা"/*lalur baba*, who is a separate person from লালু/*lalur* and doesn't exist by any other name. This is identified by a rule that captures name of a person ends with case marker র/এর and followed by a nominal relation (this comes from the linguistic resource). The alias name is of two types, the first one refers to the case where a person's name is mentioned once and then in subsequent discourse he/she is referred by his/her surname or some qualifier. For example: *Dr. Utpal Garain, Dr. Garain, UG* etc. generally represent same person in the discourse. This is identified by a string matching approach. The second one is due to the spelling variation. The spelling variation is quite natural in Bengali. A person may be addressed by names with different form of utterances in different context. For example, the person name নালু/*lalu* in our data (extended data set) having two forms, নালু/*lalu* and লেলো/*lelo*. This is done by split names character wise and then excluding vowel modifiers. For example, excluding the vowel modifier from নালু = {ন া ল ু}, and লেলো = {ন ে ল ো} is the same {ন ল} and hence নালু and লেলো are alias name.

Conflict resolution module takes care of conflict during resolution. This module uses a set of agreement/constraints which are used in the following order: (i) pronouns are number compatible, (ii) pronouns are honorific compatible, and (iii) pronouns are person compatible. If it is still not resolved then we choose the most recent one from the antecedent list as the referent.

```
0   ব্যারেট -> { তাঁর, তাঁকে, তিনি, তাঁরই, তিনিই, ⋯ }
1   বলে     VM    B-VGF o   -
2   উঠলেন   VAUX I-VGF o   -
3   যে      CC    B-CCP o   -
4   তিনিই   PRP   B-NP  o   -
5   বড়     JJ    B-NP  o   -
6   হাতীটাকে        NN    I-NP B-LIVTHINGS -
7   গুলি    NN    B-NP  o   -
8   করবেন   VM    B-VGF o   -
9   |       SYM   I-VGF o   -
```

Fig. 1. The pronoun resolution process

Resolution process: while found a noun (possible antecedent) in the text it emits its permissible pronouns (from the linguistic resource). Subsequently when the system attempts to resolve pronoun it checks the nearest antecedent object. In the above (Table 2) example, when goes to resolve the pronoun "তিনিই"/*tini-i*, it is obvious that "ব্যারেট"/*Byarat* is the antecedent, since pronoun "তিনিই"/*tini-i* is in the permissible pronoun set of "ব্যারেট"/*Byarat* (shown in Fig. 1). In case of conflict, conflict resolution module resolves such scenario.

3   Honorific Information in Bengali

The honorific agreement is very strong in Bengali language and it is used for personal nouns and pronouns. This information distinguishes people on the basis of their social status. Three types of honorificity exist in written as well as in spoken form [8, 9] in Bengali. The highest degree of honorificity (defined superior or SUP class) normally refers to people of high social status like doctors, teachers, lawyers, professors, political men, etc. or parents, grant parents, senior people of the family or society, etc. or sometimes unknown respected person. The next level of honorificity is the neutral form (defined neutral or NEU class), refers to closer members of a family, children or younger members of family, or people within peer group. The lowest level of honorificity (defined inferior or INF class) normally refers to very close friends, very close relations (who are younger like son, daughter etc.) or the people presumed to be of inferior social status like, housemaids, rickshaw-pullers, and other menial service workers.

Honorific information of a person is indicated by a word or expression with connotations conveying esteem or respect when used in addressing or referring to a person. Linguistically this word or expression is known as honorific addressing term. The most common honorifics in Bengali are usually placed immediately before the name (defined as left context) of a person (e.g. ভদ্রলোক/*bhadrolok*, ডঃ/*Dr.*, মহাশয়/*mahasaya*, মিঃ/*Mr.*, etc.). Some of the terms are also placed immediately after the name (defined as right context) of a person (e.g. বাবু/*babu*, বাবুমশায়/*babumasai*, দেবী/*debi*, etc.).

Table 3. Frequently used honorific addressing terms in Bengali

Category	Permissible terms
Left context	ভদ্রলোক/*bhadralok*, ডঃ/*dr.*, মিঃ/*mr*, শ্রী/*sri*, ...
Right context	বাবু/*babu*, মশাই/*mosai*, দেবী/*debi*, মহাশয়/*mahasaya*, ...

Table 4. Classification of Bengali pronouns based on their honorificity

Category	Permissible Pronouns
SUP	আপনি/*aapni*, ওঁর/*onr*, তাঁর/*tanr*, এঁর/*enr*, তিনি/*tini*, ...
NEU	তার/*tar*, সে/*se*, ও/*o*, ওর/*or*, ওরা/*ora*, তুমি/*tumi*, ...
INF	তুই/*tui*, তোর/*tor*, তোকে/*toke*, তোরই/*torei*, ...

Table 3 shows such a list of frequently used honorifics in Bengali. Another additional way for identifying the honorific information is to look at the inflection of the main verb associated with the noun. For example, verb খাওয়া/*eat* having three forms, খা/*eat* (refers to a person having the lowest level of honorificity, i.e. INF class), খাও/*eat* (refers to a person having the medium level of honorificity, i.e. NEU class), and খান/*eat* (refers to a person having the highest level of honorificity, i.e. SUP class).

Identification of honorific information for personal pronouns is easy as separate forms of pronouns exist in Bengali to convey honorific information [10, 11]. The second and third person pronouns have distinct forms for different degrees of honorificity. Table 4 shows such classification of frequently used pronouns in Bengali.

4   Maximum Entropy Modeling

The feature function is a binary valued function defined on the training data of the form (x, y), where x is a sentence or a phrase containing a person with its context (honorific context) information and y is the honorific information of that person. Formally the function $f(x, y)$ defined over data (X, Y) as $f: X \times Y \rightarrow \{0, 1\}$, where X is the data and Y is its class label (i.e. INF, NEU or SUP). Our feature selection model generates a 10-dimensional feature vector as described below.

$f_1(x, y) = 1$ when y is SUP and x is the person and honorific addressing term is in the left position;
$f_1(x, y) = 0$ otherwise.

$f_2(x, y) = 1$ when y is SUP and x is the person and honorific addressing term is in right position;
$f_2(x, y) = 0$ otherwise.

$f_3(x, y) = 1$ when y is SUP and x is the person and honorific addressing terms are both in left and right positions;
$f_3(x, y) = 0$ otherwise.

$f_4(x, y) = 1$ when y is SUP and x is the person and honorific addressing term is in left position and the main verb associated with x is in SUP class;
$f_4(x, y) = 0$ otherwise.

$f_5(x, y) = 1$ when y is SUP and x is the person and honorific addressing term is in right position and the main verb associated with x is in SUP class;
$f_5(x, y) = 0$ otherwise.

$f_6(x, y) = 1$ when y is SUP and x is the person and honorific addressing terms are both in left and right positions and the main verb associated with x is in SUP class;
$f_6(x, y) = 0$ otherwise.

$f_7(x, y) = 1$ when y is SUP and x is the person and the main verb associated with x is in SUP class;
$f_7(x, y) = 0$ otherwise.

$f_8(x, y) = 1$ when y is NEU and x is the person and the main verb associated with x is in NEU class;
$f_8(x, y) = 0$ otherwise.

$f_9(x, y) = 1$ when y is NEU and x is the person and there is neither left honorific terms nor right honorific terms and the main verb is absent;
$f_9(x, y) = 0$ otherwise.

$f_{10}(x, y) = 1$ when y is INF and x is the person and no honorific term is either in left or in right position and the main verb is in INF class;
$f_{10}(x, y) = 0$ other-wise.

5   Training of the Model

The above feature function is computed locally, i.e. within a sentence. We have used the Stanford University maximum entropy classifier [12] (version 3.3.0). We have defined the underlying feature function as above and trained the model using data taken from a large Bengali corpus (35 million words) [13]. The annotation format of the sample training data is in column format shown in Table 5

Table 5. Format of training data

6	মিস		XC	B-NP	B-PERSON	-
7	আগাথা		XC	I-NP	I-PERSON	SUP
8	হ্যারিসন		NNP	I-NP	I-PERSON	SUP
9	প্রথমে		NN	B-NP	o	-
10	শান্তিনিকেতনে	NNP	B-NP	o	-	
11	আসেন		VM	B-VGF	o	SUP
12	1930		XC	B-NP	o	-
13	সালে		NNP	I-NP	o	-
14	।		SYM	I-NP	o	-

Table 6. Description of training data

Column	Description
1	Token number
2	Token
3	POS
4	IOB-POS
5	Name Entity
6	Honorific information i.e. classification

Table 7. Coverage of the training data

Data description	Number
#text	25
#words	48,177
#persons	1,661
#SUP category	1,227
#NEU category	288
#INF category	146

The training data is defined of the form (x, y) where x is the data and y is its class label. In our experiment x is a sentence or a phrase containing a person with its context information and y is the honorific information of that person. The description of the annotated column format training data shown in Table 6 and Table 7 shows the coverage of the training data.

The annotation required some other NLP tools specially POS tagger, chunker and dependency parser. We retrained the Stanford POS tagger using about ten thousand POS tagged sentences collected from Linguistic Data Consortium (LDC), University of Pennsylvania and the NLP Tool Contest at [1]. The chunker described in [14] has been used for text chunking. For identifying the main verb in a sentence we used a Malt-parser based implementation of Bengali dependency parser [15].

Several information is taken from analysis of the corpus [13]. From this analysis, twenty-four (28) honorific terms are identified which may appear on the left and nine (13) honorific terms are identified which may appear on the right position of honorific nouns. These lists of terms are used to compute the feature function described above. Classification of verbs as SUP/NEU/INF is done by following a simple observation. When the verb is inflected with ন/n, we consider it as SUP category, otherwise NEU or INF category. The verb inflection does not help much to distinguish NEU or INF category. As there is no work on this verb classification task, we used the above classification technique for classifying the verbs as an unsupervised manner. We experience that SUP categories are often rightly identified but NEU and INF verb categories require some manual corrections.

6   Evaluation

For evaluating the model we used an extended version of the ICON 2011 [2] annotated data for anaphora resolution in Bengali. The test data contains thirteen texts from different domains (Tourism, Story, News article, Sports); nine of these texts are part of ICON 2011 [2] data which has been augmented (by us) by adding four more texts. The distribution of test data is shown in Table 8. The test data format corresponds to the format of training data excluding the classification information (i.e. column 6 in Table 5).

Table 8. Coverage of test data

Data description	Number
#text	13
#words	27,454
#persons	1,243
#SUP category	901
#NEU category	236
#INF category	106

The experiment has been done in two phases. At first, the accuracy of identification of honorificity of personal nouns is computed. The test data contains 1,243 personal (including the common nouns like ম/*ma* (mother), বাবা/*baba* (father), etc.) nouns and it is seen that the method can correctly identify honorificity for 1104 (89%) personal nouns [SUP: 90% (810/901); NEU: 88% (207/236) and INF: 82% (87/106) shown in Table 9]. Table 10 presents the recall, precision and F-score of this identification task.

Table 9. Classification/misclasification of the system on honorificity identification

Category	SUP	NEU	INF	Total
SUP	810	91	0	901
NEU	29	207	0	236
INF	5	14	87	106
Total	844	312	87	1243

Table 10. Performance of the system on honorificity identification

Category	P	R	F1
SUP	95.97	89.90	92.83
NEU	66.34	87.71	75.54
INF	100.00	82.07	90.15

Table 11. Comparison of results in anaphora resolution with respect of the use of honorificity

Metric		Baseline	System I	System II	System III
	P	0.437	0.477	0.489	0.538
MUC	R	0.426	0.426	0.462	0.605
	F1	0.431	0.450	0.475	0.569
	P	0.577	0.667	0.678	0.740
B^3	R	0.676	0.786	0.832	0.842
	F1	0.608	0.721	0.747	0.787
	P	0.614	0.614	0.654	0.786
CEAFM	R	0.602	0.602	0.646	0.695
	F1	0.608	0.607	0.650	0.737
	P	0.661	0.771	0.797	0.804
CEAFE	R	0.601	0.601	0.642	0.555
	F1	0.630	0.675	0.712	0.656
	P	0.480	0.500	0.542	0.765
BLANC	R	0.582	0.628	0.678	0.771
	F1	0.526	0.556	0.603	0.767
Avg.	F1	0.560	0.602	0.637	0.703

The second experiment addresses the effect of this effort of honorific agreement in anaphora resolution. For this purpose the rule base system of Senapati et al. [4] (described in section 2) has been considered.

The experiment is done on the test data set as described above and the evaluation has used five metrics namely, MUC, B^3, CEAFM, CEAFE and BLANC. The comparisons of experimental results are reported in Table 11. The column corresponding Baseline (*last noun approach*: the last noun or the noun closest to the pronoun of same category is assumed to be the correct antecedent),   System I show the result after removing the honorific agreement from the system, the column corresponding System II shows the result with honorific agreement (identified by the co-occurring approach describing in section 2) and the column corresponding System III shows when the honorific information identified by maximum entropy approach. It shows that from System-I (no honorific information used) to System-II (honorificity is captured by occurrence) absolute improvement in F-score is 6%, from System-II to System-III (honorificity is captured by maximum entropy model) the improvement is 10% whereas from System-I to System-III the improvement is 17%.

All these improvements are statistically significant ($p < 0.01$ in a two-sided t-test). This improvements show two things (i) the importance of honorific information in pronominal anaphora resolution in Bengali and (ii) maximum entropy is a good model to capture the honorificity information. Note that, the baseline result of this system is differing from that of the other system [5], though both the systems use the same dataset. This is because, the system [5] is the implementation of MARS [16] which considered only the personal pronouns whereas the present experiment considered the other pronouns too while reporting results.

7   Conclusion

This paper presents a pioneering attempt in automatically identifying honorific information for Bengali personal nouns. A maximum entropy based approach is used for this purpose and the result shows the method gives significant accuracy for doing the task. The importance of this work is also shown in the context of pronominal anaphora resolution in Bengali.

A similar approach can be followed for developing several NLP tools like identification of number information, gender identification (for instance, in Hindi or English) where such information plays crucial role in many NLP applications including pronoun resolution, machine translation, etc.

References

1. ICON. NLP Tools Contest on Dependency Parsing in Indian Languages. In: 7th Int. Conf. on Natural Language Processing, Hyderabad, India (2009)
2. ICON. NLP Tools Contest on Anaphora Resolution in Indian Languages. In: 9th Int. Conf. on Natural Language Processing, Chennai, India (2011)
3. ICON. NLP Tools Contest on Named Entity Recognition in Indian languages. In: 10th Int. Conf. on Natural Language Processing, Noida, India (2013)
4. Senapati, A., Garain, U.: Anaphora Resolution in Bangla using global discourse knowledge. In: Int. Conf. of Asian Language Processing, Hanoi, Vietnam (2012)

5. Senapati, A., Garain, U.: GuiTAR-based Pronominal Anaphora Resolution in Bengali. In: ACL, pp.126–130 (2013)
6. Berger, A., Pietra, S., Pietra, V.: A maximum entropy approach to natural language processing. Computational Linguistics 22(1), 39–71 (1996)
7. Ge, N., Hale, J., Charniak, E.: A Statistical Approach to Anaphora Resolution. In: 6th Workshop on Very Large Corpora, pp. 161–170 (1998)
8. Comrie, B.: The Major Languages of South Asia, the Middle East and Africa, pp. 61–65 (1987)
9. Bengali, T.H.: John Benjamins Publishing Company, vol. 18, pp. 75–80 (1992)
10. Majumdar, A.: Studies in the Anaphoric Relations in Bengali. Subarnarekha, India (2000)
11. Sengupta, G.: Lexical Anaphors and Pronouns in Selected SouthAsian Languages: A Principled Typology, pp. 277–280 (2011); Ed. by Lust, Barbara C. / Wali, Kashi /Gair, James W. / Subbarao, K.V.
12. Manning, C., Klein, D.: Optimization, Maxent Models, and Conditional Estimation without Magic. Tutorial at HLT-NAACL 2003 and ACL 2003 (2003)
13. Bengali Corpus. TDIL Corpus in Unicode,
    http://www.isical.ac.in/~lru/downloadCorpus.html
14. Biswas, S., Dhar, A., De, S., Garain, U.: Performance Evaluation of Text Chunking. In: Proc. of the 8th Int. Conf. on Natural Language Processing ICON, India (2010)
15. Das, A., Shee, A., Garain, U.: Evaluation of Two Bengali Dependency Parsers. In: Proc. COLING Workshop on Machine Translation and Parsing in Indian Languages, pp. 133–142 (2012)
16. Mitkov, R.: Anaphora Resolution. Longman (2002)

Statistical Relational Learning
to Recognise Textual Entailment

Miguel Rios[1], Lucia Specia[2], Alexander Gelbukh[3], and Ruslan Mitkov[1]

[1] University of Wolverhampton,
Research Group in Computational Linguistics,
Stafford Street, Wolverhampton, WV1 1SB, UK
{M.Rios,R.Mitkov}@wlv.ac.uk
[2] University of Sheffield,
Department of Computer Science,
Regent Court, 211 Portobello, Sheffield, S1 4DP, UK
L.Specia@sheffield.ac.uk
[3] Centro de Investigación en Computación,
Instituto Politécnico Nacional,
Mexico City, Mexico
www.gelbukh.com

Abstract. We propose a novel approach to recognise textual entailment (RTE) following a two-stage architecture – alignment and decision – where both stages are based on semantic representations. In the alignment stage the entailment candidate pairs are represented and aligned using predicate-argument structures. In the decision stage, a Markov Logic Network (MLN) is learnt using rich relational information from the alignment stage to predict an entailment decision. We evaluate this approach using the RTE Challenge datasets. It achieves the best results for the RTE-3 dataset and shows comparable performance against the state of the art approaches for other datasets.

1 Introduction

Recognising Textual Entailment (RTE) consists in deciding, given two text segments, whether the meaning of one segment (the (H)ypothesis) is entailed from the meaning of the other segment (the (T)ext) [7]. In order to address the task of RTE, most methods rely on machine learning algorithms. For example, a baseline method proposed by Mehdad and Magnini [18] measures the word overlap between the T-H pairs. An overlap threshold is computed over some training data.

Another approach for RTE is to determine some sort of alignment between the T-H pairs. Since T is usually longer, H is aligned to a portion of T, and the best alignment is used to compute a similarity score. A limitation of such approaches is that instead of recognising a non-entailment, an alignment that fits an optimisation criterion will be returned [17], and thus the alignment by itself is a poor predictor for non-entailment. To solve this problem, de Marneffe et al. [17] divide the RTE task such that the alignment and the entailment decision are separate processes. The alignment phase is based on matching graph representations (i.e. dependency relations) of the T-H pair. For the

A. Gelbukh (Ed.): CICLing 2014, Part I, LNCS 8403, pp. 330–339, 2014.

entailment decision, rules which strongly suggest implications are designed. A specific rewrite rule between T and H can be positive if they represent entailment or negative otherwise.

Except for Garrette et al. [8], previous work using machine learning is based on propositional representations with simple attribute-value pairs as features. Garrette et al. [8] combines first order logic and statistical methods for RTE. The approach uses discourse structures to represent T-H pairs, and a Markov Logic Network (MLN) model to perform inference in a probabilistic manner over implicativity and factivity, word meaning, and coreference. A threshold on the entailment decision given the MLN model output is manually set. Since their phenomena of interest are not present in the standard RTE datasets, they use handmade datasets. For other related work in the field, we refer the reader to [1].

In this paper we describe an RTE approach following a multi-stage architecture. In contrast to de Marneffe et al. [17], both stages are based on semantic representations in an attempt to measure entailment based on the similarity of answers to the questions *Who did what to whom, when, where, why and how*. This is done through shallow semantic parsing using a Semantic Role Labelling (SRL) tool. Furthermore, instead of using simple similarity metrics to predict the entailment decision, we use rich relational features extracted from the output of the predicate-argument alignment structures between T-H pairs. These are fed to an MLN framework, which learns a model to reward pairs with similar predicates and similar arguments, and penalise pairs otherwise. Different from [8], we do not use a manually set threshold for the entailment decision and we evaluate our method on the standard RTE Challenge datasets, which are larger and contain naturally occurring linguistic constructions that can have an effect on the entailment decision. We compare our approach to previous works for RTE based on alignment techniques, and on probabilistic modelling. Our approach achieves the best performance on the RTE-3 dataset, and competitive results on other datasets.

2   Experimental Design

Our approach to RTE is based on a two-stage architecture: i) alignment, where predicate-argument structures of H and T are aligned; and ii) entailment decision, where the alignments are considered to extract features (i.e. first order logic predicates) and these are used to build an MLN model.

2.1   Alignment Stage

We represent the T-H pair with SRLs as generated by SENNA [6] and use TINE [20, 21] to align any number of predicates and arguments between T and H. Instead of simply matching surface forms, TINE performs a flexible alignment of verb predicates by measuring (i) how similar their arguments are ($argScore$), (ii) and how related the predicates realisations are ($lexScore$). Both scores are combined as shown in Equation 1 to measure the similarity between the two predicates (Av, Bv) from a pair of sentences (A, B).

$$sim(Av, Bv) = wlex \times lexScore(Av, Bv)$$
$$+ warg \times argScore(Aarg, Barg) \tag{1}$$

where $wlex$ and $warg$ are the weights for each component, $argScore(Aarg, Barg)$ is the similarity between the arguments, computed as the cosine distance between the bag-of-words of the predicates' arguments Av, Bv. $lexScore(Av, Bv)$ is the similarity score of the predicates extracted using Dekang Lin's thesaurus [14]. The pair of predicates that maximise Equation 1 produces an alignment with an one-to-one verb-arguments relation.

2.2 Entailment Decision Stage

In the entailment decision stage we use an MLN to predict the entailment relation of a given T-H pair. Statistical relational learning [9], as opposed to a propositional formalism, is focused on representing and reasoning over domains with a relational and probabilistic structure. These models use first-order representations to describe the relations between the domain variables and probabilistic graphical models to reason over uncertainty.

MLN [19] provides a natural choice for this task as it unifies first order logic and probabilistic graphical models in a framework that enables the representation of rich relational information (such as syntactic and semantic relations) and inference under uncertainty. This framework learns weights for first order logic formulas, which are then used to build Markov networks that can be queried in the presence of new instances.

As an inherently semantic task, RTE should naturally benefit from knowledge about the relationships among elements (variables) in a text, in particular to check whether (some of) these relationships are equivalent in both T and H. It is extremely difficult to fully capture relational knowledge using standard propositional formalisms (attribute-value pairs), as it is hard to predict how many elements are involved in a relationship (e.g., a compound argument) or all possible values of these elements, and it is not possible to represent the sharing of values across attributes (e.g. the agent of a predicate which is also the object of another predicate).

The basis for our first order logic formulas are the alignments produced in the previous stage. At inference time, an aligned pair with similar situations and similar participants will likely hold an entailment relation. An alignment consists of a pair of verbs and their corresponding arguments. Several features extracted from these alignments are used as predicates to build a Markov Network. We formulate a relational model based on these predicates along with shallow features used to support the decision when there is no evidence of an alignment for a T-H pair.

Relational Model. Our model takes advantage of MLN's ability to handle relational information, and it also takes into consideration the semantic relations between the arguments and verbs. The motivation to design the relational formulas is based on how the alignment stage works. The alignment is performed via heuristics, which means that some of the decisions may introduce incorrect or poor information about the relations between the participants and situations of the entailment candidate pair. In order to alleviate this problem, the relational features reward or penalise each of the aligned verbs from the first stage by making explicit their semantic relation. In addition, the relational features generalise each of the arguments aligned by TINE.

The following variables are created to represent this information: Arg and $Verb$. Figure 1 shows the relationships between these variables in a Markov network.

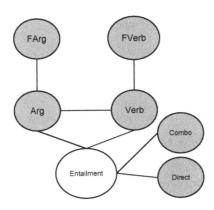

Fig. 1. Markov network of our RTE model

The value of Arg is the label given by the SRL parser for the aligned arguments (e.g., ARG1). The value of $Verb$ is the lexical realisation of the verbs, i.e., the aligned verbs themselves. Furthermore, the aligned arguments and the aligned verbs have features: $FArg$ is the set of features related to the arguments, and $FVerb$ is the set of features related to the verbs.

The features for each token of aligned arguments are as follows:

Lexical. Word, lemma and PoS of each token.
Similar Words. The 20 most similar words from Dekang Lin's thesaurus for each token. A predicate is created for each similar word.
Hypernyms. The first three levels of the hypernym tree above each noun in its first sense in WordNet. A predicate for each hypernym is created.

These argument features are represented by the following formula:

$$Token(aid, pid, +tfeature) \land Arg(aid, vid, pid) \Rightarrow Entailment(+d, pid)$$

where $tfeature$ takes the value of each of the previous features, aid and vid are the values of the Arg and $Verb$ variables
For the aligned verbs, the following features are extracted:

Bag-of-words VerbNet. $bowfeature$ is the lexical realisation of the classes shared between the verbs in VerbNet. Looking at the semantic classes of the aligned verbs brings extra information about how similar they are:

$$BowVN(vid, +bowfeature) \wedge Verb(vid, pid) \Rightarrow Entailment(+d, pid)$$

Strong Context. $strfeature$ compares components in Equation 1. If the value of $argScore(Aarg, Barg)$ is larger than that of $lexScore(Av, Bv)$, this feature is set to 1, i.e., the similarity of the context of the aligned verbs is stronger than the relationship between them; it is 0 otherwise:

$$StrongCon(vid, +strfeature) \wedge Verb(vid, pid) \Rightarrow Entailment(+d, pid)$$

Similarity VerbNet. $simvnfeature$ is set to 1 if the verbs share at least one class in VerbNet; 0 otherwise:

$$SimVN(vid, +simvnfeature) \wedge Verb(vid, pid) \Rightarrow Entailment(+d, pid)$$

Similarity VerbOcean $simvofeature$ is 1 if the verbs have the *similar* relation as given by VerbOcean [5];[1] 0 otherwise:

$$SimVO(vid, +simvofeature) \wedge Verb(vid, pid) \Rightarrow Entailment(+d, pid)$$

Token Verbs. The predicate contains the lemmas of the aligned verbs:

$$TokenVerb(vid, +tokenvfeature) \wedge Verb(vid, pid) \Rightarrow Entailment(+d, pid)$$

Finally, the relation between Arg and $Verb$ is defined by the formula:

$$Arg(aid, vid, pid) \wedge Verb(vid, pid) \Rightarrow Entailment(+d, pid)$$

The formulas sharing variables vid and aid indicate relationships between the aligned arguments and the aligned verbs, as well as their corresponding features given the SRL structure. pid relates the previous predicates to the decision of an entailment pair. Many of these formulas can take up multiple values through multiple groundings (e.g. the hypernyms of nouns). The predicate $Entailment(+d, pid)$ takes two possible values for the decision d: $true$ or $false$. The $+$ operator indicates that a weight will be learned for each grounding of the formula. The entailment decision is a hidden variable in the MLN model and it is used to query the MLN.

In the alignment stage, sometimes TINE cannot align a T-H pair, mostly because SENNA does not produce any SRL structure for certain T-H pairs. To be able to make a decision for these pairs using MLNs, we add the variables $Combo$ and $Direct$ as shallow supporting features for the entailment decision in Figure 1. $Combo$ holds the value $cfeature$, which consist of all the combinations of unigrams between the H-T pair. The following predicate is defined for each unigram combination:

$$Combo(pid, +cfeature) \Rightarrow Entailment(+d, pid)$$

---

[1] VerbOcean contains different relations between verbs.

The *Direct* variable holds the value $simdfeature$ with 1 if the verbs hold an entailment relation as given by the Directional Database [13];[2] 0 otherwise:

$$Direct(pid, +simdfeature) \Rightarrow Entailment(+d, pid)$$

The Markov network built from these formulas can then be queried for an entailment decision. For a new T-H pair, the model predicts a decision based on the type of arguments it has, the features of the words in the arguments, the alignment between its verbs, the relations between such verbs, and the shallow support features.

3   Experimental Results

We use the Alchemy[3] toolkit and the datasets from the RTE challenges 1-3 [7, 2, 10], which are publicly available, to evaluate our MLN model. To predict the entailment decision we take the marginal probabilities that Alchemy outputs for a given query, i.e., the $Entailment$ predicate. The query with the highest probability gives the entailment decision.

For a fair comparison, we evaluate our approach against previous work for RTE that is also based on alignment techniques. de Marneffe et al. [17] use a two-stage alignment similar to ours, but with dependency trees instead of SRLs. In addition, the entailment decision problem is represented with a vector of 54 features, where these features try to capture entailment and non-entailment by focusing on negations and quantifiers. Training and is performed using a logistic regression classifier. Chambers et al. [4] improve the alignment stage in [17] and combine it with a logical framework for the second stage [16]. The inference in the logical framework is expressed by a sequence of edits over texts expressions, where the edits represent operations that affect monotonicity over texts expressions. The logical framework maps alignments into a sequence of edits that defines the entailment decision. MacCartney et al. [15] propose a phrase-base alignment that uses external lexical resources. They improve the first stage via knowledge about semantic similarity and an extra, specific dataset for the training of the alignment stage.

Table 1. Accuracy against previous work based on alignment over the RTE datasets

Method	RTE-1	RTE-2	RTE-3
de Marneffe et al. [17]	-	60.5%	60.5%
Chambers et al. [4]	-	-	63.62%
MacCartney et al. [15]	-	60.3%	-
Relational Model	57%	55%	65%

Table 1 shows that our approach outperforms previous work for the RTE-3 dataset. However, the results are less positive for RTE-2. A possible reason for this error is the

---

[2] It contains directional lexical entailment rules.

[3] http://alchemy.cs.washington.edu/

low performance of our alignment technique. TINE only finds alignments for a subset of the test sets: 162 pairs (out of 287) for RTE-1, 463 pairs (out of 800) for RTE-2, and 385 pairs (out of 800) for RTE-3. Therefore, the proportionally fewer noisy alignments obtained for RTE-3 could have contributed to the better performance of the approach on this dataset. Another reason for the differences in performance across datasets can be the way the RTE datasets were built. RTE-3 contains longer T parts, with longer contexts, and therefore our method can find good quality alignments. This also seem to affect the overall performance of the participating systems, since the average accuracy (across all participating systems) for RTE-1 is 55%, while it is 59% for RTE-2, and 61% for RTE-3.

Our approach predicts a larger proportion of the *TRUE* class for RTE-3 than for RTE-2. There is a big gap between precision (54%) and recall (70%) for the RTE-3 dataset. Whereas for the RTE-2 this gap is smaller, with 52% precision and 57% recall. This behaviour could be because TINE finds more alignments for the *TRUE* pairs.

To further analyse the impact of poor alignment decisions, we test our model on the subsets of the datasets for which TINE produced an alignment. We compare the relational model only with the alignment features (i.e. without the shallow features) against a Support Vector Machine (SVM)-based approach. For the SVM algorithm, we compute a common and strong RTE baseline: the overlap of lemmas between T-H pairs as features, and use a linear kernel to learn the binary entailment decision [18]. Table 2 shows the results, where the relational model clearly outperforms the SVM model, and by a large margin on the RTE-3 dataset. This shows the potential of the relational features and MLNs for RTE.

Table 2. Accuracy on a subset of RTE 1-3 where an alignment is produced by TINE for T-H

Algorithm	RTE-1	RTE-2	RTE-3
SVM	50%	51%	56%
Relational model	57%	55%	78%

For a comparison covering the other main aspect of our approach – its probabilistic nature –, in a second evaluation experiment we compare our approach against other methods based on probabilistic modelling.

Glickman and Dagan [11] model entailment via lexical alignment, where the web co-occurrences for a pair of words are used to describe the probability of the hypothesis given the text. Harmeling [12] propose a model that, with a given sequence of transformations over a parse tree, keeps entailment decisions with a certain probability. Wang and Manning [22] merge the alignment and the decision into one step, where the alignment is a latent variable. The alignment is used into a probabilistic model that learns tree-edit operations on dependency parse trees. Beltagy et al. [3] extend the work in [8] to be able to process large scale datasets such as those from the RTE challenges. The method transforms distributional similarity judgments to weighted inference formulas, where the distributional similarity (i.e. If X and Y occur in similar contexts they describe similar entities) describes the degree of entailment between pairs.

Table 3. Accuracy against previous work based on probabilistic modelling over the RTE datasets

Method	RTE-1	RTE-2	RTE-3
Glickman and Dagan [11]	59%	-	-
Harmeling [12]	-	-	59.3%
Wang and Manning [22]	-	63%	61.1%
Beltagy et al. [3]	57%	-	-
Relational Model	57%	55%	65%

Table 3 shows a similar behaviour as the previous comparison: our approach leads to considerably better results on RTE-3, but lower performance for RTE-2. In addition, for the RTE-1 dataset, which has also been used by most of these other approaches, our relational model shows very competitive performance. In particular, it achieves the same performance as Beltagy et al. [3], which also use a MLN for the entailment decision.

4   Conclusions

We have described a proposal on using a relational statistical learning framework for the RTE task. Our experiments showed promising results. The main source of errors was found to be the alignment step, which has low coverage and can produce noisy alignments. However, we showed that when an alignment is found, the relational features improve the final entailment decision.

Future work includes improvements in the alignment stage as well as incorporating a more robust set of support features, such as using syntactic structures along with the semantic structures into a combined relational model. In other words, we could use different types of alignments (e.g., monolingual word alignment, syntactic alignment) that are based on heuristics, where the objective of the MLN formulas will be to penalise or reward the decisions made by different aligners. We also plan to define formulas that relate decisions across aligners.

Acknowledgments. This work was supported by the Mexican National Council for Science and Technology (CONACYT), scholarship reference 309261, by the QT-LaunchPad (EU FP7 CSA No. 296347) project, and Governments of Mexico and India (DST-CONACYT grant 122030; SIP IPN grant 20144534; COFAA-IPN).

References

[1] Androutsopoulos, I., Malakasiotis, P.: A survey of paraphrasing and textual entailment methods. J. Artif. Int. Res. 38(1), 135–187 (2010)

[2] Bar-Haim, R., Dagan, I., Dolan, B., Ferro, L., Giampiccolo, D., Magnini, B., Szpektor, I.: The second pascal recognising textual entailment challenge. In: Proceedings of the Second PASCAL Challenges Workshop on Recognising Textual Entailment, Venice, Italy (2006)

[3] Beltagy, I., Chau, C., Boleda, G., Garrette, D., Erk, K., Mooney, R.: Montague meets markov: Deep semantics with probabilistic logical form. In: Second Joint Conference on Lexical and Computational Semantics (*SEM), Proceedings of the Main Conference and the Shared Task: Semantic Textual Similarity, Atlanta, Georgia, USA, vol. 1, pp. 11–21 (June 2013)

[4] Chambers, N., Cer, D., Grenager, T., Hall, D., Kiddon, C., MacCartney, B., de Marneffe, M.C., Ramage, D., Yeh, E., Manning, C.D.: Learning alignments and leveraging natural logic. In: Proceedings of the ACL-PASCAL Workshop on Textual Entailment and Paraphrasing, pp. 165–170. Association for Computational Linguistics, Prague (2007)

[5] Chklovski, T., Pantel, P.: Verbocean: Mining the web for fine-grained semantic verb relations. In: Proceedings of EMNLP 2004, pp. 33–40 (2004)

[6] Collobert, R., Weston, J., Bottou, L., Karlen, M., Kavukcuoglu, K., Kuksa, P.P.: Natural language processing (almost) from scratch. Journal of Machine Learning Research 12, 2493–2537 (2011)

[7] Dagan, I., Glickman, O., Magnini, B.: The pascal recognising textual entailment challenge. In: Quiñonero-Candela, J., Dagan, I., Magnini, B., d'Alché-Buc, F. (eds.) MLCW 2005. LNCS (LNAI), vol. 3944, pp. 177–190. Springer, Heidelberg (2006)

[8] Garrette, D., Erk, K., Mooney, R.: Integrating logical representations with probabilistic information using Markov logic. In: Proceedings of the Ninth International Conference on Computational Semantics (IWCS 2011), pp. 105–114 (2011)

[9] Getoor, L., Taskar, B.: Introduction to Statistical Relational Learning (Adaptive Computation and Machine Learning). The MIT Press (2007)

[10] Giampiccolo, D., Magnini, B., Dagan, I., Dolan, B.: The third pascal recognizing textual entailment challenge. In: Proceedings of the ACL-PASCAL Workshop on Textual Entailment and Paraphrasing, Prague, pp. 1–9 (2007)

[11] Glickman, O., Dagan, I., Koppel, M.: A lexical alignment model for probabilistic textual entailment. In: Quiñonero-Candela, J., Dagan, I., Magnini, B., d'Alché-Buc, F. (eds.) MLCW 2005. LNCS (LNAI), vol. 3944, pp. 287–298. Springer, Heidelberg (2006)

[12] Harmeling, S.: An extensible probabilistic transformation-based approach to the third recognizing textual entailment challenge. In: Proceedings of the ACL-PASCAL Workshop on Textual Entailment and Paraphrasing, pp. 137–142. Association for Computational Linguistics, Prague (2007)

[13] Kotlerman, L., Dagan, I., Szpektor, I., Zhitomirsky-geffet, M.: Directional distributional similarity for lexical inference. Nat. Lang. Eng. 16(4), 359–389 (2010)

[14] Lin, D.: Automatic retrieval and clustering of similar words. In: Proceedings of the 36th Annual Meeting of the Association for Computational Linguistics and 17th International Conference on Computational Linguistics, Montréal, Canada, pp. 768–774 (1998)

[15] MacCartney, B., Galley, M., Manning, C.D.: A phrase-based alignment model for natural language inference. In: Proceedings of the 2008 Conference on Empirical Methods in Natural Language Processing, pp. 802–811. Association for Computational Linguistics, Honolulu (2008)

[16] MacCartney, B., Manning, C.D.: Natural logic for textual inference. In: Proceedings of the ACL-PASCAL Workshop on Textual Entailment and Paraphrasing, pp. 193–200. Association for Computational Linguistics, Prague (2007)

[17] de Marneffe, M.C., MacCartney, B., Grenager, T., Cer, D., Rafferty, A., Manning, C.D.: Learning to distinguish valid textual entailments. In: Proceedings of the Second PASCAL Challenges Workshop on Recognising Textual Entailment, Venice, Italy (2006)

[18] Mehdad, Y., Magnini, B.: A word overlap baseline for the recognizing textual entailment task (2009),
http://hlt.fbk.eu/sites/hlt.fbk.eu/files/baseline.pdf

[19] Richardson, M., Domingos, P.: Markov logic networks. Machine Learning 62(1-2), 107–136 (2006)

[20] Rios, M., Aziz, W., Specia, L.: TINE: A metric to assess MT adequacy. In: Proceedings of the Sixth Workshop on Statistical Machine Translation, Edinburgh, Scotland, pp. 116–122 (2011)

[21] Rios, M., Aziz, W., Specia, L.: UOW: Semantically informed text similarity. In: Proceedings of the Sixth International Workshop on Semantic Evaluation (SemEval 2012), Montréal, Canada, pp. 673–678 (2012)

[22] Wang, M., Manning, C.D.: Probabilistic tree-edit models with structured latent variables for textual entailment and question answering. In: Proceedings of the 23rd International Conference on Computational Linguistics, COLING 2010, Stroudsburg, PA, USA, pp. 1164–1172 (2010)

Annotation Game
for Textual Entailment Evaluation

Zuzana Nevěřilová

Natural Language Processing Centre, Faculty of Informatics,
Masaryk University, Botanická 68a, 602 00 Brno, Czech Republic

Abstract. Recognizing textual entailment (RTE) is a well-defined task concerning semantic analysis. It is evaluated against manually annotated collection of pairs hypothesis–text. A pair is annotated true if the text entails the hypothesis and false otherwise. Such collection can be used for training or testing a RTE application only if it is large enough.

We present a game which purpose is to collect h–t pairs. It follows a detective story narrative pattern: a brilliant detective and his slower assistant talk about the riddle to reveal the solution to readers. In the game the detective (human player) provides a short story. The assistant (the application) proposes hypotheses the detective judges true, false or non-sense.

Hypothesis generation is a rule-based process but the most likely hypotheses that are offered for annotation are calculated from a language model. During generation individual sentence constituents are rearranged to produce syntactically correct sentences.

The game is intended to collect data in the Czech language. However, the idea can be applied for other languages. The paper concentrates on description of the most interesting modules from a language-independent point of view as well as the game elements.

1 Introduction

Recognizing Textual Entailment (RTE) is defined in [6, p. 18]: "A text t entails a hypothesis h ($t \Rightarrow h$) if humans reading t will infer that h is most likely true." This definition of entailment is far more relaxed than a mathematical logic definition.

Although RTE seems to be defined loosely ("humans will infer", "most likely"), it is one of the most well defined problems in semantic analysis. RTE systems are evaluated by comparing their outputs with annotated pairs text–hypothesis (h–t pairs). Each pair is annotated either as true (if t entails h) or false (if t does not entail h).

A collection of h–t pairs can be built manually (similarly to reading comprehension tests for children and for adults[1]) but in natural language processing (NLP) automatic data gathering is preferred.

---

[1] e.g. OECD PISA http://www.oecd.org/pisa/

A. Gelbukh (Ed.): CICLing 2014, Part I, LNCS 8403, pp. 340–350, 2014.

[5] describes four scenarios leading to collecting of h–t pairs in RTE2 challenge[2]:

- IE – texts t were collected using structured template, relations tested in ACE-2004 RDR. Afterwards, hypotheses h were extracted from these texts using IE. These hypotheses have to be evaluated as positive (correct outputs) and negative (incorrect outputs) examples.
- IR – hypotheses h from evaluation datasets TREC and CLEF, texts t were selected from documents retrieved by various search engines.
- QA – transformation of answered questions to affirmations generated hypotheses h, original answers (extracted from the web by QA systems) serve as texts t.
- text summarization – a sentence occurring in summary was taken as t and simplified by removing sentence parts which leads to h.

All these retrieved h–t pairs went through manual annotation. In case of Czech language we cannot apply the same scenario since the appropriate tools are not available, so no evaluation set for recognizing textual entailment currently exist for Czech. The aim of this work is to build a considerable collection without using the above mentioned techniques.

1.1 Paper Outline

The paper is organized as follows: in section 2 we discuss the concept of collaboratively created language resources and compare our project with similar ones. In section 3 we present the game, discuss user experience w.r.t. annotation quality, and the game design. Section 4 presents the implementation and several modules that are employed to generate the hypotheses. Even though the game is in operation for a short time we present up to now results in section 5. Section 6 discusses the results and proposes future work.

2 Collaboratively Created Language Resources

Together with the rise of Web 2.0 the "collective intelligence" becomes an area of scientific interest. Non-expert users can be involved in many ways into expert tasks. [16] divides the collaboratively created language resources (CCLR) according to several criteria: motivation, annotation quality, setup effort, human participation and task character. The idea of CCLRs is based on collective "human computation" where peoples' brains are used for solving problems difficult for computer programs (such as natural language understanding or image content recognition) and relatively easy for people. Since GWAPs are games, the main motivation for contributors is the fun.

[16] split CCLRs into three categories: mechanized labor (such as Amazon Mechanical Turk), wisdom of the crowds (such as Wikipedia) and games with

---

[2] http://pascallin2.ecs.soton.ac.uk/Challenges/RTE2

a purpose (or GWAPs). Annotation GWAPs are of three basic kinds: output-agreement, input-agreement or inversion [1]. In all cases GWAPs are games for two (human) players who produce the annotation.

GWAP is a suitable model for demanding NLP tasks. Related works comprise:

- Common Sense Propositions [2] collected by *Verbosity*. One player describes a magic word to the second player whose aim is to guess the magic word only from these descriptions.
- Coreference Annotation [3] where players of *Phrase Detectives* annotate collaboratively coreferences. The game has two modes: annotation (where players select the appropriate coreferent pairs) and validation (where users validate previously annotated data).
- Paraphrase Corpora Collection [4] presents a game *1001 Paraphrases* where the doctors say something and the player has to say the same thing in other words.
- Semantic Relations Collection [14] present a categorization game collecting pairs object–category and a free association game (pairs word–associated word). The three games (*Categorilla*, *Categodzilla* and *Free Associations*) are based on real-life games. The data are available for download in text form. In the data from March 26, 2010 there are 745,030 pairs from the Free Associations and 1,199,235 pairs from Categorilla and Categodzilla.

In our case the players' task is somewhat similar to that in GWAPs. Unlike GWAPs the game is for one player, so no instant human feedback is present. The only case players receive feedback is when a proposition is annotated repeatedly. In this case the player earns points if the annotation corresponds to the majority of previous annotations.

One-player game has a great advantage over two-player games: we can cope with less participants (i.e. registered players). For collecting data in Czech language (spoken by about 10 million people) it is not easy to get a reasonably large worker base.

3   The Game

The game narrative refers to a dialogue between a detective and his/her assistant. The purpose of the dialogue is to explain the detective's reasoning to readers. Many players are familiar with this narrative pattern. In addition, the dialogue is always set in a friendly and open atmosphere even if the assistant is baffled. These conditions encourage players to annotate consciously.

The dialogue always starts with a story. The detective (human player) either provides a new story or returns back to a former story. The assistant Watsonson (application) tries to reformulate the story and entails new propositions. Afterwards, the detective can judge assistant's propositions as true or false in the given context. The basic screen with a sample dialog is shown in Figure 1.

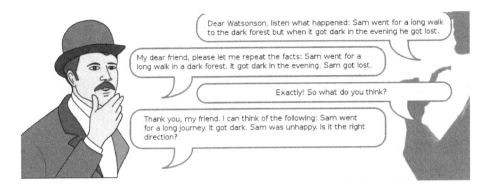

Fig. 1. The game environment is a dialogue between the detective and his assistant. N.B. the dialogue was translated in English by the author.

3.1 Data Complexity and Annotation Experience

Reading comprehension tests serve to test peoples' understanding capabilities. These tests are often considered difficult. The criticism of the game could confront the difficulty of reading comprehension tests and the lack of annotators' training. However, similarly to further semantic annotation projects, users are encouraged (by the instructions) to use their common sense to decide the annotation value. In addition, as the game advances trickiest entailments are generated. Users thus become experienced by playing the game repeatedly.

The data complexity in relation with CCLRs is discussed in [16, p.10]. According to the authors LR complexity means the data size as well as its characteristics relevant to annotation. In our case the annotation in simple yes/no decision. The data size for each h–t pair is quite small: the text consists usually of a few sentences, hypothesis a one sentence. Players are not forced to annotated every h–t pair. We suppose they prefer to annotate only clear cases.

With all this issues on mind we expect to obtain reasonable-quality annotation.

3.2 The Game Design

The game is designed as a conversational game. However, the player does not have to write much. Firstly, s/he enters a new story or obtains an old one then s/he only clicks to annotate or control the dialogue. The player can see the continuous dialogue (as shown in Figure 1) as well as popup windows with individual sentences and annotation buttons ✓ , ✗ or ☹.

The player earns points for a new story according to the number of clauses and phrases that have been identified by the syntactic parsing (story score). The player earns further points for every annotation and even more points for agreement with other annotators.

Fig. 2. Watsonson's emotions reflect the dialogue flow as well as the story score

Apart points and levels which are typical game elements two other elements are present in order to make the game fun. Firstly, the detective can encourage Watsonson to speak, appreciate him or reproach him. Secondly Watsonson's face reflect his emotions depending on the story score and the dialogue flow: he can be curious, thinking, thinking hard, happy, bored, annoyed, nosy, neutral or sad. Some of the emotions are shown in Figure 2.

4   Implementation

The game implementation is based on the integration of existing modules for natural language analysis and generation (such as morphological analyzer and syntactic parser) with new ones. It can be considered as a proof of concept of those existing "universal" software tools for processing the Czech language.

From the RTE's point of view the human player enters a text t, the computer player proposes several hypotheses h and the human player annotates the h–t pair. The hypothesis h vary from simple paraphrases (i.e. syntactic rearrangements) to real entailments (completely new sentences).

When the detective decides to return back to an older story, repeated annotations are obtained. The system encourages beginners to use this option.

4.1   Modules

For new hypotheses generation we use several modules from morphological and syntactic level processing (tokenization, disambiguation, parsing) to the semantic level. The modules for phrase re-ordering, synonym and hypernym replacement and verb frame inference are independent and are used in all possible orders to generate more hypotheses.

All semantic modules work on the phrase level (verb phrases, noun phrases, prepositional phrases, adverbial phrases, coordinations) not on word level. The stories and entailments are represented by no particular formalism (such as first order logic). Each clause is a verb phrase and a set of phrases dependent on the verb or with unknown parent (which typically applies to adverbials).

Table 1. Story representation: each sentence is divided in clauses, each clause is parsed on phrases. Phrases are marked according their syntactic roles: SUBJ(ect), VERB phrase, OBJ(ect), REFL(exive particle), ADV(erbial).

Sam šel na dlouhou vycházku do temného lesa, ale když se večer setmělo, ztratil se.										
Sam went for a long walk in a dark forest but when it got dark in the evening he got lost.										
Sam šel na dlouhou vycházku do temného lesa				ono se večer setmělo				Sam se ztratil		
Sam went for a long walk in a dark forest				it got dark in the evening				Sam got lost		
Sam	jít	(na) dlouhá vycházka	(do) temný les	on	se	večer	setmět	Sam	se	ztratit
Sam	go	(for) long walk	(in) dark forest	it		in the evening	get dark	Sam		get lost
SUBJ	VERB	OBJ	ADV	SUBJ	REFL	ADV	VERB	SUBJ	REFL	VERB

Parsing and Partial Anaphora Resolution. Players are asked to input a short story. We use syntactic parsing (SET parser [10]) to obtain phrases with known dependencies. The anaphora resolution system Saara [11] supplements unexpressed subjects and replaces demonstrative pronouns with their antecedents. Sentences are generated back from the set-of-phrases representation and they are offered for annotation. All other modules use the set-of-phrases representation. Example of the preprocessing can be viewed in Table 1.

Word Reordering. Czech is a (so called) free word order language i.e. nearly all orders of phrases are allowed. For this reason we prefer the term free phrase order. Every sentence is reformulated in all possible phrase orders. Various phrase orders never change the truth value but play a role in text cohesion. Since we generate isolated hypotheses we do not care about text cohesion. We only use the scoring module (see 4.1) to choose the most natural phrase order.

Synonym and Hypernym Replacement. We use Czech WordNet [15] for synonym replacement. The module replaces all word expressions found in Czech WordNet by their synonyms. No word sense disambiguation method is used therefore as a result false paraphrases are generated.

Since all transformations originators are recorded we can later discover WordNet synonyms unlikely in stories. For example *pes* has two senses: one corresponds to the synset `dog:1, domestic dog:1, Canis familiaris:1` in Princeton WordNet [7], the other corresponds to `martinet:1, disciplinarian:1, moralist:2`. A search in existing h–t pairs indicates the unlikely occurrence of the second sense

Table 2. Synonym replacement using Czech WordNet: "vycházka" (walk) was replaced by "výlet" (trip). N.b. the modifier "dlouhý" (long) had to be modified to fulfil the grammatical agreement with "výlet" (trip) because "vycházka" (walk) is feminine and "výlet" (trip) is masculine.

Sam	jít	(na) dlouhá vycházka	(do) temný les
Sam	go	(for) long walk	(in) dark forest
SUBJ	VERB	OBJ	ADV
Sam	jít	(na) dlouhý výlet	(do) temný les
Sam	go	(for) long trip	(in) dark forest

in stories. In fact, none of the hypotheses generated with the replacement *pes– moralista* (moralist) were judged true. An example synonym replacement is shown in Table 2.

Similarly to synonym replacement word expressions are replaced recursively by their hypernyms. In this case two restrictions apply. First, we do not replace word expression by all hypernyms but omit those from the WordNet Top Ontology. Such replacement (e.g. replace "student" by "living entity") will never generate a natural sounding expression. Second, we do not replace by hypernyms in sentences with negative polarity. While in positive sentences (such as "He came in his new coupe.") the hypernym replacement (replace "coupe" by "car") is valid, in negative sentences the replacement results always in false entailments ("He did not came in his new coupe." does not entail "He did not came in his new car."). In Czech double negative is used, so it is easier to detect correctly the sentence polarity in cases like "There was nobody in the classroom."

The hypernym replacement can generate sentences such as "Sam went for a long excursion.", "Sam went for a long journey." and "Sam went for a long travel.".

Verb Frame Inference. Word reordering and synonym replacement result in paraphrases while verb frame inference can result into new facts. In this module we take advantage of the Czech verb valency lexicon VerbaLex [9] and use verb valency frames for inferences of three types: equality, effect, precondition.

Verb frame inference is based on correct grammatical case recognition of all sentence constituents dependent on the verb or being without a parent (which applies mostly on adverbials). If the phrases and their cases are recognized correctly, the module obtains the verb plus its syntactic pattern, e.g. *be lost* + nominative:person + adverbial:non-person or *be lost* + nominative:person + *in* locative:non-person.

Subsequently the inference rules are used to transform a syntactic pattern to another pattern, e.g. *be lost* + nominative:person + adverbial:non-person → *be unhappy* + nominative:person. The inference rules were created manually, then augmented automatically using VerbaLex. The process of expansion is described in detail in [12].

For checking the phrase category constraints we use the shallow ontology Sholva. In Sholva [8] currently 154,783 words are classified into eight categories: substance, non-substance, person (incl. institutions), non-person, person-individual, non-person-individual, event, non-event. Each word can be member of more than one class. The annotation of Sholva has been done manually with multiple annotators.

The main advantages of Sholva are two: the size and the simplicity of the data. Using the category constraints we can distinguish verb frames with the same syntactic structure but distinct semantic slot categories. For example we can distinguish cases like *pass somebody on to somebody* (and infer they will communicate) and *pass something on to somebody* (and infer s/he will suffer). In many cases, distinguishing person and non-person is sufficient.

The overall process generates s from r using the following steps:

1. search for the syntactic pattern s in inference rules
2. for all rules $s \rightarrow r$: get syntactic pattern r_i
3. fill the sentence constituents from s to appropriate slots in r_i
4. check constraints with Sholva
5. if all slots are filled and constraints are satisfied generate a new sentence from r_i

An example verb frame inference is shown in Tables 3 and 4.

Table 3. The verb frame inference corresponds to the common sense inference "If someone gets lost s/he becomes unhappy"

Sam	se ztratil
Sam	got lost
SUBJ → SUBJ	ztratit se → být nešťastný
SUBJ → SUBJ	get lost → to be unhappy
Sam	byl nešťastný
Sam	was unhappy

Table 4. The verb frame inference corresponds to the common sense inference "If someone gets lost someone else will look for him"

Sam	se ztratil	
Sam	got lost	
SUBJ → OBJ(accusative)	ztratit se → hledat	
ϵ → SUBJ	get lost → look for	
někdo	hledal	Sama
somebody	looked for	Sam

Sentence Generation. When all transformations (that work with lemmata not with words) are done the generation module finds the appropriate word forms. Czech is a language with rich nominal inflection (different word forms for singular and plural as well as seven grammatical cases[3]). Verb conjugation has further intricacies (two main verb aspects, multi-word verb forms and reflexive particles). Moreover, grammatical agreements are needed between verb in past tense and the subject as well as between noun phrases and their modifiers.

The function of sentence generation module relies on correct recognition of the subject (which is always in nominative). According the subject's gender and number, the appropriate verb form is generated. Afterwards, all noun phrases' and prepositional phrases' modifiers are checked whether they fulfil the agreement on case, gender and number. For generation (i.e. finding a correct word form for a given lemma and a given tag) we use the morphological analyzer/generator majka [13].

Natural Sounding Sentences. The application produces tens to hundreds of hypotheses from each input sentence but not all of them are offered for annotation. We use a n-gram language model for calculating the most natural sounding sentence. Sentences of highest scores are offered for annotation. Apart from that a random sentence is sometimes selected for annotation to increase the collection diversity.

The appropriate n-gram frequencies were calculated using the Czes corpus[4]. The resulting score is calculated according to Equation 1 where $ngram_i$ means the i-th n-gram normalized frequency and m is the number of tokens. The normalization of each n-gram is calculated as shown in Equation 2. Here the raw frequency is normalized by corpus size and 100,000 and divided by raw frequencies of all tokens in the n-gram.

$$score = 10^2 \sum_{i=1}^{m-1} 2gram_i + 10^3 \sum_{i=1}^{m-2} 3gram_i + 10^4 \sum_{i=1}^{m-3} 4gram_i + 10^5 \sum_{i=1}^{m-4} 5gram_i \tag{1}$$

$$ngram = \frac{100000 freq_{ngram}}{corpsize \sum_{i=0}^{n} freq_i} \tag{2}$$

5   Evaluation

The project is currently in its final testing phase. Two testers inserted 275 reasonably long texts (at least one sentence, at least five words). The system generated 56,872 hypotheses. From these hypotheses 1,784 unique hypotheses were annotated (note the system strongly overgenerates) and 195 were annotated more

---

[3] Nominative, genitive, dative, accusative, vocative, locative and instrumental.
[4] 465,102,710 tokens in 2013-11-07.

Table 5. Multiple annotations

sum of annotation values	-2	-1	0	1	2	3	4	5	6	7
# hypotheses		10	351	184	1077	125	22	10	2	2	1

than once. The annotations were marked -1 when marked negative, 0 when confused and 1 when positive. The sum of repeated annotation values indicates the correctness of a hypothesis. When annotations of a particular hypothesis oscillate between true and false, the sum converges to 0 which means confused. Table 5 shows how many hypotheses received a particular sum of annotation values. Evidently, positive annotations predominate.

6   Conclusion and Future Work

We present an ongoing project of annotation game which aims to create a collection of h–t pairs for future Czech RTE system. The game is similar to GWAPs but it is only for one player. One-player games may be more suitable for collecting data in languages with minor speaker communities. Our outlook is to obtain in a few years a large collection of stories (thousands), hypotheses (tens of thousands) and their annotations as well as information about the way the hypotheses were generated.

The present results have shown which structures are preferred in the short detective stories. Some WordNet synonyms are not used in this kind of text (e.g. dog as martinet), some word orders are not used (verb in the initial position). Our future work will have two main directions. Firstly, we want to gradually reduce the generation from all possible to the most frequently annotated structures. Secondly, we need to keep the game still interesting even for experienced players. In the near future we want to add a knowledge base concerning famous detectives and their cases. We plan prospectively to add more types of inference about time and location.

Acknowledgements. This work has been partly supported by the Ministry of Education of CR within the LINDAT-Clarin project LM2010013 and by the Ministry of the Interior of CR within the project VF20102014003.

The access to computing and storage facilities owned by parties and projects contributing to the National Grid Infrastructure MetaCentrum, provided under the programme "Projects of Large Infrastructure for Research, Development, and Innovations" (LM2010005) is appreciated.

References

1. von Ahn, L., Dabbish, L.: Designing games with a purpose. Commun. ACM 51(8), 58–67 (2008), http://doi.acm.org/10.1145/1378704.1378719
2. von Ahn, L., Kedia, M., Blum, M.: Verbosity: a game for collecting common-sense facts. In: CHI 2006: Proceedings of the SIGCHI Conference on Human Factors in Computing Systems, pp. 75–78. ACM, New York (2006)

3. Chamberlain, J., Kruschwitz, U., Poesio, M.: Constructing an anaphorically annotated corpus with non-experts: Assessing the quality of collaborative annotations. In: Proceedings of the 2009 Workshop on The People's Web Meets NLP: Collaboratively Constructed Semantic Resources, People's Web 2009, pp. 57–62. Association for Computational Linguistics, Stroudsburg (2009), http://dl.acm.org/citation.cfm?id=1699765.1699774

4. Chklovski, T.: Collecting paraphrase corpora from volunteer contributors. In: Proceedings of the 3rd International Conference on Knowledge Capture, K-CAP 2005, pp. 115–120. ACM, New York (2005), http://doi.acm.org/10.1145/1088622.1088644

5. Dagan, I., Dolan, B., Magnini, B., Roth, D.: Recognizing textual entailment: Rational, evaluation and approaches. Natural Language Engineering 15(special issue 04), i–xvii (2009), http://dx.doi.org/10.1017/S1351324909990209

6. Dagan, I., Roth, D., Zanzotto, F.M.: Tutorial notes. In: 45th Annual Meeting of the Association of Computational Linguistics. The Association of Computational Linguistics, Prague (2007)

7. Fellbaum, C.: WordNet: An Electronic Lexical Database (Language, Speech, and Communication). The MIT Press (May 1998); published: Hardcover

8. Grác, M.: Rapid Development of Language Resources. Dissertation, Masaryk University in Brno (2013), http://is.muni.cz/th/50728/fi_d/

9. Hlaváčková, D., Horák, A.: VerbaLex – new comprehensive lexicon of verb valencies for Czech. In: Proceedings of the Slovko Conference (2005)

10. Kovář, V., Horák, A., Jakubíček, M.: Syntactic analysis using finite patterns: A new parsing system for Czech. In: Human Language Technology. Challenges for Computer Science and Linguistics, Poznań, Poland, November 6-8, p. 161 (2011); revised Selected Papers

11. Němčík, V.: Saara: Anaphora resolution on free text in Czech. In: Horák, A., Rychlý, P. (eds.) Proceedings of Recent Advances in Slavonic Natural Language Processing, RASLAN 2012, pp. 3–8. Tribun EU, Brno (2012)

12. Nevěřilová, Z., Grác, M.: Common sense inference using verb valency frames. In: Sojka, P., Horák, A., Kopeček, I., Pala, K. (eds.) TSD 2012. LNCS, vol. 7499, pp. 328–335. Springer, Heidelberg (2012)

13. Šmerk, P.: Towards Computational Morphological Analysis of Czech. Dissertation, Masaryk University in Brno (2010), http://is.muni.cz/th/3880/fi_d/

14. Vickrey, D., Bronzan, A., Choi, W., Kumar, A., Turner-Maier, J., Wang, A., Koller, D.: Online word games for semantic data collection. In: EMNLP 2008: Proceedings of the Conference on Empirical Methods in Natural Language Processing, pp. 533–542. Association for Computational Linguistics, Morristown (2008)

15. Vossen, P.: EuroWordNet: A Multilingual Database with Lexical Semantic Networks. Computers and the humanities. Springer (1998)

16. Wang, A., Hoang, C., Kan, M.Y.: Perspectives on crowdsourcing annotations for natural language processing. Language Resources and Evaluation 47(1), 9–31 (2013), http://dx.doi.org/10.1007/s10579-012-9176-1

Axiomatizing Complex Concepts from Fundamentals

Jerry R. Hobbs and Andrew Gordon

University of Southern California
Marina del Rey, California, USA

1 Introduction

We have been engaged in the project of encoding commonsense theories of cognition, or how we think we think, in a logical representation. In this paper we use the concept of a "serious threat" as our prime example, and examine the infrastructure required for capturing the meaning of this complex concept. It is one of many examples we could have used, but it is particularly interesting because building up to this concept from fundamentals, such as causality and scalar notions, highlights a number of representational issues that have to be faced along the way, where the complexity of the target concepts strongly influences how we resolve those issues.

We first describe our approach to definition, defeasibility, and reification, where hard decisions have to be made to get the enterprise off the ground. We then sketch our approach to causality, scalar notions, goals, and importance. Finally we use all this to characterize what it is to be a serious threat. All of this is necessarily sketchy, but the key ideas essential to the target concept should be clear.

2 Characterization and Defeasibility

In order to get started in encoding commonsense knowledge, one must build up a great deal of conceptual and notational infrastructure, and make a large number of warranted but highly controversial decisions about representation.

Among the first of these involves how tightly we can hope to define or characterize commonsense concepts. Our view is that where we can *define* a concept by necessary and sufficient conditions, that is good, but it is the exception rather than the rule. In general, the most we can hope to do is *characterize* concepts with lots of necessary conditions and lots of sufficient conditions. For example, we can't hope to define causality, but we can specify several key properties that follow from a causal relation between events, and we can list a great many pairs of causes and effects. By adding axioms, we constrain the set of possible interpretations of the predicates. It should be mentioned that this is not always done in efforts to encode commonsense knowledge. It is a common criticism that OpenCyc [14,3] is axiomatically poor; a prose description is given for a predicate that is introduced, but the set of axioms involving the predicate in general do not begin to constrain

A. Gelbukh (Ed.): CICLing 2014, Part I, LNCS 8403, pp. 351–365, 2014.

its interpretation to what the description says it means. In an early version of another popular large-scale ontology, the predicate "near" had only the property of being symmetric, which does not distinguish it from "far". In our effort, we have tried to focus on the axioms that delimit the meanings of predicates, rather than relying on the reader's intuition about the meaning of a term.

A related property of formalizations of commonsense knowledge is the defeasibility of the rules. Inferences can be drawn that subsequently must be retracted because of further information. The notation in which the axioms are expressed can be first-order logic, but there has to be a nonmonotonic proof procedure applied to them. The one we have assumed is weighted abduction [13], but our formalization could be adapted to any other approach to nonmonotonicity.

For a notation, we use a subset of Common Logic [2], essentially, textbook logic in a LISP-like format. In weighted abduction, it is possible to include "et cetera" predications in the antecedents of Horn clauses to indicate that other unspecified conditions may be relevant to the conclusion. Thus, an axiom saying that p defeasibly implies q might be written

```
(forall (x) (if (and (p x)(etc-i x)) (q x)))
```

where i is unique to this axiom. The "et cetera" predications can be thought of as the negations of McCarthy's abnormality predications in circumscriptive logic [16]. In weighted abduction, they are never proved, but they can be assumed and thereby become part of the best abductive proof of the goal expression.

In this paper, we will abbreviate an axiom like the above to

```
(forall (x) (if (and (p x)(etc)) (q x)))
```

where (etc) is understood to stand for an "et cetera" predicate unique to this axiom applied to all the universally quantified variables whose scope it is within. More generally, the reader can view (etc) as simply an indication of the defeasibility of the rule, to be dealt with by the nonmonotonic inference procedure of choice.

3   Reification and Eventuality Types and Tokens

The domain of discourse for our logical theories is the class of possible individual entities, states, and events. They may or may not exist in the real world, and if they do, it is one of their properties, expressed as (Rexist x). For example, in representing the sentence "John worships Zeus," both John and Zeus are in the universe of possible individuals, but only John *really* exists.

In a narrowly focused inquiry it is often most perspicuous to utilize specialized notations for the concepts under consideration. But our view is that in a broadbased effort like ours, this is not possible, and that it can be avoided by sufficient judicious use of reification.

For example, we treat sets as first-class individuals. Moreover, sets are taken to have "type elements", whose principal feature is that their properties are inherited by the real elements of the sets [15,8,11]. The expression (typelt x s) says that x is the type element of set s.

The term "eventuality" is used to cover both states and events [7,9]. Eventualities like other individuals can be merely possible or can really exist in the real world. A notational convention we use is that whereas the expression (p x) says that predicate p is true of x, the expression (p' e x) says that e is the eventuality of p being true of x. The relation between the primed and unprimed predicates is given by the axiom schema

```
(forall (x) (iff (p x)(exist (e)(and (p' e x)(Rexist e)))))
```

Eventuality arguments allow us to specify properties of eventualities without introducing scoping. "Pat has the goal of Chris's being happy" could be represented

```
(and (goal E P)(happy' E C))
```

That is, E is a goal of Pat's, where E is the eventuality of Chris's being happy.

Eventualities are very finely individuated. For example, Pat's walking to work and Pat's going to work are two different eventualities. The reason for this is that they may have different properties. The walking may be fast while the going isn't.

Eventualities are therefore very nearly in one-one correspondence with predications in the logic, and we can be somewhat cavalier about the distinction. For example, we can speak of the "arguments" of eventualities as a way of referring to the participants in the states or events. The expression (argn x 1 e) says that x is the first direct argument of e, and the expression (arg x e) says that x is some direct argument of e. Since eventualities can be the arguments, it is useful to define a recursive equivalent of "argument".

```
(forall (x e1)
   (iff (arg* x e1)
        (or (arg x e1)(exist (e2)(and (arg e2 e1)(arg* x e2))))))
```

Thus, in the above expression, C is an arg* of the eventuality of P's having goal E. We can think of (arg* x e) as saying that x is somehow involved in eventuality e.

We have explicitly modelled substitution in axioms (cf. [8]). The expression (subst x1 e1 x2 e2) can be read as saying that x1 plays the same role in eventuality e1 that x2 plays in eventuality e2, where e1 and e2 have the same predicate. Similarly, the expression (subst2 x1 y1 e1 x2 y2 e2) says that x1 and y1 play the same roles in eventuality e1 that x2 and y2 play in eventuality e2, respectively.

Eventualities can have type elements of sets as their arguments, and when they do, they are eventuality types. An instanceOf relation relates eventuality types and tokens. If e1 is an eventuality type whose only type element is x, the type element of set s, y is a member of s, and e2 is an eventuality such that (subst x e1 y e2) holds, then e2 is an eventuality token and an instance of e1.

Conjunctions, disjunctions, implications, and negations of eventualities are eventualities as well. The expression (not' e1 e2) says that e1 is the eventuality of eventuality e2's not really existing.

We have axiomatized a theory of time [12], and eventualities can have temporal properties. The expression (atTime e t) says that eventuality e occurs at time t. Thus, we use temporal properties rather than temporal arguments for eventualities.

The idea of reifying events is usually attributed to Davidson ([4]), although he was reluctant to reify states as well, and he did not individuate events as finely as we do. The linguist Emmon Bach ([1]) recognized the need for a concept that covered both states and events and introduced the term "eventuality". A brief exposition of eventualities as used here can be found in [7] and a more extensive exposition in [9]. The latter contains a number of arguments for the need for eventualities, ways of looking at eventualities, and arguments for very fine individuation.

4   Causality

The account of causality we employ is that of [10]. This distinguishes between the monotonic, precise notion of "causal complex" and the nonmonotonic, defeasible notion of "cause". The former gives us mathematical rigor; the latter is more useful for everyday reasoning and can be characterized in terms of the former. We begin with an abbreviated account of these concepts.

When we flip a switch to turn on a light, we say that flipping the switch caused the light to turn on. But many other factors had to be in place. The bulb had to be intact, the switch had to be connected to the bulb, the power had to be on in the city, and so on. We will use the predicate `cause` for flipping the switch, and introduce the predicate `causalComplex` to refer to the set of all the states and events that have to hold or happen for the effect to happen. The states of the bulb, the wiring, and the power supply would all be in the causal complex.

Causal complexes have two primary features. The first is that if all of the eventualities in the causal complex obtain or occur, then so does the effect. The second is that each of the members of the causal complex is relevant, in the sense that if it is removed from the set, the remainder is not a causal complex for the effect.

In practice, we can never specify all the eventualities in a causal complex for an event. So while the notion gives us a precise way of thinking about causality, it is not adequate for the kind of practical reasoning we do in planning, explaining, and predicting. For this, we need the defeasible notion of "cause".

In a causal complex, for most events we can bring about, the majority of the eventualities are normally true. In the light bulb case, it is normally true that the bulb is not burnt out, that the wiring is intact, that the power is on in the city, and so on. What is not normally true is that someone is flipping the light switch. Those eventualities that are not normally true are identified as causes. They are useful in planning, because they are often the actions that the planner or some other agent must perform. They are useful in explanation and prediction because they frequently constitute the new information. They are less useful for diagnosis, because diagnosis is employed exactly when the normal cause fails to bring about its normal effect, and the rest of the causal complex has to be examined.

In [10] the interpretation of the predicate `cause` is constrained by axioms involving the largely unexplicated notion of "presumable"; most elements of a causal complex can be presumed to hold, and the others are identified as causes.

We won't repeat that development here, but we will place some looser constraints on causes.

First, a cause is an eventuality in a causal complex.

```
(forall (e1 e2)
   (if (cause e1 e2)
      (exist (s)(and (causalComplex s e2)(member e1 s)))))
```

This allows only single eventualities to be causes, and of course many events have multiple causes. But this is not a limitation because we can always bundle the multiple causes into a single conjunction of causes. So if e1 is pouring starter fluid onto a pile of firewood and e2 is lighting a match, then the cause of the fire starting is e3 where (and' e3 e1 e2) holds.

The principal useful property of cause is a kind of causal modus ponens. When the cause happens or holds, then, defeasibly, so does the effect.

Causality is not strictly speaking transitive. Shoham ([17]) gives as an example that making a car lighter causes it to go faster, and taking the engine out causes the car to be lighter, but taking the engine out does not cause the car to go faster. In the second action, we have undone one of the presumable conditions in the causal complex for the first action. The two causal complexes are inconsistent. However, when they are consistent, cause is transitive, so it is defeasibly transitive.

```
(forall (e1 e2 e3)
   (if (and (cause e1 e2) (cause e2 e3) (etc))(cause e1 e3)))
```

Hobbs ([10]) is explicit about exactly what the content of the "et cetera" predicate is, in terms of presumable eventualities.

A causal complex consists of causes and other, presumable or nonproblematic, eventualities. The latter are frequently referred to as enabling conditions or preconditions. In the STRIPS model of Fikes and Nilsson [5] that has become the standard model for planning in artificial intelligence, the enabling conditions correspond to the preconditions and the body corresponds to the cause. The added and deleted states correspond to the effect.

5 Scales

A scale is a set of entities with a partial ordering among them.

```
(forall (s)
   (if (scale s)
      (exist (s1 e x y)
         (and (componentsOf s1 s)(partialOrdering e x y s)))))
```

The predicate componentsOf is explicated further in a theory of composite entities not discussed here [11]. The expression (partialOrdering e x y s) says that e, an eventuality type, is the relation of some x being less than some y, where x and y are components of the scale s.

```
(forall (e x y s)
   (if (partialOrdering e x y s)
      (and (scale s)(arg* x e)(arg* y e))))
```

In more conventional notation, we can think of **e** as a lambda expression and **x** and **y** as its two bound variables. However, since the **subst** predicate described above works equally well on types and tokens, we don't need to specify that **x** and **y** are variables, or types, or type elements, or anything else.

It is generally more convenient to speak directly of the partial ordering relation among elements. We can define a "less than" relation as follows, using the predicate name **lts** to indicate that it is relative to a particular scale **s**.

```
(forall (e1 x1 y1 s)
   (iff (lts' e1 x1 y1 s)
      (exist (e x y)
         (and (partialOrdering e x y s)
              (subst2 x y e x1 y1 e1)))))
```

Then the standard properties of partial orderings can be defined in terms of the predicate **lts**. The partial ordering is antireflexive, antisymmetric, and transitive. We also define the "less than or equal" relation.

We have frequent occasion to define particular scales. This is done by specifying the set of entities that are the components of the scale, and the relation that is the partial ordering of the scale.

```
(forall (s s1 e)
   (iff (scaleDefinedBy s s1 e)
      (and (scale s)(componentsOf s1 s)
           (exist (x y) (partialOrdering e x y s)))))
```

It is convenient to define a component of a scale, (**componentOf x s**), as a member of its set of components. We can define subscales and the top and bottom of a scale in the obvious way.

Suppose we have two scales with the same set of components. Then we can define a composite scale that is consistent with the two original scales. For example, suppose the set is points in the United States, in the first scale the partial ordering (in this case total) is "northOf", and in the second scale the partial ordering is "eastOf". Then in the composite scale the partial ordering is at least consistent with the "northAndEastOf" relation. We may in addition impose further structure on the composite scale, for example, by saying that the "northOf" relation takes precedence.

The loose constraints on a composite scale are as follows:

```
(forall (s s1 s2)
   (if (compositeScale s s1 s2)
      (and (exist (s3)
              (and (componentsOf s3 s1)(componentsOf s3 s2)
                   (componentsOf s3 s)))
           (forall (x y)
           (if (and (lts x y s1)(leqs x y s2))(lts x y s)))
           (forall (x y)
           (if (and (leqs x y s1)(lts x y s2))(lts x y s))))))
```

The same set **s3** is the set of components of the two original scales and the composite scale. If an entity **x** is less than an entity **y** on one of the original scales and less than or equal to **y** on the other, then it is less than **y** on the composite scale.

There is a range of structures we can impose on scales. These map complex scales into simpler scales. For example, in much work in qualitative physics the actual measurement of some parameter may be any real number, but this is mapped into one of three values – positive, zero, and negative. Where the parameter is vertical velocity, this means we are only interested in whether something is going up, staying at the same elevation, or going down.

We have introduced another sort of structure on scales, one reflected in language. What we have defined so far is adequate for characterizing the comparative and superlative forms of adjectives – "taller" and "tallest" – but not for the absolute form of adjectives – "tall". In natural language and in qualitative reasoning we often characterize something as being in the high or low region of a scale, with no more precise characterization of its location. We will call these regions the Hi and Lo regions of the scale. Each of these predicates is a relation between a scale s and one of its subscales s1 – (Hi s1 s). The top of the scale, if there is one, is the top of the Hi region of the scale, and the bottom of a scale is the bottom of its Lo region. The bottom of the Hi region and the top of the Lo region will rarely be known exactly. There is no well-established height that is the minimum height that counts as tall. Nevertheless, we can say that if a point is in the Lo region, then it is less than all the points in the Hi region.

It is often useful to go from the absolute form of an adjective to its underlying scale, for example, from "tall" to the height scale. We use the predicate scaleFor for this relation.

```
(forall (s e)
    (iff (scaleFor s e)
        (exist (s1)
            (and (Hi s1 s)
                (forall (e1 x)
                    (if (and (componentOf' e1 x s1)(argn x 1 e))
                        (iff (Rexist e)(Rexist e1)))))))))
```

For example, suppose we have (tall' e x), that is, e is the property of x's being tall. Then s is the height scale, s1 is the Hi region of the height scale, and whenever we have a relation e1 of x being in that Hi region, then e1 holds exactly when e holds. That is, some entity x is tall exactly when x is in the Hi region of the height scale. The height scale is the scaleFor the property "tall". In line 6 we specify that x must be the first argument of e, because if there are multiple arguments, we need to say which one is the relevant argument placed on the scale.

There are two primary external theories that a theory of the qualitative structure on scales should link to. The first is an as-yet-to-be-developed commonsense theory of distributions. The Hi and Lo regions usually correspond to the right and left tails of a distribution. As a first approximation, we can say that if something is in the Hi region of a scale, then defeasibly it is higher on the scale than *most* entities in some contextually deetermined comparison set.

The second is a theory of functionality or goals, as outlined in Section 6. Often when we say that an entity is tall, we mean that it is tall *enough* for something or *too* tall for something. Discovering that something is recognizing the connection between qualitative scalar judgments and functionality.

More specifically, we can say that, defeasibly, if something is in the Hi region of a scale, then that property plays a causal or enabling role in some agent's goal being achieved or not being achieved. We can state this as follows:

```
(forall (e x s1 s)
    (if (and (componentOf' e x s1)(Hi s1 s)(etc))
        (exist (c a g g1)
            (and (member e c)(goal g a)
                (or (causalComplex c g)
                    (and (not' g1 g)(causalComplex c g1)))))))
```

That is, if e is the property of x being in the Hi region s1 of some scale s, then defeasibly e is part of a causal complex c that will bring about some agent a's goal g or its negation g1. This axiom does not tell us who the agent is or what the goal is. That has to be determined from context. But it does alert us to the possible relevance of such a goal.

We axiomatize the notion of a "likelihood scale", as a qualitative, common-sense concept corresponding to standard probability and of which standard probability is one possible model. Space precludes presenting the details of this.

6   Agents and Goals

An agent is an entity that can, in the commonsense view of things, initiate a causal chain. People are agents. When someone decides to stand up and cross the room, there are neural events that are causing this, but we normally don't carry our analysis of the event to this level. We view the person's decision as the initial cause. Higher animals, organizations, and complex artifacts are also often viewed as agents.

Commonsense psychology is about people, but most of it applies more generally to agents. Agents have beliefs. We take the objects of belief to be eventualities. Because eventualities are very finely individuated, there is a straightforward translation between talking of belief in an eventuality and belief in a proposition. The expression (believe a e) can be read as saying that agent a believes the proposition that eventuality e really exists. We have developed but not included here our treatment of belief [6] because of space limitations and because it breaks no new ground in the abundant literature on logics of belief. Our use of the predicate here should be obvious and unproblematic.

Human beings are intentional agents. We have goals, we develop plans for achieving these goals, and we execute the plans. We monitor the executions to see if things are turning out the way we anticipated, and when they don't, we modify our plans and execute the new plans. The concept of a goal is central to this formulation.

The key concept in modeling intentional behavior is that of an agent a having some eventuality type e as a goal. The expression (goal e a) says that eventuality e is a goal of agent a. Normally, e will be an eventuality type that can be satisfied by any number of specific eventuality tokens, but it is entirely possible in principle for an agent to have an eventuality token as a goal, where there is only one satisfactory way for things to work out. We won't belabor the distinction here.

Agents know facts about what causes or enables what in the world, in most cases, facts of the form

```
(forall (e1 x)
    (if (p' e1 x) (exist (e2)(and (q' e2 x)(cause e1 e2)))))
```

and a similar axiom for `enable`. That is, if `e1` is the eventuality of `p` being true of some entities `x`, then there is an eventuality `e2` that is the eventuality of `q` being true of `x` and `e1` causes or enables `e2`. Or stated in a less roundabout way, `p` causes or enables `q`.

The agent uses these rules to plan to achieve goals and to infer the goals and plans of other agents. A plan is an agent's way of manipulating the causal properties of the world to achieve goals, and these axioms express causal properties.

We will work step by step toward a characterization of the planning process. The first version of the axiom we need says that if agent `a` has a goal `e2` and `e1` causes `e2`, then `a` will also have `e1` as a goal.

```
(forall (a e1 e2) (if (and (goal e2 a)(cause e1 e2))(goal e1 a)))
```

This is not a bad rule, and certainly is defeasibly true, but it is of course necessary for the agent to actually believe in the causality, and if the agent believes a causal relation that does not hold, `e1` may nevertheless be adopted as a goal. The causal relation needn't be true.

```
(forall (a e0 e1 e2)
    (if  (and (goal e2 a)(cause' e0 e1 e2)(believe a e0))(goal e1 a)))
```

We can say furthermore that the very fact that `a` has goal `e2` causes `a` to have goal `e1`. We do this by reifying the eventuality `g2` that `e2` is a goal of `a`'s, and similarly `g1`. (The e's in this axiom are the eventualities of having something; the g's are the eventualities of wanting it.)

```
(forall (a e0 e1 e2 g2)
    (if (and (goal' g2 e2 a)(cause' e0 e1 e2)(believe a e0))
        (exist (g1)(and (goal' g1 e1 a)(cause g2 g1)))))
```

That is, if agent `a` wants `e2` and believes `e1` causes `e2`, that wanting will cause `a` to want `e1`. (The belief is also in `g1`'s causal complex, but that would not normally be thought of as the cause: Why do you want `e1`? Because I want `e2`.)

Note that while the antecedent and the consequent no longer assert the real existence of having the goal (i.e., `g2` and `g1`), if we know that `g2` really exists, then the real existence of `g1` follows from the properties of `cause`.

Note also that the predicate `goal` reverses causality. For example, because flipping a light switch causes a light to go on, having the goal of the light being on causes one to want to flip the switch.

The eventuality `e1` is a "subgoal" of `e2`, and we encode this in the axiom.

```
(forall (a e0 e1 e2 g2)
    (if (and (goal' g2 e2 a)(cause' e0 e1 e2)(believe a e0))
        (exist (g1)
            (and (goal' g1 e1 a)(cause g2 g1)(subgoal e1 e2 a)))))
```

Finally, this axiom is not always true. There may be many ways to cause the goal condition to come about, and the mystery of the agent's free choice intervenes. The axiom is only defeasible. We can represent this by means of an "et cetera" proposition in the antecedent.

```
(forall (a e0 e1 e2 g2)
    (if (and (goal' g2 e2 a)(cause' e0 e1 e2)(believe a e0)(etc))
        (exist (g1)
            (and (goal' g1 e1 a)(cause g2 g1)(subgoal e1 e2 a)))))
```

That is, if agent a has a goal e2 (where g2 is the eventuality of wanting e2) and a believes e1 causes e2, then defeasibly this wanting e2 will cause a to want e1 as a subgoal of e2 (where g1 is the eventuality of wanting e1).

A similar succession of axioms can be written for enablement.

The "subgoal" relation is a relation between two goals, and implies the agent's belief that the subgoal is in a causal complex for the goal.

```
(forall (e1 e2 a)
    (if (subgoal e1 e2 a)
        (and (goal e2 a)(goal e1 a)
            (exist (e3 e4 e5 s)
                (and (causalComplex' e3 s e2)(member' e4 e1 s)
                    (and' e5 e3 e4)(believe a e5)))))))
```

In lines 5-6 of this axiom, e3 is the proposition that s is a causal complex for e2, e4 is the proposition that e1 is a member of s, e5 is the conjunction of these two propositions, and that's what agent a believes.

Goals do not have to be directly achievable by actions on the part of the agent, but successful plans have to bottom out in such actions or in states or events that will happen or hold at the appropriate time anyway.

It is formally convenient to assume that agents have one plan that they are always developing, executing, monitoring and revising, and that that plan is in the service of a single goal. We will call this goal "Thriving".

```
(forall (a) (if (agent a)(exist (e)(and (goal e a)(thrive' e a)))))
```

More specific goals arise out of the planning process using the agents' beliefs about what will cause them to thrive.

The main reason for positing this top-level goal is that now instead of worrying about the mysterious process by which an agent comes to have goals, we can address the planning problems of what eventualities the agent believes cause other eventualities, including the eventuality of thriving, and of what alternative subgoals the agent should choose to achieve particular goals. We are still left with the problem of when one goal should be given priority over another, but this is now a plan construction issue.

We will not attempt to say what constitutes thriving in general, because there are huge differences among cultures and individuals. For most of us, thriving includes staying alive, breathing, and eating, as well as having pleasurable experiences. But many agents decide they thrive best when their social group thrives, and that may involve agents sacrificing themselves. This is a common view in all cultures, as seen in suicide bombers, soldiers going into battle, and people risking death to aid others. So thriving does not necessarily imply surviving.

A good theory of commonsense psychology should not attempt to define thriving, but it should provide the materials out of which the beliefs of various cultures and individuals can be stated in a formal manner.

7   Importance

Many scales, including the scale of importance, cannot be defined precisely, but constraints can be placed on their partial ordering. That is what we will do here.

A concept, entity or eventuality is more or less important to an agent depending its relation to the agent's goals. The "more important" relation is thus a partial ordering that depends on the agent. The expression (moreImportant x1 x2 a) says that something x1 is more important than something else x2 to agent a. We place no constaints on the things x1 and x2 whose importance is being measured. They can be anything.

A plan can be thought of as a tree-like structure representing the subgoal relation. The higher a goal is in an agent's plan to thrive, the more important it is to the agent, because of the greater amount of replanning that has to be done if the goal cannot be achieved. So the first constraint we can place on the importance scale is that it is consistent with the subgoal relation.

However, this is a bit tricky to specify because an eventuality can be a subgoal of a number of different higher-level goals in the same plan, and we do not want to say an eventuality is of little importance simply because one of its supergoals is of little importance. So we first need to define the notions of an "upper bound supergoal" and a "least upper bound supergoal". An eventuality e1 is an upper bound supergoal of e2 if it is a supergoal of all of e2's immediate supergoals. More precisely, any supergoal of e2's must either be e1, be a subgoal of e1, or be a supergoal of e1.

```
(forall (e1 s a)
   (iff (ubSupergoal e1 s a)
        (and (agent a)(goal e1 a)
             (forall (e2) (if (member e2 s)(subgoal e2 e1 a)))
             (forall (e2 e) (if (and (member e2 s)(subgoal e2 e a))
                                 (or (subgoal e e1 a)(eq e e1)
                                     (subgoal e1 e a)))))))
```

The expression (ubSupergoal e1 s a) says that e1 is an upper bound supergoal of all the goals of agent a in set s. The predicate holds if and only if any eventuality e which is a supergoal of a member e1 of s is either a subgoal of e1, e1 itself, or a supergoal of e1.

A goal e1 is a least upper bound supergoal if it is an upper bound supergoal and a subgoal of all other upper bound supergoals.

```
(forall (e1 s a)
   (iff (lubSupergoal e1 s a)
        (and (ubSupergoal e1 s a)
             (forall (e)
                (if (ubSupergoal e s a)
                    (or (eq e e1)(subgoal e1 e)))))))
```

Because every goal is ultimately in the service of the top-level goal "To Thrive", every goal has a least upper bound supergoal.

Now we can say that if eventuality e1 dominates eventuality e2 on every path in the agent's plan that includes e2, then e1 is more important than e2. Every reason for wanting e2 is in the service of e1.

```
(forall (s e1 e2 a)
    (if (and (member e2 s)(lubSupergoal e1 s a))
        (moreImportant e1 e2 a)))
```

More generally,we can say that the least upper bound supergoal of a set of goals is more important than the whole set, since all the members of the set are in the service of the supergoal.

```
(forall (s e a)
    (if (lubSupergoal e s a)(moreImportant e s a)))
```

An agent's goals are important. So are eventualities that affect the agent's goals. Importance doesn't care about polarity; if passing a course is important to you, so is not passing the course. Thus, we define an eventuality as "goal-relevant" to an agent, (goalRelevant e a), if its existence implies the existence or nonexistence of one of the agent's goals. The "goal consequences" of an eventuality, (goalConsequences s e a), are those goals of the agent's whose existence or nonexistence is implied by by the eventuality. Then we can say the importance of an eventuality depends on the importance of its goal consequences. That is, if something x is more important than the goal consequences of eventuality e, then x is more important than e.

The importance of an entity depends on the importance of its properties and of the events it participates in. Thus, we define the set of "goal-relevant properties".

```
(forall (s x a)
    (iff (grProps s x a)
        (forall (e)
            (iff (member e s)(and (arg* x e)(goalRelevant e a)))))))
```

The expression (grProps s x a) says that s is the set of the set of properties of x that are relevant to a goal of qa's.

The next axiom says that the importance of an entity depends on the importance of its goal-relevant properties.

```
(forall (s x a)
    (if (and (moreImportant s x1 a)(grProps s x2 a))
        (moreImportant x2 x1 a)))
```

To summarize, x1 is more important than x2 to a if x2 is, or affects something that is, or has properties that affect something that is, in the service of x1. Note that there may be other properties constraining the moreImportant relation, but this one at least is among the most significant.

8   Threats and Seriousness

A threat situation is one in which the agent believes that things could turn out badly. To formalize this notion, we say that the agent a believes there is an eventuality e in a causal complex s whose effect e0 would cause one of a's goals not to be realized. The definition of a threat situation is as follows:

```
(forall (e s e0 a g)
   (iff (threatSituation e s e0 a g)
        (exist (e1 e2)
           (and (believe a e1)(member' e1 e s)(causalComplex s e0)
                (cause e0 e2)(not' e2 g)(goal g a)))))
```

The expression (threatSituation e s e0 a g) says that e is a threat by virtue of being a member of causal complex s with effect e0 which threatens agent a's goal g. In line 4 e1 is the eventuality of e's being in the causal complex; agent a believes this to be the case. In line 5 e2 is the eventuality of goal g not obtaining.

Threats may be real, imaginary, or just imagined. These would be further properties of e and perhaps s.

The participating entities in a threat situation can be labelled by their role. In particular, the initiating event e is the threat.

```
(forall (e s e0 a g)
   (if (threatSituation e s e0 a g)(threat e a)))
```

That is, e is the threat to agent a.

Because the threat is only one element of the causal complex, the effect is only a possibility, not an inevitability. We generally refer to something as a threat when some evasive action is still possible.

Because of space considerations, this treatment of threats is actually somewhat simplified from that presented in [6]. There rather than mere belief, we have the agent in the dynamic process of envisioning a branching causal structure, one branch of which leads to the undesirable consequence.

A threat can be more or less serious to an agent. Seriousness is a composite scale that depends on importance and likelihood. Of two equally likely threats, the more important one is more serious. Of two equally important threats, the more likely is more serious. This composition of scales is captured in the predicate compositeScale introduced in Section 5. The likelihood scale was introduced in Section 5. The importance scale was introduced in Section 7. Hence, the definition of the "seriousness" scale is as follows:

```
(forall (s a)
   (iff (seriousnessScale s a)
        (exist (s1 s2)
           (and (likelihoodScale s1)(importanceScale s2 a)
                (compositeScale s s1 s2)
                (forall (e1)
                   (if (componentOf e1 s)(threat e1 a)))))))
```

Lines 4-5 say that the seriousness scale is a composite of the likelihood scale and the importance scale. Lines 6-7 say that the elements of the scale are threats.

In terms of this we can define the partial ordering "more serious".

```
(forall (e e1 e2 a)
   (iff (moreSerious' e e1 e2 a)
        (exist (s s1)
           (and (seriousnessScale s a)(scaleDefinedBy s s1 e)))))
```

The expression (moreSerious' e e1 e2 a) says that e is the eventuality of one threat e1 being more serious than threat e2 to agent a. The predicate scaleDefinedBy, introduced in Section 5, relates a scale to its set of elements and its partial ordering.

We can define the qualitive predicate serious in terms of the Hi region of the scale, via the predicate scaleFor.

```
(forall (e e1 a)
    (iff (serious' e e1 a)
        (exist (s)
            (and (seriousnessScale s a)(scaleFor s e)))))
```

The expression (serious' e e1 a) says that e is the eventuality of threat e1 being serious to agent a. The predicate scaleFor relates a scale to the qualitative predicate of something being in the Hi region of the scale.

Thus, for a threat to an agent to be serious, it has to be high on a scale of seriousness, where "high" is related to the agent's goals, most likely in this case the goals of the agent's that are the focus of the threat. Seriousness in turn is characterized in terms of likelihood and importance of the threatened action. Importance is characterized by the action's causal impact on a goal of the agent's and that goal's place in the agent's subgoal hierarchy, or plan, for achieving the overarching goal of thriving.

9   Conclusion

Sophisticated natural language understanding will require a large knowledge base of commonsense knowledge. Much of that can be acquired automatically from large corpora and existing resources. But the very core of such a knowledge base will consist of rules that are too abstract and too complex for any automatic acquisition methods we can imagine today. It will require the manual encoding of theories such as we have described here for causality, scales, and goals. However, once these theories are explicated, some complex and deep accounts of meanings of difficult words can be constructed. We have demonstrated what this would look like for two particularlly difficult words, "serious" and "threat". These are just two representative examples of the knowledge we need and are in the position to begin to provide.

References

1. Bach, E.: On Time, Tense, and Aspect: An Essay in English Metaphysics. In: Cole, P. (ed.) Radical Pragmatics, pp. 63–81. Academic Press, New York (1981)
2. Common Logic Working Group. Common Logic Standard (2010), http://common-logic.org/
3. Cycorp (2008), http://www.cyc.com/
4. Davidson, D.: The Logical Form of Action Sentences. In: Rescher, N. (ed.) The Logic of Decision and Action, pp. 81–95. University of Pittsburgh Press, Pittsburgh (1967)

5. Fikes, R., Nilsson, N.: STRIPS: A New Approach to the Application of Theorem Proving to Problem Solving. Artificial Intelligence 2, 189–208 (1971)
6. Gordon, A., Hobbs, J.: How We Think We Think: Formalizing Commonsense Psychology (2010) (in progress)
7. Hobbs, J.: Ontological Promiscuity. In: Proceedings of the 23rd Annual Meeting of the Association for Computational Linguistics, Chicago, Illinois, pp. 61–69 (1985)
8. Hobbs, J.: Monotone Decreasing Quantifiers in a Scope-Free Logical Form. In: van Deemter, K., Peters, S. (eds.) Semantic Ambiguity and Underspecification, Stanford, California. CSLI Lecture Notes No. 55, pp. 55–76 (1995)
9. Hobbs, J.: The Logical Notation: Ontological Promiscuity (2003), http://www.isi.edu/~hobbs/disinf-tc.html
10. Hobbs, J.: Toward a Useful Concept of Causality for Lexical Semantics. Journal of Semantics 22, 181–209 (2005)
11. Hobbs, J.: Encoding Commonsense Knowledge (2005), http://www.isi.edu/~hobbs/csk.html
12. Hobbs, J., Pan, F.: An Ontology of Time for the Semantic Web. ACM Transactions on Asian Language Information Processing 3(1), 66–85 (2004)
13. Hobbs, J., Stickel, M., Appelt, D., Martin, P.: Interpretation as Abduction. Artificial Intelligence 63(1-2), 69–142 (1993)
14. Lenat, D., Guha, R.: Building Large Knowledge-Based Systems: Representation and Inference in the Cyc Project. Addison-Wesley (1990)
15. McCarthy, J.: First Order Theories of Individual Concepts and Propositions. In: Hayes, J., Michie, D., Mikulich, L. (eds.) Machine Intelligence 9, pp. 129–147. John Wiley and Sons, New York (1979)
16. McCarthy, J.: Circumscription: A Form of Nonmonotonic Reasoning. Artificial Intelligence 13, 27–39 (1980)
17. Shoham, Y.: Nonmonotonic Reasoning and Causation. Cognitive Science 14, 213–252 (1990)

A Semantics Oriented Grammar
for Chinese Treebanking

Meishan Zhang[1], Yue Zhang[2], Wanxiang Che[1], and Ting Liu[1,*]

[1] Research Center for Social Computing and Information Retrieval
Harbin Institute of Technology, China
{mszhang,car,tliu}@ir.hit.edu.cn
[2] Singapore University of Technology and Design
yue_zhang@sutd.edu.sg

Abstract. Chinese grammar engineering has been a much debated task. Whilst semantic information has been reconed crucial for Chinese syntactic analysis and downstream applications, existing Chinese treebanks lack a consistent and strict sentential semantic formalism. In this paper, we introduce a semantics oriented grammar for Chinese, designed to provide basic supports for tasks such as automatic semantic parsing and sentence generation. It has a directed acyclic graph structure with a simple yet expressive label set, and leverages elementary predication to support logical form conversion. To our knowledge, it is the first Chinese grammar representation capable of direct transformation into logical forms.

Keywords: Chinese Semantic, Semantic Representation, Chinese Treebanking.

1 Introduction

Chinese treebanking has been a much debated issue, largely due to the uniqueness of the language [1–5]. Similar to English, Chinese is an isolating language, for which meaning is defined over relatively rigid phrase structures, rather than rich morphology [6]. On the other hand, Chinese has relatively much less function words, and much more means of phrase construction, which makes its structural disambiguation a more challenging task. Much often, the resolution of syntactic ambiguities needs to resort to semantic interpretations.

Figure 1 shows an example, where the syntactic structure of "外商 (foreign capital) 投资 (investment) 企业 (business)" can be determined *only* by referring to the meaning of content words in the rest of the sentences. According to phrase structure syntax [2], the two sentences can be treated either as topicalized sentences, in which the underlined phrases serve as the topic, or as subject-predicate sentences that have a sentential (NN-VA) predication.

The example reflects the degree of flexibility in Chinese sentence construction, where patterns such as topicalization and pro-drop are quite common. As a

---

* Corresponding Author.

A. Gelbukh (Ed.): CICLing 2014, Part I, LNCS 8403, pp. 366–378, 2014.
© Springer-Verlag Berlin Heidelberg 2014

Fig. 1. Syntactic ambiguities requiring semantic knowledge to resolve

Fig. 2. An example contrast between syntactic and semantic headedness

result, the best accuracies of Chinese syntactic parsing is significantly below those of English [6–9], despite availability of large-scale syntactic treebanks. This makes the extraction of semantic information less accurate, given the fact that semantic role labeling is commonly performed on top of syntactic structures.

There have been attempts at constructing Chinese treebanks that are more semantics driven [4, 10]. By defining semantic relations directly over words, these treebanks allow statistical parsers to build semantic links directly from POS-tagged data, hence avoiding error propagation in pipeline syntactic and semantic analysis. Such treebanks typically differ from their syntactic counterparts in two ways. First, head words in dependency arcs are semantic rather than syntactic. Second, dependency labels are defined over semantic instead of syntactic relations.

Figure 2 shows an example of syntactic and semantic headedness. In (a), "美国 (American)" takes the syntactic head "的 (of)", which governs the syntactic constituent DCP. In (b), "美国 (American)" takes the semantic head "华人 (Chinese)" instead. This semantic headed link form is relatively more informative to downstream applications such as machine translation [11]. It also enjoys robustness over paraphrasing. For example, the three phrases "美国 (American) 华人 (Chinese)", "美国 (American)的(of) 华人 (Chinese)" and "在 (at) 美国 (American) 的 (of) 华人 (Chinese)" have very different syntactic structures, with the underscored modifier phrase being NP, DCP and LCP, respectively [12]. However, in the semantic headed format, the link between '美国 (American)" and "华人 (Chinese)", which bares the invariant meaning, remains the same across the three paraphrases.

These dependency treebanks, however, have two significant limitations. First, they are constructed ad-hoc over syntactic treebanks, and do not have a strict separation between syntax and semantics. Take Che et al. [4] for example, they offer 123 detailed labels to replace the original syntactic labels, making disambiguation

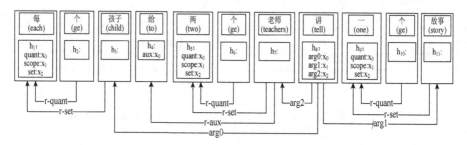

Fig. 3. The semantic representation for the Chinese sentence "每个孩子给两个老师讲一个故事 (Every child tells a story to two teachers)"

difficult[1]. On the other hand, these labels are ad hoc, and ones such as "attribute" and "coordinate" can be treated as syntactic.

Second, lack of a strict semantic formulation makes these treebanks unsuitable for wider downstream semantic tasks such as logical inference. In contrast, even syntactic formulations including CCG [13], LFG [14] and HPSG [15] allow the transformation of syntactic derivations into logical forms.

Driven by the above needs, we propose a semantics oriented grammar formalism for Chinese treebanking. The formalism uses a direct acyclic graph to represent sentential semantics, for which existing parsing technology is available [16]. In contrast to the aforementioned Chinese treebanks, the formalism is constructed following a strict semantics structure and allows transformation into logical forms. In addition, the number of arc labels is much smaller, allowing efficient parsing yet maintaining semantic expressiveness. We adopt Propbank-style predicate argument structures, yet extend elementary predications from verbs to quantifiers, adjectives and adverbs.

2   The Semantic Representation

The semantic framework we propose is lexicalized. However, instead of building semantic relations on the lexicons directly, we first project each word in the input sentence to an *elementary predication* (EP). The concept of EP is proposed in Minimal Recursion Semantics (MRS) [17] as the primary units of computational semantics, where each word can correspond to different EPs under different contexts. We adopt this notation. The semantic links in our framework are defined over EPs, as illustrated by the example in Figure 3.

2.1   Elementary Predication

EPs serve as the basic semantic frames for lexicons. By transforming words into EPs, semantic ambiguities such as predicate-argument structures, quantifier scopes and pronominal references can be resolved more easily.

---

[1] See Related Work for a more detailed discussion.

Definition 1 (Elementary Predication (EP)).
An elementary predication is composed by the following three components:

- Predicate: usually the word itself.
- Handle: the label of the EP.
- Arguments: each EP can have a list of zero or more variables arguments. These arguments denote the predicate's core semantic role. Each argument has a semantic label that describes its relation to the predicate.

Each EP can be written as $predicate:handle(label_0:x_0, \cdots, label_n:x_n)$. For example, the word "讲 (tell)" in the example sentence can be projected into the EP: "讲:h_8(agent:x_0, target:x_1, content:x_2)", as shown in Figure 3. The EP arguments can be none also, when the word does not accept any arguments at all (e.g., nouns denoting concrete objects).

EP structures are very similar to the predicate-argument structures in Propbank [18] and Chinese Propbank [19], but they have two main differences. First, the predicate of an EP is more general than that of Propbank. Propbank models only the propositions in a sentence, while EP regards every word in a sentence as a potential predicate, including quantifiers, adjectives, adverb, and expletives. In this way, EPs can handle more semantic relations than Propbank. For example, when we encode the scope attribute as a core semantic role of numeral words, numeral words' quantifier scopes can be expressed.

Second, EPs stress the integrity of their composition. A word can have only a finite number of EP structures. In a sentence, one instance of a word can take only one EP. If the value of a semantic role in an EP cannot be assigned, the semantic structure is incomplete. In contrast, without the EP structure, a semantic role labeler can neglect the framesets that Propbank defines, resulting in incomplete predicate structures or confused ellipsis phenomena.

Compared with the EPs in MRS [17], our EPs are much simpler. For example, we do not have the scope attribute as an indispensable element for every EP, since in most cases the scopes are directly reflected by the semantic relations. For those words that do have scope ambiguities, we add the scope as an argument in their EPs.

2.2 EP Arguments

There is a fixed set of EP arguments in our representation, as listed below.

- Proposition. We define the arguments of proposition words using the same method as Propbank. There are five core arguments and 14 function arguments. For more detailed descriptions, refer to Xue and Palmer [19].
- Auxiliary. We define a special argument named *aux* for the Chinese words that do not directly bare a meaning in the sentence. For example, punctation words, "的 (de, possessive marker)" and "被 (bei, passive marker)" are of this type. The aux links do not bare any meanings. However, the existence of auxiliary words influences the semantic roles of other words.

- Quantifier. We have three arguments for quantitative words: *quant, scope* and *set*. The Chinese measure words are special and seldom exist in other languages. For example, the bold words in "一**棵**树 (a tree)" and " 一**个** 人 (a person)" are measure words. We use *quant* to denote their semantics. The words with the most scope ambiguities are probably quantitative words, and we add the argument *scope* to these words. The argument *set* is used to denote the nominal word that the quantifier modifies.
- Coordination. Conjunction words such as " 和 (and)" and " 或 (or)" are not semantic heads in the formalism. We use two arguments, *conj* and *entity*, on the right-most content word of the conjunction phrase to denote its semantics. *conj* denotes the conjunction words, while *entity* denotes coordinated entities.
- Anaphora. We use the argument *refer* to denote the reference to a pronoun's semantic head.
- Interrogative. We define two arguments to represent the semantic of interrogative words in a question, namely *interrog* and *answer*. The argument *interrog* is defined on the main propositional words of a sentence, and links to the interrogative words, while the argument *answer* denotes any answer found in the current context.

The six categories cover most semantic relations of Chinese sentences. Due to inherent ambiguities and mistakes in statistical parsing, we allow some arguments to be underspecified. For example, the scope argument of quantifiers can be unfilled in a sentence. We will give detailed examples of typical arguments in Section 3.

2.3   The Sentential Structure

The building of dependencies is the assignment of arguments for EPs, where an EP may have multiple heads. Definition 2 gives a formal definition of the formalism.

Definition 2 (Sentential Structure).
The grammatical structure for a given input sentence is a labeled direct acyclic graph (DAG) $G = (V, E, R)$, which satisfies the following constraints:

- $V =< \mathrm{EP}_1, \cdots, \mathrm{EP}_n >$. The nodes of the graph G are a sequence of EPs. Each word in the sentence is mapped into an EP. Thus the number of nodes in G is equal to the number of words in a sentence.
- Each edge e_k between EP_i and EP_j $(i \neq j)$ in E is associated with an argument (label$_m$:xm). It serves as a directed semantic link from EP_i to EP_j, with EP_i being the head. The label of e_k is either label$_m$ or r-label$_m$. If the argument of the edge belongs to EP_i, then the e_k is labeled as label$_m$, otherwise it is labeled r-label$_m$, indicating an argument direction that is reverse to the head→modifier direction.

This formalism builds role-filling links between EPs. Each edge gives the true value for one argument of an EP. For example, in Figure 3, the edge "老师:h$_7$()"

$\xleftarrow{\text{arg2}}$ "讲:h_8(arg0:x_0, arg1:x_1, arg2:x_2)" assign the value "老师:h_7()" to the argument $target$:x_1 of h_8. The directed acyclic graph constraint can prevent infinite loops in the determination of the value of an EP, and allow efficient parsing algorithms to be applied.

2.4   Logic Interpretation

In the same way as syntactic grammatical relations, semantic relations from our formalism can be used as features for downstream applications, such as question answering and machine translation. As discussed in the introduction, semantic relations can potentially be more informative than their syntactic counterparts, and our grammar shares the motivation of Che et al. [4] in exploiting this advantage. One important advantage of our grammar is the support of logic interpretation, and hence it can also be used for tasks such as parsing into logical forms [20] and surface realization [21]. In this section, we illustrate how the EP-based structures can be transformed into Neo-Davidsonian first-order logic. Similar methods can be used for transformation into other logical forms.

The conversion from the DAGs into logic is rather straightforward, thanks to EPs. In particular, a propositional EP is associated with a lambda calculus expression, where the λ-free variable is used to represent an event, and a set of (zero or more) λ-bound variables are defined for its arguments. In the example in Figure 3, h_8 for "讲 (tell)" can be associated with the lambda term $\lambda x_0 \lambda x_1 \lambda x_2 \exists e_0 \; tell(e_0) \wedge arg0(e_0, x_0) \wedge arg1(e_0, x_1) \wedge arg2(e_0, x_2)$.

EPs for nominal contents are associated with only bound variables. For example, h_{11} for "故事 (story)" can be associated with the lambda term $\lambda y_0 \; story(y_0)$. Quantifier EPs are associated with logical quantifiers and a constant term 1. For example, h_5 for "两 (two)" can be associated with the term $2x.1$. EPs for measure words and auxiliaries are correlated with the constant term 1 in first-order logic. A sentential logical expression is derived by traversal of the acyclic dependency graph, performing logical conjunction and beta-reduction on each link. Scopes of quantifiers are decided by the $scope$ links, but can also be underspecified when the link is undecided. For example, the $r\text{-}set$ link between h_1 and h_3 results in $\forall x_0 \; student(x_0)$. As the $scopes$ are undecided in the example, the final logic form of the sentence can have several different interpretations, including $(\forall x_0 2x_1 \exists x_2)(\exists e_0 \; student(x_0) \wedge teacher(x_1) \wedge tell(e_0) \wedge story(x_2) \wedge arg0(e_0, x_0) \wedge arg1(e_0, x_1) \wedge arg2(e_0, x_2))))$ and $(\forall x_0 (2x_1 (\exists x_2 \exists e_0 \; student(x_0) \wedge teacher(x_1) \wedge tell(e_0) \wedge story(x_2) \wedge arg0(e_0, x_0) \wedge arg1(e_0, x_1) \wedge arg2(e_0, x_2))$. If all the $scopes$ are linked to the EP of "讲 (tell)", the former is the corresponding logic form.

3   Case Studies

In this section, we give a set of example grammatical constructions based on different sentence types. In order to analyze a given Chinese sentence, two steps must be taken.

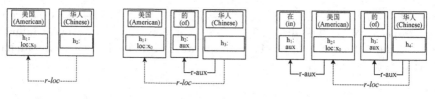

(a) "美国华人? (American Chinese)"   (b) "美国的华人? (American Chinese)"   (c) "在美国的华人? (The Chinese in American)"

Fig. 4. The semantic representation for modification

I. **EP Identification**. This process is analogous to the supertagging step in lexicalized grammar parsing [22].

II. **Edge Construction**. Given an EP, one needs to assign values to its arguments. The value of an EP argument is another EP. If found, we will link the two EPs, deciding the link direction accordingly.

Common semantic phenomena include modification, proposition, coordination, quantifier, anaphora and question. We give case studies to their representations, respectively.

3.1   Modification

As mentioned in the introduction, one benefit of using semantic formalism is the better handling of paraphrases. The example in Figure 2 can be expressed in our formalism in Figure 4. In all three phrases, "美国 (American)" is a modifier of "华人 (Chinese)", indicating the location, regardless of the function words.

3.2   Proposition

Propositional words are the most essential for sentential semantics. The EP structure of a propositional word can be denoted by

$$\text{predicate:handle}(\text{arg0:}x_0, \cdots, \text{argM:}x_M), M \leq 4$$

For each $K \leq M$, argK:x_K must be in the EP structure. The arguments arg0~arg4 refer to the core EP arguments of propositional words.

Another type of attributes for propositional words are function arguments. Different from the core arguments which are linked with content words, function arguments are defined on function words. As functional words are usually modifiers of propositional words, the edge labels between functional words and propositional words start with the mark "r-" (Sec 2.3). Functional words can have multiple heads since they can modify multiple propositional words.

Figure 5 shows an example semantic representation of this case. There are two propositional words in the example: "吃饭 (eat)" and "讨论 (discuss)".

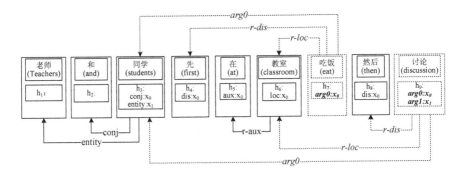

Fig. 5. The semantic representation for "老师和学生先在教室吃饭然后讨论 (Teachers and students eat in the classroom before discussion)"

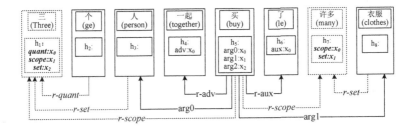

Fig. 6. The semantic representation for "三个人一起买了许多衣服 (The three persons bought many clothes together)"

The EP of the former has one argument, while that of the latter has two arguments, with the value of the second argument missing from the sentence. The function words "先 (first)" and '然后 (following)" are linked to "吃饭 (eat)" and "讨论 (discussion)", respectively, where the functional word "教室 (at classroom)" has two heads, modifying both propositional words.

3.3 Quantifier

The semantic representation of quantifier words is essential to support logic conversion.

The two essential arguments of quantifier words are *scope* and *set*. Quite often, there is a measure word following a quantifier word, and a third argument *quant* denotes this phenomenon.

Figure 6 shows an example with two quantifier words. The word "许多 (many)" has only the arguments *scope* and *set* in its EP structure, while the word "三 (three)" has three arguments in its EP due to the measure word "个 (ge)" after it. The value of argument *set* of "许多 (many)" is "衣服 (clothes)", and the

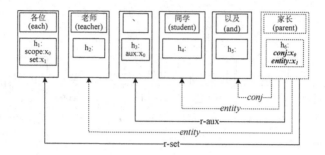

Fig. 7. The semantic representation for "各位老师、同学以及家长 (Teachers, students and parents)"

same argument for "三 (three)" is "人 (person)". The argument *scope* for the both quantifier words points to "买 (buy)", which indicates that the three persons bought many clothes together (i.e. $(3x_0\ nx_1)(\exists e_0\ person(x_0) \land n > 1 \land clothes(x_1) \land buy(e_0) \land arg0(e_0, x_0) \land arg1(e_0, x_1)))$, but not separately[2].

3.4 Coordination

The coordination structure is a very important issue in the semantic representation. Most previous work builds intermediate nodes to represent its semantics. However, additional nodes can make the dependencies between words much complicated. We choose to follow syntactic dependency treebanks and add two additional arguments *conj* and *entity* for the coordination phrase structure. All entities are linked to the last entity with the label *entity*, and all conjunction words are linked to the last entity with the label *conj*.

Figure 7 shows an example coordination structure. The three words "老师 (teacher)", "同学 (student)" and "家长 (parent)" are coordinated nouns in the sentence. We apply the two more arguments *conj* and *entity* to the EP of the last word "家长 (parent)". The two remaining words are linked to the word with the label *entity*, and the conjunction word "以及 (and)" is linked to the word with the label *conj*.

3.5 Anaphora

We adopt the semantic role *refer* (referencer) to indicate the true value of a pronoun. Thus a pronoun's EP structure can be written as predicate:handle(refer:x_0). Figure 8 shows an example, where the pronoun "他 (he)" in the sentence is a reference of the proper noun "小明 (Xiaoming)".

---

[2] For a distributive reading the argument *scope* for "三 (three)" points to "许多 (many)" and that of "许多 (many)" points to "买 (buy)", resulting in the logic meaning $(3x_0(nx_1(\exists e_0\ person(x_0) \land n > 1 \land clothes(x_1) \land buy(e_0) \land arg0(e_0, x_0) \land arg1(e_0, x_1))))$.

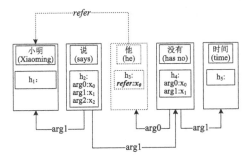

Fig. 8. The semantic representation for "小明说他没有时间 (Xiaoming said that he had no time)"

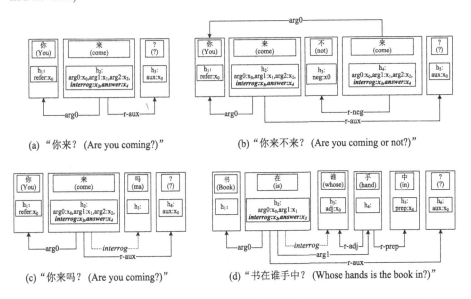

(a) "你来? (Are you coming?)"

(b) "你来不来? (Are you coming or not?)"

(c) "你来吗? (Are you coming?)"

(d) "书在谁手中? (Whose hands is the book in?)"

Fig. 9. Examples of semantic representation for questions

3.6   Question

Chinese questions are generally marked by interrogative words. However, sometimes the interrogative can be omitted from a sentence. We add the two related semantic arguments *interrog* (interrogative) and *answer* to the head propositional words of the sentence.

We show four examples in Figure 9, where Figure 9(a) and 9(b) are both questions without interrogative words, and Figure 9(c) and 9(d) are both questions with interrogative words. Figure 9(a) is a question recognized by the punctuation while 9(b) matches a common question pattern. Figure 9(c) is triggered by the interrogative particle word "吗 (ma, a question tag) " while Figure 9(d) is a question triggered by the interrogative word "谁 (whose)". To treat the these conditions consistently, the semantic arguments are imposed on the propositional words in the questions. The added arguments are in the EP of "来 (come)" for the first three examples and in the EP of "在 (be in)" for the last example.

4   Related Works

One dominant approach to sentential semantics is based on the Montagovian framework [23], adopting syntactic grammars such as CCG to build logic meaning through semantics composition. In our work, we choose not to adopt this framework for Chinese semantics as it needs resolving semantic ambiguities syntactically, which is rather difficult for Chinese. The Chinese syntax is more irregular than that of English as it is a parataxis language and lacks morphology.

Underspecification [24–26] is a useful tool for semantic representation, which allows semantic construction to be independent to syntactic structures. MRS is a representative grammar using this tool [17]. The EP structure was first introduced by this formalism, and our grammar is largely inspired by it. However, we are different in several aspects. For example, we abandon the MRS's requirement of argument *scope* for every EP, treating *scope* as a normal argument to particular EPs, but allowing underspecification, because we find that in most conditions the scope is bound to another sematic argument of the EP. For another example, we build a DAG grammar formalism for sentential semantics directly, which is independent of an extra syntactic formalism. Hence our grammar can be analyzed using existing statistical parsing algorithms.

Some researchers use Propbank and syntax dependencies for semantic representation, and the Propbank annotation has been adopted for Chinese [19]. However, this representation can only express the predicate-argument structures of propositional words conveniently. Moreover, it does not support conversion into logic forms. Debusmann et al. [27] propose Extensible Dependency Grammar (XDG) to denote sentential semantics. They classify semantic phenomena into several views; each view requires a separate structure graph.

We choose to use a single graph for the semantics of a sentence. Che et al. [4] introduces a semantics oriented dependency grammar for Chinese. They exploit the same structural representation as syntactic dependency grammars. However, they introduce 123 semantic labels to substitute syntactic labels in the dependency structures.

This fine-grained label set adds to the annotation cost, as well as difficulty in statistical disambiguation. As a result, the best performing systems as reported by Che et al. [4] gave less than 62% LAS. In addition, Che et al.'s formalism does not allow logical conversion since their dependencies are built over words.

Our formalism is designed taking consideration of and drawing inspirations from all the above work and the characteristic of the Chinese language. It abandons the Montagovian framework and adopts the underspecified framework. It denotes sentential meaning using direct acyclic graphs. It suggests the use of EPs rather than words as the basic units to build semantic relations, making the representation concise.

5   Conclusion and Future Works

We discussed several challenges to Chinese semantic treebanking and proposed a possible solution based on elementary predicates and semantic links, combining the strengths of semantic frames and light-weight grammars. Compared to existing Chinese treebanks, this formalism consists of a much simpler label set, and has a direct conversion to logical forms. Future work includes the construction of a treebank in large scale.

Acknowledgments. We thank the anonymous reviewers for their constructive comments. We gratefully acknowledge the support of the National Key Basic Research Program (973 Program) of China via Grant 2014CB340503 and the National Natural Science Foundation of China (NSFC) via Grant 61133012 and 61370164, the National Basic Research Program (973 Program) of China via Grant 2014CB340503, the Singaporean Ministration of Education Tier 2 grant T2MOE201301 and SRG ISTD 2012 038 from Singapore University of Technology and Design.

References

1. Huang, C.R., Chen, F.Y., Chen, K.J., Ming Gao, Z., Chen, K.Y.: Sinica treebank: design criteria, annotation guidelines, and on-line interface. In: Proceedings of 2nd Chinese Language Processing Workshop (2000)
2. Xue, N., Xia, F., Chiou, F.D., Palmer, M.: The penn chinese treebank: Phrase structure annotation of a large corpus. NLE 11(2), 207–238 (2005)
3. Chang, P.C., Tseng, H., Jurafsky, D., Manning, C.D.: Discriminative reordering with chinese grammatical relations features. In: Proceedings of the Third Workshop on Syntax and Structure in Statistical Translation (2009)
4. Che, W., Zhang, M., Shao, Y., Liu, T.: Semeval-2012 task 5: Chinese semantic dependency parsing. In: Proceedings of SemEval 2012, pp. 378–384 (2012)
5. Li, Z., Liu, T., Che, W.: Exploiting multiple treebanks for parsing with quasi-synchronous grammars. In: Proceedings of the ACL 2012, pp. 675–684 (2012)
6. Levy, R., Manning, C.D.: Is it harder to parse chinese, or the chinese treebank? In: Proceedings of the ACL 2003, pp. 439–446 (2003)
7. Petrov, S., Klein, D.: Improved inference for unlexicalized parsing. In: Proceedings of NAACL 2007, pp. 404–411 (2007)
8. Zhang, Y., Clark, S.: Syntactic processing using the generalized perceptron and beam search. Computational Linguistics 37, 105–151 (2011)
9. Kummerfeld, J.K., Tse, D., Curran, J.R., Klein, D.: An empirical examination of challenges in chinese parsing. In: Proceedings of the 51st ACL, pp. 98–103 (2013)
10. Li, M., Li, J., Dong, Z., Wang, Z., Lu, D.: Building a large chinese corpus annotated with semantic dependency. In: Proceedings of the SIGHAN 2003, pp. 84–91 (2003)
11. Nakazawa, T., Kurohashi, S.: Alignment by bilingual generation and monolingual derivation. In: Proceedings of COLING 2012, pp. 1963–1978 (2012)
12. Zhou, Q.: Building a large-scale annotated chinese corpus. In: Matoušek, V., Mautner, P. (eds.) TSD 2003. LNCS (LNAI), vol. 2807, pp. 106–113. Springer, Heidelberg (2003)

13. Bos, J., Clark, S., Steedman, M., Curran, J.R., Hockenmaier, J.: Wide-coverage se-
    mantic representations from a ccg parser. In: Proceedings of Coling, pp. 1240–1246
    (2004)
14. Kaplan, R., Riezler, S., King, T.H., Maxwell III, J.T., Vasserman, A., Crouch, R.:
    Speed and accuracy in shallow and deep stochastic parsing. In: Dumais, S., Marcu,
    D., Roukos, S. (eds.) Proceedings of NAACL 2004, pp. 97–104 (2004)
15. Toutanova, K., Manning, C.D., Shieber, S.M., Flickinger, D., Oepen, S.: Parse
    disambiguation for a rich hpsg grammar. In: Proceedings of TLT 2002 (2002)
16. Kenji, S., Jun'ichi, T.: Shift-reduce dependency DAG parsing. In: Proceedings of
    the Coling 2008, pp. 753–760 (2008)
17. Copestake, A., Flickinger, D., Pollard, C., Sag, I.A.: Minimal recursion semantics:
    An introduction. Research on Language and Computation 3, 281–332 (2005)
18. Daniel, G., Daniel, J.: Automatic labeling of semantic roles. In: Proceedings of the
    ACL 2000 (2000)
19. Xue, N., Palmer, M.: Annotating the propositions in the penn chinese treebank.
    In: Proceedings of the Second SIGHAN (2003)
20. Zettlemoyer, L., Collins, M.: Online learning of relaxed CCG grammars for parsing
    to logical form. In: Proceedings of the EMNLP-CoNLL, pp. 678–687 (2007)
21. Kay, M.: Chart generation. In: Proceedings of the 34th ACL, pp. 200–204 (1996)
22. Clark, S., Curran, J.R.: Wide-coverage efficient statistical parsing with ccg and
    log-linear models. Computational Linguistics 33(4), 493–552 (2007)
23. Richard, M.: Formal Philosophy: Papers of Richard Montague. Yale University
    Press, New Haven (1974)
24. Egg, M., Koller, A., Niehren, J.: The constraint language for lambda structures.
    Journal of Logic, Language and Information 10, 457–485 (2001)
25. Mel'čuk, I.: Dependency Syntax: Theory and Practice. State University of New
    York Press (1988)
26. Kahane, S.: The meaning-text theory. Dependency and Valency. An International
    Handbook on Contemporary Research. De Gruyter, Berlin (2003)
27. Debusmann, R., Duchier, D., Koller, A., Kuhlmann, M., Smolka, G., Thater, S.: A
    relational syntax-semantics interface based on dependency grammar. In: Proceed-
    ings of Coling 2004, Geneva, Switzerland, pp. 176–182 (2004)

Unsupervised Interpretation of Eventive Propositions

Anselmo Peñas[1], Bernardo Cabaleiro[1], and Mirella Lapata[2]

[1] NLP&IR group, UNED, Spain
{anselmo,bcabaleiro}@lsi.uned.es
[2] School of Informatics, University of Edinburgh
mlap@inf.ed.ac.uk

Abstract. This work addresses the challenge of automatically unfold transfers of meaning in eventive propositions. For example, if we want to interpret "throw pass" in the context of sports, we need to find the object ("ball") that transferred some semantic properties to "pass" to make it acceptable as argument for "throw". We propose a probabilistic model for interpreting an eventive proposition by recovering two additional coupled propositions related to the one under interpretation. We gather the statistics after building a Proposition Store from a document collection, and explore different configurations to couple propositions based on WordNet relations. These coupled propositions compose an actual interpretation of the original proposition with a precision of 0.57, but only for an 18% of samples. If we evaluate whether the interpretation is just useful or not for recovering background knowledge required for interpretation, then results rise up to 0.71 of precision and recall.

1   Introduction

Inherently to human communication, we use mechanisms to reduce the transmission of information able to activate more and more complex meanings in the receptor. Eventually, some of the creative constructions we are able to produce become conventions. However, if we want machines able to process and manipulate these constructions, we need to develop methods to unfold these language contractions.

The communication of events is rich in this kind of constructions, exploiting the models humans manage about the physical world, but also, exploiting the structure of events into sub-events. For example, in the US football domain, we can find frequent propositions such as "*x1 throws interception*". In this case, the requirements that the verb *throw* places onto its arguments trigger a process of inference. Since, *interceptions* aren't "*throwable*" we start the process of retrieving concepts that can be *thrown* and, at the same time, can be related to *interceptions*. We do this, because we always start assuming that the information we received is relevant, so we start a process of inference to build plausible interpretations [1].

Following with our example, if we look for the most frequent constructions used in relation to *throw* and *interception*, we find *passes* in these propositions: "*x1 throws pass, x2 intercepts pass*". So this conjunction of propositions could be a plausible interpretation for the proposition "*x1 throws interception*". To enable this retrieval, we needed to introduce the consideration of eventive nominalizations in the process.

A. Gelbukh (Ed.): CICLing 2014, Part I, LNCS 8403, pp. 379–390, 2014.

Once again, although *passes* are closer to *throwable* things, still the verb *throw* places a requirement onto its arguments that trigger the retrieval of concepts related to *throw* and *pass* in this case. Now, if look for the most frequent constructions that relate them, we find *"pass ball"* and *"throw ball"*. In this way, we have unfolded a *transfer of meaning* [2][3] between *ball* and *pass*, making *passes* *"throwable* things" by assuming some properties of *balls*. *Figure 1* draws graphically a plausible interpretation of the proposition *"x1 throw interception"*.

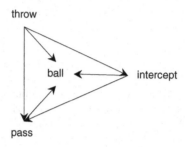

Fig. 1. Graphical interpretation *"x1 throw interception"* after two iterations

Our working hypothesis is that we can simulate this behavior gathering some probability distributions from a corpus related to the events we want to interpret. In particular, these are our research questions:

1. Is it possible to identify automatically the eventive propositions that could have a covert meaning?
2. Is it possible to automatically unfold *transfers* of meaning in eventive propositions? In particular, how the idea of interpreting an eventive proposition by recovering two coupled propositions can be modeled? What are the options to couple these propositions?
3. How effective is this approach to discover and formulate interpretation axioms? In which proportion of cases do these coupled propositions compose an actual interpretation of the original proposition?
4. What are the limitations of this approach, and conditions that affect the performance of our proposal?

In the following sections, we describe a probabilistic model to make account of this problem, the methodology to gather the propositions and their statistics, a procedure to automatically discover the axioms that express the interpretation of eventive propositions, and the findings after evaluating these axioms automatically built in an unsupervised manner.

2   Previous Work

Our work is related to type coercion [4-7] and logical metonymy [8-12], although there is an important difference with previous attempts that model their automatic interpretation. All these works start with the observation of a verb taking an argument that violates its selectional preferences and therefore, there is a need of inference for the purposes of interpretation. Most previous work on type coercion aims at determining the kind of coercion behind the surface expression. For example, [6] proposes the consideration of named entities that are used in the following coercions: place-for-people, place-for-event, place-for-product, org-for-members, org-for-event, org-for-product, org-for-facility, org-for-index, object-for-name and object-for-representation. Pustejovsky et al. [7] discussed [6] and develop a methodology to annotate coercions such as event-for-location, artefact-for-event, event-for-proposition, etc.

However, most works about building automatically an interpretation are focused on logical metonymy [8-12] where a verb takes as argument another verb that is implicit. Examples found in the literature are:

a)   *Enjoy the book → enjoy <u>reading</u> the book*
b)   *Finish the cigarette → finish <u>smoking</u> the cigarette*
c)   *Begin the sandwich → begin <u>eating</u> the sandwich*

Thus, the task of interpreting the metonymic expression is modelled as the task of recovering the implicit verb (e.g. *read, smoke, have, eat*, in the examples above).

In [6], Lapata and Lascarides address also the interpretation of constructions with adjectives such as, for example, *"fast plane"*. Once again, the goal is to recover the implicit verb *(fly)* that better fits with the expression. This is related also to the literature about the interpretation of noun compounds [13][14] by recovering their implicit predicates [15][16][17]. For example, *"malaria mosquito"* could be approximated, among many other options, by the clause *"mosquito that carries malaria"*. That is, the interpretation process relies in the recovery of the verb *carry*.

In contrast with previous work, we focus on constructions where the verb takes an *event* as argument and we want to explore the idea of recovering an implicit *argument* (the argument of the event) instead of recovering an implicit *verb*. Coming back to our example, *(x1 throw interception → x1 throw pass, x2 intercept pass)*, we are recovering *pass* as the implicit piece of information required to better interpret the original proposition. In other words, we focus on structures where the argument of the first predicate is explicitly another predicate (nominalization of an event), and we are looking for the most plausible argument for both that better explains the construction.

This idea introduces also some variations with respect to modelling. In particular, the model requires introducing a step of nominalization in an explicit manner.

3   Model

Recalling the example followed in the previous sections, given the proposition *"x1 throw interception"*, we are looking for the most probable concept c that transferred some properties to *interception* to make it *"throwable"*. More formally, given the proposition expressed as a verb-object pair *(v,o)* (e.g. *throw interception*), we are looking for the concept c (e.g. *pass*) participating in a clique with v and o (e.g. *throw pass, intercept pass*) with higher probability given the original proposition:

$$c = argmax_c \, P(c/v,o)$$

Notice that *o* has a different grammatical form in each proposition. While in proposition *(v,o)* appears as a noun (e.g. *throw interception*) in proposition *(c,o)* appears as a verb (e.g. *intercept pass*). Therefore, we also have to consider a nominalization *(v_o,o)* (e.g. *intercept, interception*) in our model. Following with our example, *Figure 2* depicts the graphical interpretation including the step of nominalization.

Fig. 2. Nominalization involved in the interpretation of *"x1 throw interception"*

Since *o* is used with the properties transferred from *c*, we can't expect to find occurrences of *c* jointly with *(v,o)*. Thus, we can assume that *v* and *o* are conditionally independent given *c*. Putting both ideas together, we will model the problem as the generative story shown in *Figure 3*.

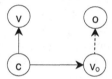

Fig. 3. Graph of the probabilistic model

where the corresponding expression is the following:

$$c = argmax_c \, P(c/v,o,v_o) \propto P(c) \cdot P(v/c) \cdot P(v_o/c) \cdot P(o/v_o)$$

We can estimate probabilities $P(c)$, $P(v/c)$ and $P(v_o/c)$ using *maximum likelihood* from the *verb-object* dependencies found after parsing a document collection.

However, $P(o/v_o)$ corresponds to the probability of the nominalization and we can't expect direct observations of both terms together. Thus we have three alternatives: (i) consider a set of deverbal suffixes, (ii) use external resources such as Nomlex [18] or the WordNet [19] derivational relation, and (iii) try to build a probabilistic model. Although worth it to explore in the future, option three is out of the scope of this work. We will address partially the problem using the derivational relation of Word-Net restricted to the *event* sub-hierarchy. However, the derivational relation of Word-Net is established between *synsets* and therefore, our results will be affected by the ambiguity of the involved words. The disambiguation in the context of this work is a very interesting research that, again, is out of the scope of this work, and we leave for future. The way we will reduce the ambiguity problem is using a collection focussed on a particular topic, where we can expect that both, candidate constructions to be interpreted, and propositions that will serve as interpretation, will be related.

On the other hand, *synsets* enable the consideration of synonyms in the retrieval of explicative propositions. For example, being *intercept* and *stop* possible synonyms, we can recover "*(x1 have run), (x2 stop run)*" as a possible interpretation for "*x1 have interception*". Furthermore, once we have a *synset*, we can explore the effect of considering its *hypernyms* and *hyponyms*. For example, a hypernym for *touchdown* is *score*. Through the derivational relation between *score-n* and *hit-v* we can recover "*(x2 hit point), (x1 score point)*" (*x1=x2* in this case) as an interpretation for "*x1 score touchdown*".

We considered that the effect of *synonyms*, *hypernyms* and *hyponyms* in the recovery of propositions deserved attention in our experimental setting and we tried different configurations.

4   Experimental Setting

As a source for discovering interpretation axioms, we will use a Proposition Store built from a collection of 30,826 New York Times articles about US football. This topic is rich in eventive expressions and language contractions and we expected to discover interesting interpretations. Following [20][21] we call propositions the tuples of words that have some determined pattern of syntactic relations among them.

To build the Proposition Store we parsed the collection using a standard dependency parser [22][23] together with TARSQI [24] and, after collapsing some syntactic dependencies, we obtained 3,022,305 raw elements in the BKB.

We have different types of propositions in our Proposition Store, coming from different syntactic structures. Out of them, we selected two types for our experimentation: NVN (e.g. *player throws pass*) and NVPN (e.g. *player scores on run*). According to the model proposed, we ignore the first argument of the selected propositions.

4.1   Identification

If we want to discover axioms that relate a proposition with an interpretation composed by two other related and coupled propositions, the first question is whether it is possible or not to identify the possible candidates that fit in this paradigm. For this purpose, we tried a very simple heuristic. TARSQI [24] provides information about whether the arguments are events or not and we can use this information to select the constructions candidate for interpretation. That is, What happens if we select all propositions NVN and NVPN where the second argument was tagged as event by TARSQI? This process gives us an initial set of 1,941 distinct candidate constructions to be interpreted. *Table 1* shows the top most frequent candidates.

We have sampled the first 100 constructions and we found that with this simple heuristic, up to 66% could be good candidates to discover axioms for its interpretation. Looking at the remaining 34% of bad candidates (see *Table 2*) we can classify the cases that don't fit. The first one is related to the constructions (41%) where the main verb is *have* or *make* (columns 1 and 2). In many cases, they are used to formulate a single event described by the object. The second one (column 3) is related to verbs (38%) denoting different states of an event. The third group is composed by the verbs (18%) related to communication (column 4). The last 3% corresponds to other cases (see column 5).

Table 1. Most frequent constructions taking an event as argument

VN		VPN	
frequency	predicate (v,o)	frequency	predicate (v,o)
6748	win, game	1820	play:in, game
4618	play, game	1271	throw:for, touchdown
4102	catch, pass	1164	run:for, touchdown
4015	score, touchdown	1009	pass:for, touchdown
2860	play, football	903	rush:for, touchdown
2698	throw, pass	865	score:on, run
2670	lose, game	784	be:in, game

Table 2. Groups of bad (v,o) candidates to be interpreted

have, game	make, mistake	win, game	say, game	play, role
have, defense	make, playoff	play, game	call, play	throw, football
have, fun	make, play	enter, game	call, game	
have, offense	make, change	miss, game	say, good	
have, plan	make, move	take, job	say, happy	
have, effect	make, deal	get, job	say, hard	
	make, adjustment	lose, job		
	make, progress	end, career		
		complete, pass		
		...		

Answering to our question, it seems feasible to consider these cases in a regular way to make accurate the process of identification. It seems there are regularities enough to address the problem with automatic classifiers. We leave this for future.

4.2   Interpretation

The second question is whether we can recover an interpretation, and if this interpretation is good or not. Recall that, given a construction (v,o), we are looking for an interpretation "(v,c), (v_o,c)" that maximizes $P(c/v,o,v_o)$. We tried 5 configurations depending on how to relate noun o with verb v_o in our model (see previous section):

1. **Str**: $P(o/v_o)=1$ if $o=v_o$, that is, o and v_o are the same string.
2. **Syn**: $P(o/v_o)=1$ if o is synonym of a nominalization[1] of v_o. We consider here that o is synonym of itself, too.
3. **Hypo**: $P(o/v_o)=1$ if o has an hyponym that is a nominalization of v_o.

4.   **Hyper:** $P(o/v_o)=1$ if o has an hypernym that is a nominalization of v_o.

5.   **Max:** $P(o/v_o)=1$ if any of the previous cases hold.

If the condition doesn't hold, then we consider $P(o/v_o)=0$.

We have evaluated the 66 candidates that passed the filter of identification in our sampling. *Table 3* shows how productive are each of these configurations to produce at least one interpretation for the given proposition (no matter if the interpretation is correct or not). It shows also the average number of interpretations per configuration that are considered in the selection of the most likely one. In one extreme, if we just consider configuration *Str*, then we find at least one interpretation (correct or not) for 30% of the input propositions. In the other extreme, configuration *Max* is able to re-cover at least one interpretation for each given proposition (once bad candidates are ignored after the identification phase).

Table 3. Percentage of eventive propositions that receive at least one interpretation for each configuration, and their average number of interpretations

Configuration	% of candidates	Average number of interpretations
Str	30%	5.67
Syn	50%	24.82
Hypo	41%	49.13
Hyper	95%	52.13
Max	100%	106.95

The average number of interpretations runs in parallel with the ability to recover at least one interpretation: from 5.67 interpretations in the more restrictive case (*Str*) up to almost 107 in the most productive one (*Max*).

4.3   Evaluation

For the purpose of evaluating if these interpretations are correct or not we have distinguished three cases in the judgment of the recovered interpretations:

1.   *Correct (C)*, if the interpretation is a concise and fairly complete explanation of the given proposition.

2.   *Useful (U)*, if the interpretation is not complete, or one of the propositions is unrelated, but still at least one proposition is related to the correct interpretation, bringing implicit lexemes related to the background knowledge that a correct interpretation should consider.

3.   *Wrong (W)*, if the interpretation is not correct and not useful.

We evaluated only the most likely interpretation for each eventive proposition. This is a very restrictive evaluation because if the correct interpretation is recovered in second or third place it will be ignored. However, the goal is to check the plausibility of our hypotheses with this preliminary experimentation. Thus, we leave for future a ranking-based evaluation.

Since we only consider the most likely interpretation, we can define *Precision* as the proportion of candidates that received a correct interpretation when an interpretation is recovered. Assuming that each candidate should receive at least one correct interpretation (which is just an upper bound since we rely on a discovery process dependent of a particular corpus), we can define *Recall* as the proportion of propositions that received a correct interpretation.

Considering just those cases evaluated as *Correct* we can have *Precision, Recall* and *F1 measure* about *Correctness*. If we add also those cases evaluated as *Useful*, then we can have *Precision, Recall* and *F1 measure* about *Usefulness*.

More formally, the measures used for *Correctness* are:

$$P = \frac{|C|}{|C| + |U| + |W|} \qquad R = \frac{|C|}{|C| + |U| + |W| + |\emptyset|} \qquad F1 = \frac{2 \cdot P \cdot R}{P + R}$$

And the measures for *Usefulness* are:

$$P = \frac{|C| + |U|}{|C| + |U| + |W|} \qquad R = \frac{|C| + |U|}{|C| + |U| + |W| + |\emptyset|} \qquad F1 = \frac{2 \cdot P \cdot R}{P + R}$$

Where $\emptyset$ is the set of candidates that didn't receive any interpretation.

4.4   Results

Tables 4, 5 and 6 show some of the interpretations that fall in each of the cases (Correct, Useful, Wrong).

Table 4. Sample of eventive propositions that received a *Correct* interpretation

(x1 be:in game) ← (x2 play position) , (x1 be:in position)
(x1 catch pass) ← (x2 throw ball) , (x1 catch ball)
(x1 drop pass) ← (x2 throw ball) , (x1 drop ball)
(x1 have interception) ← (x2 stop run) , (x1 have run)
(x1 have run) ← (x2 run ball) , (x1 have ball)
(x1 have victory) ← (x2 win game) , (x1 have game)
(x1 intercept pass) ← (x2 throw ball) , (x1 intercept ball)
(x1 lose fumble) ← (x2 fumble ball) , (x1 lose ball)
(x1 run offense) ← (x2 carry ball) , (x1 run ball)
(x1 rush touchdown) ← (x2 score point) , (x1 rush point)
(x1 stop run) ← (x2 run offense) , (x1 stop offense)
(x1 throw interception) ← (x2 intercept pass) , (x1 throw pass)
(x1 throw pass) ← (x2 snap ball) , (x1 throw ball)
(x1 watch film) ← (x2 show tape) , (x1 watch tape)

Table 7 shows *Precision, Recall* and *F1 measure* for *Correctness* and *Usefulness*. If we look at *Correctness*, the best configuration corresponds to consider just synonyms

during the nominalization process (*Syn*). While *Precision* is significantly higher, the drop on *Recall* is not so high. If we consider the most productive configuration (*Max*), we can increase *Recall*, but the drop in *Precision* is too harmful. Table 3, explains this result, since configuration *Max* is recovering more than 4 times more candidate interpretations than configuration *Syn*.

Table 5. Sample of eventive propositions that received a *Useful* interpretation

(x1 attend game) ← (x2 sport event) , (x1 attend event)
(x1 catch pass) ← (x2 snap ball) , (x1 catch ball)
(x1 drop pass) ← (x2 snap ball) , (x1 drop ball)
(x1 force fumble) ← (x2 fumble punt) , (x1 force punt)
(x1 make debut) ← (x2 get start) , (x1 make start)
(x1 make run) ← (x2 run play) , (x1 make play)
(x1 make start) ← (x2 start game) , (x1 make game)
(x1 miss attempt) ← (x2 attempt goal) , (x1 miss goal)
(x1 recover fumble) ← (x2 fumble snap) , (x1 recover snap)
(x1 return interception) ← (x2 intercept pass) , (x1 return pass)
(x1 run:for touchdown) ← (x2 hit yard) , (x1 run:for yard)
(x1 rush game) ← (x2 play defense) , (x1 rush defense)
(x1 score:on run) ← (x2 go drive) , (x1 score:on drive)
(x1 see game) ← (x2 play football) , (x1 see football)
(x1 set_up touchdown) ← (x2 score goal) , (x1 set_up goal)
(x1 win game) ← (x2 play football) , (x1 win football)

Table 6. Sample of eventive propositions that received a *Wrong* interpretation

(x1 answer question) ← (x2 question call) , (x1 answer call)
(x1 be:in playoff) ← (x2 oppose plan) , (x1 be:in plan)
(x1 do work) ← (x2 occupy spot) , (x1 do spot)
(x1 do work) ← (x2 see thing) , (x1 do thing)
(x1 give victory) ← (x2 end chance) , (x1 give chance)
(x1 go:into game) ← (x2 play season) , (x1 go:into season)
(x1 have run) ← (x2 go way) , (x1 have way)
(x1 have victory) ← (x2 end streak) , (x1 have streak)
(x1 leave game) ← (x2 play football) , (x1 leave football)
(x1 lose fumble) ← (x2 blow lead) , (x1 lose lead)
(x1 make debut) ← (x2 rush leader) , (x1 make leader)
(x1 make decision) ← (x2 take cut) , (x1 make cut)
(x1 make run) ← (x2 go way) , (x1 make way)
(x1 make start) ← (x2 commit mistake) , (x1 make mistake)
(x1 make start) ← (x2 get call) , (x1 make call)
(x1 miss attempt) ← (x2 break game) , (x1 miss game)
(x1 miss attempt) ← (x2 give chance) , (x1 miss chance)
(x1 miss practice) ← (x2 use game) , (x1 miss game)
(x1 need help) ← (x2 help team) , (x1 need team)

Table 7. Results

Configuration	Correctness			Usefulness		
	Precision	Recall	F1	Precision	Recall	F1
Str	0.43	0.09	0.15	**0.80**	0.24	0.37
Syn	**0.57**	0.18	**0.28**	0.70	0.35	0.46
Hypo	0.17	0.06	0.09	0.44	0.18	0.26
Hyper	0.21	0.17	0.19	0.67	0.64	0.65
Max	0.29	**0.23**	0.26	0.71	**0.71**	**0.71**

Looking at Usefulness, the option of considering just hyponyms during nominalization is really harmful. This behavior is consistent with its values of Correctness. However, for the rest of configurations, the more you increase the number of interpretations considered (and therefore recall), the better. Thus, the best configuration is to consider all different options to determine possible nominalizations and take the most likely interpretation (configuration *Max*).

5   Conclusions and Future Work

We have explored the idea of automatically unfold transfers of meaning in eventive propositions. Despite this is a preliminary work aiming at envisaging what is the methodology required and the challenges to be addressed, results are promising enough to explore more sophisticated approaches. Answering to our research questions, it seems feasible to identify automatically the eventive propositions that could have a covert meaning, and to unfold transfers of meaning.

With respect to identification, the keystone seems to be in detecting whether the construction contains two different (although related) events or not. We have observed that in many cases the main verb is just void of content (e.g. *have attempt, make attempt*, etc.) used to formulate a single event described by its object. In other cases, the main verb is denoting a state of an event (e.g. *enter game, play game*, etc.), and in the third main group we find propositions where the main verb is related to communication.

With respect to the interpretation, we have proposed a probabilistic model for interpreting an eventive proposition by recovering two coupled propositions related to the one under interpretation. We have built a Proposition Store from a document collection and we have considered as candidates those NVN and NVPN propositions that take an event as second argument. We explored different configurations to couple these propositions based on WordNet relations.

These coupled propositions compose an actual interpretation of the original proposition with a precision of 0.57, but only for an 18% of samples. If we evaluate whether the interpretation is just useful or not for recovering background knowledge required for interpretation, then results rise up to 0.71 of precision and recall.

This research has served us to state in which directions this work can be further developed leading us to explore more sophisticated approaches.

The first direction is to try to estimate $P(o/v_o)$. That is, try to explore if probabilistic based measures of semantic relatedness between a noun and a verb can improve our results based on a WordNet relation. We realized that this relatedness must take into account a *disambiguation* process.

According to the model proposed here, we are ignoring the first argument of the selected propositions. However, it seems reasonable to think that their consideration should help to couple propositions in a stronger way and therefore improve precision.

Another restriction in our current approach is that we are just coupling two propositions to form an interpretation. It would be really interesting to leave unrestricted the number of coupled propositions and permit them to arise from the discovery process.

About the evaluation, certainly there is a lack of resources for this kind of research. Building a larger resource of gold standard interpretations would allow us some ranking-based evaluation.

Acknowledgements. This work was partially funded by MINECO (PCIN-2013-002-C02-01) and EPSRC (EP/K017845/1) in the framework of CHIST-ERA READERS project. Anselmo Peñas received also partial support from European Commission (FP7/2007-2013) under grant agreement 288024 (LiMoSINe project). Bernardo Cabaleiro received the support of Holopedia project (TIN2010-21128-C02).

References

1. Sperber, D., Wilson, D.: Relevance: Communication and cognition, 2nd edn. Blackwell, Oxford (1995)
2. Nunberg, G.: The pragmatics of reference. Indiana University Linguistics Club, Bloomington (1978)
3. Nunberg, G.: Transfers of meaning. Journal of Semantics 12(1), 109–132 (1995)
4. Fass, D.: met*: A method for discriminating metonymy and metaphor by computer. Computational Linguistics 17(1), 49–90 (1991)
5. Utiyama, M., Murata, M., Isahara, H.: A statistical approach to the processing of metonymy. In: Proceedings of the 18th Conference on Computational Linguistics, Saarbrücken, Germany, July 31-August 04 (2000)
6. Markert, K., Nissim, M.: Data and models for metonymy resolution. Language Resources and Evaluation 43(2), 123–138 (2009)
7. Pustejovsky, J., Rumshisky, A., Plotnick, A.: SemEval-2010 Task 7: Argument Selection and Coercion. In: Proceedings of the 5th International Workshop on Semantic Evaluation, ACL 2010, pp. 27–32 (2010)
8. Lapata, M., Lascarides, A.: A probabilistic account of logical metonymy. Journal of Computational Linguistics 29(2), 261–315 (2003)
9. Lapata, M., Keller, F., Scheepers, C.: Intra-sentential context effects on the interpretation of logical metonymy. Cognitive Science 27(4), 649–668 (2003)
10. Roberts, K., Harabagiu, S.M.: Unsupervised Learning of Selectional Restrictions and Detection of Argument Coercions. In: Proceedings of the 2011 Conference on Empirical Methods in Natural Language Processing, Edinburgh, Scotland, UK, July 27–31, pp. 980–990 (2011)

11. Zarcone, A., Utt, J., Padó, S.: Modeling covert event retrieval in logical metonymy: probabilistic and distributional accounts. In: CMCL 2012 Proceedings of the 3rd Workshop on Cognitive Modeling and Computational Linguistics, pp. 70–79 (2012)
12. Shutova, E., Kaplan, J., Teufel, S., Korhonen, A.: A Computational Model of Logical Metonymy. ACM Transactions on Speech and Language Processing (TSLP) 10(3), Article No. 11 (July 2013); Special issue on multiword expressions: From theory to practice and use, part 2
13. Downing, P.: On the creation and use of English compound nouns. Language 53(4) (1977)
14. Finin, T.: The Semantic Interpretation of Compound Nominals. Ph.D. thesis, University of Illinois at Urbana Champaign (1980)
15. Nakov, P., Hearst, M.: Using verbs to characterize noun-noun relations. In: Euzenat, J., Domingue, J. (eds.) AIMSA 2006. LNCS (LNAI), vol. 4183, pp. 233–244. Springer, Heidelberg (2006)
16. Butnariu, C., Kim, S.N., Nakov, P., O Seaghdha, D., Szpakowicz, S., Veale, T.: Semeval-2 task 9: The interpretation of noun compounds using paraphrasing verbs and prepositions. In: Proceedings of the 5th International Workshop on Semantic Evaluation (2010)
17. Peñas, A., Ovchinnikova, E.: Unsupervised Acquisition of Axioms to Paraphrase Noun Compounds and Genitives. In: Gelbukh, A. (ed.) CICLing 2012, Part I. LNCS, vol. 7181, pp. 388–401. Springer, Heidelberg (2012)
18. Macleod, C., Grishman, R., Meyers, A., Barrett, L., Reeves, R.: NOMLEX: A Lexicon of Nominalizations. In: Proceedings of EURALEX 1998, Liege, Belgium (1998)
19. Miller, G.A.: WordNet: A Lexical Database for English. Communications of the ACM 38(11), 39–41 (1995)
20. Van Durme, B., Schubert, L.: Open Knowledge Extraction through Compositional Language Processing. In: Symp. on Semantics in Systems for Text Processing, STEP 2008 (2008)
21. Clark, P., Harrison, P.: Large-scale extraction and use of knowledge from text. In: The Fifth International Conference on Knowledge Capture, K-CAP 2009 (2009)
22. Marneffe, M., Manning, C.D.: The Stanford typed dependencies representation. In: COLING 2008 Workshop on Cross-Framework and Cross-domain Parser Evaluation (2008)
23. Klein, D., Manning, C.D.: Accurate Unlexicalized Parsing. In: Proceedings of the 41st Meeting of the Association for Computational Linguistics, pp. 423–430 (2003)
24. Verhagen, M., Mani, I., Sauri, R., Knippen, R., Jang, S.B., Littman, J., Rumshisky, A., Phillips, J., Pustejovsky, J.: Automating temporal annotation with TARSQI. In: Proceedings of ACLdemo 2005 (2005)

Sense-Specific Implicative Commitments

Gerard de Melo[1,*] and Valeria de Paiva[2]

[1] IIIS, Tsinghua University, Beijing
gdm@demelo.org
[2] Nuance Communications, Sunnyvale, CA
Valeria.dePaiva@nuance.com

Abstract. Natural language processing systems, even when given proper syntactic and semantic interpretations, still lack the common sense inference capabilities required for genuinely understanding a sentence. Recently, there have been several studies developing a semantic classification of verbs and their sentential complements, aiming at determining which inferences people draw from them. Such constructions may give rise to implied commitments that the author normally cannot disavow without being incoherent or without contradicting herself, as described for instance in the work of Kartunnen. In this paper, we model such knowledge at the semantic level by attempting to associate such inferences with specific word senses, drawing on WordNet and VerbNet. This allows us to investigate to what extent the inferences apply to semantically equivalent words within and across languages.

1 Introduction

Understanding a sentence requires more than just decoding its syntactic and semantic structure. Even when supplied with proper syntactic and semantic interpretations of a sentence, current natural language processing systems still lack the common sense inference capabilities required to interpret it in the way humans do. Given a sentence like "John missed that Mary had left", we are inclined to infer that Mary had indeed left. In contrast, given "John pretended that Mary had left", we are inclined to presuppose the opposite. Although both sentences share a common structure, involving verbs with sentential complements, there are clear differences in the types of implicative commitments the author is making. Such differences have been studied in detail by recent studies that have attempted to develop a classification of verbs (and verb-noun collections) that take sentential complements [1]. Similar analyses can be made with respect to other constructions, e.g. to study adjective constructions like "It is confusing that Mary has left" vs. "It is improbable that Mary has left" [2]. Likewise, one can also study implicative commitments regarding the existence of entities. For

---

* Gerard de Melo's work was supported in part by the National Basic Research Program of China Grants 2011CBA00300, 2011CBA00301, and NSFC Grants 61033001, 61361136003.

A. Gelbukh (Ed.): CICLing 2014, Part I, LNCS 8403, pp. 391–402, 2014.

instance, "She cancelled the meeting" generally leads us to believe that there was no meeting, while "They caused a strike" means that a strike did occur [3].

Certain sentences give rise to implied commitments that the author normally cannot disavow without being incoherent or without contradicting herself. Previous work has developed detailed classifications of words and their implicative commitments at the lexical level [4]. In this paper, we build on this work but attempt to model such knowledge about commitments at the semantic level instead. For this, the commitments are tied to specific word senses, drawing on WordNet and VerbNet as sense inventories. We rely on automatic disambiguation methods to produce a lexicon of sense-specific implicative commitments. Among other things, this allows us to investigate to what extent such inferences also apply to semantically equivalent words within and across languages. Our findings indicate that they are for the most part preserved when transitioning from words to synonym sets, and at least most of the English classifications collected by Nairn et al. [4] do seem to transfer to other languages such as Portuguese as well.

2 Implicative Commitments

Karttunen's seminal "The Logic of English Predicate Complement Constructions" [5] makes the point that while it is valid and helpful to classify verbal constructions according to their syntactic characteristics, e.g. whether they take propositional complements, it is also valid and even more useful to classify them according to their semantic characteristics, such as the *factivity* of complements. The term *factive verb* introduced by Kiparsky and Kiparsky [6] refers to the notion that any simple assertion using one of these verbs (e.g. "John knew that Mary had left") commits the speaker to the belief that the complement sentence ("Mary had left"), just by itself, is also true.

Karttunen observes that it is sometimes "possible to show that there is a definite connection between the semantic properties of a verb and certain syntactic characteristics", and thus goes on to investigate parts of this connection. The syntactic characteristics referred to are that these are propositional complement constructions, while the semantic characteristics are what he refers to as the "logic" in the title of his book. The latter is more general than simply factivity. In fact, 35 years later, as part of a project to extract information from text for question answering, Karttunen, together with Nairn and Condoravdi [4] presented an algorithm for detecting author commitment to the truth or falsity of complement clauses based on their syntactic type and on the meaning of their embedding predicate. They also created a small lexicon of around 300 verbs spelling out the implicative commitments that these verbs indicate. This was the starting point of our work. We contend that such implicative commitments are best described at the semantic level, as one would hope given Karttunen's choice of the word "logic" in the title of his seminal work.

The same applies to other sources and forms of implicative commitments. For instance, in some cases, nouns can induce factivity commitments. As an example, "John took the opportunity to sing" entails that he did sing [7]. Similarly, adjectives, too, can be marked for factivity and implicativity, as in a recent study [2] that collected such annotations using Amazon Mechanical Turk. Additionally, one can study implicative commitments relating to existential and temporal aspects [3].

3   Veridicity, Veracity, or Veridicality?

Throughout this paper, we are adopting a generalized, somewhat underspecified notion of implicativity that subsumes factivity, entailment, and presupposition, calling them "implicative commitments" for want of a better name. This notion seems more robust and clear-cut, as described by Nairn et al. [4], and exemplified in their small lexicon.

It remains clear that our lexicon would have to be just one of several ingredients in any natural language understanding systems aimed at fully assessing the *veridicity* of textual content. Has an event mentioned in the text really occurred? Who is the source of the information? What is the stance of the author of the text? Does the author indicate whether he or she believes the source? This more encompassing problem is discussed by Karttunen and Zaenen [8] and in Saurí and Pustejovsky's FactBank corpus and factuality studies [9,10,11].

The work in FactBank is an attempt to mark the implicative commitments of both the author and some participants of the events in some given text. As Saurí and Pustejovsky explain

> Identifying the veracity, or factuality, of event mentions in text is fundamental for reasoning about eventualities in discourse. Inferences derived from events judged as not having happened, or as being only possible, are different from those derived from events evaluated as factual. Event factuality involves two separate levels of information. On the one hand, it deals with polarity, which distinguishes between positive and negative instantiations of events. On the other, it has to do with degrees of certainty (e.g., possible, probable), an information level generally subsumed under the category of epistemic modality.

The work on FactBank has been extended by the Stanford NLP group [12], who re-annotated the same sentences using workers on Amazon Mechanical Turk, given their stance that textual entailment is to be measured in terms of the understanding of text in ordinary people, not logicians, philosophers, or linguists.

While veridicality or factuality is a possible long term goal, we instead start from the less lofty goal of *veridicity*, a level of assessment that we believe can be coded up in the lexicon. As Karttunen and Zaenen put it:

> It is useful to distinguish between two ingredients that go into determining the truth value of utterance, one is the trustworthiness of the utterer and the other is the stance of the utterer vis-à-vis the truth of the content. The latter we will call the veridicity of the content [...].

The classification chosen by Nairn et al. [4] focuses on cases in which the author's commitment to the truth of a complement clause arises solely from the larger sentence it belongs to, leaving aside other sources of information about the beliefs of the author. Note that it can be difficult to decide between *entailments*, that is, what the author is actually committed to, and the more pragmatic notion of *conversational implicatures*, that is, what a reader/hearer may feel entitled to infer. For example, "Ed did not refuse to participate" might lead the hearer to conclude that Ed participated. But the speaker could continue with "He was not even eligible" indicating the opposite.

The tags used to describe the implicative commitments of lexical items need to distinguish several types of context-specific behaviors. Some words yield an entailment in both affirmative and negative environments but there are others, "one-way implicatives", that yield entailments only in one or the other environment. Furthermore, the entailment may be either positive or negative depending on the polarity of the environment. Altogether Nairn et al. provide us with a table of "implication signatures" for a large class of complement-taking constructions. These implication signatures can be stacked together, which gives rise to their algorithmic solution for dealing with more complex phrases [4].

4   Describing Sense-Specific Implicative Commitments

In this paper we advance the state-of-the-art in that, instead of associating implicative commitments just with lexical items, we associate them with word senses, as represented by WordNet synsets. Our input is a set of resources that describe implicative commitments at the lexical level (included the Nairn et al. lexicon and further resources listed later on in Section 5.1). Our output will contain descriptions of implicative commitments at the sense level.

Generally speaking, an input entry at the lexical level consists of a partially instantiated construction and a set of tags describing the implicative commitments of that construction in particular contexts. The construction may involve one or more lexical items to be disambiguated. For instance, often, there is simply a single verb like "to find" in a syntactic frame expecting a subject and a that-complement ("John found that Mary had arrived"). Instead of describing the construction for a specific lexical item like "to find", we can instead characterize it at a more abstract semantic level, allowing an instantiation with several additional words ("notice", "discover", etc.). We thus have to disambiguate the word "to find" in order to establish which senses are relevant and know which sense-specific synonyms can be assumed to give rise to the same commitments.

In some cases, the construction may also involve two lexical items that need to be disambiguated. For instance, the construction may involve both the verb "avoid" and a description of its possible arguments, described by the word "event". These will then both be mapped to WordNet synsets, and WordNet's synset hierarchy will allow us to to recognize specific instantiations, e.g. "avoiding a strike" vs. "avoiding a road bump" – In the first case, there is no strike, whereas in the second case, the road bump remains.

4.1   Sense Mappings and Disambiguation

Our goal is thus to map one or more input lexical items $e = (l, p, f)$ on the one side, each consisting of a lemma l, part-of-speech tag p, and syntactic frame marker f, to one or more candidate WordNet synsets s on the other side. For example, our input entry could be for the lemma "admit" with part-of-speech tag `verb`, and a marker for a syntactic frame with infinitival complement (as in "The report was admitted to be incorrect"). WordNet lists a total of 8 verb senses for the input lemma "admit". For instance, the first WordNet sense for "admit" refers to the process of allowing people to enter a building, and the second one refers to admittance into a group or a community. These do not reflect the meaning that our input entry is targeting. Our system instead needs to choose the WordNet sense for *declaring to be true or admitting the existence or reality or truth of something*.

In order to achieve such disambiguated mappings in the case of verbs, we draw on another traditional lexical resource VerbNet [13]. VerbNet is a lexicon of Levin-style verb classes based on the syntactic frames of English verbs. Since the input lexicons often refer to specific syntactic frames, e.g. for sentential complements, the syntactic characterizations present in VerbNet can help us pick relevant senses.

For a given input entry e with lemma l, let $W(l)$ denote the set of WordNet senses and $V(l)$ denote the set of VerbNet senses for l. We first use the following similarity function to score the candidate VerbNet senses $v \in V(l)$:

$$\text{sim}(e, v) = \frac{1}{|V(l)|}\left(\epsilon + \max_{f_v \in F(v)} \text{sim}(f_v, f)\right) \tag{1}$$

Here, the strength of association between e and one of the verb senses $v \in V(l)$ first of all depends on the number of candidates $|V(l)|$. Additionally, we obtain the relevant VerbNet class and its syntactic frame information from VerbNet, using $F(v)$ to denote the set of syntactic frames associated with the VerbNet class of v.

We rely on a simple rule-based similarity measure between syntactic descriptors f_v from VerbNet and descriptors f from the original lexicon, which in our case are based on PARC's Bridge system [14]. For instance, VerbNet's `NP V that S` would match the `V-SUBJ-COMPEXthat` descriptor from the Bridge system but not `V-SUBJ-OBJexpl-XCOMPinf`. While this similarity function clearly prefers VerbNet senses that have at least one matching syntactic frame, we still retain the remaining VerbNet senses as possible candidates by setting ϵ to a small constant greater than 0, because VerbNet's syntactic frame descriptions are not always complete.

We then use these VerbNet similarities to help us in scoring the candidate WordNet synsets $s \in W(l)$:

$$\text{sim}(e, s) = \begin{cases} \epsilon + \sum\limits_{v \in V(l)} \dfrac{\max\limits_{s_v \in V_w(v)} \text{sim}(s_v, s)}{1 + |V_w(v)|}\, \text{sim}(e, v) & s \in W_p \\[2ex] 0 & s \notin W_p \end{cases} \tag{2}$$

Here, first of all, for a given part-of-speech tag p, W_p denotes the global set of all WordNet senses for that part-of-speech tag. If we are attempting to disambiguate an entry e for which the part-of-speech tag p denotes adjectives, for instance, then only adjective synsets $s \in W_p$ receive a non-zero similarity score.

For a given VerbNet sense v, we use $V_w(v)$ to denote the set of WordNet synsets mapped to v. For this, we make use of the existing disambiguated mappings between VerbNet and WordNet provided with the former. VerbNet's verb senses tend to be more coarse-grained, as their induction was mainly guided by syntactic considerations. Thus for a given VerbNet sense v, the set $V_w(v)$ may contain multiple WordNet synsets. We compare these synsets with our current synset of interest s, simply by using the identity function as the similarity measure $\mathrm{sim}(s_v, s)$, although there are also many existing WordNet sense similarity measures that could be plugged in here. If none of the synsets $s_v \in V_w(v)$ are similar, then the overall value of $\mathrm{sim}(e, s)$ will turn out to be just a small $\epsilon > 0$. In contrast, if there are similar synsets, we still make sure to discount using the set cardinality $|V_w(v)|$, based on the intuition that the score should be higher if there are fewer possible synset matches and thus less ambiguity.

These scoring functions thus allow us to select likely WordNet synsets for a given input entry.

4.2   Lexicon

Once we have disambiguated all the lexical items involved in the construction based on the disambiguation procedure described above, we can assign the implicative commitments tags associated with the original construction to our newly sense-disambiguated version of the construction. The latter refers to specific senses of lexical items, thus eliminating possible confusion about which senses are meant in the case of ambiguous words. At the same time, the latter is more general by enabling us to instantiate the construction with alternative synonymous words, rather than forcing us to select the original lexical item.

In addition to the new sense-specific implicative commitment markings, our final lexicon incorporates information from WordNet, VerbNet, and other lexical resources. For a given word, we list a number of (potentially overlapping) senses. Each sense corresponds either to a WordNet synset or to VerbNet sense or to both, based on the mappings between the two resources provided with VerbNet.

For some of these word senses, we now have information about constructions they can be involved in and the implicative commitments that are entailed when relevant words with those senses (or hyponym senses) are used in those specific constructions. The implicative commitments are provided as properties of the constructions, marked using tags like fact_p (based on the ones in the original lexicons). The constructions are linked to specific word senses and thus our lexicon describe the implicative commitments that words with a corresponding word sense entail when used in the corresponding syntactic constellation.

Additionally, our lexicon also provides references to other resources. For every sense corresponding to a VerbNet entry, we list the corresponding FrameNet

[15] frames, relying on the SemLink project's mappings between VerbNet and FrameNet [16]. For every sense corresponding to a WordNet entry, we also include the relevant SUMO [17] concepts based on the pre-existing WordNet-SUMO mappings [18].

We make use of the RDF standard for information interchange, but our data can be converted to any number of other formats, including simple tab-separated-value files. It is clear that this is just a first step towards a freely available open "unified" lexicon in the spirit of Crouch & King [19]. In future work, we additionally hope to add presupposition relationships between different verbs, as studied by Temper and Frank [20] (e.g., given "Spain *won* the tournament", one may presuppose that Spain *played* in the tournament). Another possible extension is to mine corpora for information about downward entailment as proposed by Danescu-Niculescu-Mizil et al. [21].

5   Results

5.1   Data

For our input, we relied on the lexicons from Nairn et al. [4] and related papers, as distributed by CSLI[1], the creation of which was largely DARPA-funded, in the context of the ACQUAINT project.

A lexicon entry consists of a lemma, a syntactic category descriptor, one or more tags to describe the implicative commitment behavior, and frequently also an example sentence for the lemma in the respective syntactic frame highlighting such a possible implicative commitment.

In addition to the simple factives data and the simple implicatives data, we also included the more recent phrasal implicatives data [1], which captures implicative behaviors that depend on the specific arguments of a verb. We likewise included the CSLI data about the factivity of adjectival statements, e.g. "It is accurate that John informed the president" vs. "It is untrue that John informed the president, as well as the reduced set of markings on nouns from the appendix of Price et al. [7].

Moreover, we incorporated the CSLI infinite temporal markings data, which provides information about possible temporal shifts in addition to factivity information. For instance, for a sentence like "Mary persuaded Ed to cook dinner", the cooking can happen after the persuasion, while for "Mary let Ed cook dinner" the cooking typically is simultaneous with the letting. Finally, we created a lexicon of entries about existential commitments, drawing on analyses in the Amaral et al. paper [3], among other sources.

To disambiguate all of this data with respect to WordNet, we made use of the algorithm described in Section 4.1, choosing the highest-ranked synset for each lexical item in the input. In the case of ties (in particular for the nouns and adjectives, where the VerbNet-based heuristics do not apply), we chose the top WordNet synset in terms of WordNet's original frequency-based sense ranking.

---

[1] http://www.stanford.edu/group/csli_lnr/Lexical_Resources/

Table 1 summarizes the results of this process. For each category of information, we describe the number of lexical items in the input and the relevant number of candidate senses in VerbNet and WordNet. We then list the number of annotated entries in the input and in our disambiguated output. Since for many data types, annotated entries only refer to a single lexical item, these numbers are often quite correlated. On a random sample of 20 simple implicatives, our automatic disambiguation had a precision of 75.0%. This is a reasonable result given that we are using fine-grained WordNet word senses and that particularly for verbs we have to choose from a very large set of candidates, as can be seen in Table 1. We could not reliably assess the recall of our disambiguation, because for many of the rarer word senses, it is in fact more difficult to establish whether the same implicatives hold as for the primary senses of those lexical items.

In any case, we are currently in the process of relying on human annotators to correct and extend our data, giving us a more complete set of reliable WordNet sense annotations.

Table 1. Implicative Commitment Data

Data Type	Lexical Items			Annotated Entries	
	Input	VN Cand.	WN Cand.	Input	Disambiguated
Simple factives	108	188	486	108	108
Simple implicatives	114	223	950	114	114
Phrasal implicatives	17	67	584	14	14
Factivity of adj. comp.	278	0	449	278	225
Factivity of extraposed adj. comp.	695	0	1534	695	623
Factivity of noun comp.	112	0	304	112	108
Temporal implicatives	86	160	811	86	86
Existential commitments	74	101	428	62	62

5.2   Discussion

One of the low hanging fruits of using our sense-specific implicative markings is the extension of coverage it affords. For example the original lexical markings repository provides the marking for "to bother" but not for its WordNet synonym "to trouble", as in "It troubled Ed that Mary was about to leave". Similarly there is a marking for "to acknowledge" as in "They acknowledged that the report was correct" but not for "to concede" in the same frame.

Obviously, not all synonyms can be used in the syntactic frame specified by the annotation entry. WordNet's synonym sets are semantically motivated, and as such may include words with different syntactic behavior. Thus the extension only applies to words with compatible syntactic behavior. For example, "to hate" is marked as factive as in "Ed hated to leave the party" or "Ed hated that Mary went home". The synonymous verb "to detest" was not marked in the original lexicon, but is a possible synonym listed by our lexicon.

Another advantage of a semantic approach is that we can more accurately describe the behavior of words when this behavior depends on the semantic context. For example, if you have "wasted your chance to go to Paris" , then you probably did not go, whereas if you "wasted your money to go to Paris", then you probably did go [1]. Similarly, "He withdrew his hand" doesn't affect the existence of the hand, while "He withdrew his offer" does reflect a change regarding the existence of the offer. Our lexicon distinguishes the two cases based on the semantic type of the complement. Given a novel example, an NLP system can make use of WordNet's hypernym hierarchy or of WordNet-based semantic relatedness measures to determine which of the two cases is more likely.

Of course, our lexicon nevertheless lacks descriptions to account for certain more involved contexts. As described in Section 3, it seems difficult to account for all possible context-specific behaviors. For example, while "to cause something" generally implies that that something comes to be, there are also examples like "The decree was causing a revolution when it was revoked", from which one is likely to conclude that a revolution was ultimately avoided [3].

5.3   Cross-Lingual Applicability

Our WordNet sense markings not only enable us to find new synonyms, but also allow us to look up non-English equivalents in sense-aligned non-English versions of WordNet, such as EuroWordNet [22] and UWN [23].

We have additionally arranged for a human-created Portuguese translation of the Nairn et al. lexicon and examples[2], and checked that most of the inferential behavior is preserved under direct translations. Some anecdotal observations can be made about this translation. First, as expected, the translation tends to indicate more English verbs, with subtle variations of meaning, going to a single Portuguese verb, for example "abhor" and "abominate" mapping to "abominar", or "acknowledge" and "recognize" to "reconhecer", or, more telling, "amaze"/"astonish"/"surprise" going to "surpreender". All the factive verbs and their examples provided by Nairn et al. seem to work in Portuguese, very much like in English. This is despite some non-direct translations, for example sometimes a single work in English like "affect" becomes a phrasal verb in Portuguese "fazer de conta" (make belief), or "to perplex" becomes "deixar perplexo". Things are not always as clear-cut on the implicatives and their examples. More work is needed here.

6   Conclusion

In this paper, we have described the creation of a freely available lexical resource that encodes sense-specific implicative commitments.

Our first contribution is to bring the remarkable information distributed by CSLI to a wider audience, so that they can be improved, and lexicographical

---

[2] The authors would like to thank Henrique Oliveira for his help with the Portuguese translations.

gaps plugged, as suggested by Nairn, Condoravdi, and Karttunen themselves. Their data, together with further data collected from other papers mentioned earlier forms the basis of our resource.

Our second contribution is a first step towards an open source "Unified Lexicon" that aligns and combines this information with main-stream semantic resources. We transfer the original information to the level of word senses, enabling applications to benefit from greater coverage as well as from finer-grained sense-specific entailment information. Our lexicon integrates this information with WordNet, VerbNet, FrameNet, and SUMO, leading to a single one-stop resource that has enormous potential as a backbone for semantic and pragmatic inference as well as for linguistically oriented ontologies.

Our third contribution is some evidence to the fact that lexical resources in English can possibly (and profitably) be used to induce lexical resources in other languages. Given that many years of work have been spent in producing a variety of lexical resources for English, we would like to channel all this effort into other useful projects for several different languages to the extent this is feasible. We appreciate that these resources will not work in multiple languages "out-of-the-box" in a satisfactory manner – completions and adaptations will be necessary – but they do seem to provide a baseline to bootstrap your work from. It is encouraging to see measurable signs that some things work the same way, if we discuss them at the level of concepts.

While some of the predictions in this sort of resource may be more clear-cut than others, individual researchers can tailor it to their applications. For example, if logicians insist that the expression "X says that Y" is always agnostic on the truth-value of "Y" no matter who or what "X" is, a more pragmatic system may decide that if "X" is "The New York Times" or the FDA (Food and Drug Administration of the US) and the author of the sentence is also a reputable source, then "Y" is to be considered true. We thus hope that our resource will be adopted for use in many systems, as the focus in natural language processing shifts from more fundamental operations to higher-level tasks requiring advanced pragmatic inferences.

Our lexicon is available from `http://lexvo.org/implicative-lexicon/`.

References

1. Karttunen, L.: Simple and phrasal implicatives. In: Proceedings of the First Joint Conference on Lexical and Computational Semantics: Proceedings of the Main Conference and the Shared Task, Proceedings of the Sixth International Workshop on Semantic Evaluation, SemEval 2012, vol. 1&2, pp. 124–131. Association for Computational Linguistics, Stroudsburg (2012)
2. Zaenen, A., Karttunen, L.: Veridicity annotation in the lexicon? a look at factive adjectives. In: Proceedings of the 9th Joint ISO - ACL SIGSEM Workshop on Interoperable Semantic Annotation, pp. 51–58. Association for Computational Linguistics, Potsdam (March 2013)

3. Amaral, P., de Paiva, V., Condoravdi, C., Zaenen, A.: Where's the meeting that was cancelled? existential implications of transitive verbs. In: Proceedings of the 3rd Workshop on Cognitive Aspects of the Lexicon, Mumbai, India, The COLING 2012 Organizing Committee, pp. 183–194 (December 2012)

4. Nairn, R., Condoravdi, C., Karttunen, L.: Computing relative polarity for textual inference. In: Proceedings of ICoS-5 (Inference in Computational Semantics) (2006)

5. Karttunen, L.: The logic of English predicate complement constructions. Publications of the Indiana University Linguistics Club, Bloomington (1971)

6. Kiparsky, P., Kiparsky, C.: Fact. In: Bierwisch, M., Heidolph, K.E. (eds.) Progress in Linguistics: A Collection of Papers, Mouton, The Hague, pp. 143–73 (1970)

7. Price, C., de Paiva, V., Holloway King, T.: Context inducing nouns. In: COLING 2008: Proceedings of the Workshop on Knowledge and Reasoning for Answering Questions, Coling 2008 Organizing Committee, Manchester, UK, pp. 9–16 (August 2008)

8. Karttunen, L., Zaenen, A.: Veridicity. In: Katz, G., Pustejovsky, J., Schilder, F. (eds.) Annotating, Extracting and Reasoning about Time and Events. Dagstuhl Seminar Proceedings 05151, Dagstuhl, Germany, Dagstuhl, Germany (2005)

9. Saurí, R., Pustejovsky, J.: Factbank: A corpus annotated with event factuality. Language Resources and Evaluation 43(3), 227–268 (2009)

10. Saurí, R., Pustejovsky, J.: From structure to interpretation: A double-layered annotation for event factuality. In: The 2nd Linguistic Annotation Workshop, LREC, Marrakech, August, May 26-27 (2008)

11. Saurí, R., Pustejovsky, J.: Are you sure that this happened? Assessing the factuality degree of events in text. Computational Linguistics 38(2) (2012)

12. de Marneffe, M.-C., Manning, C.D., Potts, C.: Veridicality and utterance meaning. In: Proceedings of the Fifth IEEE International Conference on Semantic Computing: Workshop on Semantic Annotation for Computational Linguistic Resources. IEEE Computer Society Press, Stanford (2011)

13. Kipper-Schuler, K.: VerbNet: A Broad-Coverage, Comprehensive Verb Lexicon. PhD thesis, University of Pennsylvania (2005)

14. Bobrow, D.G., Cheslow, B., Condoravdi, C., Karttunen, L., King, T.H., Nairn, R., de Paiva, V., Price, C., Zaenen, A.: PARC's Bridge and Question Answering system. In: King, T.H., Bender, E.M. (eds.) Proceedings of the GEAF 2007 Workshop, pp. 46–66. CSLI, Stanford (2007)

15. Baker, C.F., Fillmore, C.J., Lowe, J.B.: The Berkeley FrameNet project. In: Proc. COLING-ACL 1998, pp. 86–90 (1998)

16. Bonial, C., Stowe, K., Palmer, M.: Renewing and revising SemLink. In: Proc. 2nd Workshop on Linked Data in Linguistics, Pisa, Italy (2013)

17. Niles, I., Pease, A.: Toward a Standard Upper Ontology. In: Welty, C., Smith, B. (eds.) Proceedings of the 2nd International Conference on Formal Ontology in Information Systems, FOIS 2001 (2001)

18. Niles, I., Pease, A.: Linking lexicons and ontologies: Mapping WordNet to the Suggested Upper Merged Ontology. In: Proceedings of the IEEE International Conference on Information and Knowledge Engineering, pp. 412–416 (2003)

19. Crouch, D., King, T.: Unifying lexical resources. In: Proceedings of the Interdisciplinary Workshop on the Identification and Representation of Verb Features and Verb Classes, Saarbruecken, Germany (2005)

20. Tremper, G., Frank, A.: A Discriminative Analysis of Fine-Grained Semantic Relations including Presupposition: Annotation and Classification. In: Dipper, S., Zinsmeister, H., Webber, B. (eds.) Dialogue and Discourse, Special Issue on Annotating Pragmatic and Discourse Phenomena, pp. 282–322 (2013)

21. Danescu-Niculescu-Mizil, C., Lee, L., Ducott, R.: Without a "doubt"? Unsupervised discovery of downward-entailing operators. In: HLT-NAACL, pp. 137–145. The Association for Computational Linguistics (2009)

22. Vossen, P. (ed.): EuroWordNet: A Multilingual Database with Lexical Semantic Networks. Springer (1998)

23. de Melo, G., Weikum, G.: Towards a Universal Wordnet by learning from combined evidence. In: Proceedings of the 18th ACM Conference on Information and Knowledge Management, CIKM 2009, pp. 513–522. ACM, New York (2009)

A Tiered Approach to the Recognition
of Metaphor

David B. Bracewell, Marc T. Tomlinson, Michael Mohler, and Bryan Rink

Language Computer Corporation
Richardson, TX
{david,marc,michael,bryan}@languagecomputer.com

Abstract. We present a tiered-approach to the recognition of metaphor. The first tier is made up of highly precise expert-driven lexico-syntactic patterns which are automatically expended on in the second tier using lexical and dependency transformations. The final tier utilizes an SVM classifier using a variety of syntactic, semantic, and psycholinguistic features to determine if an expression is metaphoric. We focus on the recognition of metaphors in which the target is associated with the concept of "Economic Inequality" and examine the effectiveness of our approach for metaphors expressed in English, Farsi, Russian, and Spanish. Through experimental analysis we show that the proposed approach is capable of achieving 67.4% to 77.8% F-Measure depending on the language.

1 Introduction

Metaphor is a pervasive literary mechanism allowing individuals to view one concept in terms of the properties of another. As recent empirical studies have shown, metaphor is abundant in natural language occurring, as often as every third sentence [1]. Because of its abundance and often unique usage, it is important to build a system that can recognize and understand metaphor in order to aid natural language processing applications, such as authorship identification and semantic interpretation.

Metaphors enrich our conversation and provide a mechanism to map an abstract *target* domain into in a concrete *source* domain [2] allowing for the target to be discussed, understood, and affect assessed through the source. Below lists some examples of metaphor:

1. They cannot escape **poverty's grasp**.
2. The **burden of the tax** is onerous.
3. **Wages have stagnated**.

In the second example above, "tax" is being described as a "burden" transferring the the heavy weight property associated with a burden to tax.In the third example, "wages" are being described as stagnated invoking a mapping to a body of water or a volume of air which has ceased to move.

Automated methodologies for processing metaphor can be broken down into two main categories: recognition and interpretation. Interpretation of metaphor

A. Gelbukh (Ed.): CICLing 2014, Part I, LNCS 8403, pp. 403–414, 2014.

involves determining the intended, or literal, meaning of a metaphor [3,4]. The recognition of metaphor entails determining if an expression is literal or figurative. Work on automated metaphor recognition dates back to the early 1990's with the work of Fass [5] based on selectional preference violations and more recently with the work of Shutova [6] using hierarchical graph factorization clustering.

In this paper, we propose a tiered approach to the recognition of metaphor. The first tier is made up of highly precise expert-constructed lexico-syntactic patterns which recognize metaphors associated with a predefined set of source and target domains. The second tier builds off the first by automatically expanding the lexico-syntactic patterns with dependency information and a series of lexical and dependency transformations. Finally, the third tier uses a combination of highly precise identification of target elements (spans of text associated with a target domain) with an SVM classifier to determine if a target element and a candidate source span of text linked to the target through a dependency chain is metaphoric.

In this paper, we limit our focus to recognition of metaphors in which the target is associated with the abstract domain of Economic Inequality. In particular, we focus on the following sub-domains of Economic Inequality: *Poverty*, *Wealth and Social Class*, and *Taxation*. We examine the effectiveness of our approach in multiple languages: English, Farsi, Russian, and Spanish.

The paper will proceed as follows. In section 2, we will present related work in metaphor processing. Then in section 3 will layout our tiered-approach to recognizing metaphor. Next, in section 4 will give experimental results of our approach for English, Farsi, Russian, and Spanish, Finally, in section 5 we will present concluding remarks.

2   Related Work

Metaphor has been studied by researchers in many fields, including, psychology linguistics, sociology, anthropology, and computational linguistics. A number of theories of metaphor have been proposed including the Contemporary Theory of Metaphor [2], the Conceptual Mapping model [7], the Structure Mapping Model [8], and the Attribute Categorization Hypothesis [9]. Based on these theories, databases of metaphors, such as the Master Metaphor List (MML) [10] for English and the Hamburg Metaphor Database (HMD) [11] for French and German, have been constructed. The MML provides links between domains (source and target) and their conceptual mappings. The HMD fuses EuroWordNet synsets with the MML source and target domains.

An active area of research in computational linguistics has been on the recognition of figurative language [12,13]. The recognition of metaphoric expressions [5,14], one part of the more general figurative language recognition task, has in particular seen a number of advances in recent years. Much of the early work on the recognition of metaphor used hand-crafted world knowledge. The met* [5] system determined if an expression was literal or figurative using selectional preferences. Figurative expressions were then determined to be metonymic using hand-crafted patterns or metaphoric using a manually constructed database of analogies. The CorMet [14] system determined the source and target concepts of a metaphoric expression using domain-specific selectional preferences mined from Internet resources.

Exemplar-based approaches to metaphor recognition have shown good results although are often limited in the metaphoric expressions they can identify. The Metaphor Interpretation, Denotation, and Acquisition System (MIDAS) [15] employed a database of conventional metaphors that could be searched to find a match for a metaphor discovered in text. In [16] semantic signatures were utilized to expand the metaphoric expressions producing a more durable and robust system for linking into the metaphor example repository.

Clustering-based approaches have also been prominent in the recognition of metaphor. In [17] noun-verb clustering starting from a small seed of one-word metaphors were used to generate clusters representing target and source concepts connected via a metaphoric relation. The clusters were then used to annotate the metaphoricity of text. Extending upon the noun-verb clustering work, [6] examined the use of hierarchical graph factorization clustering of nouns in a fully unsupervised approach to metaphor recognition. In [18] imageability and topicality were coupled with proto source induction for the recognition of metaphor.

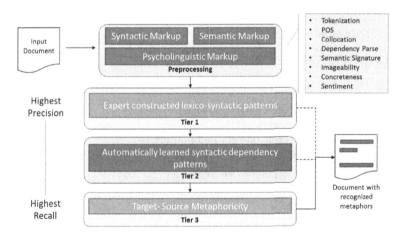

Fig. 1. Architecture of the proposed three-tiered metaphor recognition system

3 Methodology

We propose a supervised approach to the recognition of metaphor that melds human knowledge and machine learning into a single three-tiered architecture. The first tier, discussed in detail in section 3.1, is made up of high precision expert-constructed lexico-syntactic patterns for a predefined set of source and target domains. The second tier, discussed in detail in section 3.2, consists of syntactic dependency patterns which are automatically derived from the first-tier patterns. The final tier,

discussed in detail in section 3.3, uses a combination of highly accurate target domain identification using semantic signatures [3] with an SVM classifier that utilizes a variety of syntactic, semantic, and psycholinguistic features to determine if an expression is metaphoric. The overall architecture is illustrated in Figure 1.

3.1   Tier 1: Expert Constructed Lexico-Syntactic Patterns

The first tier in the proposed metaphor recognition system is made up of high precision lexico-syntactic patterns, examples are shown in Figure 2. The patterns define a source domain, target domain, and any lexical relation needed to exist for the two to be considered metaphoric. For example, In the English example in Figure 2, the pattern consists of a placeholder for a noun phrase from the source domain "BODY OF WATER" and noun phrase from the target domain of "POVERTY" and in order for the two to be metaphoric their must exist the word "of" between the source and target.

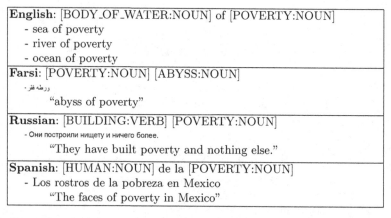

Fig. 2. Examples of high precision lexico-syntactic patterns used in the first tier of the proposed metaphor recognition system

We have defined a set of 51 source domains which either frequently occur with the target domain of Economic Inequality or one of the three sub-domains we are focusing on ("Poverty", "Wealth and Social Class", and "Taxation") in metaphoric expressions or are generic concepts often used in metaphors, e.g. "Movement" and "Vertical Scale". For each of the source domains we have defined lexical items for Nouns, Verbs, and Adjectives, which are strong exemplars of the given domain. The target domain lexical items are defined using semantic signatures of the sub-domain using the method described in [3] and enhanced using language specific lexicons for concepts not captured by the semantic signatures.

The pattern matching process allows for a gap of up to two words between any element in a pattern. For example, the English pattern listed in Figure 2 would also match "ocean of malnourished poor" and "river of unwanted and poor". The patterns in the first tier are checked against the training data and those with a precision less than 98% are discarded. This process left us with 450 patterns in English, 258 in Farsi, 65 in Russian, and 325 in Spanish covering the three sub-domains of Economic Inequality.

3.2   Tier 2: Automatically Learned Syntactic Dependency Patterns

The lexico-syntactic patterns used in the first tier are high precision, but low recall. To overcome this limitation the second tier constructs a set of automatically learned syntactic dependency patterns. The second tier leverages the first tier patterns as seed patterns. Dependency transformations are performed atop these seed patterns to discover new candidate patterns.

Following from the work of [19] we use two types of transformations. The first transformation replaces single lexical restrictions in the pattern with a wildcard. As an example, it would replace "of" in the English example shown in Figure 2 with a wildcard ("$T*$") creating the pattern:

$$[\text{BODY\_OF\_WATER:NOUN}] \ \mathbf{T*} \ [\text{POVERTY:NOUN}]$$

The second transformation works over expert defined syntactic dependency relations by replacing each dependency relation with a wildcard. Using the same example from Figure 2 with dependency information:

$$[\text{BODY\_OF\_WATER:NOUN}] \xrightarrow{prep} \text{of} \xrightarrow{pobj} [\text{POVERTY:NOUN}])$$

the pattern would be transformed into:

$$[\text{BODY\_OF\_WATER:NOUN}] \xrightarrow{T*} \text{of} \xrightarrow{pobj} [\text{POVERTY:NOUN}]$$
$$[\text{BODY\_OF\_WATER:NOUN}] \xrightarrow{prep} \text{of} \xrightarrow{T*} [\text{POVERTY:NOUN}]$$

where the relations "prep" and "pobj" get replaced by relational wildcards, meaning that other syntactic dependencies will also be considered. Using the expanded set of patterns containing wildcards, we search our training data to find matches. The matches with associated wildcards filled in are then assigned a confidence score based on the number of metaphors it matched and the number of non-metaphoric expressions it matched. Patterns matching less three times and patterns with a confidence less than 95% are discarded.

3.3   Tier 3: Target-Source Metaphoricity

The final tier of the linguistic metaphor identification system utilizes a variety of syntactic, semantic, and psycholinguistic features in an SVM classifier with highly accurate identification of target domains using semantic signatures (discussed briefly in Section 3.6 and in detail in [20]). The processing for recognizing metaphors in the third tier is as follows. First, candidate target elements (spans of text related to a target domain) are identified using semantic signatures. Semantic signatures [3] are constructed using the semantic knowledge in Wikipedia and WordNet. They have already been shown to be highly effective in identifying target domain in text and in identifying potential conceptual metaphors related to a linguistic metaphor.

Second, candidate source elements (spans of text which may be related to a source domain, known or not) are selected as all phrases within a predefined distance of the target element on a collapsed dependency tree. A collapsed dependency tree is one in which multiple dependency nodes have been merged based on a given criteria. We use Malt parser [21] in all four languages and collapse based on named entities, WordNet (and its foreign language equivalents) collocations, and noun/verb + preposition.

The final stage is to determine if each of the target element - source element pairs is metaphoric. We utilize an SVM classifier with the following features to make this judgment: (1) Imageability, (2) Concreteness, (3) Degree of dependency violation, (4) Selectional Strength, (5) Topicality, (6) Semantic Categories, and (7) Family Resemblance. Each of these features will be described in the following subsections. We optimized the C parameter of the SVM classifier by performing a grid search over the training data utilizing 10 fold cross-validation.

Imageability and Concreteness. Imageability and Concreteness are concepts from the field of psychology relating to how well an object represented by a word can be imagined (Imageability) or linked to a sensory experience (Concreteness). Imageability has been found to be a strong indicator in the recognition of metaphors [22,23].

We obtain imageability and concreteness scores by expanding the MRC [24] to full coverage of all WordNet senses across all parts-of-speech. Our methodology then moves beyond WordNet, allowing us to estimate psycholinguistics ratings in cases where such language resources are unavailable. The full details of the expansion can be found in [25].

Figure 3 lists examples of high and low imageability source elements in metaphoric and non-metaphoric phrases. The highly imageable element is "sting" and appears in the metaphoric expression "taxes will sting". The low imageability element is "outperform" and appears in the non-metaphoric expression "the rich now out perform". In this simple case, the use of an imageability feature would help to accept "taxes will sting" as being metaphoric while helping to reject "the rich now outperform".

High Imageability: If youre lucky and have itemized deductions, you will get a refund, but **taxes still sting**, especially for the middle class that pays more than its fair share.

Low Imageability: But **the rich now outperform** the middle class by as much as the middle class outperform the poor.

Fig. 3. Example of high and low imageability source terms in metaphoric and non-metaphoric phrases

3.4 Degree of Dependency Violation and Selectional Strength

The degree of dependency violation feature calculates how unexpected a source and target element are to share a given dependency relation. High degrees of violation are indicative of metaphor. For example, given the following sentence with metaphor highlighted in bold:

OK, our friends on the left have one narrow statistic that says **wealth inequality is soaring**, but to be fair this does not capture the distribution either.

"wealth inequality" (target element) and "soaring" (source element) are unlikely to share the dependency relation of subject making them appearing in this relation a high degree of violation. In contrast, in the following example:

Obama told Joe the Plumber in 2008 that its fair to **tax any income** over $250,000 at 39 percent and that when you "spread the wealth around; it's good for everybody."

"tax" (target element) and "any income" (source element) are likely to be seen with the dependency relation direct object making the combination a low dependency violation.

In a similar vein to the degree of dependency violation feature is the selectional strength feature. Selectional strength is a measure of how variable an element is in its selectional preference (here meaning between dependency relations). In simpler words, it relates to how many distinct classes of words share a given dependency relation with the source element. Source elements with low selectional strength, e.g. "is" or "think", are less likely to be metaphoric.

An example of a high selectional strength, i.e. has few semantic classes occurring in the given relation), is "tilts the field" in "But our **tax system tilts the field**". In contrast, the word "kill" has a low selectional strength in English as we tend to kill many types of things. This can be problematic as kill is often used metaphorically as is the case in "Many profitable employers argue that **taxes will kill jobs** and diminish our states competitive edge".

Both of these features are calculated using large corpora for each of the four language. Dependency relations are determine using Malt Parser and then combined as described earlier.

3.5   Topicality

Topicality is a measure of how topically related a word or phrase is to its context, i.e. hammer would topically related to a context discussing home improvement and not topically related to a context discussing legislation. Topically unrelated words are highly likely to be metaphoric.

The topical relatedness of a candidate source element is calculated by constructing a graph $G = (V, E)$ where, the vertices are the lemmatized version of the words in the candidate passage (sentence containing the candidate source and its context of two sentences before and after) and weighted edges exist between vertices whose similarity is greater than the average of all pairs. Similarity is determined using the cosine similarity between the row vectors constructed using Latent Semantic Analysis (LSA) over a large corpus.

Each lemma, l_i, is assigned a topicality score by:

$$score(l_i) = \frac{\sum_{l_j \in P(l_i)} count(l_j)}{\sum_{l_j} count(l_j)} \tag{1}$$

Where $P(l_i)$ is the set of lemmas for which a path exists l_i from l_j and $count(l_j)$ is the number of times l_j occurred in the candidate passage.

3.6   Semantic Categories

Following the work of [16] and [3], we incorporate the semantic features made available through the semantic signatures. Semantic signatures are a set of highly related and interlinked (WordNet) senses corresponding to a particular domain with statistical reliability. To generate the semantic signature we build off: (1) The semantic network encoded in WordNet; (2) The semantic structure underlying Wikipedia; and (3) Collocation statistics provided by statistical analysis of large corpora. We use the Princeton WordNet [26] for English, FarsNet [27] for Farsi, RussNet [28] for Russian, and the Multilingual Central Repository [29] for Spanish as our underlying WordNets.

We create target - source pairs of possible semantic categories using the output of the semantic signature and Wikipedia categories. A source and target element with a semantic mismatch, i.e. relating to semantic categories with little to no similarity, are more indicative of metaphor than those with no mismatch. For example, the target element "tax" and candidate source element "cow" in the expression "tax cow" represent a semantic mismatch as the corresponding semantic categories (taken from Wikipedia) "Taxation" and "Domesticated Animals" are semantically unrelated.

3.7   Family Resemblance

Family resemblance [30] based theories of categorization suggest that an item is classified based on its resemblance to the prototypical items in the category. Conceptual categories are arranged in a hierarchy in which the higher an item is in the tree the more generic it is and the lower it is in the tree the more specific it is, e.g. The concept "beagle" would be lower in the hierarchy than "canine". Each conceptual category has a set of culturally salient prototypical examples which are the items that most come to mind when imagining the given category. A prototypical example of furniture for an American would likely be "chair" whereas for a Japanese person it is likely to be "futon".

We approximate the source element's likelihood of being a prototypical example using a combination of TF-IDF and distributional semantics. We first find semantically similar concepts to the candidate source element using its associated semantic class as induced through distributional semantics. The items in the semantic class make up the examples (prototypical and not) for the candidate source element's category. We then use the TF-IDF values of the concepts in the semantic class to calculate a z-score for the candidate source element. The lower the candidate source element's z-score the less likely it is a prototype for the associated category.

4   Experimentation

For experimentation, we used a training set of roughly 1,000 metaphors per language over a wide variety of genres of data, including blog posts, forum posts, news articles, and transcripts. For testing we had a set of approximately 100 metaphors per language from the same genres as the training set. Spanish and English documents came from ClueWeb09[1], Farsi documents were gathered from web crawls, and Russian documents came from Ruwac[2].

The training and testing set were both annotated by multiple annotators. We did this not to determine inter-annotator agreement, but because we found a single annotator's recall in recognizing metaphor was poor. This problem arises based on the annotators' background and to how standard the metaphoric expression has become, e.g. "tax system" is a metaphor, but has become so standard that most people will not recognize it as one.

The results of the experimentation are shown in Table 1. We gave the system credit in recognizing a metaphor when the source and target elements it chose overlapped with the source and target elements chose by an annotator. This was because even two annotators would rarely agree on the exact same spans of text for the target and source elements.

As can be seen from Table 1, the tiered approach is able to precisely recognize metaphoric expressions in all four languages. Analyzing the errors, we found that rare words, errors in part of speech, and errors in the dependency parse caused a majority of the problems, especially in the non-English languages. In other cases the selectional strength, imageability, or concreteness of a term was too low causing a valid metaphor to be identified as non-metaphoric.

Table 2 shows the results when only the first two tiers of the system, expert-construct lexical patterns and automatically learned syntactic dependency patterns, were used. As one would expect the first two tiers resulted in high precision

---

[1] http://lemurproject.org/clueweb09/index.php
[2] http://corpus.leeds.ac.uk/ruscorpora.html

412   D.B. Bracewell et al.

Table 1. Results of the tiered system for recognizing metaphors

	Precision	Recall	F-Measure
English	73.8%	82.3%	77.8%
Farsi	60.0%	83.0%	69.6%
Russian	76.9%	51.9%	61.4%
Spanish	54.3%	88.7%	67.4%

Table 2. Results of Tier 1 and 2 at recognizing metaphors

	Precision	Recall	F-Measure
English	100.0%	7.0%	13.0%
Farsi	100.0%	10.7%	19.4%
Russian	100.0%	15.4%	26.7%
Spanish	100.0%	10.8%	19.5%

but low recall. Interestingly, Russian which had the fewest expert constructed patterns had the highest recall.

5   Conclusion

In this paper, we presented a tiered-approach to the recognition of metaphor. The first tier was made up of highly precise expert-constructed lexico-syntactic patterns for a set of 51 source domains and the three predefined sub-domains of Economic Inequality. The second tier was made up of automatically constructed syntactic dependency patterns which were learned by performing lexical and dependency transformations atop the first tier patterns. The final tier used a combination of highly accurate target domain identification using semantic signatures with an SVM classifier using a variety of syntactic, semantic, and psycholinguistic features. We examined the effectiveness of our approach for English, Farsi, Russian, and Spanish. The proposed approach was capable of achieving 67.4% F-Measure in Russian and Spanish, 69.6% F-Measure in Farsi, and 77.8% F-Measure in English.

Acknowledgments. This research is supported by the Intelligence Advanced Research Projects Activity (IARPA) via Department of Defense US Army Research Laboratory contract number W911NF-12-C-0025. The U.S. Government is authorized to reproduce and distribute reprints for Governmental purposes notwithstanding any copyright annotation thereon. Disclaimer: The views and conclusions contained herein are those of the authors and should not be interpreted as necessarily representing the official policies or endorsements, either expressed or implied, of IARPA, DoD/ARL, or the U.S. Government.

References

1. Shutova, E., Teufel, S.: Metaphor corpus annotated for source-target domain mappings. In: Proceedings of LREC (2010)
2. Lakoff, G., et al.: The contemporary theory of metaphor. Metaphor and Thought 2, 202–251 (1993)
3. Bracewell, D.B., Tomlinson, M.T., Mohler, M.: Determining the conceptual space of metaphoric expressions. In: Gelbukh, A. (ed.) CICLing 2013, Part I. LNCS, vol. 7816, pp. 487–500. Springer, Heidelberg (2013)
4. Shutova, E.: Models of metaphor in nlp. In: Proceedings of the 48th Annual Meeting of the Association for Computational Linguistics, pp. 688–697. Association for Computational Linguistics (2010)
5. Fass, D.: met*: A method for discriminating metonymy and metaphor by computer. Computational Linguistics 17(1), 49–90 (1991)
6. Shutova, E., Sun, L.: Unsupervised metaphor identification using hierarchical graph factorization clustering. In: HLT-NAACL, pp. 978–988 (2013)
7. Ahrens, K., Chung, S., Huang, C.: Conceptual metaphors: Ontology-based representation and corpora driven mapping principles. In: Proceedings of the ACL 2003 Workshop on Lexicon and Figurative Language, vol. 14. Association for Computational Linguistics,
8. Wolff, P., Gentner, D.: Evidence for role-neutral initial processing of metaphors. Journal of Experimental Psychology: Learning, Memory, and Cognition 26(2), 529 (2000)
9. McGlone, M.: Conceptual metaphors and figurative language interpretation: Food for thought? Journal of Memory and Language 35(4), 544–565 (1996)
10. Lakoff, G.: Master Metaphor List. University of California (1994)
11. Eilts, C., Lönneker, B.: The Hamburg Metaphor Database (2002)
12. Bogdanova, D.: A framework for figurative language detection based on sense differentiation. In: Proceedings of the ACL 2010 Student Research Workshop, ACLstudent 2010, pp. 67–72 (2010)
13. Shutova, E.: Computational approaches to figurative language. PhD thesis, University of Cambridge (2011)
14. Mason, Z.: CorMet: A computational, corpus-based conventional metaphor extraction system. Computational Linguistics 30(1), 23–44 (2004)
15. Martin, J.: A computational model of metaphor interpretation. Academic Press Professional, Inc. (1990)
16. Mohler, M., Bracewell, D., Hinote, D., Tomlinson, M.: Semantic signatures for example-based linguistic metaphor detection, 27 (2013)
17. Shutova, E., Sun, L., Korhonen, A.: Metaphor identification using verb and noun clustering. In: Proceedings of the 23rd International Conference on Computational Linguistics, COLING 2010, pp. 1002–1010. Association for Computational Linguistics, Stroudsburg (2010)
18. Strzalkowski, T., Broadwell, G.A., Taylor, S., Feldman, L., Shaikh, S., Liu, T., Yamrom, B., Cho, K., Boz, U., Cases, I., Elliot, K.: Robust extraction of metaphor from novel data, 67–76 (2013)
19. Rink, B., Roberts, K., Harabagiu, S., Scheuermann, R.H., Toomay, S., Browning, T., Bosler, T., Peshock, R.: Extracting actionable findings of appendicitis from radiology reports using natural language processing. AMIA Summits on Translational Science Proceedings, 221 (2013)

20. Bracewell, D.B., Tomlinson, M.T., Mohler, M.: Determining the conceptual space of metaphoric expressions. In: Gelbukh, A. (ed.) CICLing 2013, Part I. LNCS, vol. 7816, pp. 487–500. Springer, Heidelberg (2013)

21. Nivre, J., Hall, J., Nilsson, J.: Maltparser: A data-driven parser-generator for dependency parsing. In: In Proc. of LREC-2006. (2006) 2216–2219

22. Broadwell, G.A., Boz, U., Cases, I., Strzalkowski, T., Feldman, L., Taylor, S., Shaikh, S., Liu, T., Cho, K., Webb, N.: Using imageability and topic chaining to locate metaphors in linguistic corpora. In: Greenberg, A.M., Kennedy, W.G., Bos, N.D. (eds.) SBP 2013. LNCS, vol. 7812, pp. 102–110. Springer, Heidelberg (2013)

23. Turney, P.D., Neuman, Y., Assaf, D., Cohen, Y.: Literal and metaphorical sense identification through concrete and abstract context. In: Proceedings of the 2011 Conference on the Empirical Methods in Natural Language Processing, pp. 680–690 (2011)

24. Coltheart, M.: The MRC psycholinguistic database. The Quarterly Journal of Experimental Psychology 33(4), 497–505 (1981)

25. Mohler, M., Tomlinson, M., Bracewell, D., Rink, B.: Semi-supervised methods for expanding psycholinguistics norms by integrating distributional similarity with the structure of wordnet. In: Proceedings of the 9th Language Resources and Evaluation Conference (2014)

26. Fellbaum, C.: WordNet, An Electronic Lexical Database. The MIT Press (1998)

27. Shamsfard, M., Hesabi, A., Fadaei, H., Mansoory, N., Famian, A., Bagherbeigi, S., Fekri, E., Monshizadeh, M., Assi, S.: Semi automatic development of farsnet; the persian wordnet. In: Proceedings of 5th Global WordNet Conference, Mumbai, India (2010)

28. Azarova, I., Mitrofanova, O., Sinopalnikova, A., Yavorskaya, M., Oparin, I.: Russnet: Building a lexical database for the russian language. In: Proceedings of Workshop on Wordnet Structures and Standardisation and How This Affect Wordnet Applications and Evaluation, Las Palmas, pp. 60–64 (2002)

29. Atserias, J., Villarejo, L., Rigau, G., Agirre, E., Carroll, J., Magnini, B., Vossen, P.: The MEANING Multilingual Central Repository. In: Proceedings of the 2nd Global WordNet Conference (GWC), Brno, Czech Republic (January 2004)

30. Rosch, E., Mervis, C.B.: Family resemblances: Studies in the internal structure of categories. Cognitive Psychology 7(4), 573–605 (1975)

Knowledge Discovery with CRF-Based Clustering of Named Entities without a Priori Classes

Vincent Claveau and Abir Ncibi

IRISA-CNRS and INRIA-IRISA
Campus de Beaulieu, 35042 Rennes, France
vincent.claveau@irisa.fr, abir.ncibi@inria.fr

Abstract. Knowledge discovery aims at bringing out coherent groups of entities. It is usually based on clustering which necessitates defining a notion of similarity between the relevant entities. In this paper, we propose to divert a supervised machine learning technique (namely Conditional Random Fields, widely used for supervised labeling tasks) in order to calculate, indirectly and without supervision, similarities among text sequences. Our approach consists in generating artificial labeling problems on the data to reveal regularities between entities through their labeling. We describe how this framework can be implemented and experiment it on two information extraction/discovery tasks. The results demonstrate the usefulness of this unsupervised approach, and open many avenues for defining similarities for complex representations of textual data.

1 Introduction

Labeling sequences are tasks of particular interest for NLP (part-of-speech tagging, semantic annotation, information extraction, etc.). Many tools have been proposed, but in recent years, the Conditional Random Fields (CRF [1]) have emerged as the most effective for many applications. These models are supervised machine learning: examples of sequences with their labels are required.

The work presented in this paper is placed in a different context in which the goal is to bring out information from these sequences. So, we fit in a task of knowledge discovery in which supervision is not applicable: the aim is to discover how the data can be grouped into categories that make sense rather than providing these categories from expert knowledge. Therefore, these discovery tasks are based most often on clustering [2,3,4]; the crucial question is how to calculate the similarity between two interesting entities. In this paper, we propose to divert CRF by producing fake labeling problems in order to make appear entities that are regularly labeled the same way. Of these regularities is then built a notion of similarity between entities, which is thus defined by extension and not by intention.

On the application point of view, in addition to the use for knowledge discovery, the similarities obtained by our approach or the clusters produced can be used upstream of supervised tasks:

A. Gelbukh (Ed.): CICLing 2014, Part I, LNCS 8403, pp. 415–428, 2014.
© Springer-Verlag Berlin Heidelberg 2014

- it can be used to reduce the cost of data annotation. It is indeed easier to label a cluster than annotate a text instance by instance.
- it can help to identify classes difficult to discriminate, or on the contrary exhibit classes whose instances are very diverse. It then makes it possible to adapt the supervised classification task by changing the set of labels.

In the remainder of this article, we position our work in the state-of-the-art and briefly present CRF by introducing some useful concepts for the rest of the article. We then describe in Section 3 the principle of our discovery approach using supervised ML technique in an unsupervised mode for discovery tasks. Two experiments of this approach are then proposed in Sections 4 and 5, before presenting conclusions and future work in the last section.

2   Related Work

Many NLP tasks are nowadays considered as supervised ML problems: they suppose the existence of a set of pre-defined classes, and, of course, examples belonging to these classes. Yet, several studies have proposed moving to a non-supervised framework. Some of these studies are not, strictly speaking, about non-supervision but rather about semi-supervision since their goal is to limit the number of sequences to be annotated. It is particularly the case for the recognition of named entities; indeed, many studies rely on external knowledge bases (e.g. Wikipedia) or extraction rules given by an expert as a bootstrap [5,6,7]. Let us also mention the work on Part-of-Speech tagging without annotated data by HMM [8], Bayesian training [9], integer programming [10] or other approaches [11,12]. Similar work has been proposed, along with other formal frameworks for named entities [13,14]. More recently, entity linking tasks have been explored [15], their goal is to link a string mention in a document to an existing entry/category in a database. In all cases, the perspective of these studies is different than ours as they do not adopt a knowledge discovery setting: they are all based on a *tagset* already established.

The framework that we adopt in this paper is different: we aim at making categories emerge from unannotated data. Unlike previously cited work, we do not make any *a priori* on the possible label set. The task is therefore a clustering one, in which similar elements from the sequences should be grouped, as it was done for example by [4] for some named entities. Clustering words is not a new task in itself, but it relies on the definition of a representation for words (typically a context vector) and a measure of distance (or similarity, typically a cosine). Our approach aims to use the discriminative power of ML tools to provide a more effective measure of similarity. The goal is therefore to turn these techniques from supervised into unsupervised for determining the similarity between any two elements of sequences. In this paper, we report experiments using CRF, which has proved its efficiency for numerous supervised tasks [16,17,18,19, inter alia], but it is worth noting that the whole approach can be applied with other ML methods.

This way of diverting supervised machine learning techniques to bring out similarities in complex unlabeled data has been used for data mining. It was demonstrated as very useful for propositional data (i.e. described by feature-value pairs) for which defining a similarity was difficult (non numeric attributes, bias of a definition *ex nihilo*). Different ML methods have been used in this framework, including Decision Trees and Random Forests [20,21,22]. The approach consists in generating a large number of artificial learning problems, with generated synthetic data that are mixed with the real data, and then in stating what data are classified together regularly. Our approach fits into this framework, but exploits the peculiarities of CRF in order to take into account the sequential nature of textual data.

CRFs [1] are undirected graphical models that represent the probability distribution of annotations (or labels, or tags) y conditionally on observations x. More precisely, in the case of sequences like sentences, the conditional probability $P(y|x)$ is defined through the weighted sum of feature functions f and g. They are usually binary functions satisfying a certain combination of labels and attributes describing the observations and applied at each sequence position: f functions characterize the local relations between the current label in position i and observations; functions g characterize the transitions between the nodes of the graph, that is, between each pair of labels at position i and $i-1$, and the sequence of observations. These functions are defined by the user according to his knowledge of the application; their weights reflect their importance to determine the class. Learning CRF consists in estimating these weights (the vector of weights is noted θ hereafter) from training data. Indeed, from N labeled sequences, the vector θ is searched as the one that maximizes the log-likelihood of the model on these labeled sequences. In practice, this optimization problem is solved by using quasi-Newton algorithms, like L-BFGS [23]. After the learning phase, the application of CRF to new data consists in finding the most probable sequence of labels y^* given a (previously unseen) sequence of observations x. As for other stochastic methods, y^* is generally obtained with a Viterbi algorithm.

3   Principles of the Unsupervised Model

This section describes the principle of our approach. An overview is first given through an algorithm depicting the whole process. We then detail some crucial points, as well as more insights about the practical use of this method.

3.1   General Principle

As we explained above, the main idea of this approach is to derive a distance (or similarity) from repeated classifications of two objects for random learning tasks. The more often objects are detected as belonging to a same class, the closer they are supposed to be. The approach chiefly relies on the fact that CRF will make it possible to exhibit similarity between words by assigning them repeatedly identical labels in varied learning conditions. As for bagging [24], the learning process

---

Algorithm 1. Clustering by CRF

1: **input:** $\mathcal{E}_{\text{total}}$: non labeled sequences
2: **for** great number of iterations **do**
3:     $\mathcal{E}_{\text{train}}, \mathcal{E}_{\text{OoB}} \leftarrow \text{Divide}(\mathcal{E}_{\text{total}})$
4:     Randomly choose labels y_i among $\omega_1...\omega_L$ for sequences in $\mathcal{E}_{\text{train}}$
5:     Randomly generate a set of feature functions f and g
6:     Infer: $\theta \leftarrow \text{L-BFGS}(\mathcal{E}_{\text{train}}, y, f, g)$
7:     Apply: $y^* = \arg\max_y p_{(\theta,f,g)}(y|x)$ for each $x \in \mathcal{E}_{\text{OoB}}$
8:     **for all** classe ω_l among $\omega_1...\omega_L$ **do**
9:         **for all** pair x_i, x_j of $\mathcal{E}_{\text{OoB}}$ such that $y_i^* = y_j^* = \omega_l$ **do**
10:             $\mathcal{M}_{\text{co-label}}(x_i, x_j) + = \text{weight}(x_i, x_j, \omega_l)$
11: $\mathcal{M}_{sim} \leftarrow \text{Transform}(\mathcal{M}_{\text{co-label}})$
12: **return** Clustering($\mathcal{M}_{sim}$)

---

is repeated several times with different settings in order to change the learning bias. For this, several random choices are being implemented at each iteration; they concern:

- the sequences used for learning;
- the labels (number and distribution);
- the feature functions describing words.

These supervised learning tasks on artificial problems should confer, with their variety, important properties of the similarity obtained. It naturally handles complex descriptions (for instance the various attributes of the current word, the word neighbours); it operates a selection of variables by construction, and thus takes into account descriptor redundancies or ignores those of poor quality, and is robust to outliers.

Algorithm 1 gives an overview of the process. The sequential classification with CRF is repeated many times with varying data labels (the ω_i are fake classes) and learning parameters (feature functions, training set $\mathcal{E}_{\text{train}}$). The model is then applied to the data not used as training set, called 'out-of-bag' ($\mathcal{E}_{\text{OoB}}$). Pair of words (x_i, x_j) receiving same labels are memorized, and these *co-labelings*, kept in the matrix $\mathcal{M}_{\text{co-label}}$. From this matrix, similarities between each pairs can be derived (possibly with a simple normalization of the co-label counts) and then used by the clustering algorithm.

3.2 Random Learning

Of course, as we have already pointed out, the important role of randomness does not prevent the user to control the task through different bias. This is reflected for example by the provision of rich descriptions of words: part-of-speech tags of the sequences, semantic information of the words... This is also reflected in the definition of the set of feature functions from which the algorithm can draw the functions f and g at each iteration. In the experiments reported below, these functions are those usually used for Named Entity recognition : word-form, part-of-speech and upper or lower case status from the current word, the 3 preceding

Table 1. Example of sequence with observations (words, parts-of-speech) and fake labels

	l'	audience	entre	nicolas	sarkozy	et	maître	wade	...
x	DET	NC	PREP	NP	NP	COO	NC	NP	...
y	O	O	O	B-fake140	I-fake140	O	B-fake25	B-fake3	...

and the 3 following, bigrams built from these features... Concerning the sets $\mathcal{E}_{\text{train}}$ and $\mathcal{E}_{\text{OoB}}$, at each iteration, 5% sentences are randomly chosen as the training set and the remaining serves as application set.

3.3 About Random Labels

In many applications, the task of clustering is only useful for a subset of the words/phrases in the texts. In this context, it is very common to use BIO type labels that can model multi-word entities (B indicates an entity beginning, I the continuity, O is for words outside entities). Table 1 presents an example of artificial sequences derived from the data used in Section 5.

This external knowledge is part of the essential biases needed to control the process of unsupervised learning and to ensure that it applies to the specific needs of the user. But it is important to note that this knowledge about which entities have to be considered is not the same than the one that we aim to discover via clustering. In the first case, it consists in spotting the entities while in the second case, it consists in making emerge classes of entities without a priori. It is possible in this latter case to assume that we know how to delimit the interesting entities in the sequences; this is the assumption made in several studies on the classification of named entities [13,14,4]. It is also, of course, possible to consider this problem as a learning problem itself for which the user must provide some examples. In both cases, this requires expertise, provided either by intention (objective criteria to define entities) or extension (cf. Subsection 5.2). Each of the experiments reported below adopts one of these two cases.

The choice of the number of the fake labels, as well as their distribution, is important (yet, it has to be underline that the number of labels does not directly impact on the number of clusters that will be eventually generated). A very high number of fake labels may produce a model difficult to infer (CRF complexity is very dependent on the number of labels), and may also result, when applied to $\mathcal{E}_{\text{OoB}}$, in data with few entities sharing the same labels. On the opposite side, if too few random labels are used for the inference step, the model obtained may be not discriminative enough and thus may produce fortuitous co-labeling in $\mathcal{E}_{\text{OoB}}$. Of course, all of this is also dependent on the many other parameters of the inference. For instance, the feature functions may allow or not over-fitting, and thus possibly prevent or favour co-labeling of entities. The size of $\mathcal{E}_{\text{train}}$, and more specifically the number of entities that may receive the same fake label is also important : if, on a systematic basis, many training entities from probably different classes share the same fake label, the model will tend to be not discriminating enough.

In order to correctly take into account these phenomena, it would be necessary to characterize, before the labeling step and ideally before the inference step, the tendency of the model to discriminate entities enough or not. Unfortunately, such an a priori criterion is difficult to formalize. Instead, we use a simple a posteriori regularization: the co-labeling of two entities is considered as more informative if few entities have received this label. This is implemented as a weight function used when updating the $\mathcal{M}_{\text{co-label}}$ matrix. In practice, in the experiments reported below, this weighting function is defined as: $\text{weight}(x_i, x_j, \omega_l) = \frac{1}{|\{x_k | y_k = \omega_l\}|}$ and the number of labels is randomly chosen between 10 and 50 at each iteration.

For some discovery tasks, according to the particular knowledge available for them, it is also possible to bias the distribution of the random labels in the training set. For instance, if one knows that every occurrence of an entity necessarily belongs to the same class, it is important to implement this constraint in the training step. The experiment detailed in Section 4 falls within this framework.

3.4   Clustering

The final step of clustering can be implemented in various ways using different techniques and tools. The famous k-means algorithm requires centroid calculations during the process; this is of course not suitable for our non-metric space, but its variant k-medoids, which uses an object as a representative of a cluster, does not require other similarity/distance measures than those provided by $\mathcal{M}_{sim}$.

Let us underline here that in discovery tasks, the number of clusters to be produced is of course unknown. For our part, in the experiments presented in Sections 4 and 5, another clustering technique is considered, namely Markov Clustering (MCL). This technique was originally developed for partitioning large graphs [25]. Its advantage over k-medoids is that it does not require to set a priori the number of clusters expected, and it also avoids the problem of initialization of these clusters. We therefore consider only our objects (words or other entities of the sequences) as vertices of a graph whose edges are valued based on the similarity contained in $\mathcal{M}_{sim}$.

3.5   Operational Aspects

The iterative process proposed in this paper is obviously expensive but easily parallelizable. In the experiments reported below, the number of iterations was set to 1,000; with such a high number of iterations, the results obtained are stable and can be reproduced despite the several random steps of the algorithm. The main sources of cost in terms of computation time are learning CRF models and the application of these models to label data. The complexity of these steps depends on many parameters, including the size of the training sample, the variety of observations (x), the number of random classes (ω), the attributes considered (feature functions f and g)... To minimize the impact of this cost,

Table 2. Excerpt of a minute-by-minute football report

Time	Report
80mn	Zigic donne quelques frayeurs à Gallas et consorts en contrôlant un ballon chaud à gauche des 16 mètres au devant du Gunner. Le Valencian se trompe dans son contrôle et la France peut souffler.
82mn	Changement opéré par Raymond Domenech avec l'entrée d'Alou Diarra à la place de Sidney Govou, pour les dernières minutes. Une manière de colmater les brèches actuelles ?

we use an implementation of CRF WAPITI that optimizes standard inference algorithms [26].

4 Experimental Validation: Classification of Proper Names

For this first experiment, we consider the problem and data of [4]. The goal is to bring out the various classes of proper names in football (soccer) summaries. More specifically, in their experiments, the authors have attempted to classify names at the corpus level: all occurrences of a same proper name are considered to belong to a unique entity and thus to a unique class. Therefore, in this dataset, entities are not considered as polysemous; even if that point is debatable, we adopt here the same framework than [4].

4.1 Task and Data

The corpus is composed of minute-by-minute match reports in French, taken from various websites. Important events of almost every minute of a match are described (see Table 2): player replacements, fouls, goals...

These data have been manually annotated by experts according to classes defined to meet specific application requirements [27]. These annotations constitute a ground truth: it defines what could be interesting classes, and it associates each proper name to a class (see Figure 1). Note that the classes are very unbalanced with a large *player* class.

4.2 Performance Measures

Since our task of discovery relies on a final stage of clustering, it is evaluated as such. Evaluation of clustering tasks is always difficult: evaluation through external criteria requires to have a reference clustering (ground truth) whose relevance can always be discussed, but the internal criteria (e.g., a measure of cohesion of clusters) are known not to be reliable [28]. We therefore prefer the external evaluation: the clustering obtained by our process is compared with the ground truth produced by experts.

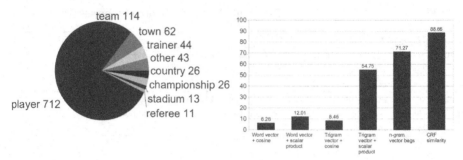

Fig. 1. Distribution of proper names in the **Fig. 2.** Clustering evaluation (ARI %); football ground truth (number of unique football dataset names)

To do this, various evaluation metrics have been proposed, such as purity or *Rand Index* [29]. These measures, however, have a low discriminating power and tend to be overly optimistic when the ground truth contains classes of very different sizes [30]. We therefore prefer the *Adjusted Rand Index* (ARI), a version of the Rand index taking into account the agreement by chance, which has become a standard measure for clustering evaluation and is known to be robust. As secondary measures, we also indicate, when possible, V-measure, normalized mutual information and adjusted purity. Their study and definitions can be found in [31].

4.3   Implementation and Results

To test our clustering method by CRF, the corpus was part-of-speech tagged; we use BIO annotation scheme to generate the fake labels. In this particular application, we take the assumption of [13,14]: entities to categorize are supposed to be known and defined. In practice, it means that the random labels are only generated for these entities; the other words in the corpus receive the label 'O'. Functions f and g are those conventionally used in information extraction: functions f bind the current label y_i to observations (word-form and part-of-speech of the current word in x_i, word-form and part-of-speech in x_{i-1}, x_{i-2} x_{i+1}, x_{i+2}, or combinations of these features), the functions g bind two successive labels (y_{i-1}, y_i). Since the task here is to classify proper nouns at the corpus level and not at the occurrence level, we force two occurrences of a same name to have the same label when generating random labels (step 4 of the algorithm). However, since the CRF models annotate at the occurrence level, the matrix $\mathcal{M}_{\text{co-label}}$ keeps track of the occurrence classifications. The transformation step (step 11) transforms this matrix into a similarity matrix $\mathcal{M}_{sim}$ of proper names in the corpus by summing the rows and columns concerning the different instances of the same names.

The results of our approach are given in Fig. 2 in terms of ARI (percentage, 0 is a random clustering or a all-in-one clustering, and 100 is for a clustering identical to the ground truth). For comparison purposes, we report the results of [4]; these were obtained using vector description of the contexts of each entity, either as a single vector, or as bags of vectors, with suited similarity functions. The clustering step is performed with the same algorithm MCL than for our system. MCL has a parameter called inflation rate that affects indirectly the number of clusters produced. For a fair comparison, the results reported for each method are those for which this setting is optimal for the evaluation measure ARI. In these experiments, it corresponds to 12 clusters with the CRF-based similarity, and 11 for the n-gram bag-of-vectors.

These results emphasize the interest of our approach compared to more standard representations and similarities. The few differences between the clusters formed by our approach and the ground truth classes focus on the class *other*. This class contains the names of individuals appearing in various contexts (personality giving the kickoff, appearing in the audience...), with too few regularity to allow CRF, no more than other methods, to succeed at bringing out a similarity. It is worth noting that the density of entity in this corpus, and the fact that any entity is very likely to appear often in various contexts makes the corpus and the discovery task particular. It may explain why some errors reported by [4] as recurrent are not made by the clustering by CRF. For example, the vector methods tend to confuse the names of cities and names of players, as they often appear close to each other and therefore share the same contexts. These mistakes are not made by the CRF approach, for which the built-in consideration of the sequentiality in the labeling process (word order and label order are taken into account) help to distinguish between these two classes.

5   Experimental Validation for Information Discovery

In order to assess the validity on another type of entity discovery task, this section presents further experiments on a news corpus, with a different definition of what are the interesting entities.

5.1   Task and Data

The data are from the ESTER2 evaluation campaign [32]. They consist of 150 hours of radio recorded between 1999 and 2003 from various sources (France Inter, Radio Classique, Africa 1...). These broadcasts have been transcribed and then annotated for named entities according to 8 categories: people, functions, places, organizations, times, human products, quantity, and a *other* category.

Unlike the previous dataset, entities are annotated at the occurrence level; so, the entity Paris can be annotated as a place or organization depending on the context. In the experiments reported below, only the *dev* part of the ESTER2 dataset; it was transcribed manually, but respects the particularities speech recognition systems: the text has no punctuation or capitalization, which makes the named entity recognition task more challenging than for well-formed written texts. Here again, the manual annotation will serve as a ground truth for our discovery task; its characteristics are given in Figure 3.

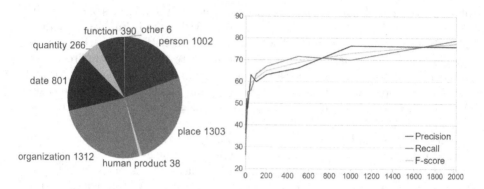

Fig. 3. Distribution of data in ESTER2 ground truth (number of occurrences)

Fig. 4. Performance of entity detection according to the number of annotated training sequences

5.2 Entity Identification

Although it is possible to adopt the same framework as above and assume that the entities to classify are known and defined, here we use a more realistic framework: a small portion of the data is annotated by an expert who defines the entities of interest (but without assigning any class). These data will serve as a first step to learn how to retrieve entities before trying to cluster them. It is therefore a supervised task with two classes (interesting entity or not), for which we use CRF in its traditional way.

Figure 4 shows the detection results, depending on the number of sequences (phrases) used for learning. The performance is evaluated in terms of precision, recall and F-score. It appears that it is possible to retrieve the named entities from relatively few training sentences with good results (compared with the published results on close tasks on this dataset [32]).

5.3 Evaluation of Clustering

The experimental framework is the same as in Section 4.3, except that the classification is done here at the occurrence level. The transformation of $\mathcal{M}_{\text{co-label}}$ in $\mathcal{M}_{\text{sim}}$ is therefore just a normalization. Entities considered are those identified by the previous step with 2,000 annotated sequences. The results, measured in terms of ARI, normalized mutual information, V-measure and adjusted purity are shown in Figure 5. As previously, we present the results of clustering techniques on the same data using more conventional similarities comparing the context and the entities (with the exception of the bags-of-bags approach that cannot apply to classification at the occurrence level). In order to assess the interest of our approach to handle complex representation of the data, we also add the results obtained by our CRF approach taking only into account the word-forms (no other features such as PoS).

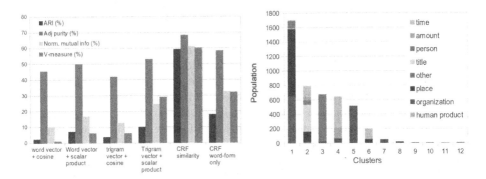

Fig. 5. Clustering evaluation; ESTER2 **Fig. 6.** Confusion between clusters and dataset   ground-truth classes

On this task, the advantage of our approach is clear. Taking into account the sequentiality appears as very important: results with n-grams are indeed better than single words for the contextual vector representations, and our approach based on CRF, which take into account more naturally this sequential aspect, are even better. It is also worth noting that using the word-forms only yields slightly better results than ngrams; it underlines the interest of our approach compared to standard ones, even when using the same set of features, but it also emphasizes the benefit of using complex representation (including PoS for instance), that are easily handled by our approach.

Clusters obtained by our approach, however, are not identical to those of the ground truth as one can see in Fig 6. Indeed, some clusters bring down the results by grouping entities belonging to distinct classes of ground truth. This is the case for 'organization' and 'place', which is a common mistake caused by polysemous names of country or town. This is also the case for 'time' and 'amount' which are difficult to distinguish without additional knowledge. Indeed, in the absence of other information than the form of words and parts-of-speech, it seems impossible to distinguish entities such as `last four days` (time) vs. `on the last fifteen kilometres` (amount).

6   Conclusion and Perspectives

Solving fake learning problems with a ML technique helps to bring out similarities between textual elements. This similarity is making the most of the richness of description that the ML method allows (typically parts-of-speech, sequential information...). This defines a similarity in a non-metric space that is expected to be robust due to repeated random choices in the inference process. The use of CRF, successfully used for many (supervised) tasks, appears as an obvious choice, but of course, the same principles may be applied with other ML methods (HMM, MaxEnt, random forests, SVM).

Evaluations conducted on two information extraction tasks[1] highlight the interest of the approach; although we are well aware of the limits of the evaluation of a discovery task which requires the establishment of a ground truth, which is what we want to avoid by using discovery techniques. It should also be noted that there is no machine learning without bias, even more when dealing with unsupervised learning [33]. These biases represent the knowledge of the user and help define the problem. The provision of information on entities to consider, the description of sequences and the definition of feature functions are pieces of information allowing the user to control the discovery task on its object of study.

Several improvements and perspectives are possible as a result of this work. From a technical point of view, the step transforming co-occurrences into similarities, which is a simple normalization in our experiences, could be deepened. Using other functions (such as those used to identify complex terms: mutual information, Jaccard, log-likelihood, χ^2...) to obtain more reliable similarities is foreseen. It may help to overcome the weakness of our clustering algorithm that can merge two clusters only because a few entities are strongly connected with many other nodes. Several variations concerning this clustering step may also be considered. It is for example possible to use hierarchical clustering algorithms. It is also possible to directly use the similarities between the words for other tasks, such as information retrieval, or for smoothing language models... From a practical point of view, it would be interesting to obtain an explicit definition of the similarity by recovering λ_i and μ_i along with their corresponding functions f and g. This would makes it possible to apply the similarity function to new texts without repeating the costly stages of learning.

References

1. Lafferty, J., McCallum, A., Pereira, F.: Conditional random fields: Probabilistic models for segmenting and labeling sequence data. In: International Conference on Machine Learning, ICML (2001)
2. Wang, W., Besançon, R., Ferret, O., Grau, B.: Filtering and clustering relations for unsupervised information extraction in open domain. In: Proceedings of the 20th ACM International Conference on Information and Knowledge Management (CIKM), Glasgow, Scotland, UK, pp. 1405–1414 (2011)
3. Wang, W., Romaric Besan, C., Ferret, O., Grau, B.: Evaluation of unsupervised information extraction. In: Proceedings of the 8th International Conference on Language Resources and Evaluation (LREC 2012), Istanbul, Turquie (2012)
4. Ebadat, A.R., Claveau, V., Sébillot, P.: Semantic clustering using bag-of-bag-of-features. In: Actes de le 9e Conference en Recherche d'information et Applications, CORIA 2012, Bordeaux, France (2012)
5. Kozareva, Z.: Bootstrapping named entity recognition with automatically generated gazetteer lists. In: Proceedings of the Eleventh Conference of the European Chapter of the Association for Computational Linguistics: Student Research Workshop, Trento, Italy, pp. 15–21 (April 2006)

---

[1] For replicability concerns, the football dataset is available from [27], ESTER2 from ELRA, WAPITI is available on `wapiti.limsi.fr`

6. Kazama, J., Torisawa, K.: Exploiting wikipedia as external knowledge for named entity recognition. In: Proceedings of the 2007 Joint Conference on Empirical Methods in Natural Language Processing and Computational Natural Language Learning, pp. 698–707. Association for Computational Linguistics, Prague (2007)

7. Wenhui Liao, S.V.: A simple semi-supervised algorithm for named entity recognition. In: Proceedings of the NAACL HLT Workshop on Semi-supervised Learning for Natural Language Processing, pp. 58–65. Association for Computational Linguistics, Boulder (2009)

8. Merialdo, B.: Tagging english text with a probabilistic model. Computational Linguistics 20, 155–171 (1994)

9. Richard, D., Benoit, F.: Semi-supervised part-of-speech tagging in speech applications. In: Interspeech 2010, Makuhari, Japan, September 26-30 (2010)

10. Ravi, S., Knight, K.: Minimized models for unsupervised part-of-speech tagging. In: Proceedings of ACL-IJCNLP 2009, pp. 504–512 (2009)

11. Smith, N., Eisner, J.: Contrastive estimation: Training log-linear models on unlabeled data. In: Proceedings of ACL (2005)

12. Goldwater, S., Griffiths, T.L.: A fully bayesian approach to unsupervised part-of-speech tagging. In: Proceedings of the ACL (2007)

13. Collins, M., Singer, Y.: Unsupervised models for named entity classification. In: Proceedings of Empirical Methods for Natural Language Processing (EMNLP) Conference (1999)

14. Elsner, M., Charniak, E., Johnson, M.: Structured generative models for unsupervised named-entity clustering. In: Proceedings of the Conference on Human Language Technology and North American Chapter of the Association for Computational Linguistics (HLT-NAACL 2009), Boulder, Colorado (2009)

15. Ji, H., Grishman, R.: Knowledge base population: Successful approaches and challenges. In: Proceedings of the 49th Annual Meeting of the Association for Computational Linguistics: Human Language Technologies, HLT 2011, vol. 1, pp. 1148–1158. Association for Computational Linguistics, Stroudsburg (2011)

16. Wang, T., Li, J., Diao, Q., Wei Hu, Y.Z., Dulong, C.: Semantic event detection using conditional random fields. In: IEEE Conference on Computer Vision and Pattern Recognition Workshop, CVPRW 2006 (2006)

17. Pranjal, A., Delip, R., Balaraman, R.: Part of speech tagging and chunking with hmm and crf. In: Proceedings of NLP Association of India (NLPAI) Machine Learning Contest (2006)

18. Constant, M., Tellier, I., Duchier, D., Dupont, Y., Sigogne, A., Billot, S.: Intégrer des connaissances liguistiques dans un CRF : Application à l'apprentissage d'un segmenteur-étiqueteur du français. In: Traitement Automatique du Langage Naturel (TALN 2011), Montpellier, France (2011)

19. Raymond, C., Fayolle, J.: Reconnaissance robuste d'entités nommeés sur de la parole transcrite automatiquement. In: Actes de la Conférence Traitement Automatique des Langues Naturelles, Montréal, Canada (2010)

20. Liu, B., Xia, Y., Yu, P.: Cltree-clustering through decision tree construction. Technical report, IBM Research (2000)

21. Hastie, T., Tibshirani, R., Friedman, J.H.: The Elements of Statistical Learning: Data Mining, Inference, and Prediction. Springer, New York (2001)

22. Shi, T., Horvath, S.: Unsupervised learning with random forest predictors. Journal of Computational and Graphical Statistics 15(1), 118–138 (2005)

23. Schraudolph, N.N., Yu, J., Günter, S.: A stochastic quasi-Newton method for online convex optimization. In: Proceedings of 11th International Conference on Artificial Intelligence and Statistics of Workshop and Conference Proceedings, San Juan, Puerto Rico, vol. 2, pp. 436–443 (2007)
24. Breiman, L.: Bagging predictors. Machine Learning 24(2), 123–140 (1996)
25. van Dongen, S.: Graph Clustering by Flow Simulation. Thése de doctorat, Universit é d'Utrecht (2000)
26. Lavergne, T., Cappé, O., Yvon, F.: Practical very large scale CRFs. In: Proceedings the 48th Annual Meeting of the Association for Computational Linguistics (ACL), pp. 504–513. Association for Computational Linguistics (July 2010)
27. Fort, K., Claveau, V.: Annotating football matches: Influence of the source medium on manual annotation. In: Proceedings of the 8th International Conference on Language Resources and Evaluation (LREC 2012), Istanbul, Turquie (2012)
28. Manning, C., Raghavan, P., Schütze, H.: Introduction to information retrieval. Cambridge University Press (2008)
29. Rand, W.M.: Objective criteria for the evaluation of clustering methods. Journal of the American Statistical Association 66(336), 846–850 (1971)
30. Vinh, N.X., Julien Epps, J.B.: Information theoretic measures for clusterings comparison. Journal of Machine Learning Research (2010)
31. Hubert, L., Arabie, P.: Comparing partitions. Journal of Classification 2(1), 193–218 (1985)
32. Gravier, G., Bonastre, J.F., Geoffrois, E., Galliano, S., Tait, K.M., Choukri, K.: ESTER, une campagne d'évaluation des systèmes d'indexation automatique. In: Actes des Journeés d'Étude sur la Parole, JEP, Atelier ESTER2 (2005)
33. Mitchell, T.M.: The need for biases in learning generalizations. Rutgers Computer Science Department Technical Report CBM-TR-117, Reprinted in Readings in Machine Learning (May 1980, 1990)

Semi-supervised SRL System
with Bayesian Inference

Alejandra Lorenzo and Christophe Cerisara

LORIA / UMR 7503, Vandoeuvre-les-Nancy, France
alelorenzo@gmail.com, cerisara@loria.fr

Abstract. We propose a new approach to perform semi-supervised training of Semantic Role Labeling models with very few amount of initial
labeled data. The proposed approach combines in a novel way supervised
and unsupervised training, by forcing the supervised classifier to over-
generate potential semantic candidates, and then letting unsupervised
inference choose the best ones. Hence, the supervised classifier can be
trained on a very small corpus and with coarse-grain features, because
its precision does not need to be high: its role is mainly to constrain
Bayesian inference to explore only a limited part of the full search space.
This approach is evaluated on French and English. In both cases, it
achieves very good performance and outperforms a strong supervised
baseline when only a small number of annotated sentences is available
and even without using any previously trained syntactic parser.

1 Introduction

1.1 Data Scarcity in Semantic Role Labeling

Semantic Role Labeling (SRL) is a major task in Natural Language Processing
which provides a shallow semantic parsing of a text. Its primary goal is to identify
and label the semantic relations that hold between predicates (typically verbs),
and their associated arguments [1]. The analysis of semantic relations and predicate argument structures has many potential applications in Natural Language
Processing (NLP). In particular, applications such as natural language understanding, machine translation, information extraction and question answering
are shown to benefit from semantically annotated text. The extensive research
carried out in this area resulted in a variety of annotated resources, which, in
time, opened up new possibilities for supervised SRL systems. Although such
systems show very good performance, they require large amounts of annotated
data to be successful. This annotated data is not always available, very expensive
to create and often language and domain specific [2].

To bypass these shortcomings, different solutions have been proposed. Unsupervised and semi-supervised learning techniques are two possible options to
address the data scarcity problem. Unsupervised learning attempts to induce
the annotations from large amounts of unlabeled data, while semi-supervised
models are trained on both a limited quantity of labeled examples and a larger

A. Gelbukh (Ed.): CICLing 2014, Part I, LNCS 8403, pp. 429–441, 2014.

unlabeled corpus. A first claass of semi-supervised systems for SRL exploit a bootstrapping approach, such as self-training and co-training [3,4]. An alternative solution to combine labeled and unlabeled data is "semi-unsupervised" systems, which start from some unsupervised model and train a small number of this model's parameters on the limited labeled corpus available [5]. One example of the application of these semi-unsupervised approaches on the semantic role labeling task is described in [6].

1.2   Proposed Semi-supervised Approach

Our proposed approach to semantic role labeling is inspired by the work of [6]. However, one of the main differences is that, while they essentially use the labeled data to build an informed prior distribution over the unsupervised model parameters, we rather use the labeled data to train a supervised classifier which role is to generate a set of potential arcs, and then introduce these labels as virtual evidence [7] to constrain unsupervised Bayesian inference. The proposed system can thus either be viewed as a semi-supervised approach, where the output of the initial bootstrapped supervised system is filtered by the unsupervised model, or as a semi-unsupervised approach, where inference in the unsupervised model is constrained and guided by the supervised solutions. It is, to the best of our knowledge, the first SRL system that does not explicitly root itself in one or the other paradigm.

Another advantage of the proposed approach is that it produces labeled semantic roles and can thus be evaluated with supervised SRL metrics. Moreover, the proposed model automatically detects all candidate argument chunks and predicates and we do not make any assumption about predicate argument structures. Instead, we let the model infer a semantic structure by detecting not only the semantic role associated with each argument but also the predicate it shall be linked to. As shown in Section 3, the proposed approach shows competitive performance even under the assumption of no pre-existing syntactic parser. Hence, as opposed to many previous works that rely on either gold syntactic trees or trees obtained with supervised parsers trained on a large training corpus, we rather train our parser on the same small training corpus than for SRL. To summarize, the only inputs needed in the proposed approach are part-of-speech tags, and a very small initial corpus labeled with syntax and semantic arcs.

2   Task Definition

Semantic role labeling is the task of automatically finding the semantic roles for each predicate in a sentence. That is, finding out which constituents in a sentence are semantic arguments for a given predicate and then determining the appropriate role for each of these arguments. Different definitions for "semantic role" have been proposed. In this work, we use the definition provided by PropBank [8], since it is commonly used in the NLP community and there are available versions of this resource in the target languages. The semantic roles in

PropBank are defined with respect to individual verb senses or predicates. Thus, each predicate has a number of roles. In general, roles A0 and A1 attempt to capture Proto-Agent and Proto-Patient roles [9], and thus are more valid across verbs and verb instances than A2-A5 roles.

In this work, we focus on both determining which constituents in a sentence are semantic arguments for each predicate, and labeling these arguments with semantic roles. For this, we decompose the overall process into two main steps:

1. Candidate arcs generation: this process generates a set of possible candidate semantic arcs (see Section 3);
2. Bayesian inference: this process selects the most likely semantic arcs from the set of candidates (see Section 4).

Figure 1 illustrates the results produced by the proposed approach on an example sentence. The inferred dependency arcs and semantic relations are shown respectively above and below the sentence.

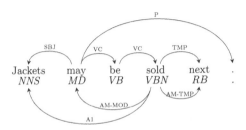

Fig. 1. Example of inferred sentence

3   Candidate Arcs Generation

The first stage is summarized in Figure 2, with the following notation:

- $\mathcal{L}$ is the small initial manually labeled corpus used to train our supervised classifiers. It is typically composed of 50 sentences that are automatically tagged with an existing POS-tagger, and manually annotated with labeled dependencies and semantic relations.
- $\mathcal{U}$ is a large unlabeled corpus, only automatically annotated with POS tags.
- $\mathcal{T}$ is the test or gold manually annotated corpus. It is only used to evalute the performances.

This first stage only exploits an initial set of 50 manually labeled sentences to train both the MATE syntactic parser [10], and a Maximum Entropy semantic model M_s with L-BFGS optimization[1]. Both these supervised models are used to produce a set of candidate semantic arcs on the unlabeled corpus.

---

[1] We use the Stanford Classifier: http://nlp.stanford.edu/software/classifier.shtml

These candidate semantic arcs are then used in the next stage to constrain Bayesian inference to only use plausible semantic arcs, hence greatly reducing the size of the search space on the unlabeled corpus. It is thus very important to maximize the recall of M_s so that the set of candidate semantic arcs comprises most if not all gold arcs. In other words, M_s must miss as few semantic arcs as possible, which imply to maximize its recall while its precision shall be kept at a reasonable level but is not as important there.

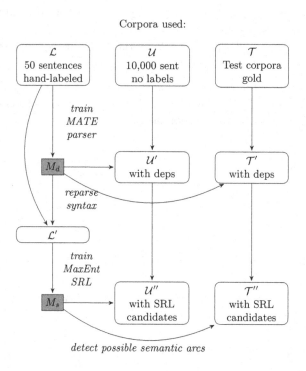

Fig. 2. Procedure to generate candidate semantic arcs, before inference

3.1 Supervised Semantic Model

Following common practice, all non-auxiliary verbs are selected as predicates[2]. Then the supervised Maximum Entropy semantic model M_s is trained on the small manually labeled corpus $\mathcal{L}'$ as detailed in Figure 3. For this training, we first run an optional pre-processing step, which consists of a very simple rule based chunker that uses word forms and POS tags to segment noun and prepositional phrases, and thus reduce the number of possible arguments. The Maximum Entropy semantic model's features $\Phi(a, p)$ computed for each predicate-argument pair are shown in Table 1.

Once the semantic model is trained on the labeled corpus $\mathcal{L}'$, it is applied on the unlabeled $\mathcal{U}'$ and test $\mathcal{T}'$ corpus as detailed in Figure 4.

---

[2] In this work we do not disambiguate between predicate senses.

Table 1. Features used in M_s for each argument-predicate pair (a, p)

- Letter 4-gram, prefix and suffix 4-grams and lengths of a, p, (a, p),
  argument context bigram (a_{t-1}, a_{t+1}) and syntactic dependents of a
- POS tags of a and p
- Distance from a to p: -4 and less, -3, $\cdots$, $+3$, $+4$ and more
- indicator that a is the syntactic head of a NP or PP
- Dependency type from a
- Boolean true iff a and p are directly linked syntactically

```
 1: for every predicate p (p is any non auxiliary verb as given by the POS tags) do
 2:     for every argument a (a is the estimated syntactic head of any NP or PP) do
 3:         Compute features Φ(a, p)
 4:         if arc (a, p) ∈ L' then
 5:             Set class c = label of (a, p) ∈ L'
 6:         else
 7:             Set class c = NOARC
 8:         end if
 9:         Add the observation (Φ(a, p), c) to the training set T_r
10:     end for
11: end for
12: Train the maximum entropy model on T_r
```

Fig. 3. Train MaxEnt semantic model

4   Bayesian Model

The candidate semantic arcs proposed by the semantic model M_s are used as constraints during inference on $\mathcal{U}'' \cup \mathcal{T}''$ of the posterior of the Bayesian model. This Bayesian model is designed to encode standard linguistic features very similar to the ones used in most other unsupervised SRL models. The factors representing these features are described next and include lexical roles preferences $p(w|a)$, arguments position $p(pos|a)$ and syntactic roles preferences $p(d|a)$, which give the following joint on $\mathcal{U}'' \cup \mathcal{T}''$:

$$P(W, A, POS, D) = \prod_{u \in \mathcal{U}'' \cup \mathcal{T}''} \prod_{w_t \in A_u} P(w_t | a_t) P(pos_t | a_t) P(d_t | a_t)$$

where u is one sentence of the corpus, A_u is the observed set of argument candidates proposed by M_s for sentence u, d_t is the observed estimated dependency type from w_t, and (a_t, pos_t) are latent and chosen during inference from the set of candidate semantic arcs proposed by M_s for w_t. Exactly one semantic arc is chosen for each w_t during inference.

4.1   Lexical Roles Preferences

$P(w|a)$ follows a Multinomial distribution smoothed with a symmetric Dirichlet with constant concentration hyper-parameter $\alpha = 0.001$. The same smoothing is

```
 1: for every predicate p (p is any non auxiliary verb as given by the POS tags) do
 2:     for every argument a (a is the estimated syntactic head of any NP or PP) do
 3:         Compute features Φ(a, p)
 4:         Use M_s to compute P(c|Φ(a, p)), where c spans all arc labels
 5:         for every arc label c do
 6:             if P(c|Φ(a, p)) > P(NOARC|Φ(a, p)) then
 7:                 Add the arc (a, p, c) to the set of candidate arcs
 8:             end if
 9:         end for
10:     end for
11: end for
```

Fig. 4. Compute candidate semantic arcs

applied to the two other factors described next. The α parameter has not been tuned at all but has been set beforehand to 0.001 in order to favor peaky distributions. w is the observed lexical form of the head of the candidate argument chunk, which is given by M_d (see Figure 2). This factor shall encode the fact that some words are more likely to play a given role than others. A typical example in French are personal pronouns "je, tu, il" (I, you, he), which are more likely to be A0, while "lui, leur" (him, them) are more likely to be A2. Another example is the preposition "de" (of), which is more likely to be AM than A0.

4.2   Argument Position

$P(pos|a)$, where pos is the position of the argument relative to its predicate, follows a similar Multinomial distribution than the previous factor. The position variable can take two values: left or right. This factor shall encode the fact that, in French and English, the relative position of the role is relevant. Hence, the active case as well as the declarative forms of sentences are largely dominant in the corpus, and thus the A0 role is more likely to occur before the verb.

4.3   Syntactic Preferences

$P(d|a)$, where d is the dependency label that governs the argument, also follows a smoothed Multinomial distribution. This factor shall encode part of the well-known correlation between syntactic dependencies, such as subject, and semantic roles, such as A0.

4.4   Inference

Inference is realized with the Metropolis-Hastings algorithm. The chosen proposal distribution proposes, for one random sentence u and argument $w_t \in A_u$, to replace its current semantic arc (a_t, pos_t) with a new one amongst the set of candidate arcs proposed by M_s for w_t, eventually attaching to a new predicate and/or with another label. Note that this proposal is based on the assumption that every argument is linked to exactly one predicate. This strong assumption is reasonable on the target French SRL corpus, in which less than 8% of the arguments are linked to more than one predicate. .

For each of these possible moves, the proposal is non-uniform, in order to speed up convergence. We rather set the proposal distribution so that 80% of the time, the move that leads to the largest posterior is chosen, and we distribute the remaining 20% probability mass uniformly over the other possible moves. It is easy to check that the Bayesian model is identifiable with discrete and finite variables and that the detailed balance condition is verified. These conditions guarantee that inference converges towards a stationary posterior distribution. Before inference, the semantic arcs are initialized by choosing for each argument w_t, the semantic arc with the maximum score given by M_s.

5   Experimental Validation

The proposed semi-supervised SRL approach is evaluated on French and English. In both cases we compare our approach with the MATE state-of-the-art supervised semantic parser [11] (called **MATEsrl**). For this comparison, both the **MATEsrl** system and the proposed approach are trained on the same 50 sentences. Furthermore, as explained in Section 3, instead of using gold syntactic dependencies, for all three corpora we use the dependencies obtained with the MATE syntactic parser (**MATEdep**), which is also trained only on the 50 sentences of the labeled corpus. The objective of this comparison is to prove that the proposed weakly-supervised approach outperforms a supervised approach when only a few number of annotated examples are available. The choice of this supervised **MATEsrl** system is motivated by its very good performances in general, and on the French corpus in particular. Indeed, we have also evaluated the performances of the **MATEsrl** system when trained on the full French corpus with 10-fold cross-validation, which then gives a labeled F1 of 98.7% and an unlabeled F1 of 98.8%. Note that cross-validation has only been used in this specific experiment, and not in any of the others, because in the other experiments, we only use 50 sentences for training and may thus use a large test set of 500 sentences.

In English, the proposed system has further been compared with the state-of-the-art semi-supervised SRL system presented in [12]. However, for this comparison we used slightly different settings, adapted to those described in [12]. For all the experiments, the scores are computed as in the CoNLL09 evaluation campaign, except for the labeled scores that assume that the gold labels of the predicates are known, because our system does not do sense detection. To simplify notations, we omit next the ′ and ″ when referring to the corpora derived from $\mathcal{L}$, $\mathcal{T}$ and $\mathcal{U}$.

5.1   Evaluations on French

Data. The data used in this evaluation is the French CLASSIK corpus [13]. The "gold" section of this corpus has been manually labeled and contains 1000 sentences in total. $\mathcal{L}$ is composed of the 50 first sentences, and $\mathcal{T}$ is composed of the 500 last sentences. We can thus make the size of $\mathcal{L}$ vary from 50 to 500 sentences when drawing Figure 5. $\mathcal{U}$ is composed of 10,000 sentences taken from the non-manually annotated part of the French CLASSIK corpus.

Comparison with MATE. In this experiment, the **MATEsrl** parser is trained on the same 50 sentences from $\mathcal{L}$ than our proposed system. The results obtained by both the **MATEsrl** system and the proposed approach are shown, respectively, in the first and last rows of Table 2. Note that, although the **MATEsrl** performances are very high when trained on the full corpus, they drop down dramatically when trained on only 50 sentences. In this case, the proposed approach largely outperforms the supervised system.

The second and third rows in Table 2 show the performances obtained when using only the first stage of our system, that is the output of the Maximum Entropy classifier, without doing Bayesian inference. The first "all links" model simply includes all of the semantic arcs produced by our supervised classifier. On the average, this classifier produces about 5 candidate semantic arcs per argument. Obviously, the F-measure is quite low here, because this classifier has been designed to produce many more arcs than necessary, so that the subsequent inference step only selects a few of them. But the recall is more interesting than the F1 in this experiment, because it shows the best performance that can be reached with this first stage of classifiers.

The second "Optimum links" shows the best possible results that can be reached with the set of candidate links proposed by the classifier, given our restriction that every argument can be linked to at most one predicate. So in this line, we select for every argument the single arc that matches a corresponding arc in the gold semantic structure, or a random arc if none of the candidates is correct. This line gives the real upper bound of performances that can be reached by our system, given our restrictions and the current setup of the deterministic and supervised parts of the model, i.e., at the exclusion of the unsupervised model. This oracle does not consider the labels of the arcs. Note that its recall is slightly lower than the "All links" recall, because of the few arguments that are linked to several predicates. Its precision is also lower than 100% because of the false alarms from both predicate and argument detection.

Comparing the last two lines of Table 2, we can note that the proposed system only adds about 15% of errors more than the oracle system, in terms of F1, precision and recall. This suggests that the inference stage is doing correctly its job and that further improvements can be obtained mainly by working on the deterministic part and restrictions of our system.

5.2   Impact of Quantity of Labeled Data

Figure 5 shows the evolution of the unlabeled F1 in function of the number of manually annotated sentences both for the supervised **MATEsrl** system and for the proposed weakly supervised system. As expected, the proposed system gives

Table 2. SRL experimental validation on French in terms of labeled and unlabeled F1-measure, Precision and Recall

System	F1 lab.	Prec. lab.	Rec. lab.	F1 unlab.	Prec. unlab.	Rec. unlab.
MATE	36.4	40.8	32.9	58.7	65.8	53.0
All links	31.4	20.6	**66.0**	33.3	21.8	**70.0**
Optimum links				76.7	86.2	69.1
Inference	**54.8**	62.7	48.7	**73.5**	84.1	65.3

much better performances than the MATE system with a small amount of manually labeled data, but still remains better for up to 450 sentences, although the difference between both systems decreases when more labeled data is included. This suggests that the proposed model may still be improved by better tuning the complexity and number of features used to train our first stage classifier, depending on the size of the available corpus.

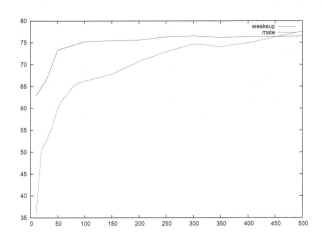

Fig. 5. Unlabeled F1 as a function of the number of manually annotated sentences

5.3   Evaluations on English

Two evaluation experiments are realized in English: first, we compare our approach with the supervised **MATEsrl** system, in a similar way as done in French. Second, we compare our system with the semi-supervised SRL presented in [12].

Data. The data used for the evaluations on English is the standard CoNLL 2008 shared task [14] version of Penn Treebank WSJ and PropBank. As done in [12], $\mathcal{T}$ corresponds to the test portion of the CoNLL 2008 corpus. $\mathcal{L}$ is composed of the first 50 (Table 3) or 400 (Table 4) sentences extracted from the training corpus of CoNLL 2008, while $\mathcal{U}$ is composed of the next 10,000 sentences.

Table 3. SRL experimental validation on English in terms of labeled and unlabeled F1-measure, Precision and Recall

System	F1 lab.	Prec. lab.	Rec. lab.	F1 unlab.	Prec. unlab.	Rec. unlab.
MATE	26.0	31.7	22.1	53.9	65.5	45.8
Inference	**47.5**	**54.7**	**42.0**	**67.2**	**77.4**	**59.4**

Comparison with MATE. Both systems are trained on only 50 sentences. The results obtained by both the **MATEsrl** semantic parser and the proposed system are shown in Table 3. Like in French, the proposed approach largely outperforms the state-of-the-art supervised semantic parser when only a few number of annotated examples are available.

Comparison with semi-supervised. We compare next our approach with the semi-supervised SRL system presented in [12], which also produces labeled semantic arcs. In [12] the authors present a Latent Words Language Model, which learns word similarities from unlabeled text and use them in different semi-supervised SRL systems as additional features or to automatically expand a small training set. They experiment with different sizes of the training corpus and show that for a small training corpus they outperform a state-of-the-art supervised baseline. We mimic their experimental setup by increasing the size of our training corpus from 50 to 400 sentences, which represents about 1% of the training corpus. The first two lines in Table 4 show the results presented in [12] when training

Table 4. Comparison with **MATEsrl** and [12] for two training corpus sizes

System	Training	F1 lab.
Supervised Baseline [12]	5%	40.49
LWFeatures [12]	5%	60.3
MATE SRL supervised	1%	54.4
Inference	**1%**	**60.6**

on 5% of the training set. The first line is their baseline, a supervised SRL system, while the second line shows the results of the same supervised system, but using extra features given by the latent words language model. These features correspond to the estimated distribution of the latent words for every word for both the training and test set. The last two lines in the table show, respectively, the results achieved by our baseline, the **MATEsrl** parser, and our proposed system when training only on 1% of the training set. We can observe that with only 1% of the training data (i.e., about 400 sentences), the proposed system matches the performances of the semi-supervised approach trained on 5% of the training corpus (i.e. with about 1900 sentences).

6   Additional Related Works

A variety of algorithms have been proposed for semi-supervised learning[3]. And there are many more examples of applications of these semi-supervised approaches to SRL other than the ones described in the introduction and evaluation sections. For instance, [3] and [4] tested self-training and co-training on SRL; [17] used a graph-alignment method to SRL; and, more recently, [18] used a graph-based label propagation semi-supervised approach to improve the coverage of a frame-semantic parsing model and reported significant improvements over a state-of-the-art baseline, both in frame identification accuracy and full frame-semantic parsing F_1. Finally, another interesting approach, also related with our work, is "prototype-based" learning [19,20]. In this approach, prior knowledge is specified declaratively, by providing "prototypes" (e.g., a list of representative words) for each label. Then they use distributional similarity between the words in the corpus and the prototypes as features in a generative model. Similarly, our proposed framework might support the inclusion of "prototypes" in the form of rules that generate candidate semantic arcs, as a replacement or in addition to the supervised Maximum Entropy model used in this work.

7   Conclusions and Future Work

We present in this work a new approach to SRL that is able to work competitively even when only a small amount of labeled data is available. The proposed approach exploits both supervised and unsupervised methods, without privileging one or the other by design. It is based on the combination of a supervised semantic role labeler that generates many more potential candidate arcs than traditional supervised systems, with a Bayesian unsupervised model that maximizes the joint posterior of several linguistic factors that are commonly used in the unsupervised SRL field. This combination is realized thanks to "virtual evidence" that acts as new types of constraints for Bayesian inference.

Because the proposed approach relies on a supervised SRL classifier, it produces labeled semantic roles and it does infer a semantic structure by detecting to which predicate each argument should be linked. This semi-supervised direction is very promising, specially for those domains and languages for which little or no annotated data is available.

We successfully evaluated the proposed model on two languages, French and English, showing, in both cases, consistent performances improvement over a state-of-the-art supervised SRL system on small amounts of labeled data. Furthermore, we showed for English that its accuracy reaches a level comparable to that of a state-of-the-art semi-supervised SRL systems even when the amount of labeled data is smaller.

The system could be improved in many ways, and in particular the proposed unsupervised model. We could, for instance, include some penalization term for sampling the same role for several arguments of a verb instance (at least for core roles).

---

[3] We refer the reader to [15] or [16] for an overview on semi-supervised methods.

Acknowledgments. This work has been partially funded by the French ANR project ContNomina.

References

1. Màrquez, L., Carreras, X., Litkowski, K.C., Stevenson, S.: Semantic role labeling: An introduction to the special issue. Comput. Linguist. 34, 145–159 (2008)
2. Pradhan, S.S., Ward, W., Martin, J.H.: Towards robust semantic role labeling. Comput. Linguist. 34, 289–310 (2008)
3. He, S., Gildea, H.: Self-training and Cotraining for Semantic Role Labeling: Primary Report. Technical report, TR 891, University of Colorado at Boulder (2006)
4. Lee, J.Y., Song, Y.I., Rim, H.C.: Investigation of weakly supervised learning for semantic role labeling. In: ALPIT, pp. 165–170 (2007)
5. Daumé III, H.: Semi-supervised or semi-unsupervised? In: Proc. NAACL Wokshop on Semi-supervised Learning for NLP (2009)
6. Titov, I., Klementiev, A.: Semi-supervised semantic role labeling: Approaching from an unsupervised perspective. In: Proceedings of the International Conference on Computational Linguistics (COLING), Bombay, India (2012)
7. Jain, D., Beetz, M.: Soft evidential update via markov chain monte carlo inference. In: Dillmann, R., Beyerer, J., Hanebeck, U.D., Schultz, T. (eds.) KI 2010. LNCS, vol. 6359, pp. 280–290. Springer, Heidelberg (2010)
8. Palmer, M., Gildea, D., Kingsbury, P.: The proposition bank: An annotated corpus of semantic roles. Comput. Linguist. 31, 71–106 (2005)
9. Dowty, D.: Thematic proto-roles and argument selection. Language 67, 547–619 (1991)
10. Bohnet, B.: Top accuracy and fast dependency parsing is not a contradiction. In: Proc. International Conference on Computational Linguistics, Beijing, China (2010)
11. Björkelund, A., Hafdell, L., Nugues, P.: In: Proceedings of the Thirteenth Conference on Computational Natural Language Learning: Shared Task, CoNLL 2009, pp. 43–48. Association for Computational Linguistics, Stroudsburg (2009)
12. Deschacht, K., Moens, M.F.: Semi-supervised semantic role labeling using the latent words language model. In: EMNLP, pp. 21–29 (2009)
13. van der Plas, L., Merlo, P., Henderson, J.: Scaling up automatic cross-lingual semantic role annotation. In: Proceedings of the 49th Annual Meeting of the Association for Computational Linguistics: Human Language Technologies: Short Papers, HLT 2011, vol. 2, pp. 299–304. Association for Computational Linguistics (2011)
14. Surdeanu, M., Johansson, R., Meyers, A., Màrquez, L., Nivre, J.: The conll-2008 shared task on joint parsing of syntactic and semantic dependencies. In: Proceedings of the Twelfth Conference on Computational Natural Language Learning, CoNLL 2008, pp. 159–177. Association for Computational Linguistics, Stroudsburg (2008)
15. Zhu, X.: Semi-Supervised Learning Literature Survey. Technical report, Computer Sciences, University of Wisconsin-Madison (2005)
16. Pise, N.N., Kulkarni, P.: A survey of semi-supervised learning methods. In: Proceedings of the 2008 International Conference on Computational Intelligence and Security, CIS 2008, vol. 2, pp. 30–34. IEEE Computer Society, Washington, DC (2008)
17. Fürstenau, H., Lapata, M.: Graph alignment for semi-supervised semantic role labeling. In: EMNLP, pp. 11–20 (2009)

18. Das, D., Smith, N.A.: Semi-supervised frame-semantic parsing for unknown predicates. In: Proceedings of the 49th Annual Meeting of the Association for Computational Linguistics: Human Language Technologies, HLT 2011, vol. 1, pp. 1435–1444. Association for Computational Linguistics, Stroudsburg (2011)
19. Haghighi, A., Klein, D.: Prototype-driven learning for sequence models. In: Proceedings of the Main Conference on Human Language Technology Conference of the North American Chapter of the Association of Computational Linguistics, HLT-NAACL 2006, pp. 320–327. Association for Computational Linguistics, Stroudsburg (2006)
20. Haghighi, A., Klein, D.: Prototype-driven grammar induction. In: Proceedings of the 21st International Conference on Computational Linguistics and the 44th Annual Meeting of the Association for Computational Linguistics. ACL-44, pp. 881–888. Association for Computational Linguistics, Stroudsburg (2006)

A Sentence Similarity Method Based on Chunking and Information Content

Dan Ştefănescu, Rajendra Banjade, and Vasile Rus

Department of Computer Science, The University of Memphis
Memphis, TN, 38152, USA
{dstfnscu,rbanjade,vrus}@memphis.edu

Abstract. This paper introduces a method for assessing the semantic similarity between sentences, which relies on the assumption that the meaning of a sentence is captured by its syntactic constituents and the dependencies between them. We obtain both the constituents and their dependencies from a syntactic parser. Our algorithm considers that two sentences have the same meaning if it can find a good mapping between their chunks and also if the chunk dependencies in one text are preserved in the other. Moreover, the algorithm takes into account that every chunk has a different importance with respect to the overall meaning of a sentence, which is computed based on the information content of the words in the chunk. The experiments conducted on a well-known paraphrase data set show that the performance of our method is comparable to state of the art.

Keywords: Similarity, Semantic Similarity, Sentence Similarity, Paraphrase Identification, Short Text Similarity, Information Content.

1 Introduction

Sentence similarity is a well-studied topic in Natural Language Processing because of its use in many tasks such as question-answering, text mining, text summarization, plagiarism detection, assessing the correctness of student answers in education technologies, or assessing the translation quality of automatic translation systems. Due to this wide applicability, the research literature is abundant in methods for detecting or assessing sentence similarity, which are often presented as methods for identifying paraphrases. The performance of such methods was (and still is) commonly evaluated using the Microsoft Research Paraphrase Corpus (MSRP) [4]. MSRP contains 5,801 sentence pairs overall, out of which 3,900 (67.23%) are considered paraphrases. However, MSRP does not seem to be the ideal data set for testing paraphrase identification systems, because of the high degree of word overlap: about 96% of word level alignments are between identical words. The implication of this would be that for a sentence similarity method that exploits similarity at word level, the results would not vary significantly just by switching between word-to-word similarity measures. Yet, sentence similarity is more than word overlap and as such, MSRP should provide the means of testing what various methods are exploiting more than just word-to-word

A. Gelbukh (Ed.): CICLing 2014, Part I, LNCS 8403, pp. 442–453, 2014.

similarity. As a consequence, in the case of MSRP, word overlap similarity could be considered the baseline and so, researchers' role is to come up with improvements over this baseline. On the other hand, word overlap methods are sensitive to preprocessing [12, 18]. For example, just using lemmas instead of stems or occurrence forms (tokens) would give different results for word overlapping measures. Anyhow, on the MSRP test data, one can obtain 64.3% accuracy by using word overlap (cf. [1]), which is less than what can be obtained with a fake system that would consider each pair as a paraphrase: 66.55%. Thus, such a virtual system is usually considered to be the baseline.

In this research work, our objective is to find an efficient and yet a robust solution to the problem of quantifying the semantic similarity between sentences, with the purpose of using it in a real-world application. The software application we target is an on-line Intelligent Tutoring System prototype, already tested by hundreds of students. The experiments presented in this paper are motivated by our continuous efforts to improve the semantic similarity function that is responsible for assessing the correctness of the answers given by the students interacting with our system. For practical reasons, this solution should be simple enough so that its implementation is feasible and reproducible, while achieving state of the art results.

Beside this introductory section, the content of the paper is organized as follows. Section 2 briefly discusses the related work, while section 3 describes how the input sentences are processed and represented. Section 4 explains in detail how the similarity of two sentences is computed, followed by section 5, which discusses the performance of the proposed method. The paper ends with a section of conclusions.

2   Related Work

At the moment of writing this paper, the ACL wiki's entry for paraphrase identification[1] lists no less than 17 systems ordered based on their performances on MSRP. This is in itself a statement about the high volume of research work dedicated to this topic, but still, it is a small figure compared to the number of the systems that have been implemented along the years.

Early applications of text similarity were in the field of information retrieval. These early developments were essentially "bag-of-words" strategies developed for solving well-known problems such as selecting the documents most relevant to a given query [23] or text classification [17]. The most basic methods rely on lexical matching and return scores based on the number of lexical units that occur in both fragments (whether sentences or not). Salton and Buckley [22] were among the first to try to improve these methods by applying various weighting and normalization schemes based on word frequencies. Also, certain pre-processing steps such as stemming, tagging or short-word removal were shown to improve the results of the systems. In fact, it was recently proven that different pre-processing variations can be responsible for large differences in the performance of a system [12, 18].

---

[1] http://aclweb.org/aclwiki/index.php?title=Paraphrase_
Identification_%28State_of_the_art%29

In time, the methods moved beyond lexical matching to using corpus and know-ledge-based word-to-word similarity measures [27, 15, 6, 9] and machine learning techniques were employed to further improve results by learning from lexical and semantic features [10] or dependency-based features [29]. More recent approaches exploit Machine Translation evaluation measures [7, 13], graph structures [20] or vector space models [8, 21].

Given a sentence pair, our method exploits the parse trees of the sentences to ex-tract the principal syntactic constituents (chunks) and their dependencies. Based on their mapping, which is done on phrase/chunk level semantic similarity criteria, the pair is assigned an overall similarity score. Among other approaches that are also exploiting the parse tree information, but are substantially different than ours, we mention those of Rus et al. [20], Das and Smith [3] and Socher et al. [25].

3   Sentence Representation

To assess the semantic similarity between sentences, we propose a method that ex-ploits the definition of a paraphrase as "a *statement that expresses something that somebody has written or said using different words, especially in order to make it easier to understand*" (cf. Oxford on-line dictionary). In other words, a paraphrase is an alternative for expressing the same meaning with different words (in the same language). This means that the elements that are making up the meaning of the origi-nal sentence must be preserved in the paraphrase. In our approach, we consider these elements to be the principal phrasal chunks forming the sentence, and the dependen-cies among them. The rationality of considering chunks instead of words as the basic meaning constituents is backed-up by the linking theory and relies on the fact that very often in a text, the syntactic constituents correspond to the semantic constituents of that text. Van Valin Jr. [28] argues that in terms of Role and Reference Grammar theory of semantic representation, the constructionist approach on the syntax-semantics interface fits naturally with the linking from syntax to semantics (representing the hearer's perspective). Moreover, this linking is governed by the Completeness Constraint:

"*All of the arguments explicitly specified in the semantic representation of a sentence must be realized syntactically in the sentence, and all of the referring expressions in the syntactic representation of a sentence must be linked to an argument position in a logical structure in the semantic representation of the sentence.*"

Text chunking consists of dividing a text into phrases or chunks so that syntactically related words become members of the same phrase. In this paper, our generalized assumption is that these syntactic constituents are the actual manifestation of the se-mantic constituents of the sentence. In this context, by dependencies we mean the direct links or connections between the considered chunks, as given by the hierarchic-al relations of the parse tree. Consider the following sentence from MSRP:

Example 1: In Soviet times the Beatles' music was considered propaganda of an alien ideology.

We process the above sentence with the Stanford NLP Parser [24], which outputs the corresponding full parse tree. Starting from this tree, our algorithm extracts the principal syntactic constituents of the sentence, considering all noun and adverbial phrases of maximum length, as long as there is no change in the type of the phrase. Thus, from an annotation such as (NP1 (NP2 ...) (NP3 ...)), our algorithm would select NP1 as a principal chunk, while from an annotation like (NP1 (NP2 ...) (PP (...) (NP3 ...))), NP2 and NP3 would be considered principal chunks. Each verb is considered a singular verb phrase (VP), but the auxiliaries are removed. Again, our assumption is that the selected chunks represent the principal semantic constituents of the sentence. For the above given example, the parse tree returned by Stanford's parser and the selected principal chunks are the following:

(ROOT (S (PP (IN In) **(NP (JJ Soviet) (NNS times)))** **(NP (NP (DT the) (NNP Beatles) (POS '))** **(NN music))** (VP (VBD was) (VP **(VBN considered)** (NP **(NP (NN propaganda))** (PP (IN of) **(NP (DT an) (JJ alien) (NN ideology))**)))))) (. .)))

1. [NP Soviet/JJ/soviet times/NNS/time]
2. [NP the/DT/the Beatles/NNP/Beatles '/POS/' music/NN/music]
3. [VP considered/VBN/consider]
4. [NP propaganda/NN/propaganda]
5. [NP an/DT/a alien/JJ/alien ideology/NN/ideology]

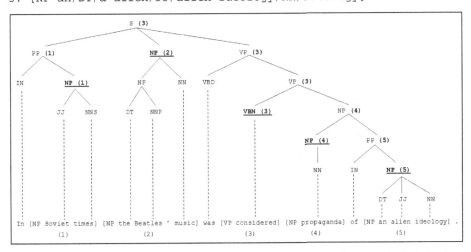

Fig. 1. Parse tree for the given sentence. Highlighted nodes correspond to the principal chunks. The propagated chunk index is shown in brackets

It is important to stress that our algorithm works directly with the output provided by the Stanford NLP Parser. Our algorithm exploits the tree structure to obtain the dependencies between the considered principal chunks (see the underlined nodes in Figure 1). To find the dependencies, the algorithm propagates the chunk index information from these nodes to their immediate parents. The chunk index information of a parent node is the chunk index of the leftmost child corresponding to a verb phrase if such a child exists, or that of the leftmost child corresponding to a noun phrase, otherwise. In Figure 1, this information is shown in brackets.

In this current version of our algorithm, we decided not to consider prepositions and complementizers (e.g. *if, although, while, even though, in case, so that*), even if though they may have their own contribution to the meaning of a sentence. Nevertheless, for the purpose of computing sentence similarity, we believe that their role is not crucial. We also got rid of any existing punctuation, such as (POS ') and (. .), in the above given example.

Based on the propagation of chunk indexes, the algorithm simplifies the parse tree presented in Figure 1. The newly generated tree is depicted in Figure 2.

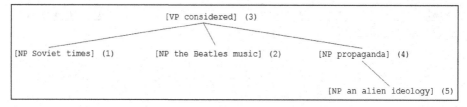

Fig. 2. Simplified version of the tree in Fig. 1

The dependencies between chunks can now be easily identified in the simplified parse tree, by exploiting all the hierarchical links between the nodes. In this example, the dependency links are: (1)-(3), (2)-(3), (4)-(3) and (5)-(4).

4   Computing the Similarity of Two Sentences

As stated at the beginning of the previous section, we consider the basic elements that are contributing to the meaning of a sentence to be the chunks in that sentence and the dependencies among them. This is in contrast to previous approaches which used individual words as basic elements [20]. Under our assumption, we consider two sentences to have the same meaning if the majority of their chunks can be semantically aligned (i.e. they mean the same thing) and if the dependencies existing between chunks in one sentence can be found between the corresponding aligned chunks in the other sentence. For example, let us consider another sentence, which is a paraphrase of the sentence in Example 1:

Example 2: Ex-KGB agent Putin added that the Beatles were considered 'propaganda of an alien ideology'.

For this sentence, our algorithm generates the following simplified parse tree:

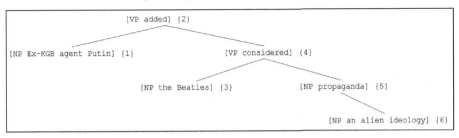

Fig. 3. Simplified tree corresponding to the paraphrase in our example

In this case, the dependency links are {1}-{2}, {4}-{2}, {3}-{4}, {5}-{4}, {6}-{5}. Notice that because of the different indexing, the dependencies of the two sentences (the ones in Example 1 and Example 2, respectively) are not directly comparable.

The next step of our algorithm is to align the chunks of the two sentences. This is accomplished by employing the Hungarian algorithm, also called Kuhn-Munkres method [11]. This approach is inspired by the work of Rus and Lintean [19], who proposed a solution for text-to-text similarity based on the optimal assignment problem. This is a fundamental combinatorial optimization problem which consists of finding a maximum weight matching in a weighted complete bipartite graph, $G = \{X \cup Y; X \times Y\}$. The classic example is about assigning a group of workers, i.e. X, to a set of jobs, i.e. Y, based on the expertise level of each worker at each job, i.e. the weight of an xy edge, $w(x,y)$. Given two sentences, the text-to-text similarity method proposed by Rus and Lintean considers X to be the set of words in the first sentence, Y to be the set of words in the second sentence, while the weights $w(x,y)$ are word-to-word similarity scores that can be obtained by using any word-to-word similarity measure. After the words in X are aligned to the words in Y, the final score is essentially computed as a normalized sum of the weights corresponding to the edges that were selected in the optimal alignment:

$$sim(X,Y) = \frac{2 * \sum_i w_i(x,y)}{|X| + |Y|} \tag{A}$$

In our case, X is the set of chunks in the first sentence and Y is the set of chunks in the second sentence. Thus, $w(x,y)$ values are similarity scores computed between the chunks in X and those in Y. Our algorithm does these computations by recursively employing Rus and Lintean's method and applying formula (A). We experimented with various word-to-word similarity methods based on WordNet [5] and an LSA model generated using the whole Wikipedia [26]. Only $w(x,y)$ values higher than a certain threshold are deemed valid.

Returning to our example, after running the Hungarian algorithm at the chunk level, we obtain the following alignment: (3)-{4}, (2)-{3}, (4)-{5}, (5)-{6}, which means that chunks (1), {1} and {2}, respectively, remain unpaired. This alignment allows us to compare chunk dependencies in the two sentences and so, in the next step, our algorithm computes two different similarity scores for the sentence pair: one is a chunking similarity score based on the chunk alignment, while the other one is a dependency similarity score, based on the comparability of the dependencies. The next sub-sections will further detail these computations.

4.1   Chunking Similarity and Information Content

For each sentence pair, the chunking similarity is calculated as a chunk overlap score, exploiting the chunk alignment and essentially trying to apply equation (A) at the chunk level. However, every chunk has a different importance in the sentence and consequently, we modified the formula in equation (A) so as to take the importance into account. We define the importance of a chunk based on the information content of the words it contains. We were inspired by the work of Resnik [16], which provides a way of computing the information content for concepts in a taxonomy such as WordNet.

According to Resnik [16], the information content (*ic*) of a concept *c*, can be computed as the negative of log likelihood: -log $p(c)$ (as explained in what follows). The lower the probability of a concept, the higher its informativeness. However, this model cannot be applied directly to words based on their estimated probabilities of occurrence, because in numerous texts informative words can be more frequent than non-informative ones. For example, in the Wikipedia (as of January 2013), "*England*" has more occurrences than "*entity*", but intuitively, the former should be more informative. Resnik however, is not working with words, but with concepts in the hierarchical taxonomy WordNet, in which links represent IS-A relations. In WordNet, more general concepts are at the top of the hierarchy while the specific ones are at the bottom. Each concept in WordNet (called *synset*) has a well-defined meaning that can be represented by any of the specific senses of certain words (e.g. *literals*). Let us consider *words*(c), the set of all words that can represent c and all the other concepts for which c is an ancestor (any concept higher up) in the hierarchy of the taxonomy. For example, in Figure 4, $words(c_3) = \{w_4\}$; $words(c_2) = \{w_2, w_3, w_4\}$ and $words(c_1) = \{w_1, w_2, ..., w_8\}$.

Resnik calls *words*(c), the *set of words subsumed by concept c* and computes the frequency of c as the sum of the number of occurrences of all words subsumed by c in a reference corpus:

$$freq(c) = \sum_{w \in words(c)} count(w)$$

Consequently, $\hat{p}(c) = freq(c)/T$, where T is the total number of words observed, and $ic(c) = -log\ \hat{p}(c)$. In his research, Resnik computes the information content only for nominal concepts (nouns). We reproduced his work on the 3.0 version of WordNet, getting the word counts from an already mentioned January 2013 version of Wikipedia, and we further extended it for the other grammatical categories. The only problem is that the WordNet does not have a hierarchical structure for adjectives or adverbs and we do not consider them as informative as the nouns or verbs. This is why we decided to halve the information content values computed for adjectival and adverbial concepts. To be clear, the total number of words observed, T, includes all the words of WordNet, no matter their grammatical category.

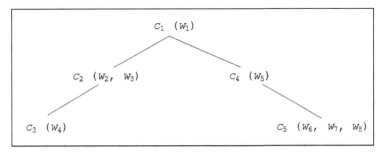

Fig. 4. Generic synsets (c) and their literals (w) in the hierarchical structure of the WordNet

Because Word Sense Disambiguation is beyond the scope of this research work, we consider the information content of a word to be equal to the information content of the most general concept it can represent, which is the concept having the minimum information content value: $ic(w) = \min_{c \mid w \in c} ic(c)$. Thus, we want to make sure that potentially non-informative words will always have low information content values assigned to them. However, the input data may contain words that are not literals in WordNet. For such a word w_i, our algorithm selects a WordNet literal w_j, of the same grammatical category, for which the similarity with w_i is sufficiently high (>= 0.8 or otherwise the highest), according to the above mentioned LSA model generated on the whole Wikipedia. Once w_j is found, the algorithm transfers the information content value to w_i: $ic(w_i) = ic(w_j)$.

We consider the importance of a chunk to be equal to the sum of the information content values assigned to the words in that chunk:

$$i(chunk) = \sum_{w \in chunk} ic(w)$$

Now, equation (A) can be modified so as to be applied at the chunk level and to include the importance of chunks. Consequently, the chunking similarity (cs) for the sentence pair is computed as:

$$cs(X, Y) = \frac{\sum_x i(x)w(x, y_x) + \sum_y i(y)w(x_y, y)}{\sum_{x \in X} i(x) + \sum_{y \in Y} i(y)} \tag{B}$$

4.2 Dependency Similarity

At this point, the chunk alignment between two given sentences is available, as generated by the Hungarian algorithm. Consequently, the chunk dependencies can now be compared. Returning to our example, we have the following:

```
alignment:  (3)-{4}, (2)-{3}, (4)-{5}, (5)-{6}
dependencies 1 (D₁):  (1)-(3), (2)-(3), (4)-(3), (5)-(4)
dependencies 2 (D₂):  {1}-{2}, {4}-{2}, {3}-{4}, {5}-{4}, {6}-{5}
```

Using the chunk alignment, we transfer the indexing for the dependencies in sentence 2, so as to match the indexing in sentence 1 (or the other way around), excluding (for both sentences) those dependencies for which one of the ends is not in the alignment. In our case, the new sets of dependencies are:

```
dependencies 1 (D₁): (2)-(3), (4)-(3), (5)-(4)
dependencies 2 (D₂): (2)-(3), (4)-(3), (5)-(4)
```

The dependency similarity (*ds*) is computed as the Dice coefficient over the sets:

$$ds(X,Y) = \frac{2|D_1 \cap D_2|}{|D_1| + |D_2|} \tag{C}$$

We prefer Dice coefficient over Jaccard index because the latter is penalizing too much for different dependencies that are sharing only one chunk.

It is important to specify that the dependency score is not penalized for the dependencies that have been excluded, because they correspond to the chunks that remained unaligned. The chunking similarity function already takes this information into account and so, we see no reason in penalizing twice for the unaligned chunks. We should also note that an (i)-(j) dependency will match a (j)-(i) dependency and this will increment the numerator in equation (C).

4.3   Sentence Similarity

The final sentence similarity score is mainly based on the chunking similarity function, while the dependency similarity score (*sim*) is acting as a bonus activator, having the role of increasing the score only when its value is higher than a certain threshold:

$$sim(X,Y) = \begin{cases} cs(X,Y), & \text{if } ds(X,Y) < \theta \\ \min(\alpha \cdot cs(X,Y), 1) & otherwise \end{cases} \tag{D}$$

In our experiments, α and θ values were not optimal, but empirically chosen so as to maximize the performance of the algorithm on the training data. The values we ended up with are $\alpha = 1.3$ and $\theta = 0.7$. Moreover, when more than a certain number of chunks remain unaligned (4 in our experiments), our algorithm reduces the value of $sim(X,Y)$ to its half.

5   Results

We exemplify how to compute the sentence similarity score for the two sentences in examples 1 and 2, which form a pair in MSRP, by successively applying equations (B), (C) and (D). It is important to mention that the information content (*ic*) values, calculated as described in section 4.1, are usually in the interval [1, 10]. Since $w(x,y)$ values are in the interval [0, 1], in order for the equation (B) to function as intended, our algorithm brings *ic* values to the interval [0, 1] simply by dividing them by 10. Looking at the chunk alignment, we see that most of the aligned chunks are identical and so, their associated $w(x,y)$ values are 1, as calculated by using equation (A). The only difference is for the pair (2)-{3}: "*the Beatles music*" vs. "*the Beatles*", for which we have that $w((2), \{3\}) = 0.8$. The chunk importance values turned out to be the following: $i("the\ Beatles\ music") = 0.89$; $i("the\ Beatles") = 0.594$; $i("considered")$ $= 0.312$; $i("propaganda") = 0.587$ and $i("an\ alien\ ideology") = 0.734$. Thus, we have:

$$cs(X,Y) = \frac{(0.89 * 0.8 + 0.312 + 0.587 + 0.734) + (0.594 * 0.8 + 0.312 + 0.587 + 0.734)}{(0.89 + 0.312 + 0.587 + 0.734) + (0.594 + 0.312 + 0.587 + 0.734)}$$
$$= 0.94$$

$$ds(X,Y) = 1$$

The sentence similarity score computed using equation (D) would be $1.3 * 0.94 = 1.22$, which is finally trimmed to **1**, due to the *min* operator.

We tested our approach with the MSRP test data. Since our algorithm assigns values in the [0,1] interval for each sentence pair instead of doing just a binary classification, we used the training data to figure out the optimal threshold value for which the sentences in a pair should be considered paraphrases. Other than for this threshold and also for the values of α and θ, our method does not need any training. The paraphrase threshold value turned out to be 0.19 (which might indicate that our assigned similarity scores are too low), with an overall accuracy of 0.742. Table 1 shows our results in comparison to those obtained by state of the art approaches.

Table 1. Our results compared to the state of the art in terms of Accuracy and F-measure

Reference	Algorithm	Acc.	F
this paper	syntactic constituents & dependencies	.742	.821
Madnani et al. [13]	combination of MT metrics	**.774**	**.841**
Socher et al. [25]	recursive autoencoder	.768	.836
Das and Smith [3]	product of experts	.761	.827
Fernando and Stevenson [6]	WordNet similarity with matrix	.741	.824
Rus et al. [21]	Latent Semantic Space	.736	.818
Rus et al. [21]	optimal Latent Dirichlet Allocation	.733	.809
Rus et al. [20]	graph subsumption	.706	.805
Hassan [8]	explicit semantic space	.670	.793
Mihalcea et al. [15]	cosine sim. with tf-idf weighting	.654	.753

6   Conclusions

This paper introduces a method for assessing the semantic similarity between sentences which operates under the generalized assumption that the syntactic constituents of a sentence are the actual manifestation of its semantic constituents. This view is influenced by linking theory and the completeness constraint. Thus, we consider the basic elements that compose the meaning of a sentence to be its syntactic constituents and the dependencies between them. Our algorithm considers that two sentences have the same meaning if there is a good match in terms of similarity between their chunks and also between their dependencies. In the final score, every chunk is considered to have a different importance, computed based on the information content of its words. The final output is a value in the [0,1] interval, representing the similarity score for the sentences given as input. The described method is straightforward and relatively easy to implement and reproduce, compared to others, while achieving state of the art results.

Our future work will focus on better exploiting the role of dependency similarity in computing the final similarity score and on assessing the results of our method on other data sets for semantic textual similarity, such as STS [2] or ULPC [14].

Acknowledgments. This research was supported in part by Institute for Education Sciences under awards R305A100875. Any opinions, findings, and conclusions or recommendations expressed in this material are solely the authors'.

References

1. Achananuparp, P., Hu, X., Shen, X.: The evaluation of sentence similarity measures. In: Song, I.-Y., Eder, J., Nguyen, T.M. (eds.) DaWaK 2008. LNCS, vol. 5182, pp. 305–316. Springer, Heidelberg (2008)
2. Agirre, E., Diab, M., Cer, D., Gonzalez-Agirre, A.: Semeval 2012 task 6: A pilot on semantic textual similarity. In: Proceedings of the First Joint Conference on Lexical and Computational Semantics, Volume 2: Proceedings of the Sixth International Workshop on Semantic Evaluation, pp. 385–393. Association for Computational Linguistics (2012)
3. Das, D., Smith, N.A.: Paraphrase identification as probabilistic quasi-synchronous recognition. In: Proceedings of the Joint Conference of the 47th Annual Meeting of the ACL and the 4th International Joint Conference on Natural Language Processing of the AFNLP, vol. 1, pp. 468–476. ACL (2009)
4. Dolan, B., Quirk, C., Brockett, C.: Unsupervised construction of large paraphrase corpora: Exploiting massively parallel news sources. In: Proceedings of the 20th International Conference on Computational Linguistics, p. 350. ACL (2004)
5. Fellbaum, C.: WordNet: An Electronic Lexical Database. MIT Press, Cambridge (1998)
6. Fernando, S., Stevenson, M.: A semantic similarity approach to paraphrase detection. In: Computational Linguistics UK (CLUK 2008) 11th Annual Research Colloquium (2008)
7. Finch, A., Hwang, Y.S., Sumita, E.: Using machine translation evaluation techniques to determine sentence-level semantic equivalence. In: Proceedings of the Third International Workshop on Paraphrasing (IWP2005), pp. 17–24 (2005)
8. Hassan, S.: Measuring semantic relatedness using salient encyclopedic concepts. PhD, Thesis. University of North Texas (2011)
9. Islam, A., Inkpen, D.: Semantic similarity of short texts. In: Recent Advances in Natural Language Processing V, vol. 309, pp. 227–236 (2009)
10. Kozareva, Z., Montoyo, A.: Paraphrase identification on the basis of supervised machine learning techniques. In: Salakoski, T., Ginter, F., Pyysalo, S., Pahikkala, T. (eds.) FinTAL 2006. LNCS (LNAI), vol. 4139, pp. 524–533. Springer, Heidelberg (2006)
11. Kuhn, H.W.: The Hungarian Method for the assignment problem. Naval Research Logistics Quarterly 2, 83–97 (1955)
12. Lintean, M.: Measuring Semantic Similarity: Representations and Methods (Doctoral dissertation). The University of Memphis, Memphis, TN (2011)

13. Madnani, N., Tetreault, J., Chodorow, M.: Re-examining machine translation metrics for paraphrase identification. In: Proceedings of the 2012 Conference of the North American Chapter of the Association for Computational Linguistics: Human Language Technologies, pp. 182–190. Association for Computational Linguistics (2012)
14. McCarthy, P.M., McNamara, D.S.: The user-language paraphrase challenge (2008)
15. Mihalcea, R., Corley, C., Strapparava, C.: Corpus-based and knowledge-based measures of text semantic similarity. In: AAAI, vol. 6, pp. 775–780 (2006)
16. Resnik, P.: Using information content to evaluate semantic similarity in a taxonomy. arXiv preprint cmp-lg/9511007 (1995)
17. Rocchio, J.J.: Relevance feedback in information retrieval. Prentice Hall, Ing. Englewood Cliffs, New Jersey (1971)
18. Rus, V., Banjade, R., Lintean, M.: On Paraphrase Identification Corpora. LREC (2014)
19. Rus, V., Lintean, M.: A Comparison of Greedy and Optimal Assessment of Natural Language Student Input Using Word-to-Word Similarity Metrics. In: Proceedings of the Seventh Workshop on Building Educational Applications Using NLP, The 2012 Conference of the North American Chapter of the Association for Computational Linguistics: Human Language Technologies, Montreal, Canada (2012)
20. Rus, V., McCarthy, P.M., Lintean, M.C., McNamara, D.S., Graesser, A.C.: Paraphrase Identification with Lexico-Syntactic Graph Subsumption. In: FLAIRS Conference, pp. 201–206 (2008)
21. Rus, V., Niraula, N., Banjade, R.: Similarity measures based on latent dirichlet allocation. In: Gelbukh, A. (ed.) CICLing 2013, Part I. LNCS, vol. 7816, pp. 459–470. Springer, Heidelberg (2013)
22. Salton, G., Buckley, C.: Term-weighting approaches in automatic text retrieval. Information Processing & Management 24(5), 513–523 (1988)
23. Salton, G., Lesk, M.E.: Computer evaluation of indexing and text processing. Journal of the ACM (JACM) 15(1), 8–36 (1968)
24. Socher, R., Bauer, J., Manning, C.D., Ng, A.Y.: Parsing With Compositional Vector Grammars. In: Proceedings of ACL 2013 (2013)
25. Socher, R., Huang, E.H., Pennin, J., Manning, C.D., Ng, A.: Dynamic pooling and unfolding recursive autoencoders for paraphrase detection. In: Advances in Neural Information Processing Systems, pp. 801–809 (2011)
26. Ştefănescu, D., Banjade, R., Rus, V.: Latent Semantic Analysis Models on Wikipedia and TASA, LREC (2014)
27. Turney, P.: Mining the Web for Synonyms: PMI-IR versus LSA on TOEFL (2001)
28. Van Valin Jr, R.D.: Lexical representation, co-composition, and linking syntax and semantics. In: Advances in Generative Lexicon Theory, pp. 67–107. Springer (2013)
29. Wan, S., Dras, M., Dale, R., Paris, C.: Using dependency-based features to take the "para-farce" out of paraphrase. In: Proceedings of the Australasian Language Technology Workshop (2006)

An Investigation on the Influence of Genres and Textual Organisation on the Use of Discourse Relations

Félix-Hervé Bachand, Elnaz Davoodi, and Leila Kosseim

Department of Computer Science & Software Engineering
Concordia University
Montreal, Canada
{f_bachan,e_davoo,kosseim}@encs.concordia.ca

Abstract. In this paper, we investigate some of the problems associated with the automatic extraction of discourse relations. In particular, we study the influence of communicative goals encoded in a given genre against another, and between the various communicative goals encoded between sections of documents of a same genre. Some investigations have been made in the past in order to identify the differences seen across either genres or textual organization, but none have made a thorough statistical analysis of these differences across currently available annotated corpora. In this paper, we show that both the communicative goal of a given genre and, to a lesser extend, that of a particular topic tackled by that genre, do in fact influence in the distribution of discourse relations. Using a statistically grounded approach, we show that certain discourse relations are more likely to appear within given genres and subsequently within sections within a genre. In particular, we observed that *Attributions* are common in the newspaper articles genre while *Joint* relations are comparatively more frequent in online reviews. We also notice that *Temporal* relations are statically more common in the methodology sections of scientific research documents than in the rest of the text. These results are important as they give clues to allow the tailoring of current discourse taggers to specific textual genres.

1 Introduction

Consider the simple discourse: *Writing a scientific paper takes time; we wrote this one in two months.* In a coherent text, textual units are not understood in isolation but in relation with each other through discourse relations that may or may not be explicitly marked. The fact that "we" wrote this paper in two months, *illustrates* that writing a scientific paper takes time. Research on discourse analysis tries to model the coherence relations that hold between textual units, and these allow us to interpret the text and understand the communicative purpose of its units. This, in turn, is useful for many Natural Language Processing (NLP) applications such as automatic summarisation, question answering and text simplification. The objective of this paper is to explore the relationship between genre and textual organisation and the use of discourse relations.

A. Gelbukh (Ed.): CICLing 2014, Part I, LNCS 8403, pp. 454–468, 2014.
© Springer-Verlag Berlin Heidelberg 2014

The task of automatic discourse relation extraction is a particularly difficult one. One important difficulty stems from the need for the system to be aware of the rhetorical purpose of the discourse on several levels. The rhetorical structure of a document can be divided into several levels of abstraction, from the general, down to the more specific. Discourse parsers available today (eg. [1] [2] [3]) attempt to extract rhetorical relations between Elementary Discourse Units (EDUs) without trying to build to the highest level of discourse relations schemas, namely the textual genre. For our purpose, we consider three levels of abstractions related to rhetorical structures: the genre, the sections, and the relations between individual EDUs. We argue that in order to extract discourse relations effectively, a system should consider the higher level rhetorical structures that we describe here as genre and section. By genre we mean that texts can have a variety of communicative goals [4]. Examples of genres include: instructional texts, reviews, scientific papers, newspaper articles, etc. At a lower level, we consider the textual organization of the document. By that we mean that the documents could be separated into different sections and subsections, each emphasising a lower-level communicative purpose. For example, given a scientific paper, the sections will typically include: abstract, introduction, methodology, results, etc. It should be noted that different genres will typically exhibit different textual organisations. Compare our previous example of a scientific paper to a review of a film. The likelihood of the appearance of a methodology section in such a document is very low. Instead we are expecting sections such as plot description, criticism, conclusion, etc. This shows that the higher level genre distinction can be used to better identify the more fine grained textual organisation categorizations. This is in line with the hierarchical view of discourse analysis presented in [5]. Based on this, it seems intuitive that the distribution of the various types of discourse relations be influenced by its occurrence in a given section of the document as opposed to another. Both the genre and, subsequently, the textual organization are important features to be considered in the automatic tagging of discourse relations.

2   Previous Work

Currently available discourse parsers do not take genre and textual organisation into consideration when extracting discourse relations (eg. [1] [2] [3]). To some extend, [2] and [3] estimate the influence of textual organisation by using the distance between a relation and the beginning of the text. Neither, however, take the genres or sections of texts into account in a definitive sense.

A few attempts have already been made at investigating the relation between genres and textual organisation and their influence on the distribution of discourse relations.

Bonnie Webber's investigation [6] shows that genre does in fact appear to play a role in the distribution of discourse relations. In order to reach this conclusion, the author performed a frequency analysis of the PDTB corpus [7]. She split the corpus into four distinct sections, each identifying a specific genre. The *news* section accounts for the largest portion of the corpus, with 1902 documents. The remaining 208 documents are split into *essays*, *summaries*, and *letters*. The author observed that in the case of labelling implicit relations, which are those that are not marked explicitly by expressions such as *therefore* and *in order to*, especially when such relations appear in between sentences, the genre appears to be a worthwhile feature to investigate. Another interesting point relates the overall structure of discourse relations across a given document. For example, a news article might start by giving an effective summary of its contents, while an essay is less likely to do so. This leads to the hypothesis that we should not only consider the distribution of relations one at a time, but we should also consider sequences of such relations and the influence of genre on the observed patterns. This is similar to the notion of rhetorical schemas described in [5].

Another interesting research deals with the concept of "stages" [8]. These are similar in nature to our notion of textual organisation. The author studied a corpus of movie reviews written by non-professionals and aggregated from various web-sites such as *Rotten Tomatoes* and *Epinions*, and found that such reviews are typically organized in five sections: *subject matter, plot description, character descriptions, background* and *evaluation*. These sections could be segmented in two larger communicative goals: description and evaluation, that usually appear in this order. In addition, [8] observed that the evaluation sections tend to contain more evaluative and subjective words. On the other hand, descriptive sections tend to contain more temporal connectives, as well as more causal-type connectives. These observations are relevant to our purpose as the appearance of such connectives hints towards the existence of certain discourse relations.

Another interesting work is that of [9] which argues that discourse relations can be used as a feature to segment a given document on various topics. This shows that there does exist a relation between discourse relations and document sections. In order to evaluate their hypothesis, the authors constructed a corpus of 140 texts in Brazilian Portuguese picked from a number of sections of mainstream news agencies. These were manually annotated using the Rhetorical Structure Theory (RST) framework [5] and split into various topics [10] (or what we refer to as sections). Their conclusion is that some relations tend to be more frequent around topic boundaries, while others were never recorded to occur around these same boundaries. The authors evaluated topic segmentation based on RST type annotation and noticed an improvement over their baseline implementation. This again appears to show that a relation between the distribution of discourse relations and sections does in fact exist.

However, to our knowledge, no previous work has attempted to measure empirically the influence of genre and textual organisation on the distribution of discourse relations using today's large scale annotated discourse corpora.

3   Methodology

In order to measure the influence of genre and textual organisation in the task of automatic discourse relation extraction, we analysed the distribution of discourse relations across various corpora spanning over various genres. Within some of these corpora, we also identified sections and analysed the distribution of discourse relations across these sections.

3.1   Corpora

The surge of interest in computational discourse analysis could not have happened without the availability of large-scale annotated corpora. The first major effort was the RST Discourse TreeBank (RST-DT) [11], which uses then followed by: Graphbank [12], the Discourse Relations Reference Corpus (DRRC) [13] and the Penn Discourse TreeBank (PDTB) [7] which included 3165 documents (50,000 sentences) tagged using [14]'s model. Through strict annotation guidelines, these resources attained a high inter-annotator agreement, which made them usable for training machine learning techniques. In addition, the field of BioNLP became interested in the extraction of the *causality* relation in bio-medical texts and developed the BioCause corpus [15] and their own shared-tasks. In 2011, [16] took a larger view of the problem by tagging all relations in bio-medical texts and developed the Biomedical Discourse Relation Bank (BioDRB).

Most work on computational discourse analysis are based on two principal frameworks for the annotations of discourse level relations. The first, Rhetorical Structure Theory (RST) [5], was conceived in order to fill the need for a framework that could be used in tasks related to natural language generation. A detailed set of annotation guidelines based on the RST framework was later created by Marcu, et al. [17]. More recently, the Penn Discourse Tree Bank [7], makes use of a new annotation framework for discourse structures. Guidelines for the purpose of creating corpora within this framework were penned by [14]. PDTB has now become one of the most widely used corpora due to its size and annotation that attempt to remain framework agnostic.

For the purpose of our work, we used the following corpora:

RST Discourse Treebank. The RST Discourse treebank [11] consists of 385 articles from *The Wall Street Journal* annotated on discourse relations, with over 20,000 EDUs which are related together and tagged with RST's 78 relations. Given the source of these documents, they are generally written in a formal language.

Maite Taboada's Review Corpus. A second corpus we considered which also uses the RST framework, as well as Marcu's annotation guidelines is Taboada's Review corpus [18] [19]. This corpus is composed of 400 reviews gathered from the *Epinions* website, with over 12,000 discourse relations identified. These reviews are authored by non-professionals. As a result, the type of language used in these documents tend to be more informal and the overall structures seem to be somewhat more liberal.

Penn Discourse Tree Bank. The Penn Discourse Tree Bank (PDTB) is a large scale corpus which, much like RST, annotates discourse level relations [7].

The PDTB annotation style is an attempt at providing annotations which are theory-neutral. The annotation guidelines [14] were used in the creation of a number of corpora. They describe 43 discourse relations, some of which are hierarchically related. The original corpus covers the entire *Wall Street Journal* section of the Penn Treebank. The corpus is composed of 2304 texts, which are marked with over 40,000 relations.

Biomedical Discourse Relation Bank. The last corpus we investigated in our work is the Biomedical Discourse Relation Bank (BioDRB) [16]. It is composed of 24 open-access research papers in the biomedical field. Nearly 6000 relations were marked using the PDTB framework. An interesting feature of this corpus is that each document is split into several sections. These sections can be used as the basis of our investigation on textual organization (see Section 4.2)

3.2   Inter-framework Discourse Mapping

Since the annotation guidelines used for the corpora we used ([8] [14] [16] [20]) differ in some manners, some work had to be performed in order to map the different discourse relations across corpora. The RST-DT contains 78 discourse relations, grouped into 18 meta-relations, while the PDTB is built from 43 discourse relations which can be grouped overall into 4 broad categories. Although both the RST-DT [11] and the online reviews [19] [18] are based on the RST framework, they do not use exactly the same set of relations. Mapping between the RST Discourse Treebank and the Online review corpus was performed by only considering the *meta-relations* of each. Since the major differences in annotations between RST-DT and the reviews corpus are found in the finner grained relation types, using these higher level *meta-relations* allowed us to perform a sensible mapping between annotations. For example, both RST-DT and Taboada's Review Corpus include relations that can be grouped under the *Contrast* relation, even if some of these exact relations differ in both corpora. Similarly, although both the PDTB and the BioDRB corpora are grounded on the PDTB annotation guidelines [14], the BioDRB annotations differ in a number of ways. Some modifications were made by the authors of this corpus as they found that certain aspects of the original framework were inappropriate for their task. [16] provides a mapping between their new relations set. Because of this, we converted the data from the original PDTB corpus into the new relations of the BioDRB corpus, following the descriptions provided. In order to adequately compare the PDTB and BioDRB corpora, we relied on the descriptions given in [16] which detail the changes made to the original PDTB guidelines in order to obtain those used in the creation of the more recent corpus. Since this description shows how the authors converted the original PDTB annotation guidelines to the ones used in the creation of the bio-medical documents corpus, we have followed the same path and used the relations described in [16] while comparing these two corpora. Details on how to map the original PDTB relations to those we used are given in this same paper.

3.3   Log Likelihood Ratio

In order to identify statistically significant differences between genres and textual organization, we performed frequency profiling using the log likelihood ratio described in [21]. This measure allows us to compare the distribution of discourse relations across multiple corpora and sort them according to the importance of their relative frequencies. It then allows to identify the most relevant data points, but qualitative examination must subsequently be performed. The resulting numbers themselves only provide a measure of which discourse relations are statistically more informative. As described in [21] the log likelihood ratio for a given relation between corpora a and b is computed as:

$$ LL = 2 \times \left(\left(O_a \times log \left(\frac{O_a}{E_a} \right) \right) + \left(O_b \times log \left(\frac{O_b}{E_b} \right) \right) \right) \tag{1} $$

where:

$$ E_a = \frac{N_a \times (O_a + O_b)}{N_a + N_b}, E_b = \frac{N_b \times (O_b + O_a)}{N_b + N_a} \tag{2} $$

N_i corresponds to the total count of all relations in a given corpus, and O_i corresponds to the count of the relation for which we are currently making calculations in that corpus. The second and third formulas gives us the *expected values*, which are then used, as E_i, in the first formula.

4   Analysis

4.1   Distributions of Discourse Relations across Genre

We first studied the influence of genre on the distribution of discourse relations. To do so, we split the corpora described in Section 3.1 into two categories: RST framework corpora and PDTB framework corpora.

RST Framework Corpora. The first two corpora we analysed both use the RST framework [5] and guidelines [17]: RST-DT and the Taboada Review Corpus (see Section 3.1). From the RST-DT corpus, we only selected the documents that have been identified as newspaper articles in [6], leaving out "erratas", "letters", and "summaries", in order to limit our investigations to documents that are news stories. On top of the genres themselves being different, it should be noted that the newspaper articles of the RST-DT corpus are written in a very formal language, while the online reviews of the Taboada's corpus tend to be much more informal, as indicated in Section 3.3. In order to view the differences in terms of discourse relations between these two corpora, we calculated the log-likelihood ratio [21]. These results are shown in Table 1.

Table 1 shows the relations which appear to vary in a statistically significant manner while comparing the two corpora. The most obvious statistical differences stem from *Joint, Attribution, Enablement, Same-Unit, Background,* and *Elaboration*.

Table 1. Log Likelihood Ratio between RST-DT and Taboada's Reviews Corpus Using RST's Meta-Relations

Relation	RST-DT	Reviews	LL ratio
Joint	10.55%	33.35%	1,637.57
Attribution	11.07%	0.01%	1,546.38
Enablement	18.31%	3.02%	1,321.68
Same-Unit	9.36%	0.01%	1,305.96
Background	2.99%	13.23%	940.39
Elaboration	25.79%	9.25%	915.17
Contrast	4.90%	16.59%	888.74
Cause	2.55%	9.59%	578.64
Condition	1.04%	5.92%	517.30
Comparison	1.49%	0.01%	199.42
Explanation	3.41%	1.13%	134.44
Topic-Change	1.01%	0.01%	132.42
Textual-organization	0.88%	0.01%	114.62
Temporal	2.37%	4.73%	101.87
Summary	0.71%	0.14%	45.14
Topic-Comment	0.84%	0.26%	35.03
Manner-Means	0.73%	0.54%	3.41
Evaluation	1.97%	2.17%	1.25

A number of observations can be made from Table 1. First, *Elaborations* appear much more frequently in newspaper articles than in online reviews (25.79% vs. 9.25%). In fact, this single relation accounts for a quarter of the relations of the first corpus. It does not seem surprising that the *Elaboration* relation be used so often in news paper articles, as we would expect texts to bring forward an idea and then elaborate on it. This type of pattern can probably be expected regardless of genre. What we notice, however, is that while *Elaboration* is frequent in newspaper articles, the *Joint* relations are roughly three times more likely in the online review corpus (10.55% vs. 33.35%). The *Joint* relation is described in the annotation guidelines [17] as a pseudo-relation which should be used by annotators when no other relation seems appropriate. What these two distributions seem to indicate has more to do with the fact that the review corpus is written less formally than the newspaper articles of the *The Wall Street Journal*. What we mean by that is that, not only is the task of identifying discourse relations a difficult one, but using such relations appropriately is also difficult. The flow of a discourse redacted by a professional writer and with the help of an editorial staff is likely to be more easily identifiable than that of an amateur reviewer posting online, without any sort of peer review. For these reasons, we believe that the occurrence of such an elevated number of the *Joint* relation in the review corpus is likely due to the differences in the writers capacity to make an adequate use of discourse relations and the inherent ability at writing in a coherent manner. A similar conclusion can be reached when observing the distribution of *Same-Unit* relations, more frequent in the RST-DT corpus (9.36% vs. 0.01%). This particular pseudo-relation is intended to represent embedded relations. For example, consider this excerpt from wsj_1362 where the EDUs marked are related as *Same-Unit*, while the first EDU contains a *Enablement* relation.

[a reserve it is establishing to cover expected pollution cleanup costs at an Ohio plant][reduced its third-quarter net income by $1.9 million.]

The use of such embedded discourse relations can once again be attributed to better stylistic choices made by authors with a better grasp of the language used. *Attribution* relations are much more frequent in the newspaper corpus than in the review corpus (11.07% vs. 0.01%). This comes as no surprise as reported news should include a number of statements that are later attributed to their authors. On the other hand, this is not the type of discourse structure we would expect from a review. The *Enablement* relation serves to provide a description of a condition which enables a subsequent occurrence. Finding this relation type in a corpus of newspaper articles is not surprising as events are often described in order to introduce more recent occurrences. Examples of *Background* relations are more frequent in the Online Reviews corpus. This again seems logical, as we would expect observations of reviewers to be justified by providing background information. Such relations are therefore useful in the case of reviews.

PDTB Framework Corpora. We now turn our attention to corpora making use of the PDTB framework for the annotation of discourse relations. We compare here the Penn Discourse Tree Bank to the Biomedical Discourse Relation Bank (BioDRB).

Table 2. Log Likelihood Ratio between PDTB and Bio-medical corpora, using PDTB Relations

Relation	PDTB	BioDRB	LL ratio
Circumstance	0.05%	4.09%	354.86
Contrast	16.00%	5.01%	228.97
Background	0.29%	2.36%	104.12
Condition	3.45%	0.39%	99.42
Purpose	6.05%	10.96%	66.73
Instantiation	4.28%	1.56%	50.45
Concession	3.62%	5.85%	24.18
Temporal	10.75%	13.86%	17.43
Restatement	6.98%	9.47%	16.89
Conjunction	23.03%	18.92%	16.80
Alternative	1.17%	0.66%	5.87
Continuation	13.48%	15.19%	4.42
Exception	0.04%	0.16%	4.20
Reinforcement	1.27%	1.79%	4.01
Similarity	0.14%	0.09%	0.51
Cause	9.40%	9.63%	0.12

Table 2 shows the differences in the distribution of discourse relations between our two corpora. Once again, we see a number of statistical differences between our corpora. The most noticeable being *Circumstances, Contrast, Background, Condition,* and *Purpose.* First, *Circumstance* relations are favored in the BioDRB corpus. These relations are often used in order to explain the specific conditions of a given experiment and subsequently describe the observed results within said conditions. The *Contrast* relation appears to be favored within the PDTB corpus (16.00% vs. 5.01%). A common way of using such a relation when dealing with text of the newspaper article genre is to compare divergent opinions. For example, consider the following from ws j_0047:

> [A majority of an NIH-appointed panel recommended late last year that the research continue under carefully controlled conditions] [but the issue became embroiled in politics as anti-abortion groups continued to oppose federal funding]

This type of discourse is quite common when dealing with news items in the social sphere, as divergence of opinions is generally what make the news. On the other hand, the *Background* relations are more frequent in the BioDRB corpus (0.29% vs. 2.36%). Such relations are used when background information is provided in order to allow the reader to fully grasp the arguments being made. It is not surprising to find such a relation in scientific papers where claims are often made based on background knowledge or previous work. *Condition* relations appear in the PDTB corpus at a greater frequency (3.45% vs. 0.39%). These relations are often used to put forward conditions necessary for certain predictions to become reality. Such conditions do not need to be realized, they can simply be hypothetical. This is not an uncommon strategy for journalists, as exemplified in ws j_0664:

> [If the exchange falters in these moves,][it might once again fall behind its chief New York competitor, the Commodity Exchange.]

Finally, we notice that the distributions of relations of *Cause* and *Similarity* seem to be constant across the two textual genres studied.

4.2 Distributions of Discourse Relations across Sections

We now turn our attention to the influence of textual organisation on the distribution of discourse relations. Our hypothesis is that, just like with the higher-level communicative goal of the textual genre, the organisation of discourse into sections play a role in influencing the discourse relations employed in the lower-level communicative goals expressed though sections. For example, we believe that the distribution of discourse relations encountered in the *abstract* section of scientific papers should differ from those found in the *methodology* section. In order to evaluate this claim, we once again analysed the BioDRB corpus which is already split into sections. Each of these section refers to the usual sections found in scientific papers: *introduction, methodologies, results, abstracts,* and *discussions.* In order to discover statistically significant data, we again computed

Table 3. Distributions of Discourse Relations Across Sections in BioDRB

	Overall	Introduction		Methods	
Relation	Distribution	Distribution	LL ratio	Distribution	LL ratio
Alternative	0.92%	0.79%	0.07	0.59%	0.88
Background	3.07%	3.94%	1.07	1.17%	13.84
Cause	12.32%	12.76%	0.16	2.49%	**89.83**
Circumstance	4.10%	1.89%	**11.24**	1.17%	23.33
Concession	7.11%	9.45%	5.31	0.59%	**77.58**
Condition	0.61%	0.31%	1.33	1.02%	1.94
Conjunction	18.81%	16.69%	2.98	14.79%	9.73
Continuation	11.20%	12.76%	2.55	17.86%	33.30
Contrast	6.63%	4.25%	6.36	1.46%	43.86
Exception	0.28%	0.31%	0.03	0.44%	0.63
Instantiation	2.01%	2.83%	2.71	0.15%	21.82
Purpose	11.96%	14.8%	4.83	12.45%	0.16
Reinforcement	2.37%	2.36%	0.01	0.59%	14.42
Restatement	8.44%	10.08%	1.94	7.03%	2.41
Similarity	0.25%	0.16%	0.33	0.15%	0.45
Temporal	9.92%	6.61%	8.46	38.07%	**478.55**

	Results		Abstracts		Discussions	
Relation	Distribution	LL ratio	Distribution	LL ratio	Distribution	LL ratio
Alternative	0.68%	0.75	0.33%	1.39	1.46%	5.42
Background	3.15%	0.04	5.35%	3.77	3.65%	0.77
Cause	11.5%	0.78	11.37%	0.21	19.07%	**53.27**
Circumstance	9.63%	**108.90**	4.68%	0.22	1.28%	37.70
Concession	6.3%	1.64	7.36%	0.02	10.68%	24.93
Condition	0.09%	10.62	0.67%	0.01	1.09%	5.06
Conjunction	23.17%	11.93	26.42%	**7.46**	17.88%	1.84
Continuation	9.63%	2.22	6.02%	**8.08**	7.85%	13.19
Contrast	10.14%	32.58	4.35%	2.50	7.48%	2.41
Exception	0.26%	0.05	0.33%	0.03	0.18%	0.60
Instantiation	0.85%	12.22	1.34%	0.70	3.92%	26.70
Purpose	12.35%	0.21	12.04%	0.00	9.58%	7.58
Reinforcement	1.28%	8.75	1.67%	0.64	4.65%	31.80
Restatement	10.65%	8.16	10.03%	0.77	6.02%	12.39
Similarity	0.26%	0.00	0.33%	0.07	0.36%	0.64
Temporal	0.09%	**260.37**	7.69%	1.51	4.84%	43.42

the log likelihood ratio [21]. This time, this measure was computed for each section with respect to the overall distribution of relations in the corpus.

The results shown in Table 3 show the relations that have a more statistically significant difference in distribution across sections. The most striking values of the log likelihood ratio are seen with *Temporal* relations which are significantly more frequent in the **Methods** (38.07% vs. 9.92%) and **Results** (0.09% vs. 9.92%) sections. This seems intuitive as we would expect a description of methodologies to include experimental steps to be taken in succession through time. Other values are also worth noting. The most statistically significant difference observed in the **Introduction** section shows a slight tendency to disfavor *Circumstance* relations (1.89% vs. 4.10%). Such a relation is used to describe the conditions in which an event occurs, without the need for the event and the circumstances to influence each other. It seems relevant to use this relation in the **Results** section, where circumstances are first given in order to set the stage for the results observed. Such a discourse schema is not as useful in an introduction. The **Methods** section's second and third most significant distinctions are seen with the *Cause* (2.49% vs. 12.32%) and *Concession* (0.59% vs. 7.11%) relations. It should be noted that both these relations are less likely to appear in the **Methods** section. Again, the **Methods** tend to favor *Temporal* relations, describing successive steps in an experiment. The **Results** section shows *Temporal* (0.09% vs. 9.92%) and *Circumstance* (9.63% vs. 4.10%) relations to be the most significant differences. Seeing a higher frequency of *Circumstance* relations in this section seems again intuitive as the presentation of results are often made in the context of the circumstances observed during experimentation. In the **Abstracts** section, the *Conjunction* (26.42% vs. 18.81%) and *Continuation* (6.02% vs. 11.20%) relations are the most statistically different. *Conjunction* relations are used in order to link EDUs as part of a list. This seems appropriate, especially in an **Abstract**, where statements are compressed due to constraints on length. In the **Discussion** section, *Cause* (19.07% vs. 12.32%) is statistically the most different in its distribution. Again, this appears intuitive as the discussion should explain the causes for the observed results. Finally, relations such as *Alternative*, *Exception*, and *Similarity* do not seem to be used differently across the sections studied.

Overall, Table 3 shows that even through a small investigation of the distributions of discourse relations, sections do in fact appear to play an important role and that these differences can be justified fairly naturally. One interesting observation is that the result of the log-likelihood ratio calculations shown in Table 3 are generally much lower than those seen in Tables 1 and 2. This suggests that the communicative goal of the textual genre has a larger influence over these distributions than the communicative goal of given sections in documents of a same genre.

Since currently, only the BioDRB corpus is segmented into sections, in order to further evaluate our claims on the influence of textual organisation, we proceeded with the following investigation: using the RST-DT corpus [11], we clustered the discourse relations used according to how far within the document they occur. Specifically, we counted the number of discourse relations in each document, and separated them in five pseudo-sections, each containing 20% of the total discourse relations found in the document. For example, a document with 10 discourse relations would have its first two grouped together, followed by the next two, and so on. With this simple heuristic, and assuming that the documents of our corpus all share the same general pattern, as dictated by the genre, we were able to identify in which portion of the documents certain relations are more likely to occur and approximate the notion of textual organisation. Once again, the log likelihood ratio is calculated by comparing a given pseudo-section to all the overall distribution.

Table 4. Distributions of Discourse Relations Per Pseudo-Sections in RST-DT

Relation	Overall Distribution	0-20% Distribution	0-20% LL ratio	20-40% Distribution	20-40% LL ratio
Attribution	12.00%	10.86%	2.65	11.09%	1.39
Background	3.64%	4.76%	**7.71**	3.81%	0.26
Cause	3.04%	3.49%	1.59	2.75%	0.57
Comparison	1.68%	1.60%	0.08	1.37%	1.26
Condition	1.29%	1.01%	1.45	1.42%	0.38
Contrast	5.94%	5.90%	0.01	5.66%	0.24
Elaboration	31.28%	30.45%	0.53	32.91%	2.63
Enablement	2.30%	2.50%	0.43	2.55%	0.72
Evaluation	2.33%	2.25%	0.06	2.12%	0.39
Explanation	3.88%	3.72%	0.17	4.06%	0.25
Joint	13.45%	11.00%	**11.29**	13.14%	0.08
Manner-means	0.88%	0.74%	0.50	0.88%	0.00
Same-unit	11.10%	11.88%	1.28	11.65%	0.86
Summary	0.84%	2.14%	**37.90**	0.65%	0.99
Temporal	2.97%	2.88%	0.06	2.79%	0.19
Textual-organization	1.23%	1.74%	4.50	0.72%	5.59
Topic-change	1.19%	1.78%	6.36	0.74%	4.30
Topic-comment	0.98%	1.28%	2.16	0.95%	0.01

Relation	40-60% Distribution	40-60% LL ratio	60-80% Distribution	60-80% LL ratio	80-100% Distribution	80-100% LL ratio
Attribution	11.76%	0.04	13.14%	2.94	14.60%	1.39
Background	3.18%	1.27	3.40%	0.28	3.47%	2.92
Cause	2.73%	0.67	3.47%	1.59	3.11%	0.96
Comparison	1.92%	0.87	1.71%	0.03	1.98%	0.06
Condition	1.06%	0.92	1.31%	0.02	1.78%	1.56
Contrast	6.04%	0.08	6.04%	0.08	6.78%	0.00
Elaboration	31.87%	0.52	29.19%	2.69	35.74%	0.00
Enablement	2.30%	0.00	1.74%	3.33	2.68%	0.03
Evaluation	2.32%	0.00	2.55%	0.56	2.68%	0.00
Explanation	4.37%	1.62	4.15%	0.53	3.58%	4.29
Joint	13.91%	0.54	14.38%	1.84	16.48%	1.90
Manner-means	0.77%	0.30	0.97%	0.27	1.13%	0.38
Mame-unit	10.89%	0.04	10.23%	1.36	12.19%	0.48
Summary	0.52%	3.21	0.34%	**8.66**	0.65%	2.66
Temporal	3.34%	1.18	3.27%	0.80	2.93%	1.64
Textual-organization	0.56%	**10.18**	1.13%	0.18	2.16%	**9.05**
Topic-change	0.92%	1.40	1.08%	0.20	1.56%	0.68
Topic-comment	0.81%	0.65	1.17%	0.96	0.79%	2.59

Table 4 shows the distribution of discourse relations, along with their log likelihood ratio, for each pseudo-section of the documents. A first observation is that a *Summary* is statistically more likely to occur at the onset of the document, and unlikely past half of the document, especially in the 60% to 80% pseudo-section. This seems to make sense given our newspaper article corpus, where documents are likely to start with a very brief summary of the news item, followed by the detailed explanation. The *Background* relation is more frequently seen at the beginning of a document as well. This, again, seems intuitive as providing a background in order to contextualize a news item is a typical writing strategy. The *Joint* relation, on the other hand, is less likely to occur at the beginning of documents. We assume here that since the bulk of the information provided in such documents should be located towards the middle, it seems more likely that such a relation, which joins together said information, should occur in the body of a newspaper article. The *Textual-organization* relation, which links a sub-heading with its associated section, is noted to be unlikely towards the middle of such newspaper articles, but more likely towards the end. This is likely due to the use of sub-headings to introduce new items which are related to the news being covered in the article.

5   Conclusion and Future Work

In this paper, we have performed an analysis of the distributions of discourse relations across various genres and sections. Using currently available annotated corpora, which themselves use different discourse relation frameworks, we have studied how both genre and textual organisation affects the distribution of discourse relations at the unigram level. As the RST framework suggests, discourse analysis is a hierarchical process and the construction of a discourse starts at the top with the communicative goal of the textual genre, and subsequently trickles down to the sections and sub-sections and finally between individual EDUs. In particular, we observed that *Attributions* are much more common in newspaper articles than in online reviews, and that newspaper articles favor *Enablement* while online reviews favor *Joint* relations. *Circumstance* relations are favored in scientific papers compared to newspaper articles, while the opposite is true of the *Contrast* relation. Our investigation of lower-level communicative goals across sections shows that the *Temporal* relation is significantly different across sections, while a number of other relations provide significant statistical differences across specific sections, to a lesser degree. Our observations therefore suggest that a worthwhile approach to extracting discourse relations should take into account both the genre of the text, hopefully with access to annotated corpora within that same genre, and should then partition the text into sections which themselves provide an influence on the distributions of the low-level discourse relations. In addition, while the task of identifying discourse relations is difficult, the use of those same relations appropriately is difficult for the authors themselves. We believe that some of our results, especially when comparing newspaper articles to online reviews, hints towards how the use of a more formal language usually comes with a better use of discourse relations.

As future work, it would be interesting to study the distribution of specific sequences of discourse relations, by observing discourse bigrams and trigrams, as opposed to the distribution of unigrams alone. This would be a step towards the automatic creation of discourse schemas described within the RST framework. Future work also includes the analysis of discourse relations across other types of textual genres such as poetry, political speech, but doing so is difficult to do objectively without properly annotated corpora.

Acknowledgement. The authors would like to thank the anonymous reviewers for their comments on an earlier version of the paper. This work was financially supported by an NSERC grant.

References

1. Soricut, R., Marcu, D.: Sentence level discourse parsing using syntactic and lexical information. In: Proceedings of the 2003 Conference of the North American Chapter of the Association for Computational Linguistics on Human Language Technology, Edmonton, Canada, vol. 1, pp. 149–156 (2003)
2. Hilda: A discourse parser using support vector machine classification
3. Feng, V.W., Hirst, G.: Text-level discourse parsing with rich linguistic features. In: Proceedings of the 50th Annual Meeting of the Association for Computational Linguistics, Jeju Island, Korea, vol. 1, pp. 60–68 (2012)
4. Swales, J.: Genre analysis: English in academic and research settings. Cambridge University Press (1990)
5. Mann, W.C., Thompson, S.A.: Rhetorical structure theory: A framework for the analysis of texts. IPRA Papers in Pragmatics 1, 79–105 (1987)
6. Webber, B.: Genre distinctions for discourse in the Penn Treebank. In: Proceedings of the Joint Conference of the 47th Annual Meeting of the ACL and the 4th International Joint Conference on Natural Language Processing of the AFNLP, Suntec, Singapore, vol. 2, pp. 674–682 (2009)
7. Prasad, R., Dinesh, N., Lee, A., Miltsakaki, E., Robaldo, L., Joshi, A.K., Webber, B.L.: The Penn Discourse TreeBank 2.0. In: Proceedings of the 6th International Conference on Language Resources and Evaluation (LREC), Marrakech, Morocco, pp. 2961–2968 (2008)
8. Taboada, M.: Stages in an online review genre. Text & Talk-An Interdisciplinary Journal of Language, Discourse & Communication Studies 31(2), 247–269 (2011)
9. Cardoso, P.C., Taboada, M., Pardo, T.A.: On the contribution of discourse structure to topic segmentation. In: Proceedings of the Special Interest Group on Discourse and Dialogue (SIGDIAL), Metz, France, pp. 92–96 (2013)
10. Cardoso, P.C., Maziero, E.G., Castro Jorge, M., Seno, E.M., Di Felippo, A., Rino, L.H., Nunes, M.: Cstnews-A discourse-annotated corpus for single and multi-document summarization of news texts in Brazilian Portuguese. In: Proceedings of the 3rd RST Brazilian Meeting, Brazil, pp. 88–105 (2011)
11. Carlson, L., Okurowski, M.E., Marcu, D.: RST Discourse Treebank. Linguistic Data Consortium, University of Pennsylvania (2002)
12. Wolf, F., Gibson, E., Fisher, A., Knight, M.: Discourse Graphbank. Linguistic Data Consortium, Philadelphia (2004)
13. Taboada, M., Renkema, J.: Discourse relations reference corpus. Simon Fraser University and Tilburg University (2008),
http://www.sfu.ca/rst/06tools/discourse_relations_corpus.html
14. Prasad, R., Miltsakaki, E., Dinesh, N., Lee, A., Joshi, A., Robaldo, L., Webber, B.L.: The Penn Discourse Treebank 2.0 annotation manual. Technical Report, Institute for Research in Cognitive Science, University of Pennsylvania (2007),
http://www.seas.upenn.edu/pdtb/PDTBAPI/pdtb-annotation-manual.pdf

15. Mihaila, C., Ohta, T., Pyysalo, S., Ananiadou, S., et al.: Biocause: Annotating and analysing causality in the biomedical domain. BMC Bioinformatics 14(2) (2013)
16. Prasad, R., McRoy, S., Frid, N., Joshi, A., Yu, H.: The biomedical discourse relation bank. BMC Bioinformatics 12, 188
17. Marcu, D.: Instructions for manually annotating the discourse structures of texts (1999), http://www.isi.edu/marcu
18. Taboada, M., Anthony, C., Voll, K.: Methods for creating semantic orientation dictionaries. In: Proceedings of the 5th International Conference on Language Resources and Evaluation (LREC), Genova, Italy, pp. 427–432 (2006)
19. Taboada, M., Grieve, J.: Analyzing appraisal automatically. In: Proceedings of AAAI Spring Symposium on Exploring Attitude and Affect in Text, Stanford University, CA, pp. 158–161 (2004)
20. Carlson, L., Marcu, D., Okurowski, M.E.: Building a discourse-tagged corpus in the framework of rhetorical structure theory. In: Proceedings of the Second SIGdial Workshop on Discourse and Dialogue, SIGDIAL 2001, Aalborg, Denmark, vol. 16, pp. 1–10 (2001)
21. Rayson, P., Garside, R.: Comparing corpora using frequency profiling. In: Proceedings of the Workshop on Comparing Corpora, Hong Kong, pp. 1–6 (2000)

Discourse Tagging for Indian Languages

Sobha Lalitha Devi, S. Lakshmi, and Sindhuja Gopalan

AU-KBC Research Centre, MIT Campus of Anna University, Chennai, India
sobha@au-kbc.org

Abstract. Indian Language Discourse Project is to develop large corpus annotated with various types of discourse relations which are explicit and implicit. As an initial step towards it we have annotated corpus in three languages, Hindi, Tamil and Malayalam belonging to the two major language families in India-Indo Aryan and Dravidian. In this paper we describe our initial experiments in annotating all the three language corpus and the domains of the corpus belongs to health. The initial experiment brought out various types of discourse connectives in the three languages and how they vary amongst the languages. The preliminary study itself revealed that there is cross linguistic variation among the three languages. We have shown the inter annotator agreement for all the three languages.

Keywords: Discourse Connectives, Indo-Aryan, Dravidian, Inter-annotator Agreement.

1 Introduction

Discourse connectives are cohesive links that makes a discourse coherent. Coherence makes a text semantically meaningful. Coherence relation can be explicitly marked by connectives, which marks the presence of relationship between two discourse units. Annotation of discourse connectives have become a prime task as discourse level annotated corpus plays an important role in performing tasks such as text summarization, question-answering systems and knowledge mining [4]. Researchers believe that richer the annotated linguistic information available, the better the discourse analysis [8]. Hence tagging of discourse connectives is necessary. In the below example 1(b) the connective "so" marks the coherence relation between two sentences explicitly. This shows that the first clause in 1(a) is the cause for the contingency explained in second clause. The introduction of discourse marker establishes the relation between the two sentences that keeps the discourse coherent.

1. (a). There is a depression in Bay of Bengal. It is raining.
   (b). [There is a depression in Bay of Bengal.]/arg1 **so** [It is raining.]/arg2
2. He eats ice cream. He likes it.

The connectives are used to show how the ideas are logically connected but in some cases these logical relations are so apparent that they remain implicit as in Example (2).

A. Gelbukh (Ed.): CICLing 2014, Part I, LNCS 8403, pp. 469–480, 2014.

Work on discourse connectives have been explored in various languages like French [11], Turkish [2], Czech [6], Arabic [1], English [8] etc. Penn Discourse Tree Bank (PDTB) is the large scale annotated corpora of linguistic phenomena in English. The PDTB is the first to follow the lexically grounded approach to annotation of discourse relations. It is unique in adopting a theory-neutral approach to annotation. PDTB provides argument structure of discourse relations and sense labels for each relation following a hierarchical classification scheme. Various works have been carried out based on PDTB annotation [15, 14, 7].

There are very less work done on discourse connectives for Indian language. Published works on Discourse tagging is available only for Indian languages like Hindi [3, 9, 12, 13] and Tamil [5, 10] where PDTB is used as the base for annotation. The question is whether the PDTB format is the right method for the annotation for discourse connectives in Indian Languages.

The main goal of our work is to develop a large scale corpus for Indian languages annotated with various types of discourse relations. Our early effort is to develop the corpus for Indo-Aryan and Dravidian languages such as Hindi, Malayalam and Tamil. In this initial work, we have taken a corpus of 3000 sentences from health domain for all three languages. Annotation of the corpus with various types of discourse relation like explicit and implicit discourse relations and other relations are done. The arguments are marked by arg1 (initial argument) and arg2 (the second argument) and the arg2 always follows the connective. Various types of connectives and its arguments are described in this paper. We have validated our annotation using Inter annotator agreement. In the following section we have described about the corpus used in our work. Section 3 explains the discourse connectives and arguments in Hindi, Malayalam and Tamil. The annotation task work flow and inter annotator agreement is presented in section 4. The results and observation are discussed in section 5. Section 6 presents the challenges and future work. The paper ends with the conclusion.

2   Corpus Used

For the purpose of analysing the discourse connectives and its arguments we have used the health domain corpus for all the three languages. Large-scale corpus for English in the Penn Discourse Tree Bank (PDTB) schema [9] and for Hindi in Hindi Discourse Relation Bank [14] schema is available. As there is no large corpus for Hindi, Malayalam and Tamil, we developed the corpus by collecting health related articles from the web. The inconsistency such as hyperlinks from the data was removed and a corpus of 3000 sentences was chosen for annotating the connectives and its arguments.

3   Discourse Connectives and Arguments in Hindi, Tamil and Malayalam

All the three languages share certain similarities such as verb final language, free word order but structurally they are different. Hindi, an Indo-Aryan language is an ergative language. Tamil and Malayalam, belonging to the Dravidian family of languages, are nominative-accusative languages. In Tamil and Malayalam nouns are inflected with case markers by suffixation, but this phenomenon does not occur in Hindi except for pronouns. The pronouns are suffixed with case markers. In Tamil and Malayalam some connectives remain agglutinated where as in Hindi they always occur as a free word. Syntactically the clausal constructions in Hindi are using correlative relative whereas Tamil and Malayalam has nonfinite verb form for subordination. The difference at the syntactic structure level gives rise to the variations in discourse connectives.

3. Hi: [faast food khaane se bhookh to shaanta ho jaatii hai.]/arg1
    fast   food   eat   by hunger   get   reduce
    kintu [shariira ko inse koyii laabh   nahiiM pahuncataa.]/arg2
    but   body   for this any benefit   no   bring
    (Eating fast food may reduce the hunger. But it does not bring any benefits for the body.)
4. Ta: [vayiRRil   kutalpuN irun**taal**]/arg1 [vayiRu   valikkum.]/arg2
    stomach+in ulcer   is+there   stomach   pains
    (If there is ulcer in stomach, stomach pains.)

The above examples show how the connectives are formed in Hindi and Tamil languages. In example (3) the connective is an independent element "kintu" which is a coordinate conjunction. Where as in (4) they are bound morphemes which inflect the nonfinite verb and same type of construction occur in Malayalam. In (4) verb "iru" is found agglutinated with the connective "-aal".

In Hindi the connectives occur in paired form, i.e. two connectives share same arguments. This type of construction does not occur in Tamil and Malayalam.

5. (a). Hi: agar [bacche ko bukhaar hai]/arg1 **to**   [kyaa   karen?]/arg2
        if   child   for fever   is   then   what   do
  (b). Ml: [kuttikk   paniyund**engil**]/arg1 [enthu cheyyum?]/arg2
        child+dat   fever_has if   what   do
        (If the child has fever, what to do?)

The sentence 5(a) has two connectives "agar" and "to" which shares same arguments. In Malayalam and Tamil the equivalent to Hindi paired connective "agar-to" is bound morpheme "-engil" and "-aal" as in 5(b) and (4).

3.1   Discourse Markers in Hindi, Malayalam and Tamil

The discourse markers fall into two major categories, Explicit and Implicit relations. While tagging, we also observed that there are other types of relations. The following examples give the various types of discourse connectives and their occurrence.

3.1.1   Explicit Connectives

The explicit connectives are free words or morphemes that trigger discourse relations. They explicitly signal the presence of discourse connectives between sentences or clauses. Explicit connectives can occur within a sentence or between two sentences, which need not be adjacent sentences. The connectives can occur at the initial, final or medial position in an argument. In example (3) the connective "kintu", occurs inter sententially by connecting the two sentences. The connective occurs at the initial position of the second argument.

6. Hi: [koyii soocanaa yaa rahasya **yadii** ek   klik kii doorii par ho]/arg1 **to**

　　any idea   or secret if one click   distance on is
　　[koyii kyoN apne dimaag ko kaSta   denaa caahegaa]/arg2
　　any why your brain to trouble give   want
　　(If any idea or secret is there in one click distance, then why anyone wants to give trouble to your brain.)

The above example shows that the explicit connective "yadii-to" occurs intra sententially, sharing same arguments. The connective "yadii" is present at the medial position of the first argument and "to" at the initial position of the arg2. In example (4) it is observed that the conditional marker "-aal" occur at the final position of the first argument. In the above examples we see that the relations between two discourse units are explicitly realized. Six types of explicit connectives have been observed.

3.1.1.1 Subordinators. Subordinators are subordinate conjunctions which join two clauses together, thereby making one clause dependent upon another. They connect the main clause with the adverbial clause and in certain case with noun or adjectival clause. From the corpus we observed that most commonly occurring subordinators in all the three languages are since, because and when.

7. Ml: [choot   adikamaayaalum   thaNuppadikamaayaalum kunjungaLkk
　　　Hot   excessive_and   chillness_excessive   baby-pl
　　sahikkaan kazhiyate varumpoL]/arg1 [avar karanjuthutangum.]/arg2
　　tolerate_cannot   come+when   they crying+start
　　(When babies cannot tolerate excessive hot and chillness, they start crying.)

In the above example "poL=appoL (this change of appoL to poL is due to morpho-phonemic changes which is called as Sandhi in Indian languages)" occur as subordinators. The subordinators in Hindi occur as free word whereas in Tamil and Malayalam both lexical and morpheme can become the connectives.

3.1.1.2 Coordinators. Coordinators give equal emphasis for two clauses or sentences. They connect two words, phrases, clauses and sentences. The most commonly observed Coordinators in the corpus are "but" and "and".

In example (3) Coordinator is "kintu" that connect two clauses. The two clauses remain independent. Coordinators occur in similar way in Tamil and Malayalam. The intra sentential coordinators can occur between the clauses but not at the beginning or end of the sentences in all the three languages.

3.1.1.3 Conjunct Adverbs. Conjunct adverbs are said to modify the clauses or sentences in which they occur. They join independent clauses together. They are part of adverbs and conjunction. Given below is an example of such a relation.

8. Ml: [kazhuth, mukham, kaiviralukal ennivitangalil   karuthaniramuNtaakaan
       Neck,   face,   fingers   all+these+palces   black+color+come
       kozhuppu   kaaraNamaakum.]/arg1 **athinaal** [eNNayil varutha
       fat   reason+will+be   Therefore oil   fried
       aahaaram, kozhuppulla Bakshanam enniva   ozhivaak kaNam.]/arg2
       food   fatty   food   all+these   avoid.
       (Fat can make the neck, face and fingers turn to black color. Therefore we
       have to avoid oily foods and fatty stuffs.)

In the above example "athinaal" is the adverbial conjunct which shows cause and effect relationship, where arg1 is effect and arg2 is the cause. Conjunct adverbs join ideas together in an emphatic way. This type of conjunctions occurs also in Tamil and Hindi.

3.1.1.4 Correlative Conjunction. Correlative conjunctions are simple pair of conjunctions that is used in a sentence to join different words or group of words. They are not used to link sentences themselves; instead they link two or more words of equal importance within a sentence itself. They always occur within a sentence.

9. Ta: [kaalaraa intiyaavil **maTTum alla**]/arg1 [**aanaal** ulakam muzhuvatum
       cholera India+in   not only   but   world   whole
       oru periya piracanaiaka uruvaakkirukiratu.]/arg2
       one big   issue   has become
       (Cholera has become a big issue not only in India, but also in the whole
       world.)

In the above example "maTTum alla aanaal" acts as correlative marker connecting equal structures. In Hindi correlative conjunction is "na keval balkii", where "na keval" occurs in the medial position of the argument1 whereas in Tamil and Malayalam "maTTum alla" and "maathramalla" occurs in the final position of the argument1.

While analyzing we have come across clausal markers which behaves as connectives. In the following section 3.1.1.5 we discuss two such clausal markers which can be considered as connective markers.

3.1.1.5 Relativizer and Complementizer. Relativizer pronoun does not refer to a preceding noun, rather it comments on the whole preceding clause or sentence. A relative pronoun makes the relative clause by having the same referent as the element of the main clause which the relative clause modifies. It provides link between main clause and relative clause.

10. Hi: [stana kainsara kaa ilaaja   mahangaa hai,]/arg1 **jo** [eka laakha taka
    breast cancer gen treatment costly   is   which one lakh upto
    hotaa hai.]/arg2
    can   be
    (The treatment for Breast cancer is costly, which can be upto one lakh.)

In the above example (10) "jo" is the sentential relative verb that connects the relative clause with the main clause. The relative clause uses "jo" as the grammatical device to indicate the connection with the main clause. This type of relation is true for Tamil and Malayalam.

Complementizer can also be considered as a special type of connectives. It is a conjunction which marks a complement clause. It can turn a clause into subject or object.

11. Ta: [mazhai varum]/arg1 **enRu** [raaju ninaitaan.]/arg2
    Rain   will come   that   Raju thought
    (Raju thought that rain will come.)

In Indian languages, there is clause inversion where subordinate and matric clause can be swapped. So the arguments move according to the position of the clauses and the connective markers.

3.2   Implicit Connectives

If there exist a relationship between two adjacent pairs of sentences and if no explicit connective is present then an implicit relation can be inferred. We have marked the "Implicit" label where an implicit relation can be inferred.

12. Hi: [isa game ke sare khiladi sachin tendulkar se bhI mahaan hE.]/arg1**Implicit**
    [Inko klIna bOld karna kisI ke basa kI bath nahI.]/arg2
    (All players in this game are greater than even Sachin Tendulkar. It is not possi
    ble for anyone to get them clean bowled.)

Here implicit connections like hence (ataha), so (isliye), because (kyonkI), also (bhI), and (aur), as (jIse hI), for (ki), further (ke athirikth), in addition (ke alaava) can be inferred. Hence annotating implicit connectives depends on the annotators. It is difficult to assign a connective for implicit relation.

In the above example it is shown that two sentences are not explicitly connected but a relationship can be inferred implicitly. The implicit relation can be inferred within a sentence and also between sentences. Here we are not posting a connective as done in PDTB because we find that this can lower inter annotator agreement. There are connectives with same lexical meaning but the contextual meaning can be different. Hence if annotators use different connectives with same lexical meaning the contextual meaning is changed.

3.3   Other Types of Relations

We have observed a type of relation where the connective is a compound word with an anaphoric entity and a lexical connective. The anaphoric entity can be a pronominal or a wh word and the connectives are the subordinate connectives. The compounding can also take an emphatic in Dravidian languages such as "taan". The anaphoric entity refers the connective to the sentence preceding it or the clause preceding it. The second argument is always anaphoric that belong to the category of explicit connectives. Some examples below show such relations.

13. Ml: [varkkala maRRu theerapradesathil ninnu vyathyasthamaayi madhya keralathinte BooprakrithiyaaN.]/arg1 **athukoNtaaN** [arabikkatalinot valare cern uyarnna kunnukaL kaaNaan kazhiyunnath.]/arg2
(Varkkala has a different land from other seashores in middle of Kerala. That is why we are able to see lot of big hills near to the Arabian Sea.)

In the above example "athukoNtaaN" signals a relationship between two sentences. Same type of relation is observed in Tamil.

In the example given below there is no discourse relation between two adjacent sentences but pragmatic knowledge is required for connecting the sentences. We call this as noun-noun relation. These types of constructions are frequent in corpus. Consider the following examples.

14. Ta: [tayir nam uTalukku oru aru maruntu.]/arg1 **Noun-Noun relation** [tayir nalla jiiraNa caktiyay tarum]/arg2
(Curd is a good medicine for our health. Curd gives good digestion power.)

In the above examples it is seen that there is no explicit or implicit relation between two sentences. But an entity based relation is observed between two sentences. Same type of relation is found in Hindi and Malayalam.

We maintained a list of connectives that occurred while analysing the data. Table 1 shows the commonly occurring connectives and its type.

Table 1. Commonly occurring connectives classified according to its type

Type of Connectives	Hindi	Malayalam	Tamil	English
Subordinator	cuuNkI/kyuuNkI	ethennaal/ ithinaal	enenil/ataal	because
Coordinator	aur/evam	um/mattum	maRum/um	and
	isliye	athinaal	athanaal	so
Conjunct Adverbs	isliye	atinaal	aakaiyaal	therefore
Relativizer	jisse	ennathinaal	ethanaal	because of which
Correlative conjunction	na keval balki	maathramalla- pakshe	mattumalla aanaal	not only but also
Complementizer	ki	enn	enRu	That

4   Annotation Task

4.1   Details of Tagged Data for Each Language

Annotation of all the relation types and its arguments were carried out sequentially across the text. We tagged the connectives and its argument syntactically in the corpus for all the three languages. We have listed below the count of different types of connectives obtained from the three corpora.

Table 2. Count of different types of connectives

Connective	Hindi	Malayalam	Tamil
Explicit	841	1192	936
Implicit	90	52	63
Other Relations	152	85	103

There were totally 1192 explicit connectives in Malayalam in which 190 connectives occur as morphemes and 1002 connectives occurred as free word. In Tamil there were 936 explicit connectives in which 448 connectives occurred as morphemes and 488 connectives occurred as free words. There are 841 explicit connectives in Hindi corpus. One single sentence can serve the purpose of connective and its arguments. Sometimes, one of the preceding sentence acts as an argument. Also the argument can be a non-adjacent sentence. But the text span follows the minimality-principle. According to this principle, the parts of clauses, clauses or sentences that are minimally necessary and sufficient for the interpretation of the relation are marked in the text span of arguments. As discussed above the connectives of Hindi are free words and connectives in Malayalam and Tamil are morphemes and free words. When free words occur we tag them as connectives and the discourse units between which the relation is inferred is marked as arg1 and arg2. When there are morphs we keep them along with the word to which it is attached.

4.2   Inter-annotator Agreement

The inter annotator agreement has been calculated to get the reliability of the annotation. The study was conducted on 3k sentences and the annotation was done by two annotators based on the guidelines. Cohen's-kappa was chosen for obtaining the inter annotator rate for explicit connectives in the corpus. The kappa coefficient is generally regarded as the statistic of choice for measuring agreement on ratings made on a nominal scale. The kappa statistic k is a better measure of inter-annotator agreement which takes into account the effect of chance agreement.

$K = (p_o - p_c)/(1- p_c)$, where p_o is agreement rate between two human annotators and pc is chance agreement between two annotators. The kappa coefficient is calculated by looking into the agreement among each argument boundary by the two annotators and the result of the agreement is shown in the table below.

Table 3. Inter-Annotator Agreement

	Hindi	Malayalam	Tamil
Connective	0.84	0.89	0.86
Arg1 start	0.73	0.82	0.72
Arg1 end	0.76	0.74	0.75
Arg2 start	0.77	0.72	0.75
Arg2 end	0.79	0.9	0.78
	Hindi	Malayalam	Tamil

The results on inter annotator agreement between the annotators for Hindi language shows that there is almost perfect agreement in tagging the connectives and substantial agreement in tagging the arguments for the connectives. In Malayalam, there is almost perfect agreement in tagging connectives, start of arg 1 and end of arg 2 and substantial agreement in tagging the end of arg 1 and start of arg 2. In Tamil, there is almost perfect agreement in tagging the connectives and substantial agreement in tagging its arguments between the annotators. From the above results it shows that the variation in agreement rate was particularly noted at arg 1 end and arg 2 start because various types of embedded structures and combined connectives were seen in all languages which produced a deviation in agreement.

5   Results and Observations

In an attempt to develop annotated corpora for the three languages Hindi, Tamil and Malayalam, we observed that there are cross linguistic variations among the three languages. We found that all relations can either be between sentences or clauses and not with phrases. In Hindi Discourse Relation Tree Bank [13] they had done a semantic labeling of arguments for cause-effect relationship. As this type of relationship depends on the main and subordinate clause the labeling of arguments must be syntactic. We have found six types of connectives in the corpus. But there are many other types of connectives that can be classified into a predefined category.

5.1   Results of Tagging

In our annotation for implicit connectives instead of adding a connective, an "Implicit" label is used. The context of the sentence changes when two different connectives are added even if the connectives have same meaning as shown in example in (12). While considering the text span we noted that implicit connectives also existed among non-adjacent sentences in all the three corpora. PDTB has mentioned an implicit relation among adjacent sentences only.

In Malayalam and Tamil paired connectives does not exist. Only a single connective serves this purpose as in 5(b) and 5(c). Whereas in Hindi paired connectives like "agar-to", yadii-to" exists as in 5(a). While tagging the Hindi corpus we also found that there exist some paired connectives other than the one that occurs generally. These types of connectives include "yooN-lihaajaa", "haalaaNkii-parantu", "cuuNkii-ataha", "cuuNkii-isliye". In PDTB, paragraphs were chosen as self-contained units and they never observed an implicit relation between paragraph boundaries. When Indian languages are considered, implicit relations were seen between paragraph boundaries.

15. Ml: kochi raajyathe thrippoonnithara,chandra gupthante kaalathe ----- [kristhuvinu munp kochi thuramugam illayirunnu ennum pinnit katalil ninn   uyarnnu vanna-thaan ennathinu thelivukal unt.]/arg1
Implicit [aadhyammayi kochiye patti vivarikkunnath chineese yaathrikarraya mahvanum   fayseenumaan.]/arg2

In the above example (15) it is seen that the implicit connective is realized between the paragraphs.

Generally in the case of paired connective "agar-to", agar occurs implicitly but "to" is always explicitly marked. In our corpus we found that in certain sentences both "agar" and "to" occur implicitly.

16. Hi: **Implicit**[baccoM ko ghamorii ho rahii hai,]/arg1 **Implicit** [kyaa kareN?]/arg2
(If children get prickly heat then what to do?)

The above example (16) show that the connective "agar" and "to" in the sentence remain implicit.

We also found that certain connectives occur together and share same arguments. In those cases we have tagged both the connectives as a single one. For example the connectives "lekin" and "baad meM" occur together and share same arguments. While considering the discourse relations some embedded structures were seen like the below examples.

17. Hi: **yadii** [[raatrii ke bhojana meM roTii kama khaaeN]/arg1 **aur** [caaval pratidin khaaeN]/arg2]/arg1 **to** [yah halkaa bhojana aapkaa svaasthya Tiika rakhe-gaa.]/arg2
(For dinner if one eats less roti and eats rice daily then this light diet will keep our body healthy)

In the above example (17) there are two types of connectives "yadii-to" and "aur". We see that "yadii-to" and aur shares the arguments. This is true for all three languages.

In our analysis of the tagged data for Tamil and Malayalam we have found that the causative marker "kaaraNam" is prefixed or preceeded with pronominal "athu" or "ithu" and sometimes suffixed with morpheme "aal". In Hindi the causative marker is "ke kaaraN" which also in some cases is seen prefixed or preceded with pronominal. "-aal" is the predominantly used morpheme in Tamil and Malayalam that acts as a connective. This produces a cause effect relation between clauses. The marker -aal in Malayalam and Tamil is highly polysemous. There are various senses. When it attaches to a verb it act as conditional marker. It also denotes an instrumental case when attached to a noun. It is found that the verb of the clause in conjunction with the noun to which the suffix -aal attaches, gives the information whether the -aal is instrumental or causal. This ambiguity is resolved by the verb phrases.

6   Challenges and Future Work

In this approach the assignment of labels is done purely syntactically and hence it is first such approach towards developing an annotated corpus for Indian languages such as Hindi, Malayalam and Tamil. Tagging of implicit connectives is a major challenge as it mainly depends on annotators choice of inferring a relation which may affect the inter annotator rate. In example (12) it is shown that different types of connectives can be implicitly marked between the sentences. The morpheme "aal" in Tamil and Malayalam acts as a conditional as well as instrumental marker. Hence disambiguation is needed while annotating.

The overlapping of arguments because of the presence of two or more connectives in a discourse unit also poses difficulty while tagging. As this is the initial step towards developing an annotation schema for the three languages we have to improve our work by annotating more implicit connectives and other relation types. We have to go in deep into the sense classification of the three languages and the implementation of the system has to be done.

7   Conclusion

In our initial step towards developing an annotated corpus a detailed analysis of connectives and its arguments are presented. The distribution of the connectives in the corpus and their syntactic pattern are discussed. We have outlined the issues that occur when adapting to PDTB type of annotation. Some specific approaches are described, which involved reasonably large corpora, highlighting the inter-annotator agreement obtained while following these guidelines.

References

1. Al-Saif, A., Markert, K.: The Leeds Arabic Discourse Treebank: Annotating Discourse Connectives for Arabic. In: LREC (2010)
2. Zeyrek, D., Webber, B.: A Discourse Resource for Turkish: Annotating Discourse Connectives in the METU Corpus. In: IJCNLP, Hyderabad, India (2008)
3. Kolachina, S., et al.: Evaluation of Discourse Relation Annotation in the Hindi Discourse Relation Bank. In: LREC (2012)
4. Korbayova, K.I., Webber, B.: Information structure and the formal presuppositions of discourse connectives. In: ESSLLI Workshop on Information Structure, Discourse Structure and Discourse Semantics, The University of Helsinki, Helsinki (2001)
5. Menaka, S., Rao, P.R.K., Devi, S.L.: Automatic identification of cause-effect relations in tamil using CRFs. In: Gelbukh, A.F. (ed.) CICLing 2011, Part I. LNCS, vol. 6608, pp. 316–327. Springer, Heidelberg (2011)

6. Mladová, L., Zikánová, S., Hajičová, E.: From Sentence to Discourse: Building an Annotation Scheme for Discourse Based on Prague Dependency Treebank. In: LREC (2008)
7. Patterson, G., Kehler, A.: Predicting the Presence of Discourse Connectives: EMNLP 2013, Seattle, October 18-21 (2013)
8. Prasad, R., Dinesh, N., Lee, A., Miltsakaki, E., Robaldo, L., Joshi, A., Webber, B.: The Penn discourse Treebank 2.0. In: LREC (2008)
9. Prasad, R., Husain, S., Sharma, D.M., Joshi, A.: Towards an Annotated Corpus of Discourse Relations in Hindi. In: IJCNLP (2008)
10. Rachakonda, T.R., Sharma, D.M.: Creating an Annotated Tamil Corpus as a Discourse Resource. In: Linguistic Annotation Workshop (2011)
11. Roze, C., Danlos, L., Muller, P.: LEXCONN: A French lexicon of discourse connectives. In: MAD 2010 (Multidisciplinary Approaches to Discourse), Moissac, France, pp. 114–125 (2010)
12. Sobha, L., Patnaik, B.N.: Discourse Connectives and Their Arguments in Malayalam. In: 24th South Asian Language Analysis, November 19-21. University of Stony Brook, New York (2004)
13. Oza, U., et al.: The Hindi discourse relation bank. In: Third Linguistic Annotation Workshop, Association for Computational Linguistics (2009)
14. Webber, B., Knott, A., Joshi, A.: Multiple discourse connectives in a lexicalized grammar for discourse. In: Third International Workshop on Computational Semantics, Tilberg, Netherlands, pp. 309–325 (1999)
15. Versley, Y.: Towards finer-grained tagging of discourse connectives. In: Workshop beyond Semantics: Corpus-based Investigations of Pragmatic and Discourse Phenomena (2011)

Classification-Based Referring Expression Generation

Thiago Castro Ferreira and Ivandré Paraboni

School of Arts, Sciences and Humanities, University of São Paulo (USP / EACH)
Av. Arlindo Bettio, 1000 - São Paulo, Brazil
{thiago.castro.ferreira,ivandre}@usp.br

Abstract. This paper presents a study in the field of Natural Language Generation (NLG), focusing on the computational task of referring expression generation (REG). We describe a standard REG implementation based on the well-known Dale & Reiter Incremental algorithm, and a classification-based approach that combines the output of several support vector machines (SVMs) to generate definite descriptions from two publicly available corpora. Preliminary results suggest that the SVM approach generally outperforms incremental generation, which paves the way to further research on machine learning methods applied to the task.

Keywords: Natural Language Generation, Referring Expressions, Classification, SVM.

1 Introduction

Referring expressions such as definite descriptions, pronouns, proper names etc. are ubiquitous in language use. In Natural Language Generation (NLG) systems, Referring Expression Generation (REG) is known as the computational task of producing these linguistic forms in order to describe a target object in a given context [1].

REG generally involves at least two relatively independent tasks: the content determination (or content selection) of definite descriptions (e.g., 'the large black dog') [2] and surface realisation [3–5]. In this work we focus on the former, that is, the task of deciding which semantic properties should be selected to compose a description of the target (or 'what to say' as opposed to 'how to say it'). When there is no risk of confusion, we hereby use the term 'REG' in the particular sense of content selection of definite descriptions[1].

A typical REG algorithm takes as an input a target object t to be described and a context containing a number of distractor objects. The goal of the algorithm is to compute a uniquely identifying set of semantic properties of t, so that t can be distinguished from every other object in the context. For instance, in a visual scene representing a class room we may refer to a particular student as in (a) below.

(a) *'the blonde girl on the second row, on the left, wearing a blue jacket'*

---

[1] Not to be mistaken with the NLP reference resolution task, e.g., [6].

A. Gelbukh (Ed.): CICLing 2014, Part I, LNCS 8403, pp. 481–491, 2014.
© Springer-Verlag Berlin Heidelberg 2014

This is however only one among many possibilities. Discourse participants may choose widely different reference strategies to uniquely describe the same target object (i.e., the student) as in, e.g., (b-d) below:

(b) *'the tall girl wearing a blue jacket'*
(c) *'the girl next to Robert's youngest sister'*
(d) *'the new student'*

At the most basic level, a REG algorithm is expected to address the issue of how to compute an 'optimal' attribute set to prevent the generation of ambiguous or overly long descriptions, among other issues that have been extensively discussed in the literature [2]. By contrast, the possibly more challenging task of selecting the 'right' attribute set (i.e., those attributes that humans speakers would actually choose, as in (a-d) above), still remains an open research question[2].

Existing approaches to REG have mainly focused on algorithmic solutions to the task [1, 8, 9]. This contrasts, for instance, the use of machine learning techniques that are mainstream in many other NLP fields. With the introduction of large-scale, publicly available corpora for REG such as TUNA [10], GRE3D3 [11] and GRE3D7 [12], however, this scenario has started to change. Besides the recent series of REG Shared Tasks in [7, 13, 14], examples of machine-learned and corpus-based REG algorithms have began to emerge [15–17].

As a step further in the use of machine learning techniques applied to REG, our own work presents a novel classification-based approach that combines the output of a set of support vector machines (SVMs) to generate relational definite descriptions. In addition to that, we also discuss an extended version of the well-known Incremental algorithm in [1] that has been modified for the same purpose. Both approaches are applied to the generation of definite descriptions found in the GRE3D3/7 corpora.

The rest of this paper is structured as follows. Section 2 discusses related work in the field and the training and test data. Section 3 presents our extended incremental and classification-based algorithms. Section 4 describes an experiment to evaluate both algorithms and its results. Finally, Section 5 presents additional remarks and suggests future work.

2   Related Work

2.1   Referring Expression Generation

The computational task of attribute selection for referring expression generation (REG) is easily illustrated by the work in [1], which introduces the well-known Incremental algorithm. The algorithm takes as an input a context D comprising a set of domain objects with their corresponding properties represented as attribute-value pairs (e.g., $\langle type, dog \rangle$ or $\langle colour, black \rangle$). The following is an example of context conveying four domain objects (three dogs and a cat) of various sizes and colours.

---

[2]   See for instance the discussion on the 'humanlikeness' of referring expressions in [7].

$e_1 : \langle type, dog \rangle, \langle size, small \rangle, \langle colour, black \rangle$
$e_2 : \langle type, dog \rangle, \langle size, large \rangle, \langle colour, white \rangle$
$e_3 : \langle type, cat \rangle, \langle size, small \rangle, \langle colour, black \rangle$
$e_4 : \langle type, dog \rangle, \langle size, small \rangle, \langle colour, brown \rangle$

One particular object t is the target to be described by means of a referring expression (i.e., a uniquely identifying set of properties), and the remaining objects in D are assumed to be distractors. The goal of the algorithm is to produce a list of attributes L such that L uniquely describes the target t and no other distractor object in D.

The Incremental algorithm iterates over a preference list P representing the order in which the target attributes should be considered for selection. Each attribute a in P is selected for inclusion in the output description L if a effectively help ruling out at least one distractor object in D. This procedure is repeated until the point in which the output description L allows the target object t to be uniquely identified.

In the above example, assuming a preference order $P = \{type,\ size,\ colour\}$, we may refer to e_1 as $L = \{\langle type, dog \rangle, \langle size, small \rangle, \langle colour, black \rangle\}$, which could be realised as 'the small black dog'. The reference to the $size$ attribute rules out e_2, which is large (i.e., not small), the reference to $type$ rules out e_3, which is a cat, and the reference to $colour$ rules out e_4, which is brown. Similarly, e_2 may be described as 'the large dog', e_3 simply as 'the cat' and e_4 as 'the small brown dog'.

Attributes that do not rule out any distractors are best avoided since they lead to overspecified descriptions, which may be prone to false conversational implicatures [18]. For instance, in a context in which there is only one cat, 'the small cat' is overspecified in the sense that $size$ is not strictly required for disambiguation. For reasons of computational complexity, however, the Incremental algorithm does not perform backtracking (hence the name 'incremental'). If an attribute included in L is made redundant by a subsequent selection, the output description L will remain overspecified. For instance, in the reference to e_4 as 'the small brown dog', the $size$ attribute was made redundant by the subsequent inclusion of $colour$.

The Incremental algorithm does not explicitly handle relational properties as in 'the dog *next to* the small cat', but many others do [2, 19–22]. These algorithms, however, tend to assign higher priority to the atomic properties of the target, using relations to other objects only as a last resort. Our present work will attempt to establish a more balanced use of atomic and relational attributes by making use of frequency estimates obtained from corpora.

2.2   Machine-Learned REG

The recent availability of REG corpora such as TUNA [10], GRE3D3 [11], GRE3D7 [12] and others has lead to a number of corpus-based approaches to the task, as in [15–17]. In [15], for instance, attribute selection makes use of rule induction techniques to predict patterns in referring expressions extracted from the Coconut dialogue corpus [23].

Our own work makes use of descriptions produced in the GRE3D3/7 online experiments described in [17]. GRE3D3/7 descriptions refer to 3D objects (e.g., boxes, spheres etc.) in simple visual scenes, and they make frequent use of relational properties as in 'the small red ball on top of the cube'. GRE3D3 contains 630 referring expressions produced by 63 speakers, and GRE3D7 contains 4480 produced by 287 speakers.

GRE3D3/7 data have been applied to a series of REG experiments described in [16, 17] and others. These experiments made use of decision-tree induction to determine whether a particular reference pattern is applicable. In the work in [17] decision trees also help decide whether each attribute should be included in a particular referring expression.

In our present work we attempt to improve this by using SVMs instead of decision trees. More importantly, we will go one step further and use the classifiers output to assemble the actual referring expressions as seen in each corpus.

3   Current Work

Besides using the existing GRE3D3/7 annotated descriptions[3], we also annotated the stimuli images used in each experiment with their atomic and relational properties, so that this could be taken as an input to our REG algorithms.

We consider two alternative approaches to referring expression generation: a straightforward, frequency-based extension of the original Incremental approach [1] to handle atomic and relational properties alike, and a classification-based approach that makes use of support vector machines. Each alternative is discussed in turn in the next sections.

3.1   Extended Incremental Approach (EIA)

As a first approach to the generation of GRE3D3/7 descriptions, the Incremental algorithm [1] was extended so as to allow the use of atomic and relational attributes alike. In our so-called Extended Incremental Algorithm (EIA), a relational description is represented as a set of properties of a target and related landmark objects. For instance, 'The cube in the left of the red ball' conveys a reference d_1 to the main target object (i.e., the cube) and an additional reference d_2 to the related landmark object (ball) as follows:

$$\{d_1 = \{\langle type, cube\rangle, \langle left\text{-}of, d_2\rangle\}, d_2 = \{\langle colour, red\rangle, \langle type, ball\rangle\}\}$$

An overview of EIA is illustrated by Algorithm 1. As in the original Incremental approach, EIA takes as an input a target object t, a context D and a list P of preferred attributes. Depending on whether the object being described is the target or a landmark, however, the algorithm will compute different preference lists as discussed below.

---

[3] http://www.jetteviethen.net/downloads.html

Algorithm 1. Extended Incremental Approach (EIA)

---

function $getDescription(t, L, D, preference, activeObjects, isTarget)$
    $L[t] \leftarrow \{\}$
    $C \leftarrow D - \{t\}$
    $activeObjects \leftarrow activeObjects \cup t$
    $Pref \leftarrow getPreferenceList(preference, isTarget)$
    for $A_i \in Pref$ **and** $|C| > 0$ **do**
        $V \leftarrow value(t, A_i)$
        if $|C \cap rulesOut(A_i, \langle A_i, V \rangle)| > 0$ **then**
            if $relationalAttribute(A_i) = true$ **then**
                if $V \notin activeObjects$ **then**
                    $L[t] \leftarrow L[t] \cup \langle A_i, V \rangle$
                    $C \leftarrow C - rulesOut(A_i, \langle A_i, V \rangle)$
                    $L \leftarrow getDescription(V, L, D, preference, activeObjects, false)$
                end if
            else
                $L[t] \leftarrow L[t] \cup \langle A_i, V \rangle$
                $C \leftarrow C - rulesOut(A_i, \langle A_i, V \rangle)$
            end if
        end if
    end for
    return L
end function

EIA starts by assigning an empty list to the L output description, defining the set of distractors C and inserting t in the list of $activeObjects$, which contains all objects that have already been mentioned in L. As in the original Incremental approach, the core of the content selection procedure consists of iterating over the list of preferred attributes and selecting those attributes that help ruling out distractors, up to the point in which the set of distractors C is empty. To this end, the auxiliary function $rulesOut$ is assumed to return, for a given property $\langle A_i, V \rangle$, the set of all distractor objects whose attribute A_i has a value different from V.

The $getPreferenceList$ function (not shown) is assumed to return the appropriate list of preferred attributes for describing either the target or landmark. These lists are ordered by attribute frequencies as seen in the training data, beginning with the most frequent attribute. In the case of relational attributes, all their frequencies are summed up to form a single entry in the preference list for each object (i.e., target or landmark). Thus, a typical preference order for describing a target cube object may be represented as, e.g., $P(target) = \{type, colour, relation, size\}$.

When the *relation* attribute is selected, the most frequent relational attribute for that particular object type (i.e., a cube) is applied. Attributes that do not appear in the training data (i.e., with zero frequency estimates) are not included in the preference list, and are therefore never selected.

If the selected attribute denotes a relation to another object (represented as the value V of the relational attribute), a recursive call is made to describe V as well. The use of the *activeObjects* list guarantees that an object already included in the description will not be mentioned twice, hence avoiding circularity [19] as in 'the bowl on the table that supports the bowl'.

3.2   Classification-Based Approach (SVM)

As in many other NLP fields, the use of machine learning techniques in REG has become more widespread as large-scale resources (i.e., REG corpora) become available. In particular, the work in [17] makes use of decision-tree induction to learn whether to select each attribute individually, and also to decide whether to follow a particular reference pattern in general.

In what follows we discuss a similar classification-based approach to decide whether to select each attribute individually. Instead of decision-trees, however, we will make use of support vector machines with radial basis function kernel. In addition to that, we will not only attempt to predict individual attribute selection, but we will also present an algorithm that combines the classifiers output to assemble an actual referring expression.

We intend to learn eight individual binary classifiers corresponding to the target and landmark attributes to be predicted, and a multi-class classifier for the *relation* attribute prediction. Possible values for *relation* are *no relation*, *right-of*, *left-of*, *next-to*, *on-top-of* and, in the case of GRE3D3, also *in-front-of*. These classifiers are summarized in Table 1.

Table 1. Target and landmark referential attribute classifiers (based on [16])

Id	Description
tg_type	type of the target object
tg_colour	colour of the target object
tg_size	size of the target object
tg_location	location of the target object
lm_type	type of the landmark object
lm_colour	colour of the landmark object
lm_size	size of the landmark object
lm_location	location of the landmark object
relation	relation between the target and landmark

All target attribute classifiers (i.e., those beginning with 'tg_' in Figure 1) and also the *relation* classifier are trained using all descriptions in each corpus. Landmark attribute classifiers (beginning with 'lm_' in Figure 1) are trained using relational descriptions only.

All classifiers make use of the same basic feature vector input as proposed in [17]. These features are summarized in Table 2.

Table 2. Feature Vector taken from [16]

Id	Description
TG_Size	size of the target object
LM_Size	size of the landmark object
Relation_Type	type of relation between target and landmark
Num_TG_Size	number of objects of same size as the target
Num_LM_Size	number of objects of same size as landmark
TG_LM_Same_Size	target and landmark share size
Num_TG_Col	number of objects of same colour as target
Num_LM_Col	number of objects of same colour as landmark
TG_LM_Same_Col	target and landmark share colour
Num_TG_Type	number of objects of same type as target
Num_LM_Type	number of objects of same type as landmark
TG_LM_Same_Type	target and landmark share type

We use Python *Scikit-learn* software [24] for the actual SVM implementation. For the multi-class prediction of the *relation* attribute, we followed the 'one-against-one' approach [25].

In Algorithm 2, the actual REG module is represented by the *getDescription* function. This function receives as an input the target object t, the attribute set to be considered (*attributes*) and the attribute predictions (*predictions*) obtained from the SVMs. The output is a description L of the target t, which may or may not include relations to other objects.

Algorithm 2. Classication-based Approach (SVM)

function *getDescription*(t, *attributes*, *predictions*)
  $L \leftarrow getObjectDescription(t, \emptyset, attributes[target], predictions[target])$
  if *predictions*[*relation*] <> *norelation* **then**
    $V \leftarrow value(t, predictions[relation])$
    if $V <> null$ **then**
      $L[t] \leftarrow L[t] \cup \langle A_i, V \rangle$
      $L \leftarrow getObjectDescription(V, L, attributes[landmark], predictions[landmark])$
    end if
  end if
  return L
end function

function *getObjectDescription*(t, L, *attributes*, *objectPredictions*)
  $L[t] \leftarrow \{\}$
  for $A_i \in attributes$ **do**
    $V \leftarrow value(t, A_i)$
    if $objectPredictions[A_i] = positive$ **then**
      $L[t] \leftarrow L[t] \cup \langle A_i, V \rangle$
    end if
  end for
  return L
end function

The algorithm starts by considering the target attribute predictions (tg_-). For each positive class prediction, *getObjectDescription* will include the corresponding attribute in target description.

Next, the *relation* attribute is considered. If *relation* prediction equals the 'no relation' class, the algorithm terminates, returning an atomic description L of the target object. If not, the related landmark object is included in the output description and *getObjectDescription* is called once more to describe the landmark object. L in this case will correspond to a relational description.

4   Evaluation

4.1   Procedure

We carried out two separate evaluation procedures making use of the GRE3D3 [11] and GRE3D7 [12] data, respectively. In both cases, our goal was to verify whether the classification-based approach (SVM) may outperform incremental generation (EIA) as discussed in the previous section.

Besides comparing SVM against EIA, both algorithms were also compared against a *Random* baseline strategy. *Random* is a variation of EIA that always selects the attribute *type* in the first place, and then considers the remaining attributes in random order, up to the point in which a uniquely identifying description is obtained, or once all available attributes have been attempted.

Both SVM and EIA were trained and tested using 10-fold cross validation. The number of folds was chosen so that the number of referring expressions of each speaker was kept balanced within each fold. In the case of the SVM approach, kernel parameters for the GRE3D3 data were set as $C = 1.0$ and $\gamma = 0.1$. For GRE3D7 we used $C = 10.0$ and $\gamma = 0.1$.

Evaluation was carried out by comparing the resulting referring expression produced by each system with the reference description found in the corpus. As in [7] and others, overall precision for each algorithm was computed by measuring Accuracy scores (i.e., the number of exact matches between System and Reference description pairs).

Also as in [7], the degree of overlap between each System-Reference description pair was measured by computing both Dice [26] and MASI [27] scores. However, given that Dice and MASI scores usually co-relate, MASI scores are provided for illustration purposes only, and are not further discussed in the analysis to follow.

4.2   Results

Table 3 summarizes the results for the REG task using EIA, SVM and *Random*. In the case of SVM, Table 4 shows prediction accuracy scores for each individual attribute[4].

---

[4] The   software   used   to   obtain   the   results   is   available   in
http://www.CICLing.org/2014/data/15

Table 3. REG results

	GRE3D3			GRE3D7		
Algorithm	Dice	MASI	Accuracy	Dice	MASI	Accuracy
Random	0.54	0.31	0.16	0.51	0.22	0.10
EIA	0.68	0.46	0.24	0.88	0.75	0.61
SVM	0.78	0.61	0.46	0.89	0.77	0.64

Table 4. Attribute prediction accuracy

Attribute	GRE3D3	GRE3D7
type	1.00	1.00
colour	0.78	0.99
size	0.90	0.74
location	0.98	0.98
relation	0.58	0.87
lm_type	1.00	1.00
lm_colour	0.70	0.87
lm_size	0.87	0.54
lm_location	0.90	0.98

We applied Wilcoxon's Signed-rank test over Dice scores, and the Chi-squared test over accuracy scores. Differences between the algorithms under evaluation are significant as summarized in Table 5.

Table 5. Significance test results

	GRE3D3				GRE3D7			
	Dice		Accuracy		Dice		Accuracy	
Algorithm Pairs	W	p	χ^2	p	W	p	χ^2	p
Random x EIA	828.0	< .0001	30.0	< .0001	68158.0	< .0001	12947.2	< .0001
EIA x SVM	33626.5	< .0001	167.10	< .0001	1319418.5	0.060	16.95	< .0001

4.3 Discussion

Both Incremental (EIA) and Classification-based (SVM) approaches outperformed the *Random* baseline algorithm in both corpora, according to both Dice and Accuracy scores. Moreover, SVM outperforms EIA in all situations except for the comparison based on Dice scores for the corpus GRE3D7, in which case there was no significant difference between EIA and SVM. This generally confirms our main research hypothesis.

In the case of individual attribute prediction, SVM results were generally similar to decision-tree induction in [17]. However, as the work in [17] does not fully address the REG task, a more direct comparison to our current work is not possible.

5   Final Remarks

This paper has sketched an algorithm for referring expression generation based on a series of SVM classifiers. Our approach extended the work in [17] by combining the classifiers output to assemble referring expressions.

Our approach has been tested against an extended version of the Incremental algorithm [1] on two publicly available REG corpora. In both cases, our preliminary results suggest that the classification-based approach is indeed superior to incremental generation.

As future work, we intend to train additional machine learning models and improve the classification-based approach by including speaker-dependent features to account for the human variation in the generation of referring expressions. In addition to that, we are currently collecting a large corpus of definite descriptions on a new, more realistic domain, to further asses the current proposal.

Acknowledgments. The authors acknowledge support by CAPES and FAPESP.

References

1. Dale, R., Reiter, E.: Computational interpretations of the Gricean maxims in the generation of referring expressions. Cognitive Science 19(2), 233–263 (1995)
2. Krahmer, E., van Deemter, K.: Computational generation of referring expressions: A survey. Computational Linguistics 38(1), 173–218 (2012)
3. Pereira, D.B., Paraboni, I.: A language modelling tool for statistical NLP. In: 5th Workshop on Information and Human Language Technology (TIL 2007), pp. 1679–1688. Anais do XXVII Congresso da SBC. Rio de Janeiro (2007)
4. Pereira, D.B., Paraboni, I.: Statistical surface realisation of portuguese referring expressions. In: Ranta, A., Nordström, B. (eds.) GoTAL 2008. LNCS (LNAI), vol. 5221, pp. 383–392. Springer, Heidelberg (2008)
5. de Novais, E.M., Paraboni, I.: Portuguese text generation using factored language models. Journal of the Brazilian Computer Society 19(2), 135–146 (2013)
6. Cuevas, R.R.M., Paraboni, I.: A machine learning approach to portuguese pronoun resolution. In: Geffner, H., Prada, R., Machado Alexandre, I., David, N. (eds.) IBERAMIA 2008. LNCS (LNAI), vol. 5290, pp. 262–271. Springer, Heidelberg (2008)
7. Belz, A., Gatt, A.: The attribute selection for GRE challenge: Overview and evaluation results. In: Proceedings of UCNLG+ MT: Language Generation and Machine Translation, Copenhagen, MT Summit XI, pp. 75–83 (2007)
8. Krahmer, E., van Erk, S., Verleg, A.: Graph-based generation of referring expressions. Computational Linguistics 29(1), 53–72 (2003)
9. de Lucena, D.J., Pereira, D.B., Paraboni, I.: From semantic properties to surface text: The generation of domain object descriptions. Inteligencia Artificial. Revista Iberoamericana de Inteligencia Artificial 14(45), 48–58 (2010)
10. van Deemter, K., van der Sluis, I., Gatt, A.: Building a semantically transparent corpus for the generation of referring expressions. In: Proceedings of the Fourth International Natural Language Generation Conference, INLG 2006, pp. 130–132. Association for Computational Linguistics, Stroudsburg (2006)
11. Viethen, J., Dale, R.: The use of spatial relations in referring expression generation. In: Proceedings of the Fifth International Natural Language Generation Conference, INLG 2008, pp. 59–67. Association for Computational Linguistics, Stroudsburg (2008)
12. Viethen, J., Dale, R.: GRE3D7: A corpus of distinguishing descriptions for objects in visual scenes. In: Proceedings of the UCNLG+Eval: Language Generation and Evaluation Workshop, pp. 12–22. Association for Computational Linguistics, Edinburgh (2011)

13. Gatt, A., Belz, A., Kow, E.: The TUNA challenge 2008: overview and evaluation results. In: Proceedings of the Fifth International Natural Language Generation Conference, INLG 2008, pp. 198–206. Association for Computational Linguistics, Stroudsburg (2008)
14. Gatt, A., Belz, A., Kow, E.: The TUNA-REG challenge 2009: Overview and evaluation results. In: Proceedings of the 12th European Workshop on Natural Language Generation, ENLG 2009, pp. 174–182. Association for Computational Linguistics, Stroudsburg (2009)
15. Jordan, P.W., Walker, M.A.: Learning content selection rules for generating object descriptions in dialogue. J. Artif. Int. Res. 24(1), 157–194 (2005)
16. Viethen, J., Dale, R.: Speaker-dependent variation in content selection for referring expression generation. In: Proceedings of the Australasian Language Technology Association Workshop 2010, Melbourne, Australia, pp. 81–89 (December 2010)
17. Viethen, J.: The Generation of Natural Descriptions: Corpus-Based Investigations of Referring Expressions in Visual Domains. PhD thesis, Macquarie University, Sydney, Australia (2011)
18. Grice, H.P.: Logic and conversation. In: Cole, P., Morgan, J.L. (eds.) Syntax and Semantics, vol. 3, Academic Press, New York (1975)
19. Dale, R., Haddock, N.: Generating referring expressions involving relations. In: Proceedings of the Fifth Conference on European Chapter of the Association for Computational Linguistics, EACL 1991, pp. 161–166. Association for Computational Linguistics, Stroudsburg (1991)
20. Paraboni, I.: An algorithm for generating document-deictic references. In: Procs. of Workshop Coherence in Generated Multimedia, Associated with First Int. Conf. on Natural Language Generation (INLG 2000), Mitzpe Ramon, pp. 27–31 (2000)
21. Krahmer, E., Theune, M.: Efficient context-sensitive generation of referring expressions. In: Information Sharing: Reference and Presupposition in Language Generation and Interpretation, vol. 143, pp. 223–263. CSLI Publications, California (2002)
22. Kelleher, J.D., Kruijff, G.J.M.: Incremental generation of spatial referring expressions in situated dialog. In: Proceedings of the 21st International Conference on Computational Linguistics and the 44th Annual Meeting of the Association for Computational Linguistics, ACL-44, pp. 1041–1048. Association for Computational Linguistics, Stroudsburg (2006)
23. Eugenio, B.D., Jordan, P.W., Thomason, R.H., Moore, J.D.: The agreement process: An empirical investigation of human–human computer-mediated collaborative dialogs. International Journal of Human-Computer Studies 53(6), 1017–1076 (2000)
24. Pedregosa, F., Varoquaux, G., Gramfort, A., Michel, V., Thirion, B., Grisel, O., Blondel, M., Prettenhofer, P., Weiss, R., Dubourg, V., Vanderplas, J., Passos, A., Cournapeau, D., Brucher, M., Perrot, M., Duchesnay, E.: Scikit-learn: Machine learning in Python. Journal of Machine Learning Research 12, 2825–2830 (2011)
25. Knerr, S., Personnaz, L., Dreyfus, G.: Single-layer learning revisited: a stepwise procedure for building and training a neural network. In: Soulié, F., Hérault, J. (eds.) Neurocomputing. NATO ASI Series, vol. 68, pp. 41–50. Springer, Heidelberg (1990)
26. Dice, L.R.: Measures of the amount of ecologic association between species. Ecology 26(3), 297–302 (1945)
27. Passonneau, R.: Measuring agreement on set-valued items (MASI) for semantic and pragmatic annotation. In: Proceedings of the Fifth International Conference on Language Resources and Evaluation (LREC), Valletta, Malta, pp. 831–836 (2006)

Generating Relational Descriptions Involving Mutual Disambiguation

Caio V.M. Teixeira, Ivandré Paraboni,
Adriano S.R. da Silva, and Alan K. Yamasaki

School of Arts, Sciences and Humanities, University of São Paulo (USP / EACH)
Av. Arlindo Bettio, 1000 - São Paulo, Brazil
ivandre@usp.br

Abstract. This paper discusses the generation of relational referring expressions in which target and landmark descriptions are allowed to help disambiguate each other. Using a corpus of referring expressions in a simple visual domain - in which these descriptions are likely to occur - we propose a classification approach to decide when to generate them. The classifier is then embedded in a REG algorithm whose results outperform a number of naive baseline systems, suggesting that mutual disambiguation is fairly common in language use, and that this may not be entirely accounted for by existing REG algorithms.

Keywords: Natural Language Generation, Relational Referring Expressions, Underspecification.

1 Introduction

In Natural Language Generation (NLG), the computational task of referring expression generation (REG) consists of producing a set of semantic properties to uniquely distinguish an intended referent from other objects in the same context. Consider for instance Figure 1, which illustrates a simple context set containing five objects: a sphere and two cones on the left side, and a cube and a second sphere on the right side.

Fig. 1. A visual domain conveying simple geometric objects

Consider the goal of referring to the target $r = Obj2$. This may be accomplished, for instance, by making use of atomic properties of r, as in (a-b) below.

A. Gelbukh (Ed.): CICLing 2014, Part I, LNCS 8403, pp. 492–502, 2014.
© Springer-Verlag Berlin Heidelberg 2014

(a)The green cone
(b)The small green cone

Alternatively, we may also make use of relational properties, i.e., we may refer to a second object - hereby called a *landmark* - as in (c-d) below, and possibly in combination with some atomic properties as well.

(c)The cone *next to the sphere*
(d)The small cone *on the right side of the large one*

The generation of relational descriptions as in (c-d) has been the focus of a number of studies in REG [1,2,3,4,5] and, of particular interest to our present discussion, there is the issue of how much information is desirable - or necessary - to convey in order to describe each individual objects (i.e., target and landmark).

Algorithms such as [1] implicitly assume that target and landmark descriptions are allowed to disambiguate each other, as in previous example (c), in which both 'cone' and 'sphere' would be ambiguous had we interpreted each description in isolation. This contrasts, for instances, studies such as [5], in which mutual disambiguation is shown to disrupt the search for the target in more complex situations of reference.

In this paper we ask when relational descriptions involving mutual disambiguation should be produced as opposed to more (e.g., fully) distinguishing alternatives. To this end, we will focus on the particular case of mutual disambiguation in which the landmark description is left underspecified. Using a corpus of referring expressions in which mutual disambiguation is ubiquitous, we propose a classifier approach to decide when to generate under or fully-specified landmark descriptions. The classifier is embedded in a REG algorithm and compared to a number of baseline systems.

The rest of this paper is structured as follows. Section 2 introduces the computational problem of REG and related work. Section 3.1 describes the training and test data and Section 3.2 describes the proposed classifier. Section 3.3 presents our REG approach and Section 4 discusses evaluation and its results. Finally, Section 5 draws a number of conclusions and suggests future work.

2 Basic Concepts

In this section we introduce the computational problem of referring expression generation, and related work on the generation of (possibly underspecified) relation descriptions.

2.1 Referring Expression Generation

The content selection for referring expressions generation (REG) has received a great deal of attention in NLG research [6]. As a prominent example of REG approach, in what follows we will discuss the Incremental algorithm in [7].

The Incremental algorithm takes as an input an intended referent (or target object) r to be described as a referring expression L, and a context C containing a set of distractor objects. Objects are represented as (*attribute-value*) pairs (or properties), as in (*size-large*). The goal of the algorithm is to produce description of the target r represented by a set L of semantic properties of r such that L distinguishes r from every distractor in C, and which may be subsequently realised as a surface string in a given target language [8,9,10].

Each attribute is considered for inclusion in L in turn according to a preference order P. An attribute is selected only if it helps disambiguate the target r (i.e., only if it helps ruling out distractors in C). The algorithm terminates when a uniquely distinguishing description L is obtained (i.e., when L denotes r and no other distractor) or when all possible attributes in P have been considered (in which case L remains ambiguous). In the first case, L may be realised as a definite description as in 'the green cube', and in the second case as an indefinite description as in 'a red sphere'.

For instance, the context in Figure 1 may be represented by a knowledge base as follows.

Obj1: (*type-sphere*), (*size-large*), (*colour-red*), (*position-left*).
Obj2: (*type-cone*), (*size-small*), (*colour-green*), (*position-left*).
Obj3: (*type-cone*), (*size-large*), (*colour-red*), (*position-left*).
Obj4: (*type-cube*), (*size-large*), (*colour-green*), (*position-right*).
Obj5: (*type-sphere*), (*size-large*), (*colour-red*), (*position-right*).

Let $r = Obj2$, $C = \{Obj1, Obj3, Obj4, Obj5\}$, and $P = \{type, position, colour, size\}$. The algorithm starts with an empty list L and iterates over the preference order P considering whether to include each attribute in turn.

The first attribute to be considered (according to P) is *type*. Since te corresponding property (*type-cone*) rules out *Obj1*, *Obj4* and *Obj5*, whose *type* is not *cone*, this property is added to L and these distractors are removed from C.

Next, the *position* attribute is considered. Since C now contains only *Obj3*, which shares the same (*position-left*) property with the target, this attribute does not rules out any distractor objects in C, and it is thus disregarded.

Finally, the *colour* attribute is considered. Since (*colour-green*) rules out *Obj3*, which is *red*, this property is also added to L, and *Obj3* is removed from C. This leaves the context C now empty, so the algorithm terminates and returns the expression $L = \{(type, cone), (colour, green)\}$, which could be realised as, e.g., 'the green cone'. Similarly, *Obj1* would have been described as 'the sphere on the left'; *Obj3* as 'the red cone', *Obj4* as 'the cube' and *Obj5* as 'the sphere on the right'.

The order P in which attributes are considered is assumed to be domain-dependent, and it may have a great impact on the outcome of the algorithm. For instance, had we considered $P = \{position, type, size, colour\}$ in the above example, *Obj2* would have been described as 'the small (object) on the left'[1]. Similarly, *Obj3* would have been described as 'the large cone on the left' and *Obj4* as 'the cube on the right'.

2.2 Relational Descriptions Involving Mutual Disambiguation

Besides having atomic properties (e.g., *type, colour* etc.), objects may also hold relations (e.g., *next to, behind* etc.) between themselves, as in 'the small cone next to the large one'. Not surprisingly, many well-known REG algorithms, including a number of extensions of the Incremental approach in [7], are capable of handling relations between target and landmark objects, e.g., [1,2,3].

Mutual disambiguation in REG is however an elusive research topic. Although some of the existing approaches to relational REG do allow mutual disambiguation between target and landmark descriptions, this seems to be largely a side-effect of the main reference strategy implemented by each system. This seems to be particularly the case when brevity or minimality are among the goals under consideration, e.g., [11,12].

The few existing accounts of mutual disambiguation in relational REG are actually made in the context of the complimentary discussion on reference over-specification [13,14,5]. In some cases, landmark underspecification is explicitly described as a situation to be avoided in favour of fully or overspecified descriptions. In [5], for instance, landmark underspecification is shown to make search for domain objects more difficult in certain spatial domains, as in 'the man behind the door' in a context with several identical doors to be inspected by the hearer in order to identify its target.

Situations in which mutual disambiguation is best avoided contrast simpler situations of reference as in [1] and others. Algorithms such as [1] allow mutual disambiguation between target and landmark descriptions, as in 'the bowl on the table' in a context with several bowls and tables, but only one bowl supported by a table. Descriptions of this kind are assumed to be felicitous, at least in simple visual domains as in this example, but no further discussion is presented. Similarly, the design of algorithms such as [3] also suggests the ability to generate underspecified landmark descriptions, although the issue is not explicitly discussed.

3 Current Work

Our work consists of a REG algorithm that makes use of a classifier to decide when to generate a relational description involving mutual disambiguation. In this section we describe the training and test data under consideration (Section 3.1), the computational model of mutual disambiguation (Section 3.2) and the REG algorithm proper (Section 3.3).

---

[1] The algorithm in [7] forces the inclusion of the attribute *type* in L regardless of its discriminatory power.

3.1   Training and Test Data

Standard text corpora for NLP [15,16] are ubiquitously available but, as pointed out in [17], they lack the necessary semantic 'transparency'. In what follows we investigate the generation of underspecified descriptions as seen in the Stars corpus of referring expressions[2]. The corpus contains 803 descriptions and their corresponding contexts. Descriptions were produced by 64 participants in a controlled experiment focused on reference involving relations, who were requested to identify targets in 11 images containing simple geometric objects. An example of stimulus used in the experiment is illustrated in Figure 2.

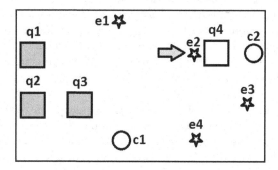

Fig. 2. A situation of reference to a target (e2) via a landmark object (q4), adapted from the stimuli used in the corpus

In this domain, referring to a target object as $e1$ in Figure 2 will usually involve referring to the nearest landmark object ($q4$). In what follows we will focus on a subset of 6 situations of reference (making 384 descriptions in total) in which mutual disambiguation was always possible, but with varying degrees of difficulty.

Unambiguous reference may be successfully accomplished either by making use of a underspecified landmark descriptions as in (a) below, or by making use of a non-underspecified landmark description as in (b). Depending on the situation (i.e., depending on whether the landmark colour helps disambiguation or not), landmark descriptions as in (c) may be underspecified or not.

   (a) the star next to the box
   (b) the star next to the box, on the right
   (c) the star next to the white box

In the selected data, 43% of descriptions show mutual disambiguation, and 53% do not. Further 14% were either ambiguous or non-relational descriptions. By comparison, about 15.5% of descriptions in the GRE3D corpus [18] involve

---

[2] The Stars corpus is freely available for download for research purposes from http://each.uspnet.usp.br/ivandre/Stars-corpus/

mutual disambiguation and, in the more complex situations of reference found in the GRE3D7 corpus [4], mutual disambiguation occurs in 9% of all descriptions.

The Stars data was randomly split into training (300 descriptions) and test (84 descriptions) sets. Using the training data we built a model of reference underspecification described in the next section. Attribute frequencies as required by the algorithm in Section 3.3 were also extracted from the training data, as in [19]. The test data will be applied in the evaluation work in Section 4.

3.2   A Computational Model of Mutual Disambiguation

Using the training data set described in the previous section, we built a Naive-Bayes classifier to determine when a reference to a target r via a landmark object o in a given context C should allow mutual disambiguation between target and landmark descriptions.

The goal of the classifier is to learn a binary class *underspecify* that represents, for a particular situation of reference, whether an underspecified landmark reference strategy is applicable. The learning features under consideration are summarized in Table 1.

Table 1. Learning features for the binary class *underspecify* given target r and landmark o in a context C

Feature	Description
context-size	The number of objects in C.
same-landmark-type	The number of objects of the same type as o.
hpos	The target position on screen (left/right).
same-hpos	The number of objects horizontally aligned with r.
same-vpos	The number of objects vertically aligned with r.
absolute-attrib	True if o has a uniquely distinguishing colour.

Decisions made by the classifier will be taken into account by the REG algorithm discussed in the next section.

3.3   Generating Relational Descriptions Involving Mutual Disambiguation

The proposed algorithm makes use of the *underspecify* classifier and a straightforward implementation of the basic Incremental approach [7] modified so as to handle both atomic and relational properties.

The algorithm is divided into two main blocs: the *Describe* procedure that invokes the classifier to decide whether to use mutual disambiguation or not, and the *Full-specification* procedure that generates a standard description (i.e., not involving mutual disambiguation) if necessary. These procedures are summarised as follows.

Describe(C, r):
1 $lm = \text{nearestLandmark}(C, r)$
2 IF underspecify(C)
    3 $L = type(r) + relation(r, lm) + type(lm)$
    4 RETURN L
5 ELSE
      6 $L = \text{Full-specification}(C, r)$
      7 RETURN L

Full-specification(C, t):
8 $Lt = \text{MakeReferringExpression}(C, t)$
9 $L = L + Lt$
10 $lm = \text{nearestLandmark}(C, t)$
11 IF $lm \in Lt$ AND $P \neq \phi$
    12 $t = lm$
    13 $L = L + \text{Full-specification}(C, t)$
14 RETURN L

Given a context C and a target r, the algorithm starts (step 1) by generating a test instance of the class *underspecify* and by invoking the underspecify classifier (step 2) described in the previous section. If landmark underspecification is recommended, a mutually disambiguating description conveying the target type, the relation to the landmark object and the landmark type is returned (steps 3-4). In the Stars domain this may produce, for instance, descriptions as in 'the star next to the box'.

If landmark underspecification is not recommended, the algorithm will make a series of fully-specified descriptions of r (step 8) and related landmarks (step 12), if necessary. This procedure is similar to [20,21] and others.

The core of the algorithm is the *MakeReferringExpression* function (step 8), which produces a set Lt of minimally distinguishing properties of the current target (which may be r itself, or a related landmark lm). *MakeReferringExpression* implements a standard REG algorithm as [7] which selects only discriminatory attributes. In any individual description, attributes are considered in order of preference as seen in the training data, starting with the most frequent attribute. From the training portion of the Stars corpus, two distinct preference lists P_{target} and $P_{landmark}$ for each object type were computed:

$$P_{target} = \{next, left, right, others\}$$
$$P_{landmark} = \{others, next, left, right\}$$

If an individual description contains a relational attribute (step 11), a subsequent call to the algorithm (step 13) will produce a separate list of attributes to describe the landmark object, and so forth. The algorithm terminates when there is no further landmark objects to be described, or when the list of attributes P is empty (step 14).

4   Evaluation

4.1   Procedure

The algorithm proposed in Section 3.3 - hereby called *proposal* - was evaluated intrinsically by comparing its results to those produced by three baseline systems: a *random* strategy that sorts the list P in random order; a *fully-specifying* algorithm that always produces a fully-specified description of each individual object (i.e., target and landmarks alike), and a *minimally distinguishing* algorithm that always generates minimal relational descriptions in the form 'the star next to the box'.

Each of the four systems was individually applied to the generation of the descriptions found in the test data - hereby called *Reference* set) - as described in Section 3.1. As a result, four sets of *System* descriptions were obtained. Following [22] and others, evaluation was carried out by comparing each *System* set to the *Reference* set, and by measuring overall accuracy (Acc), Dice [23] and MASI [24] scores.

Accuracy (*Acc*) values represent the proportion of situations in which the *System* description is identical to the *Reference* counterpart. Accuracy values range from 0 to 1, in which 1 indicates total coincidence between the two sets. However, given that descriptions in the Stars corpus are not fully annotated (e.g., low-frequency attribute values are simply annotated as 'others'), total coincidence with the *Reference* description is expected to be rare for all systems under evaluation.

Dice scores measure the degree of overlap between two sets. *Dice* scores range from 0 (total dissimilarity) to 1 (total similarity). *MASI* tends to co-relate with *Dice*, but it gives more weight to situations in which one set is a subset of the other [24]. In the analysis to follow we will mainly focus on *Dice* scores, which are accompanied by *MASI* for illustration purposes only.

4.2   Results

Before discussing the evaluation of the REG algorithm proper, Table 2 shows the results of the Naive-Bayes classifier over the (84-instances) test data set[3].

Table 2. Reference strategy classification

Underspecify	Precision	Recall	F-measure
no	0.91	0.75	0.82
yes	0.69	0.91	0.78

---

[3] The software used to obtain the results is available in
http://www.CICLing.org/2014/data/14

The corresponding confusion matrix is illustrated in Table 3.

Table 3. Confusion matrix for the class *underspecify*

classified as ->	no	yes
no	39	13
yes	3	29

We applied the four systems to the generation of the 84 test descriptions taken from the corpus. Table 4 summarizes our findings.

Table 4. Results over test data

Algorithm	Acc.	Dice	MASI
Random	0.00	0.42	0.12
Fully-specifying	0.06	0.54	0.21
Minimally-distinguishing	0.05	0.55	0.21
Proposal	0.11	0.61	0.28

The *proposal* algorithm outperforms the baseline alternatives according to all criteria. The difference in *Dice* scores is highly significant according to Wilcoxon's signed rank test (W=-793, Z=-4.95, p<0.0001). The choice made by the proposed algorithm between under and full specification was correct in 67 cases (79.8%).

5   Conclusions

This paper discussed the computational generation of relational descriptions involving mutual disambiguation. We collected a corpus of referring expressions in a simple visual domain in which these descriptions are likely to occur, and produced a classifier to decide when to generate under or fully-specified landmark descriptions. The classifier was embedded in a REG algorithm whose results outperform a number of naive baseline systems.

Despite the simplicity of the data set and the proposed computational model, our experiments suggest that mutual disambiguation between target and landmark descriptions is fairly common in language use, and that it may not be entirely accounted for by existing REG algorithms. Thus, as future work we will attempt to generalise the present findings by making use or larger, more complex corpora. Moreover, we intend to take into account the issue of variation between speakers, as individuals may follow distinct (e.g., under or fully-specified) reference strategies.

Acknowledgments. The authors acknowledge support by FAPESP and the University of São Paulo.

References

1. Dale, R., Haddock, N.J.: Content determination in the generation of referring expressions. Computational Intelligence 7, 252–265 (1991)
2. Krahmer, E., Theune, M.: Efficient context-sensitive generation of referring expressions. In: van Deemter, K., Kibble, R. (eds.) Information Sharing: Reference and Presupposition in Language Generation and Interpretation, pp. 223–264. CSLI Publications, Stanford (2002)
3. Krahmer, E., van Erk, S., Verleg, A.: Graph-based generation of referring expressions. Computational Linguistics 29(1), 53–72 (2003)
4. Viethen, J., Dale, R.: GRE3D7: A corpus of distinguishing descriptions for objects in visual scenes. In: Proceedings of the UCNLG+Eval: Language Generation and Evaluation Workshop, Edinburgh, Scotland, pp. 12–22 (July 2011)
5. Paraboni, I., van Deemter, K.: Reference and the facilitation of search in spatial domains. Language and Cognitive Processes online (2013)
6. Krahmer, E., van Deemter, K.: Computational generation of referring expressions: A survey. Computational Linguistics 38(1), 173–218 (2012)
7. Dale, R., Reiter, E.: Computational interpretations of the Gricean maxims in the generation of referring expressions. Cognitive Science 19 (1995)
8. Pereira, D.B., Paraboni, I.: A language modelling tool for statistical NLP. In: 5th Workshop on Information and Human Language Technology (TIL 2007). Anais do XXVII Congresso da SBC. Rio de Janeiro, pp. 1679–1688 (2007)
9. Pereira, D.B., Paraboni, I.: Statistical surface realisation of portuguese referring expressions. In: Nordström, B., Ranta, A. (eds.) GoTAL 2008. LNCS (LNAI), vol. 5221, pp. 383–392. Springer, Heidelberg (2008)
10. de Novais, E.M., Paraboni, I.: Portuguese text generation using factored language models. Journal of the Brazilian Computer Society, 1–12 (2012)
11. Dale, R.: Cooking up referring expressions. In: Proceedings of the 27th Annual Meeting of the Association for Computational Linguistics, pp. 68–75 (2002)
12. Gardent, C.: Generating minimal definite descriptions. In: Proceedings of the 40th Annual Meeting of the Association for Computational Linguistics, pp. 96–103 (2002)
13. Arts, A., Maes, A., Noordman, L.G.M., Jansen, C.: Overspecification facilitates object identification. Journal of Pragmatics 43, 361–374 (2011)
14. Koolen, R., Gatt, A., Goudbeek, M., Krahmer, E.: Factors causing overspecification in definite descriptions. Journal of Pragmatics 43, 3231–3250 (2011)
15. Aziz, W.F., Pardo, T.A.S., Paraboni, I.: An experiment in spanish-portuguese statistical machine translation. In: Zaverucha, G., da Costa, A.L. (eds.) SBIA 2008. LNCS (LNAI), vol. 5249, pp. 248–257. Springer, Heidelberg (2008)
16. Cuevas, R.R.M., Paraboni, I.: A machine learning approach to portuguese pronoun resolution. In: Geffner, H., Prada, R., Machado Alexandre, I., David, N. (eds.) IBERAMIA 2008. LNCS (LNAI), vol. 5290, pp. 262–271. Springer, Heidelberg (2008)
17. Gatt, A., van der Sluis, I., van Deemter, K.: Evaluating algorithms for the generation of referring expressions using a balanced corpus. In: 11th European Workshop on Natural Language Generation, ENLG 2007 (2007)
18. Dale, R., Viethen, J.: Referring expression generation through attribute-based heuristics. In: Proceedings of the 12th European Workshop on Natural Language Generation, ENLG 2009, pp. 58–65. Association for Computational Linguistics, Stroudsburg (2009)

19. de Lucena, D.J., Pereira, D.B., Paraboni, I.: From semantic properties to surface text: The generation of domain object descriptions. Inteligencia Artificial. Revista Iberoamericana de Inteligencia Artificial 14(45), 48–58 (2010)

20. Paraboni, I., van Deemter, K.: Issues for the generation of document deixis. In: Procs. of Workshop on Deixis, Demonstration and Deictic Belief in Multimedia Contexts, in Association with the 11th European Summers School in Logic, Language and Information (ESSLLI 1999), pp. 44–48 (1999)

21. Paraboni, I.: Generating references in hierarchical domains: The case of Document Deixis. PhD thesis, University of Brighton (2003)

22. Gatt, A., Belz, A.: The attribute selection for GRE challenge: Overview and evaluation results. In: Proceedings of UCNLG+MT: Language Generation and Machine Translation (2007)

23. Dice, L.R.: Measures of the amount of ecologic association between species. Ecology 26(3), 297–302 (1945)

24. Passonneau, R.: Measuring agreement on set-valued items (MASI) for semantic and pragmatic annotation. In: Proceedings of the International Conference on Language Resources and Evaluation, LREC (2006)

Bayesian Inverse Reinforcement Learning for Modeling Conversational Agents in a Virtual Environment

Lina M. Rojas-Barahona[1] and Christophe Cerisara[2]

[1] Université de Lorraine/LORIA, Nancy
[2] CNRS/LORIA, Nancy
{lina.rojas,christophe.cerisara}@loria.fr

Abstract. This work proposes a Bayesian approach to learn the behavior of human characters that give advice and help users to complete tasks in a situated environment. We apply Bayesian Inverse Reinforcement Learning (BIRL) to infer this behavior in the context of a serious game, given evidence in the form of stored dialogues provided by experts who play the role of several conversational agents in the game. We show that the proposed approach converges relatively quickly and that it outperforms two baseline systems, including a dialogue manager trained to provide "locally" optimal decisions.

1 Introduction

Reinforcement Learning (RL) has been widely used for learning dialogue strategies [1–5]. Dialogues are modeled as an optimization problem, simulating the inherent dynamic behavior of conversations in order to find the globally optimal policy. However, the RL problem assumes the reward function is known. Indeed the reward function is usually handcrafted, as pointed out in [6], "the reward function is almost always set by intuition, not data". Inverse reinforcement learning (IRL) has been defined in [7] as the problem of recovering the reward function from experts' demonstrations. It tries to find an optimal reward, which leads to a decision policy that follows as closely as possible the examples provided by experts maximizing the expected cumulated reward in the long-run.

In this work we explore Bayesian Inverse Reinforcement Learning (BIRL) [8] to infer the reward function from humans who perform the task of instructing players in a serious game. We also apply the improvements to BIRL proposed in [9], namely the Modified BIRL (MBIRL), in order to reduce the computational complexity in large state spaces. This work covers a first step towards dialogue optimization with user simulation. Therefore, instead of designing in advance the reward function to "properly instruct players", which is a difficult and subjective task, we rather propose to learn it from humans. Once we have found the reward function we can apply classical reinforcement learning with user simulation for building a dialogue system and afterwards testing it with real users.

The adapted Bayesian approach is evaluated in terms of policy loss [9] and is compared against two baselines. The first one uses random rewards, while the second one exploits corpus-estimated locally-optimal rewards (i.e., supervised learning). The results show that the proposed approach converges relatively quickly and consistently

A. Gelbukh (Ed.): CICLing 2014, Part I, LNCS 8403, pp. 503–514, 2014.

outperforms both baselines, which confirms that taking into account the dynamic properties of the environment leads to virtual characters that better reproduce the behavior of experts. Qualitatively, our models have thus learned to adequately inform users and provide help when needed.

2   Reinforcement Learning for Dialogue Management

We focus on Markov decision processes for modeling dialogues because we aim to model unimodal conversations (i.e., a chatbot), thus we do not tackle speech recognition uncertainty. We first introduce Markov decision processes, then we present some reward functions commonly used in dialogue systems.

2.1   Markov Decision Processes

A finite Markov decision process (MDP) is a tuple $M = (S, A, T, \gamma, R)$ where:

- S: A set of possible states that represent the dynamic environment.
- A: A set of possible actions.
- $T : S \times A \times S \rightarrow [0, 1]$ is a transition probability function. For any action $a \in A(s)$ taken in a state $s \in S$, the probability of transiting to the next state s' is given by $T(s, s')$.
- γ: A discounting factor in the range of $[0, 1)$, which controls the prediction horizon of the algorithm.
- R: The reward function that specifies the reward gained at every state. It contains the information that guides the agent towards the goal. R is a function of the state that is bounded in absolute value by R_{max}.

A stationary *policy* is a map $\pi : S \rightarrow A$ and the discounted infinite-horizon expected reward for starting in state s and following policy π thereafter is given by the value function $V^\pi(s)$ that satisfies the following *Bellman Equation*:

$$V^\pi(s) = R(s) + \gamma \sum_{s'} T(s, \pi(s), s')V^\pi(s') \tag{1}$$

The discounted infinite-horizon expected reward for starting in state s, taking action a and following policy π thereafter is given by the *Q-function* $Q^\pi(s, a)$ that satisfies the following equation:

$$Q^\pi(s, a) = R(s) + \gamma \sum_{s'} T(s, a, s')V^\pi(s') \tag{2}$$

A policy π is optimal in M iff, for all $s \in S$:

$$\pi(s) = \arg\max_{a \in A} Q^\pi(s, a) \tag{3}$$

$Q^*(s, a, \mathbf{R})$ is the optimal Q-function of the optimal policy π^* for a known reward function $\mathbf{R}$.

2.2 Reward Functions for Dialogue Systems

Previous work on RL for learning dialogue strategies typically use reward functions that penalize long dialogues, returning a final positive reward for task completion or user satisfaction [1, 2, 10, 11]. This might be an intuitive reward function for slot-filling applications, such as train ticket or restaurant reservation, in which usually customers know exactly what they want and they expect to be accurately informed by the system as fast as possible. However, this reward function might be inappropriate in other domains or even for some other user profiles in the same domain. This is especially true in tutorial dialogues, where learners usually have to complete a task and may not know exactly how to do it. In such tutoring situations, the reward function might be designed according to the student-learning gains [12]. However, it is usually difficult even for tutors to write down the correct formula for being a good tutor. In our case, we are interested in building conversational virtual humans for a serious game. Although virtual characters can been seen as tutors because they provide information and help players to successfully complete different tasks, the learning-gain is relaxed since not only some conversations are optional, but also besides asking for help, players may also talk with virtual characters just for fun.

3 Related Work

Two dialogue systems were built in [13] for the same game scenario presented in this paper: (i) combining an information-state dialogue manager and a supervised model for interpretation; and (ii) using a supervised model for dialogue management. Both systems were evaluated with real users reaching a relatively low user satisfaction. These results motivated our interest to explore IRL because it leverages the local learning of supervised models in the context of MDPs, optimizing the cumulated reward at long term. IRL has been first introduced by [7], then it has been applied in car driving simulation [14] and autonomous helicopter aerobatics [15]. In dialogue systems, IRL has been first proposed as one strand of dialog research by [6]. It has been applied to user simulation in [16], learning the behaviour of users for simulating iterations in a RL dialogue manager with a known reward. Instead, we are applying IRL for dialogue management, learning the tutor (i.e. system) reward function from experts. In addition, their user simulator learns from human-computer data, which contains iterative turns between humans and a rule-based dialogue manager, while we are using human-human data, thus avoiding possible incoherent or unusual turns due to system errors. Unlike previous work [17, 16], we are using a Bayesian refined IRL algorithm instead of the original IRL algorithm proposed in [14] (the reader is referred to [18] for a review of IRL algoritms). Finally, [17, 16] applied IRL for slot-filling dialogue systems while we are applying it for building twelve distinct conversational agents in a serious game.

4 Conversational Agents in a Serious Game

In this section, we introduce the serious game and the dialog scenario. We then describe the dialogue states, the actions as well as the transition probability function.

Table 1. Description of the 12 dialogs in the game

Dialog Id	VC	Player	Goals	Location
1	Lucas	Ben	Find the address of the enterprise.	Uncle's place.
2	M.Jasper	Lucas	The manufacturing first step	Enterprise reception
3	Samir	Julie	Find the plans of the joystick *Optional: job, staff, studies, security policies*	Designing Office
4	Samir	Julie	Find out what to do next *Optional: jobs in the enterprise, staff in the enterprise*	Designing Office
5	Melissa	Lucas	Find the mould *Optional: where are the moulds*	Plant
6	Melissa	Lucas	Find the right machine	Plant
7	Melissa	Lucas	Confirm you have found the right mould and machine and find out what to do next	Plant
8	Operator	Julie	Knowing about the material space and about the job *Optional: find out what to do in the case of failure* *helping to feed a machine with the right material*	Material Space
9	Serge	Ben	Perform quality tests. *Optional: VC's job*	Laboratory Tests
10	Serge	Ben	Find out what to do next. *Optional: know what happens with broken items*	Laboratory Tests
11	Sophia	Julie	Find the electronic components, knowing about VC's job	Finishing
12	Sophia	Lucas	Finishing process *Optional: know about conditioning the product*	Finishing

4.1 Scenario and Demonstrations from Experts

The objective of the virtual agents is to engage the player in a conversation in the context of a serious [1] game called *Mission Plastechnologie*[2] (MP). In this game, the player seeks to build a joystick in order to free their uncle trapped in a video game. To build this joystick, the player must explore a factory and interact with different virtual humans through twelve distinct dialogs (i.e., chatbots), each of them occurring in a specific place of the virtual world with various mandatory goals to be achieved and optional goals to be discussed (See Table 1). Note that defining the reward for each of these dialogues is not as simple as giving a positive reward when the joystick is built by the player. Instead, virtual characters (i.e., tutors) have to instruct players, providing valuable information and supporting spontaneous conversations through a sort of fun relaxed tutoring (as mentioned in Section 2.2).

To learn human behavior, we are taking the experts' (i.e., seven subjects performing the Wizard of Oz) demonstrations from the corpus of the MP dialog scenario [19]. It contains 1250 Human-Human dialogues involving 6845 Wizard of Oz turns and 3610 player turns.

4.2 States, Actions and Transitions

As shown in Table 1, there are 12 distinct conversations in the game between 7 virtual characters (VC) and 3 player characters. Each of these dialogues talks about mandatory

---

[1] A serious game is a game designed for a primary purpose other than pure entertainment.

[2] The game is designed to promote careers in the plastic industry, is French speaking and was created by Artefacto, http://www.mission-plastechnologie.com/

and optional goals. The player either asks for information about these goals or asks for help. Accordingly, the virtual human either informs about the goals or provides help. It can also handle out of domain topics, misunderstandings or request information. The following example shows an excerpt of the third dialogue in the MP Game between Samir (the system) and Julie (the player) annotated with dialogues acts as in the corpus. The goal to be achieved by the player is to find the plans of the joystick, the goals to be discussed (which are optional) are: learn about the virtual character's job, his studies, his colleagues as well as the security policies of the enterprise.

Samir: Hello my name is Samir, the product designer {greet}
Samir: What are you doing here young people? {ask(task(X))}
Julie: We come to build the joystick of Professor Geekman {find_plans}
Samir: You are in the right place, the plans are in the closet ... {inform(do(find_plans))}
Samir: Before leaving would you like to hear about my job, the studies I did or my colleagues? {ask(domore(X))}
Julie: Ok, tell me about your job. {inform_job}

We use MDPs to model virtual humans in the game. We designed coarse-grained states containing user and system contributions to the dialogue; either by explicitly asking about the domain specific tasks (i.e. the dialogue goals) or by producing general dialogue acts (e.g., greeting, asking for help, acknowledgments, etc). A binary variable that indicates whether or not the dialogue has finished is also included. With this state representation we have 32 states for the shortest dialogue (the first dialogue in Table 1), and 432 states for the longest dialogue (i.e., the third dialogue in Table 1 with 5 goals).

State variables

1. Has any of the characters ended the dialogue with a farewell action ? : 1 for setting a terminal state, 0 otherwise.
2. The last goal either informed or requested by the system: 0 when the system has not informed/requested about any goal, otherwise the goal id (e.g., from 1 to up to 5 for the longest dialogue).
3. The last goal either asked or confirmed by the player: 0 when the user has not yet asked/confirmed about any goal, otherwise the id of the goal (e.g., from 1 to up to 5 for the longest dialogue).
4. The last general dialogue act produced by the system: 0 for absence of general dialog act, 1 when providing help, and 2 when asking the player about the task to be solved (e.g., "How may I help you").
5. The user has asked for help: 0 if the user has not asked for help, 1 otherwise.

Actions. We are considering only the following actions in our experiments.

- *quit*: farewell greeting.
- *inform(do(g_i))*: informing about how to achieve goal g_i.
- *inform(help)*: providing help
- *ask(task(X))*: Asking the player about the task, it corresponds to a general welcome sentence (e.g., "How may I help you"). Note that this action neither occurs in dialogue 1 nor in dialogue 7.

- *WAIT*: the system gives the turn back to the user.
- *ack*: the system acknowledges understanding.
- *other*: the system answers to out of context turns.

Virtual characters always greet the player at the beginning, thus we do not need to learn this behaviour.

Transition Function. The transition function is not deterministic when the next state reflects an (unpredictable) user action. This is typically the case after the WAIT system action. However, BIRL requires this transition function to be given, and we have thus estimated such non-deterministic transition probabilities using smoothed counts from the observed corpus as follows:

$$P(s'|s, a) = \frac{N(s,a,s')+\alpha}{N(s,a)+N_\chi\alpha}$$

Where $N(s, a, s')$ and $N(s, a)$ are respectively the number of times the transition (s, a, s') and the state-action pair (s, a) have been seen in the corpus, and N_χ is the number of observed state-action pairs. α is a smoothing constant arbitrarily set to 0.1.

The other transitions that reflect a system action are deterministic and have been defined as:

$$P(s'|s, a) = \begin{cases} 1, & \text{if } s' = next\_s(s, a) \\ 0, & \text{otherwise} \end{cases}$$

Where $next\_s(s, a)$ is a function that computes the next state given a system action a. For instance, when the system informs about the first goal, g_1, the action $a_t = inform(do(g_1))$ yields the next state s' to have the state variable 2 set to 1.

5   Bayesian Inverse Reinforcement Learning

The IRL problem as defined in [7] is described as follows: given a finite state space S, a set of actions $A = \{a_1, a_2, ...a_k\}$, a transition probability $P_{ss'}^a$, a discount factor γ, and a policy π, determine a set of possible reward functions R such that π is the optimal policy for the given MDP. The IRL problem is an ill-posed problem [14], because potentially an infinite number of rewards may be optimal. Bayesian IRL approaches model this uncertainty by inferring the posterior distribution of the reward vector $\mathbf{R}$, treating the demonstration sequences as the evidence and relying on a prior on the reward function [8].

The IRL agent receives a sequence of observations of the expert's behaviour $O_\chi = \{(s_1, a_1), (s_2, a_2), ..., (s_k, a_k)\}$, which means that at time step i, the virtual character χ that mimics the expert is in state s_i and takes the action a_i. After applying Bayes Theorem, the posterior can be written as:

$$Pr(\mathbf{R}|O_\chi) = \frac{Pr(O_\chi|\mathbf{R})Pr(\mathbf{R})}{Pr(O_\chi)} \qquad (4)$$

We model next the reward function by a simple n-dimensional real vector, where n is the number of different states. Then, $Pr(\mathbf{R}|O_\chi)$ is the posterior distribution of the reward vector given the observed state-action pairs of the expert. $Pr(O_\chi|\mathbf{R})$ is the likelihood of the observed expert state-action pairs given the reward vector $\mathbf{R}$. This likelihood is modeled in [8] with a parameter α representing the degree of confidence we have in the expert's ability to choose a good action as follows:

$$Pr(O_\chi|\mathbf{R}) = \frac{1}{Z}e^{\alpha \sum_i Q^*(s_i,a_i,\mathbf{R})} \tag{5}$$

$Pr(\mathbf{R})$ is the prior distribution and $Pr(O_\chi)$ is the probability of the evidence over the entire space of reward vectors $\mathbf{R}$, which is not needed in the BIRL algorithm. The original BIRL algorithm, namely PolicyWalk, follows a Markov Chain Monte Carlo (MCMC) technique iterating as follows: Given a reward vector $\mathbf{R}$, it performs random walks over the neighbors of $\mathbf{R}$ on a grid of length δ, finding a new proposal $\bar{\mathbf{R}}$, such that: $\bar{\mathbf{R}}(s) = \mathbf{R}(s) \pm \delta$. The proposal is accepted with probability $\min\{1, \frac{Pr(\bar{\mathbf{R}}|O)}{Pr(\mathbf{R}|O)}\}$, where the posterior is given by Eq (4).

The expected value of the reward given this posterior is then computed over all these samples. Note that the normalizing constants cancel out in the ratio used to accept the proposed $\bar{\mathbf{R}}(s)$ and that finding Q^* in Eq (5) requires to solve the MDP at every MCMC iteration. This can be done for example with the policy iteration (PI) algorithm [20].

BIRL converges slowly when applied to large state spaces. One reason for this is that it infers the reward of every state, although many states have little expert evidence. Second, searching over a reward function space easily increases the number of MCMC iterations needed to approximate the mean of the posterior. To solve these limitations, [9] proposed a modified BIRL (MBIRL) that:

- infers only those states that are similar to the observed ones according to a *kernel-based relevance function*.
- uses simulated annealing to focus the sampled distribution around its maximum, hence reducing the number of samples needed to converge. Therefore, they use a modified acceptance probability of $\left(\frac{Pr(\bar{\mathbf{R}}|O)}{Pr(\mathbf{R}|O)}\right)^{\frac{1}{T_i}}$ where T_i is a decreasing *cooling schedule*.

6 Experiments

In this section we introduce the baselines, the evaluation metrics and the experiment setup for 12 dialogues in the game. We have defined for each dialogue the state and action space as explained in Section 4.

6.1 Baselines

We evaluate the performances of the proposed system by comparing it with two baselines:

- Using random rewards (RR);

- Exploiting "locally-estimated" rewards (LR), i.e., rewards that are trained on the corpus with the additional assumptions that the reward prior $Pr(\mathbf{R})$ is uniform, that the states are conditionally independent given the reward $P(O_\chi|\mathbf{R}) = \prod_i P(s_i|\mathbf{R})$ and that the state likelihood is multinomial with parameters representing the reward $P(s = k|\mathbf{R}) = R_k$, so that the path that maximizes the cumulated reward also maximizes the likelihood. Then:

$$\arg\max_R Pr(\mathbf{R}|O_\chi) = \arg\max_R Pr(O_\chi|\mathbf{R})$$

$$= \arg\max_R \prod_i P(s_i|\mathbf{R}) \tag{6}$$

Let n_k be the number of times the kth state of the states space occurs in the expert observations: $n_k = |\{(s_i = k, a_i)\}_{i \in O_\chi}|$
Then we want to maximize the likelihood $\prod_k P(s = k|\mathbf{R})^{n_k}$ under the constraint $\sum_k R_k = 1$, which gives the locally optimum reward:

$$\hat{R}_k = \frac{n_k}{N_\chi}$$

with $N_\chi = |\{(s_i, a_i)\}_{i \in O_\chi}|$ the number of observed state-action pairs.

6.2 Evaluation Metrics

We consider two evaluation metrics: the policy loss [9] and the system training time.

- *Policy loss*: The policy loss is the ratio $\frac{n_{\neq}}{N_\chi}$, where $n_{\neq} = |\{(s_i, a_i \neq \pi(s_i))\}_{i \in O_\chi}|$ is the number of expert state-action pairs that disagree with the learned policy π and $N_\chi = |\{(s_i, a_i)\}_{i \in O_\chi}|$ is the number of observed state-action pairs.
- *Elapsed time*: The time in milliseconds it takes to MBIRL and to the policy iteration algorithms to finish.

6.3 Experimental Setup

We used 2000 MCMC iterations and 20 policy iterations (PI) with a discount factor $\gamma = 0.9$. The experiments were run 5 times (except for dialogues 3 and 8 that were run only once) and the averaged measures are reported in Table 2. Dialogue 3, which has the largest state and action spaces ($|S| = 432, |A| = 11$), was run with 1000 MCMC iterations and 10 PI iterations.

Parameters for BIRL. For solving the BIRL posterior defined in Eq (4), we set the parameter α of Eq(5), representing the degree of confidence we have in the expert , to $\alpha = 0.85$, based on the Inter-Annotator Agreement between experts. We use the *Beta* distribution as prior, where $R_{max} = 6$ and $R_{min} = -6$, R_{max} was set taking into account the maximum number of goals that can be discussed in a dialogue (i.e., 5) plus the action of providing help when requested.

$$P_{Beta}(R(s) = r) = \frac{1}{(\frac{r - R_{min}}{R_{max} - R_{min}})^{\frac{1}{2}}(\frac{R_{max} - r}{R_{max} - R_{min}})^{\frac{1}{2}}}$$

Table 2. Results of MBIRL. The dialogues marked with (*) are optional. The first four columns stand for dialogue id (Id), number of dialogues ($ND.$), number of states $|S|$ and number of actions $|A|$. The next three columns represent the policy loss of the baselines and MBIRL. The last two columns shown the policy iteration (PI) and MCMC elapsed time.

Id	$ND.$	$|S|$	$|A|$	LR (p-loss)	RR (p-loss)	MBIRL (p-loss)	PI (ms)	MCMC (ms)
1	105	32	6	0.79	0.73	**0.66**	559	177062
2	112	48	7	0.82	0.82	**0.66**	852	415988
3	113	432	11	0.80	0.95	0.85	112231	590935186
4	106	192	9	0.74	0.85	0.79	15693	50988645
5	107	108	8	0.81	0.86	**0.72**	7661	5129894
6*	102	48	7	0.94	0.75	**0.69**	876	397773
7	105	72	7	0.95	0.88	**0.64**	8649	5183773
8	105	300	10	0.86	0.92	0.89	109863	351388062
9	104	108	8	0.79	0.82	**0.73**	6513	5620255
10*	93	108	8	0.79	0.80	0.82	7625	4770796
11	115	108	8	0.81	0.80	**0.71**	6949	4938122
12*	82	108	8	0.80	0.83	**0.79**	6614	4994690

Parameters for MBIRL. The state relevance kernel exploits a radial basis kernel that uses the Euclidean distance as a measure of similarity as follows:

$$k(s, s') = e^{\frac{-||s-s'||_2}{2\xi^2}},$$

where $\xi = 100$. The cooling schedule parameter is set to $\frac{1}{T_i} = 25 + \frac{i}{50}$ where i is the MCMC iteration.

7   Results and Discussion

In this section we present the results of the quantitative evaluation measured in terms of policy loss and elapsed time as explained in Section 6.2. We also present a qualitative evaluation in which we compare the trajectories of the experts and the trajectories of the optimal policy π obtained by MBIRL.

Performance. Table 2 shows the performances of MBIRL in the context of our serious game. Two important issues affect performance: the size of the state-space and the limited number of expert observations. In general MBIRL outperforms both locally-optimal and random rewards, reaching a policy loss of 0.66 ± 0.10 for the shortest dialogue, dialogue 1 ($|S| = 32, |A| = 6$) and around 0.72 for dialogues 5,9 and 11 ($|S| = 108, |A| = 8$). However, with a larger state-action space such as in dialogue 3 ($|S| = 432, |A| = 11$), 4 ($|S| = 192, |A| = 9$) and 8 ($|S| = 300, |A| = 10$), the models do not improve over the locally-optimal reward, which suggests that the number of samples that are generated is not large enough. Moreover, for state spaces greater than 300 states, MBIRL takes a prohibitively long running time to finish. For instance, dialog 3 took 590935186 milliseconds (around 7 days) to finish when running with 10 policy iterations and 1000 MCMC iterations. Similarly, dialogue 8 took 351388062 milliseconds, which is equivalent to 4 days, to finish.

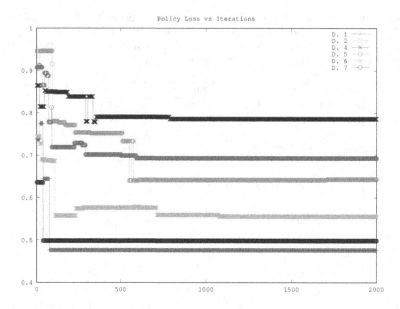

Fig. 1. 0-1 policy loss as a function of the number of MCMC iterations for the first six dialogues that finished with 20 PI

The huge computational expense for large state spaces is an important limitation of MBIRL since it needs to solve one RL problem per iteration. A potential solution to this issue might be to use appropriate function approximation both for the Q function and for modeling the reward function **R**, but this is left for future work.

Unsurprisingly, the MBIRL policy loss is higher for optional dialogues such as dialogue 10 and 12 than for mandatory dialogues within the same state-action space size (i.e., dialogues 5,9 and 11). Indeed, MBIRL does not improve significantly over either the locally-optimum or the random reward in the optional dialogue 10, showing that the scarce number of observations in this dialogue significantly affects performance (see column (ND.), number of dialogues, in Table 2).

Convergence. Figure 1 shows that MBIRL converges in around 600 iterations when using 20 PI, towards a policy loss that is better than the one obtained with the initial random rewards.

Trajectories. Figure 2 shows two dialogue trajectory excerpts with both the gold (or expert) trajectory and the trajectory inferred by MBIRL. Interestingly, in most of the dialogues, both trajectories coincide in the first state and in those states where the system has to inform about mandatory goals just after explicitly requested by the user. This is also the case of the states where the system properly provides help as requested by the user. On the other hand, the learned policy usually fails to close the dialogue and it sometimes contains repetitions e.g., once it has informed about a goal, it may inform again later on.

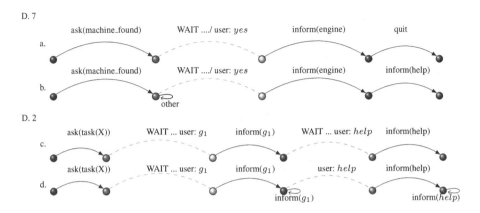

Fig. 2. Comparison of expert vs. MBIRL trajectories for dialogues 7 (top) and 2 (bottom). (a) and (c) depict expert trajectories, while (b) and (d) show the trajectories of MBIRL optimal policy π.

8 Conclusion

In this work we applied a Bayesian algorithm for apprenticeship learning to model the behaviour of distinct conversational agents in a virtual environment. The reward function is then learned from expert demonstrations. Most noticeably, the learned reward tends to reproduce quantitatively and qualitatively the expert decisions, in particular all models learned to provide information and help as requested. We conclude that the proposed approach is a viable option to learn a reward that leads to human-like policies while still benefiting from the interesting dynamic properties of Markov Decision Processes. We found that two main factors affect performance, the size of the state-action space and the limited number of expert demonstrations. Certainly, the computational expense drastically increases when dealing with complex dialogues in a large state-action space ($|S| >= 192$, $|A| >= 9$), without improving the locally optimum.

A potentially interesting future work concerns the study of function approximation to model both Q and $\mathbf{R}$ functions. Such an approach may provide an elegant way to solve the states space dimensionality issue, as well as give the possibility to model more complex behaviour such as handling misunderstandings, repetitions and out of context inputs as well as proposing new topics to be discussed. Finally, exploiting BIRL for both dialogue management and user simulation might constitute an interesting strand of research to better model uncertainty at every stage of dialogue modeling.

Acknowledgments. This work was partially supported by the French ANR (*Agence Nationale de la Recherche*) funded project ContNomina.

References

1. Levin, E., Pieraccini, R., Eckert, W.: A stochastic model of human-machine interaction for learning dialog strategies. IEEE Transactions on Speech and Audio Processing 8(1), 11–23 (2000)

2. Rieser, V., Lemon, O.: Reinforcement learning for adaptive dialogue systems. Springer (2011)
3. Pietquin, O., Dutoit, T.: A probabilistic framework for dialog simulation and optimal strategy learning. IEEE Transactions on Audio, Speech, and Language Processing 14(2), 589–599 (2006)
4. Cuayáhuitl, H.: Hierarchical reinforcement learning for spoken dialogue systems. PhD thesis, Citeseer (2009)
5. Williams, J.D., Young, S.: Partially observable markov decision processes for spoken dialog systems. Computer Speech & Language 21(2), 393–422 (2007)
6. Paek, T., Pieraccini, R.: Automating spoken dialogue management design using machine learning: An industry perspective. Speech Communication 50(8), 716–729 (2008)
7. Ng, A.Y., Russell, S.J.: Algorithms for inverse reinforcement learning. In: Icml, pp. 663–670 (2000)
8. Ramachandran, D., Amir, E.: Bayesian inverse reinforcement learning. Urbana 51, 61801 (2007)
9. Michini, B., How, J.P.: Improving the efficiency of bayesian inverse reinforcement learning. In: IEEE International Conference on Robotics and Automation (ICRA), pp. 3651–3656. IEEE (2012)
10. Walker, M.A.: An application of reinforcement learning to dialogue strategy selection in a spoken dialogue system. Journal of Artificial Intelligence Research 12, 387–416 (2000)
11. Young, S., Gašić, M., Keizer, S., Mairesse, F., Schatzmann, J., Thomson, B., Yu, K.: The hidden information state model: A practical framework for pomdp-based spoken dialogue management. Computer Speech & Language 24(2), 150–174 (2010)
12. Tetreault, J.R., Litman, D.J.: A reinforcement learning approach to evaluating state representations in spoken dialogue systems. Speech Communication 50(8), 683–696 (2008)
13. Rojas Barahona, L.M., Lorenzo, A., Gardent, C.: An end-to-end evaluation of two situated dialog systems. In: Proceedings of the 13th Annual Meeting of the Special Interest Group on Discourse and Dialogue, pp. 10–19. Association for Computational Linguistics, Seoul (2012)
14. Abbeel, P., Ng, A.Y.: Apprenticeship learning via inverse reinforcement learning. In: Proceedings of the Twenty-First International Conference on Machine Learning, p. 1. ACM (2004)
15. Abbeel, P., Coates, A., Ng, A.Y.: Autonomous helicopter aerobatics through apprenticeship learning. The International Journal of Robotics Research 29(13), 1608–1639 (2010)
16. Chandramohan, S., Geist, M., Lefevre, F., Pietquin, O., et al.: User simulation in dialogue systems using inverse reinforcement learning. In: Proceedings of the 12th Annual Conference of the International Speech Communication Association, pp. 1025–1028 (2011)
17. Boularias, A., Chinaei, H.R., Chaibdraa, B.: Learning the reward model of dialogue pomdps from data. In: NIPS Workshop on Machine Learning for Assistive Techniques, Citeseer (2010)
18. Zhifei, S., Joo, E.M.: A review of inverse reinforcement learning theory and recent advances. In: 2012 IEEE Congress on Evolutionary Computation (CEC), pp. 1–8. IEEE (2012)
19. Rojas-Barahona, L.M., Lorenzo, A., Gardent, C.: Building and exploiting a corpus of dialog interactions between french speaking virtual and human agents. In: Proceedings of the 8th International Conference on Language Resources and Evaluation (2012)
20. Sutton, R.S., Barto, A.G.: Reinforcement Learning: An Introduction. MIT Press, Cambridge (1998)

Learning to Summarize Time Series Data

Pranay Kumar Venkata Sowdaboina[1], Sutanu Chakraborti[1], and Somayajulu Sripada[2]

[1] Department of Computer Science, Indian Institute Technology Madras
svpranay@gmail.com, sutanuc@cse.iitm.ac.in
[2] Computing Science, University of Abeerdeen

Abstract. In this paper we focus on content selection for summarizing time series data using Machine Learning techniques. The goal is to exploit a parallel corpus to predict the appropriate level of abstraction required for a summarization task. This is an important step towards building an automated NLG (Natural Language Generation) system to generate text for unseen data. Machine learning approaches are used to induce the underlying rules for text summarization, which are potentially close to the ones that humans use to generate textual summaries. We present an approach to select important points in a time series that can aid in generating captions or textual summaries. We evaluate our techniques on a parallel corpus of human generated weather forecast text corresponding to numerical weather prediction data.

1 Introduction

The focus of our work is content selection for time series data, where the task is to summarize the input data, rejecting unimportant portions of the input. Our aim is to identify points or observations in time series data that are relatively more important than other points in the input data. These identified points can be used in generating textual summaries while ensuring that the summary is small and also the information lost is minimal. While traditional systems have emphasized solely on linguistic processing, an interesting recent direction is the application of Machine Learning (ML) techniques to Natural Language Generation (NLG).

In this paper, we focus on NLG systems that process numeric datasets (e.g time series data) and produce human readable output text. For example, an existing software to generate the pollen forecast for Scotland produced a short textual summary of pollen levels mentioned below.

Grass pollen levels for Friday have increased from the moderate to high levels of yesterday with values of around 6 to 7 across most parts of the country. However, in Northern areas, pollen levels will be moderate with values of 4.

In contrast, the actual forecast (written by a human meteorologist) for the same data was,

Pollen counts are expected to remain high at level 6 over most of Scotland, and even level 7 in the south east. The only relief is in the Northern Isles and far northeast of mainland Scotland with medium levels of pollen count.

Automatic text generation systems are generally categorized into either template based systems or NLG systems. Template based systems accept an input and produce an output by performing a sequence of string manipulation steps. On the other

A. Gelbukh (Ed.): CICLing 2014, Part I, LNCS 8403, pp. 515–528, 2014.

hand, NLG systems have a deeper understanding of linguistics and domain knowledge. NLG systems can work with morphology, punctuation and can produce multi-lingual output with guaranteed conformance to standards. Such NLG applications are widely used today across multiple domains. Apart from weather forecast text generation, STOP (Reiter et al., 2003b) is an NLG system that generates tailored smoking-cessation letters. There have also been efforts in generating textual summaries of medical data to aid medical professionals in taking quick decisions. It has already been demonstrated that statistical techniques could be used to generate high-quality summaries using a consistent terminology.

Text generation can be viewed as a 3-stage pipeline architecture (Reiter, 2003c). The process begins with document planning (also called content planning) which performs content selection, followed by micro-planning which analyses ways of expressing the information linguistically and lastly the surface realization of grammatical text.

One common bottleneck in building an NLG system is that it takes extensive resources of time and labour to build the initial prototype. Lack of sufficient domain knowledge can lead to difficulties in content selection, which has adverse effects on downstream modules like sentence formation, structuring and final realization. Typically, NLG systems cannot generate text unless they have a syntactic representation of the expected output text or declarative domain knowledge base.

Our interest is in building hybrid text generation systems, which could combine the advantages of both NLG and template based systems. ML techniques can aid NLG systems in improving the quality of text generated by selecting or generating fragments that can fit into templates. ML techniques can be used for high-level content selection and templates can be used for low-level realization of the text. To the best of our knowledge, this is the first paper that demonstrates the applicability of Machine Learning techniques to the problems of content selection and lexical choice.

2   Background

An important problem in generating textual summaries of time series data is identifying the appropriate level of abstraction. Any typical time series data has minor variations, which are often regarded as noise by humans when summarizing the data. If each of these minor and unimportant variations have to be reported in the text, then the summary becomes huge and irrelevant. In this section, we position our work in the context of research related to content generation using ML techniques.

2.1   Sumtime-Mausam

SUMTIME-MAUSAM (Sripada et al, 2003), a research project at University of Aberdeen, UK aimed to develop better technology for producing summaries of time series data by integrating leading-edge time series and NLG technology. The goal of their project was to automatically generate weather forecast texts from numerical weather prediction (NWP) data to be used by an oil rig company. They deployed an end-to-end NLG system, which generates weather forecasts in English text that are then post-edited by human forecasters and sent to oil-rig employees working off-shore to inform them about weather conditions. Our work is an attempt to use ML for addressing issues related to knowledge engineering overheads encountered in building NLG systems. We focus on time series segmentation, verb selection and point selection.

The segmentation problem (Sripada et al., 2002) can be stated as follows (refer Figure 1) : Given a time series T, produce the best representation either (a) using exactly K segments or (b) such that the maximum error for any segment does not exceed some user specified threshold, or (c) such that the combined error of all segments is less than some user specified threshold.

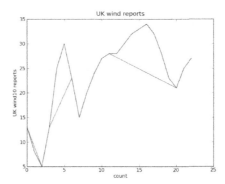

Fig. 1. Figure depicting the segmentation problem

It has been demonstrated that humans tend to use similar rules to summarize time series data (Reiter, 2003c). Since our goal is to generate text as close as possible to human forecasts, solving the segmentation problem is an important step towards content identification. An important parameter is the stopping criterion, which dictates when the segmentation process should stop. This suggests that the stopping criterion is dependent on the range of values in the input data. Preferably, the stopping criterion must also accommodate end user preferences to generate personalized summaries.

The developers of SUMTIME-MAUSAM investigated several issues empirically, by analyzing how people made choices while writing forecasts. The various issues brought up by them are as follows.

- Should *west* be reported as W or $W'ly$?
- Should 8 knots be represented as 08 or 8 ?
- Should *backing* or *becoming* be used to describe the change in wind ?
- should *by evening* or *by late evening* be used to replace time $00Hours$?

The following is the human written forecast corresponding to the weather data in Table 1 for gust10M.

NNE SOON 22-26 GUSTS 36 THEN GRADUALLY DECREASING NE 10-14 BY LATE EVENING

SUMTIME-MAUSAM developers have identified a set of rules (refer Table 2) based on end user preferences that indicate the stopping criterion (a threshold value). These rules are used for segmenting wind speed and wind direction data. In addition they made

Table 1. Weather data as observed on April 11, 2001

date	timetext	winddir	windspeed	gust10m	gust50m
10/4/2001	19:00:00	NNE	20	25	31
10/4/2001	22:00:00	NNE	22	27	33
11/4/2001	01:00:00	N	21	26	32
11/4/2001	04:00:00	N	20	25	31
11/4/2001	07:00:00	N	18	22	28
11/4/2001	10:00:00	N	16	20	24
11/4/2001	13:00:00	N	14	17	21
11/4/2001	16:00:00	NNW	9	11	14
11/4/2001	19:00:00	WNW	7	8	10

several interesting observations such as (a) channels need coordination, (b) stopping criterion should be sensitive to end user, (c) importance being relative, etc.

Table 2. Stopping criteria for time series segmentation acquired from K.A.

Wind speed	Direction threshold (degrees)	Speed threshold (magnitude)
0 - 15	44	5
15 - 40	22	5
40 - 65	22	5
> 65	22	5

In this paper, we want to explore opportunities to learn desirable mappings from a parallel corpus, which can then be used in generating summaries for a new time series.

2.2   Experiments with a Classifier

Researchers working on SUMTIME-MAUSAM (Sripada et al, 2003) evaluated the classifier approach for identifying the right choice of time phrase. They built classifiers that predicted the usage of time-phrases in the forecast. An error rate of 48% was reported in predicting the usage of the time phrase **IN THE MORNING** with the use of a J48 classifier trained on features associated with the time phrase. Feature values and the class label for a specific report are listed below.

- Class: IN-THE-MORNING
- Time: 0600 (from alignment)
- Wind-Speed: 18
- Author: F5
- Previous-Word: Number (for this feature, all numbers were replaced by a generic Number token)
- Previous-Time-Phrase: None
- Phrase-Position: 2 (that is, this is the second phrase in the sentence)
- How-Far-in-Future: 1 day (that is, this forecast is for the day following the day the forecast was issued)

2.3   Stopping Criteria Based on Statistical Measures

A technique was proposed for estimating number of segments for a time series based on permutation tests in (Vasko et al, 2002). The algorithm works by segmenting the time series data continuously (using a top down approach) and reducing the error until the reduction is solely due to noise. They propose a statistical measure by which we can identify whether the reduction in the error is due to noise or some structure in the data. Lavrenko (Lavrenko et al., 2000) proposed another approach during their work on mining concurrent time series. They used a piecewise fitting algorithm that uses a top down approach with an automatic stopping criteria based on the t-test.

3   Our Approach

We estimate the number of segments by learning from a parallel corpus. A parallel text is text placed alongside data from which it was generated and a collection of parallel texts constitutes our parallel corpus. SUMTIME-METEO is a parallel corpus of 1045 weather forecast texts and the numerical data that human forecasters examined when writing the texts. Numerical Weather Prediction (NWP) models generate predicted values of various weather parameters such as wind speed, wind direction and precipitation for various time points. Human forecasters use the time series data sets generated by NWP models as the major source of information when writing the forecast texts. The training data is obtained by extracting a set of features from the time series and the corresponding output class label is obtained by parsing the human written forecast text.

The motivation behind the ML approach is the observation that the stopping criteria is dependent on multiple channels and also on interactions between these channels. For example, a change in the wind direction at very high wind speeds may necessarily be reported but a change in wind direction at slow wind speeds can be ignored. Our trained models should be capable of automatically detecting these chennels and accurately predicting the output.

An important question that needs to be addressed is the following: how does a system that learns from parallel corpus handle changes in content selection, sentence formation, and change of vocabulary over time? We observe that the ML approaches described in this paper are flexible; as and when the end users of the system observe that the content generated no longer reflects the changes in the domain knowledge or terminology, we need to re-train our models that perform content selection or micro-planning. A more elegant option is incremental learning, which is outside the scope of the paper. Output of the system can be provided for human manual edits and the edited text can be used as the training corpus.

In our initial experiments, we trained a classifier to identify the appropriate number of segments for a given time series. With the classifier output and the time series data a segmentation algorithm generates the segments required for an NLG system to generate appropriate weather forecasts.

We use simple statistical measures and characteristic features derived from the wind and wave related weather forecasts to build the classifier. This makes the model induced by the ML system understandable to humans. The features used are as follows.

- *regdev* : Sum of the deviations for all points between the original time series and the time series reconstructed from the output segments by the segmentation algorithm. This sum also acts as a measure of information loss between the original time series and the segmented time series.
- *range* : Difference between the minimum and the maximum values (speed and direction) in the time series.
- *variance* : Variance of the wind speed values in the time series.
- *max* : Maximum value in the time series.
- *endtoendslope* : Slope between the first and last points in the time series.
- *meanslope* : Average of slopes for all the initial $n - 1$ segments.
- *mean* : Mean of the wind speed values in the time series.
- *windchanges* : Number of changes in wind direction in the entire time duration.
- *median* : Median of the wind speed values in the time series.
- *min* : Minimum value in the time series.
- *meanDeviation* : Deviation of a point from the mean in the time series.
- *regDeviation* : Deviation of a point from the regression curve in the time series.
- *slopeChange* : Change of slope between adjacent segments to be merged.
- *windChange* : Change in the wind direction at a point in time series. This feature takes a non-zero value if there is a change in the wind direction.

3.1   Estimating Segment Count

As our first approach we build a decision tree classifier based on the above mentioned features to predict the number of segments to be reported for a new time series. For training we use the number of segments reported in the forecast text as the class label. An advantage of decision trees over other classifiers in the current context is that the rules induced can easily be interpreted.

For example, consider the forecast "NNE SOON 22-26 GUSTS 36 THEN GRADUALLY DECREASING NE 10-14 BY LATE EVENING" we observe two indications of wind speed values reported as ranges, 22-26 and 10-14. Hence we assign a value of 2 for the class label (corresponding to number of segments) for this time series. An alternative approach would be to view this as a regression problem, but it is not discussed in this paper.

With the segment count prediction and the input data, a segmentation algorithm generates the segments required for an NLG system to generate appropriate weather forecasts. The segmentation algorithm recursively merges low cost segment pairs until the required number of segments is reached. The cost function can be imagined as using sum of squares, furthest point, difference in slopes between a pair of segments or any other measure.

One observation from our initial experiment was that the segments generated by the above approach result in continuous segments but we observe that humans use discontinuous segments. We present an alternative approach of point selection in section 3.3 to solve this problem.

3.2   Identifying Points of Interest

Once the number of segments is identified by our learner, we must identify a point from each segment at which the wind speed and wind direction should be reported in the forecast.

We view the entire process of content selection from time series as a two-step process. In the first step we identify the segment count, then estimate segments using a segmentation algorithm. Secondly, we need to identify a point in each segment that is to be reported. An error in the system could occur because of an error in the estimation of the number of segments or an improper choice of points from the segments. To overcome these problems, we propose a technique to directly identify the points from the time series that are relatively more important than other points. From Table 3, we observe that a few important points are chosen from the time series and are used for generating the summary.

Table 3. Forecast : E 02-06 VEERING SE 28-32 BY MIDDAY, LATER SSE 14-18

Wind speed	6	10	18	26	30	28	22	18	16
Wind direction	"E"	"SE"	"SE"	"SE"	"SE"	"SE"	"SE"	"SE"	"SSE"
Reported points	E 02-06				SE 28-32				SSE 14-18

Our approach is to use a neural network that has an input node for each channel and an output node for each recording in the time series data (refer Figure 2). The value of the output node determines whether the specific recording will be reported in the summary. The model is trained using the parallel corpora and the output for each data point is computed by assigning a value of 1 for reported recordings and a value of 0 otherwise.

In our experiment we considered two input channels for each data point in the time series. One of them is the numerical speed value and the other is wind direction, which takes a non-zero value if there is a change in the wind direction. With this choice of input channels, our model presupposes that point selection happens based on wind speed and wind direction. The basic approach can be extended to included more complex features. Each output node returns a value in the range $[0, 1]$ which indicates the relative importance of that particular point in the time series. This can be used to identify whether to report an observation from the time series or to ignore it.

From corpus analysis we observe that human forecasts typically contain a maximum of four segments for a time series. Since the output from the neural network is a real value from $[0, 1]$ by adjusting the threshold we can control the number of segments that qualify to be reported. We observe that as we decrease the threshold more points qualify to be reported resulting in a fall in the precision and a rise in recall.

Finally, we could either select a fixed number of data points from the neural network scores by taking the top k scores or by model prediction we can choose a variable number of data points based on the input time series. In section 4, we present our results for both fixed and variable counts. In the case of variable segment counts, we use our ML model to estimate the number of segments required and then choose the k nodes associated with the top k scores.

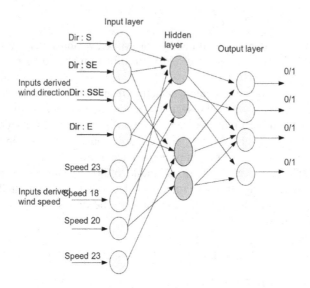

Fig. 2. Neural network for point selection with multiple input nodes for each observation

3.3 Verb Selection

Once the number of segments is identified, we need to identify the right verb that describes each segment. Our approach is to formulate the verb selection as another ML problem.

Observe the following two weather forecasts. Along with the indication of the speed and/or direction, a verb is used to quantify the wind speed and/or wind direction. In our experiments, we focus on the choice of verb that describes the wind speed.

- SSE 10-14 SLOWLY EASING AND BACKING ESE 06-10.
- ESE 6-10 VEERING SSE AND RISING 18-22 LATER IN THE MORNING.

In the above two forecasts, the verbs used are EASING, BACKING, VEERING, RISING. The words VEERING and BACKING are used to indicate the change in the direction of wind. The word BACKING refers to change in the wind direction in anti-clockwise direction and the word VEERING refers to the change in wind direction in the clockwise direction. Here the choice of BACKING and VEERING is determined by the wind direction changes whereas verbs such as RISING, INCREASING are used to describe the kind of rise or fall in the wind speed. From Table 4, we can observe that the verb INCREASING is used to describe different magnitudes of change depending on the context. The choice of RISING versus INCREASING is largely dependent on the text producer.

A walkthrough over the corpus revealed that the magnitude change in wind speed does not directly map to a particular choice of verb. We decided to leave it to the learner to arrive at the right usage of these verbs by learning from the parallel corpus. We identify six verbs from the forecast texts and try to learn their usage from the corpus. We cast the problem of verb selection as a classification task, to identify the class label

Table 4. Weather reports with verb 'INCREASING'

Forecasts with 'INCREASING' verb	Change in wind speed value
ENE 04-08 INCREASING 14-18 BY MIDDAY THEN EASING 08-12 BY EVENING	10
ENE 08-12 VEERING SE-S 02-06 BY MIDDAY THEN INCREASING SSE 20-24 BY LATE EVENING	18
S'LY 24-28 EASING SE 12-16 BY MIDDAY THEN INCREASING 16-20 BY EVENING	4

(a verb) for a time series data segment. The decision tree classifier experiments reported here were conducted on the SUMTIME dataset to identify the right verb to indicate the variations in wind speed and wind direction.

4 Experimental Results

SUMTIME-MAUSAM dataset is a parallel corpus of numerical weather data and human written forecasts. It contains non-linguistic input data and output texts written from this data. Marine models generate predictions for wind and wave parameters at three hourly intervals. Similarly other weather parameters such as cloud, precipitation are also predicted hourly intervals.

4.1 Estimating Segment Count

We build a decision tree (a fragment of which is shown in Figure 3) by training the classifier for 941 time series with 11 attributes. The reduced error pruning algorithm is

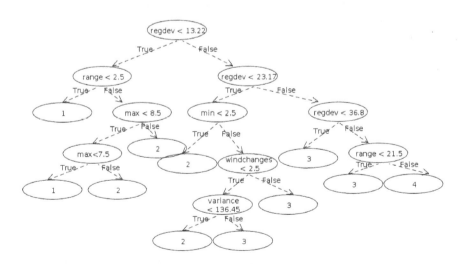

Fig. 3. A decision tree learned to estimate the number of segments

Table 5. Arriving at the time series segments using the predicted segment count

Time series	['22', '13', '10', '9', '9', '7', '6', '3']
Forecast	E-NE 20-25 VEERING SE 08-12 BY MIDDAY, BECOMING N'LY LESS THAN 08 LATER
Prediction	3
Segments	22 - 13 , 13 - 6 , 6 - 3
Time series	['6', '10', '12', '14', '16', '14', '12', '10']
Forecast	ENE 04-08 INCREASING 14-18 BY MIDDAY THEN EASING 08-12 BY EVENING
Prediction	3
Segments	6 - 10 ,10 - 16 ,16 - 10
Time series	['18', '18', '16', '14', '12', '10', '6', '4']
Forecast	SSE 16-20 BACKING ESE-E 04-08 BY LATE AFTERNOON AND N'LY BY LATE EVENING
Prediction	2
Segments	18 - 10 ,10 - 4

used to learn the decision tree. We have used Weka (Mark Hall et al, 2009) and Matlab to train models for our experiments.

From the decision tree, it is easy to observe that as the deviation from regression curve increases the number of segments also increases. For example, the left subtree at the root node with $regdev < 13.22$ has its leaves with segment count selecting as 1 or 2 whereas the right subtree with $regdev > 13.22$ has leaves with segment counts ranging from 2 to 4 with majority leaves having value of 3.

Another interesting observation that can be made from this decision tree is the co-ordination across channels (refer Figure 4). From the subtree shown in Figure 4 we can observe the interactions between the following two parameters: the count of changes in the wind direction and the variance of the wind speed values in the time series. It is interesting to note that the rules learnt by the Reduced Error Pruning algorithm are very intuitive and humans tend to do similar reasoning in making decisions.

We achieved a precision of 0.708 and recall of 0.697 (averaged over 10-fold test-train splits) for our classifier with eight class labels, each indicating the number of segments required to summarize the time series.

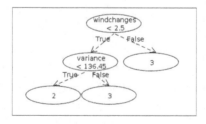

Fig. 4. A subtree depicting the coordination between wind speed and wind direction parameters

Table 6. Confusion matrix for verb selection. The rows in the confusion matrix represent the correct choice of verb and the columns represent the verb that was actually selected by the system.

FRESHENING	INC	RISING	EASING	DEC	FALLING	System/*Actual*
7	2	3	0	0	0	*FRESHENING*
27	37	24	1	0	0	*INCREASING*
9	10	14	0	1	0	*RISING*
4	0	0	62	19	20	*EASING*
0	0	0	20	6	5	*DECREASING*
0	0	0	4	3	10	*FALLING*

For an unseen time series, the classifier identifies the number of segments needed to report it. The output from the classifier becomes the input to the segmentation algorithm unit to identify the segments. It can be observed that the generated segments either overlap or map to the range in the manual forecast (refer Table 5 for some examples).

4.2   Point Selection Using Neural Network Approach

In our initial neural network configuration, we kept the number of points to be selected as a fixed input parameter. This parameter 'limit' can be varied from 1 through 4 to choose the desired number of segments that are reported in the forecast. We use precision and recall values to measure the performances of our approaches. Precision is the fraction of retrieved instances that are reported, while recall is the fraction of reported instances that are retrieved. These values are reported in table 7.

The above configuration has the drawback of a fixed number of segments being generated irrespective of the time series; this surely is not reflective of how humans identify pivotal points for text generation. To overcome this problem we extend the above framework to use our segment count estimator, so that the appropriate number of segments would be generated for a given time series. In the variable segment configuration, our neural network model achieved a precision of 0.672 and a recall of 0.678.

4.3   Verb Selection

The results of our experiments on verb selection as a classification task are summarized in Table 6. Here are some observations on the confusion matrix. Firstly, for majority (5 out of 6) classes the diagonal contains the maximum value in the row; thus our approach passes the first sanity check. Secondly, for the class label RISING, the model

Table 7. Performance of various point selection approaches

Approach	Precision	Recall
Neural Network (variable segments)	0.672	0.678
Neural Network (2 segments)	0.68	0.60
Neural Network (3 segments)	0.60	0.77
Neural Network (4 segments)	0.51	0.89

correctly predicts RISING for 14 instances and misclassifies RISING as INCREASING and FRESHENING for 10 and 9 instances respectively. These misclassifications could be due to inconsistent labeling in the training corpus by humans. From empirical analysis it has been observed that for the same time series different lexical choices could be made by different authors.

5   Discussion

Natural Language Generation has largely been viewed as a task that needs significant acquisition of knowledge from domain experts. Most earlier work (Reiter et al, 2003a) for example) have advocated this line of thinking. However, over the last one decade, there have been interesting efforts that examined the effectiveness of machine-learning over parallel corpora, with the primary goal of addressing the knowledge acquisition costs. (Belz, A, 2005) addresses the use of Machine Learning for idiolect, but the focus is not so much on content selection. In contrast, our work shows that Machine Learning approaches can aid in both content selection (segmentation) and micro-planning (verb selection, in our domain). Using ML for content selection is a new research direction that has gained serious attention over the last few years. (Colin Kelly et al, 2009) address the problem of content selection in the domain of generating text from structured reports of cricket matches. However the problem here is of identifying informative content from a set of available fields, which is a considerably simpler problem than identifying a variable number of segments from a time series. In an earlier work, (Duboue et al, 2003) used ML to mine simple rules from graphs, which were then used to generate text. The work reported in this paper is expected to have implications in other complex NLG applications where the alignments of the data points to the textual contents is not straightforward, and knowledge acquisition overheads in both content selection and lexical choice are substantial. The evaluation results seem promising in the light of the observation in (Sripada et al, 2003) that despite elaborate human intervention in Knowledge Acquisition, SUMTIME MAUSAM developers report that they failed to detect 17% of phrases because of segmenting errors, and 40% of the phrases did not match, possibly due to inappropriate lexical choice. This attests to the complexity of the task and the difficulty in modelling human processes in summary generation from time series.

We need to be cautious in overdoing Machine Learning in NLG tasks. The early pioneers in NLG (Reiter et al, 1997) advocated that NLG systems be built by careful analysis of the target text corpus, and by talking to domain experts. Our argument is that while corpus-driven ML clearly cannot replace the human effort involved, they can lessen it substantially. In order to compliment and not compete against traditional approaches, it is important that experts are able to read into the knowledge induced by learners and tweak them to address domain specific needs if necessary. Black box learners like Support Vector Machines are thus not appropriate choices in NLG content selection and idiolect problems. Decision trees may start off with slightly lower accuracies compared to SVM, but are expected to outperform black-box approaches over time, since the rules read out from decision trees are interpretable, and experts can adapt them easily. Changes in domain needs can also be addressed conveniently.

We have shown in this paper that simple interpretable features that are easily understood by experts can act as good starting points for learning, in comparison to a complex linear combination of features, for example. Once humans identify a simple set of features, approaches inbuilt within decision tree learners, such as those based on Information Gain, can identify those that play a more significant role in either content selection or lexical choice.

6 Conclusion and Outlook

In the current work, we proposed ML techniques to learn the underlying rules by which humans generate weather forecast texts. We framed content selection and verb selection problems in time series summarization as classification tasks, and our empirical results show the effectiveness of our approaches in content generation in the context of a real world NLG task. ML was used in our approach for three distinct tasks : identifying the number of segments, selecting the number of points to be used for generating the textual summary, and the choice of verbs corresponding to the segments.

NLG system have the ability to vary content in fine-grained and flexible way. In large demographic countries, the literacy rate varies significantly. Hence the choice of vocabulary can affect the readability of the article. A NLG system could produce news reports which differ in structure and vocabulary to suit the particular audience. User preferences can be learnt from parallel corpus and used in the NLG system to control the level of readability in the generated text and generate user-preference based reports. Also, it has been shown in Human Computer Interface (HCI) studies that a textual caption accompanying a graphical image helps in better understanding than a plain image. Automated text generation applications might also find good use in mobile technology where downloading an image could demand a significantly higher bandwidth compared to a text download.

References

[Belz, A, 2005]Belz, A.: Corpus-driven generation of weather forecasts. In: Proceedings of the 3rd Corpus Linguistics Conference (CL 2005) (2005)

[Colin Kelly et al, 2009]Kelly, C., Copestake, A., Karamanis, N.: Investigating content selection for language generation using machine learning. In: Proceedings of the 12th European Workshop on Natural Language Generation (ENLG 2009), Athens, pp. 130–137 (2009)

[Duboue et al, 2003]Duboue, P.A., McKeown, K.R.: Statistical Acquisition of Content Selection Rules for Natural Language Generation (EMNLP 2003), pp. 121–128 (2003)

[Goldberg et al, 1994]Goldberg, E., Driedger, N.: Using natural-language processing to produce weather forecasts. In: Proceedings of the IEEE Expert (1994)

[Lavrenko et al., 2000]Lavrenko, V., Schmill, M., Lawrie, D., Ogilvie, P., Jensen, D., Allan, J.: Mining of concurrent text and time series. In: Proceedings of the 6 th ACM SIGKDD Intl Conference on Knowledge Discovery and Data Mining Workshop on Text Mining (2000)

[Mark Hall et al, 2009]Hall, M., Frank, E., Holmes, G., Pfahringer, B., Reutemann, P., Witten, I.H.: The WEKA Data Mining Software: An Update. SIGKDD Explorations 11(1) (2009)

[Reiter et al, 1997]Reiter, E., Dale, R.: Building applied natural language generation systems. Natural Langauge Engineering 3(1), 57–87 (1997)

[Reiter et al, 2003a]Reiter, E., Sripada, S., Robertson, R.: Acquiring correct knowledge for natural language generation. Journal of Artificial Intelligence Research 18, 491–516 (2003a)

[Reiter et al., 2003b]Ehud Reiter, R.R., Osman, L.M.: Generating tailored smoking cessation letters. In: Artificial Intelligence (2003b)

[Reiter, 2003c]Reiter, E.: Learning the meaning and usage of time phrases from a parallel text-data corpus. In: Proceedings of the HLT-NAACL 2003 Workshop on Learning Word Meaning from Non-Linguistic Data (2003c)

[Sripada et al., 2002]Sripada, S., Reiter, E., Hunter, J., Yu, J.: Segmenting time series for weather forecasting. In: Applications and Innovations in Intelligent Systems X. Springer (2002)

[Sripada et al, 2003]Sripada, S.G., Reiter, E., Davy, I.: SUMTIME-MOUSAM: Configurable Marine Weather Forecast Generator (2003)

[Sripada et al., 2001a]Somayajulu, S.G., Reiter, E., Hunter, J., Yu, J.: Segmenting time series for weather forecasting. University of Aberdeen, U.K. (2001a)

[Sripada et al, 2001b]Somayajulu, S.G., Reiter, E., Hunter, J., Yu, J.: Modelling the task of Summarising Time Series Data using KA Techniques. University of Aberdeen, U.K. (2001b)

[Sripada et al, 2003]Sripada, S.G., Reiter, E., Hunter, J., Yu, J.: Exploiting a parallel text-data corpus. In: Proceedings of Corpus Linguistics (2003)

[Vasko et al, 2002]Vasko, K.T., Toivonen, H.T.: Estimating the number of segments in time series data using permutation tests. In: IEEE International Conference on Data Mining (2002)

Author Index